# Display and Interface Design

## Subtle Science, Exact Art

# Display and Interface Design

## Subtle Science, Exact Art

Kevin B. Bennett
John M. Flach

CRC Press
Taylor & Francis Group
Boca Raton London New York

CRC Press is an imprint of the
Taylor & Francis Group, an **informa** business

CRC Press
Taylor & Francis Group
6000 Broken Sound Parkway NW, Suite 300
Boca Raton, FL 33487-2742



Printed in the United States of America on acid-free paper
10 9 8 7 6 5 4 3 2 1

International Standard Book Number: 978-1-4200-6438-4 (Hardback)



**Visit the Taylor & Francis Web site at**
**http://www.taylorandfrancis.com**

**and the CRC Press Web site at**
**http://www.crcpress.com**

*To Jens Rasmussen for his inspiration and guidance.*

# Contents

# Preface

This book is about display and interface design. It is the product of over 60 years of combined experience studying, implementing, and teaching about performance in human–technology systems. Great strides have been made in interface design since the early 1980s, when we first began thinking about the associated challenges. Technological advances in hardware and software now provide the potential to design interfaces that are both powerful and easy to use. Yet, the frustrations and convoluted "work-arounds" that are often still encountered make it clear that there is substantial room for improvement. Over the years, we have acquired a deep appreciation for the complexity and difficulty of building effective interfaces; it is reflected in the content of this book. As a result, you are likely to find it to be decidedly different from most books on the topic.

We view the interface as a tool that will help an individual accomplish his or her work efficiently and pleasurably; it is a form of decision-making and problem-solving support. As such, a recurring question concerns the relation between the structure of problem representations and the quality of performance. A change in how a problem is represented can have a marked effect on the quality of performance. This relation is interesting to us as cognitive psychologists and as cognitive systems engineers.

As cognitive psychologists, we believe that the relationship between the structure of representations and the quality of performance has important implications for understanding the basic dynamics of cognition. It suggests that there is an intimate, circular coupling between perception–action and between situation–awareness that contrasts with conventional approaches to cognition (where performance is viewed as a series of effectively independent stages of general, context-independent information processes). We believe that the coupling between perception and action (i.e., the ability to "see" the world in relation to constraints on action) and between situation and awareness (i.e., the ability to make sense of complex situations) depend critically on the structure of representations. Thus, in exploring the nature of representations, we believe that we are gaining important insight into human cognition.

As cognitive engineers, we believe that the relation between representations and the quality of performance have obvious implications for the design of interfaces to support human work. The design challenge is to enhance perspicuity and awareness so that action and situation constraints are well-specified relative to the demands of a work ecology. The approach is to design representations so that there is an explicit mapping between the patterns in the representation and the action and situation constraints. Of course, this implies an analysis of work domain situations to identify these constraints, as well as an understanding of awareness to know what patterns are likely to be salient. We believe that the quality of representing the work domain constraints will ultimately determine effectivity and efficiency; that is, it will determine the quality of performance and the level of effort required.

## Intended Audience

The primary target audience for this book is students in human factors and related disciplines (including psychology, engineering, computer science, industrial design, and industrial/organizational psychology). This book is an integration of our notes for the courses that we teach in interface design and cognitive systems engineering. Our goal is to train students to appreciate basic theory in cognitive science and to apply that theory in the design of technology and work organizations. We begin by constructing a theoretical foundation for approaching cognitive systems that integrates across situations, representations, and awareness—emphasizing the intimate interactions between them. The initial chapters of the book lay this foundation.

We also hope to foster in our students a healthy appreciation for the value of basic empirical research in addressing questions about human performance. For example, in the middle chapters we provide extensive reviews of the research literature associated with visual attention in relation to the integral, separable, and configural properties of representations. Additionally, we try to prepare our students to immerse themselves in the complexity of practical design problems. Thus, we include several tutorial chapters that recount explorations of design in specific work domains.

Finally, we try to impress on our students the intimate relation between basic and applied science. In fact, we emphasize that basic theory provides the strongest basis for generalization. The practical world and the scientific/academic world move at a very different pace. Designers cannot wait for research programs to provide clear empirical answers to their questions. In order to participate in and influence design, we must be able to extrapolate our research to keep up with the demands of changing technologies and changing work domains. Theory is the most reliable basis for these extrapolations.

Conversely, application can be the ultimate test of theory. It is typically a much stronger test than the laboratory, where the assumptions guiding a theory are often reified in the experimental methodology. Thus, laboratory research often ends up being demonstrations of plausibility, rather than strong tests of a theory.

We believe this work will also be of interest to a much broader audience concerned with applying cognitive science to design technologies that enhance the quality of human work. This broader audience might identify with other labels for this enterprise: ergonomics, human factors engineering, human–computer interaction (HCI), semantic computing, resilience engineering, industrial design, user-experience design, interaction design, etc. Our goals for this audience parallel the goals for our students: We want to provide a theoretical basis, an empirical basis, and a practical basis for framing the design questions.

Note that this is not a "how to" book with recipes in answer to specific design questions (i.e., *Interfaces for Dummies*). Rather, our goal is simply to help people to frame the questions well, with the optimism that a well-framed question is nearly answered. It is important for people to appreciate that we are dealing with complex problems and that there are no easy answers. Our goal is to help people to appreciate this complexity and to provide a theoretical context for parsing this complexity in ways that might lead to productive insights. As suggested by the title, our goal is to enhance the subtlety of the science and to enhance the exactness of the art with respect to designing effective cognitive systems.

**Kevin Bennett**

**John Flach**

# *Acknowledgments*

We would like to thank Don Norman, Jens Rasmussen, James Pomerantz, and a host of graduate students for their comments on earlier drafts. Larry Shattuck, Christopher Talcott, Silas Martinez, and Dan Hall were co-developers and co-investigators in the research projects for the RAPTOR military command and control interface. Darby Patrick and Paul Jacques collaborated in the development and evaluation of the WrightCAD display concept. Matthijs Amelink was the designer of the Total Energy Path landing display in collaboration with Professor Max Mulder and Dr. Rene van Paassen in the faculty of Aeronautical Engineering at the Technical University of Delft.

A special thanks from the first author to David Woods for his mentorship, collaborations, and incredible insights through the years. Thanks as well to Raja Parasuraman, who was as much friend as adviser during graduate school. Thanks to Jim Howard who introduced me to computers and taught me the importance of being precise. Thanks to Herb Colle, who created a great academic program and an environment that fosters success. Finally, thanks to my family and their patience during the long process of writing.

The second author wants especially to thank Rich Jagacinski and Dean Owen, two mentors and teachers who set standards for quality far above my reach, who gave me the tools I needed to take the initial steps in that direction, and who gave me the confidence to persist even though the goal remains a distant target. Also, special thanks to Kim Vicente, the student who became my teacher. I also wish to acknowledge interactions with Jens Rasmussen, Max Mulder, Rene van Paassen, Penelope Sanderson, Neville Moray, Alex Kirlik, P. J. Stappers, and Chris Wickens, whose friendship along the journey have challenged me when I tended toward complacency and who have encouraged me when I tended toward despair. Finally, I want to thank my wife and sons, who constantly remind me not to take my ideas or myself too seriously.

This effort was made possible by funding from a number of organizations. The Ohio Board of Regents and the Wright State University Research Council awarded several research incentive and research challenge grants that funded some of the research projects described in the book and provided a sabbatical for intensive writing. The army project was supported through participation in the Advanced Decision Architectures Collaborative Technology Alliance Consortium, sponsored by the U.S. Army Research Laboratory under cooperative agreement DAAD19-01-2-0009. Development of the WrightCAD was supported through funding from the Japan Atomic Energy Research Institute; the basic research on optical control of locomotion that inspired it was funded by a series of grants from the *Air Force Office of Scientific Research* (AFOSR).

Finally, thanks to the software specialists. Daniel Serfaty and Aptima kindly provided us with their DDD® simulation software and Kevin Gildea helped us use it effectively. Randy Green implemented (and often reimplemented) many of the displays, interfaces, and environments that are described here.

# *The Authors*

**Kevin B. Bennett** is a professor of psychology at Wright State University, Dayton, Ohio. He received a PhD in applied experimental psychology from the Catholic University of America in 1984.

**John M. Flach** is a professor and chair of psychology at Wright State University, Dayton, Ohio. He received a PhD in human experimental psychology from the Ohio State University in 1984.

Drs. Flach and Bennett are co-directors of the Joint Cognitive Systems Laboratory at Wright State University. They share interests in human performance theory and cognitive systems engineering, particularly in relation to ecological display and interface design. Their independent and collaborative efforts in these areas have spanned three decades.

# 1

# *Introduction to Subtle Science, Exact Art*

## 1.1 Introduction

> *Science and art* have in common intense seeing, the wide-eyed observing that generates empirical information. (Tufte 2006, p. 9; emphasis added)
>
> The purpose of an evidence presentation is to assist thinking. Thus presentations should be constructed so as to assist with the fundamental intellectual tasks in reasoning about evidence: describing the data, making multivariate comparisons, understanding causality, integrating a diversity of evidence, and documenting the analysis. (Tufte 2006, p. 137)

The topic of this book is display and interface design or, in more conventional terms, human–computer interaction. Given its popularity and the fact that it impacts the majority of us on a daily basis, it is not surprising that many books have been written on the subject. A search on Amazon.com™ for the term "human–computer interaction" confirms this intuition, since over 10,000 books are identified. Given this number, even those who are not inherently skeptical may be inclined to ask a simple question: "Do we really need yet another one?" Since we have taken the time and effort to write this book, our answer is obviously "yes."

One reason underlying our affirmative response is a relatively pragmatic one. Consider the extent to which the interfaces with which you interact (a) are intuitive to learn and (b) subsequently allow you to work efficiently. If your experience is anything like ours, the list of interfaces that meet these two simple criteria is a short one. To state the case more bluntly, there are a lot of bad displays and interfaces out there. We own scores of applications that we have purchased, attempted to learn how to use, and essentially given up because the learning curve was so steep. This is despite the fact that we are computer literate, the applications are often the industry standard, and we know that they would be useful if we could bring ourselves to invest the time. For other applications, we have invested the time to learn them, but are constantly frustrated by the ways in which interface design has made it difficult to accomplish relatively simple and straightforward tasks.

There is no reason for this situation to exist. Advances in computational technology have provided powerful tools with the potential to build effective interfaces for the workplace. However, this potential is rarely realized; interfaces that are both intuitive and efficient (and therefore pleasurable to use) are the exception, rather than the rule. The organizations responsible for building these interfaces did not begin with the goal of making them unintuitive and inefficient. There are profits to be made when applications have effective interfaces; there are costs to be avoided in terms of decreased productivity and safety. All things considered, only one logical conclusion can be drawn: Display and interface design is a surprisingly complicated endeavor; the difficulty in getting it right is grossly underestimated, even by researchers and practitioners who are experts in the field.

Given the sheer number of books written on the topic and the current state of affairs described earlier, one might be tempted to conclude that something very important is missing in these books. This conclusion is consistent with our experiences in teaching courses on display and interface design. Over the last 20 years we have searched for, but never found, a single book that addresses the topic in a way that meets our needs. Although researchers and practitioners from various disciplines have treated some pieces of the puzzle quite admirably, no one has yet synthesized and integrated these puzzle pieces into a single coherent treatment that meets our needs. As a result, the syllabi of our courses contain an assortment of book chapters and articles rather than a primary text.

In the end, we have arrived at the conclusion that our perspective on display and interface design may just be a fairly unique one. Our goal in writing this book is to share that perspective with you: to describe those pieces of the puzzle that have been treated well, to fill in the gaps that are missing, and to provide a coherent synthesis and integration of the topic. To accomplish this goal we have drawn upon a wealth of experience accrued over decades; this includes thinking about the issues, conducting research on various topics in the field, implementing a large number of displays and interfaces in a variety of work domains, and conveying the resulting insights to our students and colleagues. We will now convey why we feel this perspective is both unique and useful. In the process we will describe how this book is positioned relative to others that have been written on the topic.

## 1.2 Theoretical Orientation

> [C]omputer scientists and engineers should have no problem in understanding the nature of interface design as *science and art;* … the contingent nature of each interface, is a reflex of design's dual nature as science (in respect to scientific principles of design applied to computers) and art (in respect to a particular, original way of designing). (Nadin 1988, p. 53, emphasis original and added)

One broad category of books written on display and interface design addresses the problem from what might be referred to as a "user-driven" approach. These books are typically written by social scientists (e.g., psychologists, anthropologists, educators, etc.). The primary emphasis is on the contributions of the user: understanding broad capabilities and limitations (e.g., memory, perception), how users think about their work domain (e.g., mental models), situated evaluations of display and interface concepts (usability studies), and user preferences and opinions about design interventions (iterative design). These books also consider interface technology (e.g., widgets, menus, form-filling, etc.), although it is typically a secondary emphasis. Important insights can be gained from this perspective, particularly when considered in light of the historical backdrop where users were not considered as an integral part of computer system design.

A second broad category of books addresses the problem from what might be considered a "technology-driven" approach. These books are typically written by computer scientists and engineers. The primary emphasis is on understanding interface technology (e.g., menus, forms, dialog boxes, keyboards, pointing devices, display size, refresh rate, system response time, error messages, documentation, etc.) and how it can be used in the interface. These books also typically consider the user (see topics outlined in the previous paragraph), although it is a secondary emphasis. Once again, it is necessary to think about interface technology when considering human–computer interaction, and important insights can be gained from this perspective.

Collectively, these two complementary perspectives constitute conventional wisdom about how the process of the design and evaluation of displays and interfaces should proceed. This emphasis is readily apparent in the label that is typically applied to this endeavor: human–computer interaction. The vast majority of books written on the topic share one of these two orientations or a combination of the two.

### 1.2.1 Cognitive Systems Engineering: Ecological Interface Design

In our opinion, although these two perspectives are certainly necessary, they are not sufficient. The human is interacting with the computer for a reason, and that reason is to complete work in a domain. This is true whether the work is defined in a traditional sense (e.g., controlling a process, flying an airplane) or the work is more broadly defined (e.g., surfing the Internet, making a phone call, finding a book of fiction).

This represents one fundamental dimension that differentiates our book from others written on the topic. Our approach is a problem-driven (as opposed to user- or technology-driven) approach to the design and evaluation of interfaces. By this we mean that the primary purpose of an interface is to provide decision-making and problem-solving support for a user who is completing work in a domain. The goal is to design interfaces that (1) are tailored to specific work demands, (2) leverage the powerful perception-action skills of the human, and (3) use powerful interface technologies wisely.

This can be conceptualized as a "triadic" approach (domain/ecology, human/awareness, interface/representation) to human–computer interaction that stands in sharp contrast to the traditional "dyadic" (human, interface) approaches described before. Ultimately, the success or failure of an interface is determined by the interactions that occur between all three components of the triad; any approach that fails to consider all of these components and their interactions (i.e., dyadic approaches) will be inherently and severely limited.

The specific approach that guides our efforts has been referred to as cognitive systems engineering (CSE; Rasmussen, Pejtersen, and Goodstein 1994; Vicente 1999; Flach et al. 1995; Rasmussen 1986; Norman 1986). CSE provides an overarching framework for analysis, design, and evaluation of complex sociotechnical systems. In terms of interface design, this framework provides analytical tools that can be applied to identify important characteristics of domains, the activities that need to be accomplished within a domain, and the information that is needed to do so effectively. In short, it allows decisions to be made about display and interface design that are informed by the characteristics of the underlying work domain.

We also believe that this book is uniquely positioned relative to other books written from the CSE perspective. Three classic books written on CSE (Rasmussen et al. 1994; Vicente 1999; Rasmussen 1986) have focused on descriptions of the framework as a whole. It is a complicated framework, developed to meet complicated challenges, and this was a necessary step. Display and interface design were not ignored; general principles were discussed and excellent case studies were described. However, the clear focus is on the framework, the tools, and the analyses; display and interface design is a secondary topic.

Our book reverses the emphasis: display and interface design is the primary focus while CSE provides the orienting underlying framework. It was specifically designed to complement and build upon these classic texts and related but smaller scale efforts that apply CSE to the interface (i.e., ecological interface design; Rasmussen and Vicente 1989, 1990; Vicente and Rasmussen 1990).

### 1.2.2 With a Psychological Twist

There is another way in which our book is fundamentally different from other books written from the CSE perspective. CSE is an inherently interdisciplinary endeavor, as indicated by the simultaneous concern with work domains, humans, and interface technologies. However, the vast majority of the books on the topic of CSE have been written by engineers. Our training and experience are in psychology, a discipline that has produced substantive literatures that address general human capabilities and limitations as well as specific treatments of how this knowledge can be used to inform display and interface design. This perspective has been incorporated into the book, once again providing a treatment that is complementary to previous efforts in CSE. For example, this perspective differentiates our book from an excellent text on ecological interface design (Burns and Hajdukiewicz 2004) that shares both the general orientation (CSE) and the specific focus on interface and display design.

## 1.3 Basic versus Applied Science

The final differentiating feature of our book is a more general one that will require more detailed explanations and examples. Conventional wisdom divides researchers into two camps: those who practice basic research and those who practice applied research. Our approach, as evidenced by the content of this book, is characterized by both (each type of research is valuable) and yet by neither (each type of research by itself is inadequate). In our opinion, the distinction between basic and applied research is an artificial one; they are only different facets of scientific research that are complementary and that should provide mutually reinforcing results. As we have found out the hard way, the problem in adopting this approach is that one invites criticism from not one, but both of these two camps.

### 1.3.1 Too Theoretical!

The criticism from the applied research camp will be introduced through an example. We recently participated in the formation of a consortium composed of the oil refinery industry and university researchers. One of the projects subsequently solicited by the oil industry involved the control systems and displays in oil refining plants. These interfaces are outdated and the industry is beginning to contemplate a conversion to newer digital control and graphical display systems.

We were very excited about the opportunity to apply some of the lessons learned from innovations in the nuclear power and aviation industries to the design of displays and interfaces in this new domain. We immediately began searching online to learn about the processes of oil refining (e.g., the cracking process) and began gearing up to visit some regional plants to talk to the experts and to learn about the processes and their strategies for managing them. We wrote a proposal describing how we would approach the problem—beginning with a study of the domain and the domain experts to identify the task and information demands and the alternative strategies used to manage these demands and ending with the design of new interfaces that provided decision-making and problem-solving support to the operators.

Unfortunately for us, the proposal was not funded. The feedback that we received was, in essence, that our approach was far too theoretical in nature and that the industry was not interested in "basic science." We were told that this was an engineering problem (i.e., applied research) rather than a scientific one and that our job was to answer the customer's question—not to build a scientific research program. Actually, the customer's question was how to compare graphical displays to the conventional digital displays. They wanted experiments to answer this question.

We have seen this general attitude in a variety of guises. Industry and government organizations often want the product (e.g., an interface) but balk at the suggestion that any activities not directly related to implementation are necessary. The attitude can be paraphrased in the following manner: Theoretical analyses and fancy principles may be fine for the academics in their white towers, but they have no place in the applied science of building an interface (just build it and test it!).

This attitude is not just restricted to nonacademics. We sometimes hear from prospective students that other professors have suggested that we are too theoretical. Many years ago, a colleague in our department questioned whether the time and effort required by the CSE approach was justified (this colleague is now an advocate). One of our students, a terse and laconic army officer, may have summed it up best: "Is the juice really worth the squeeze?" Our work does have strong theoretical underpinnings because we believe the well-worn maxim that nothing is as practical as a good theory. This is a point to which we will return later.

### 1.3.2 Too Applied!

We have also fielded criticisms from those in the basic research camp. Both of us are trained as experimental cognitive psychologists. However, the majority of our colleagues in cognitive psychology dismiss our work as "applied science." Of course, this translates to "not science at all." For some reason, to study visual search, reaction time, compensatory tracking, crypto-arithmetic, etc., is considered to be basic science, but to study perception, decision making, control, or problem solving in the context of aviation or process control is applied science. There seems to be a sense that research is only basic if it deals with general abstractions from nature. However, if it attempts to delve deeply into any specific natural phenomenon, then it gets labeled as "applied."

The attitude from the basic camp can be further refined. Researchers in this camp believe in reductionism. The goal is to conduct research on basic activities in controlled laboratory settings. These settings are purposefully devoid of the complexities of the real world because these complexities will confound conclusions that can be drawn, via the scientific method, about the basic underlying processes. Understanding these basic processes is the key to the generalization of results since they constitute the common underlying threads that are woven across more complex settings. The goal is to find the truths that exist at the microlevel and then recombine them in an effort to explain more complex activities. From this perspective, any research conducted in more complex settings is unscientific; the inherent confounds will limit the conclusions that can be drawn and the principles that can be devised. Furthermore, any research conducted in these complex settings will produce results that are particular to that setting only and that cannot be generalized to others.

We fully appreciate the benefits of basic research. In fact, some of our best insights regarding display design have their origin in results obtained in controlled laboratory research settings (Bennett and Flach 2008). We also fully appreciate the need for controlled experimentation in evaluating the effectiveness of our designs. Where we differ is with regard to the settings within which those evaluations must take place. The context, which the basic scientists want to strip away, is the most informative piece of the puzzle (e.g., situated action—Suchman 1987).

As a result, we believe that evaluations must take place using realistic simulations and scenarios that capture the essential characteristics of the associated real-world contexts. This allows controlled, yet relevant, experimental results to be obtained. We also differ with regard to the prospects for generalization of results. It is only under these conditions of evaluation (i.e., ones in which fundamental demands of real-world settings are captured) that the results are likely to generalize. We are unapologetically skeptical about approaches that trivialize complex problems.

## 1.4 Pasteur's Quadrant

As this discussion indicates, our beliefs about science place us somewhere between a rock and a hard place, at least with respect to conventional wisdom. We believe that there are general principles to be learned about cognitive systems and that the search for these general principles is a legitimate scientific enterprise. Additionally, we believe that to discover these general principles requires that we become immersed in the particulars of nature's full complexity.

We recognize and appreciate the need for controlled experimental tests of hypotheses that emerge from observations of nature's complexity. But we also maintain a healthy respect for the limitations of any single experimental setting for tapping into the full complexity of nature. We share the desire to find general principles, but we believe that these principles must be grounded in life as it is lived.

We believe that the success or failure of a design can be an important test of our hypotheses and theories. The feedback from such tests can reflect back and inform the design of more representative experimental settings. Design success seldom depends on a well articulated theory; in fact, design innovation typically precedes scientific insight. However, explaining the success or failure of a design is a critical challenge for a basic science of cognition. It can directly inform theory and it can have important implications for the design of more controlled empirical observations. We envision an ideal world where science and design are tightly coupled so that each shapes and is shaped by the other in the search for a deeper understanding of nature that helps us to adapt more effectively.

Only recently have we discovered that this niche, where research is motivated by a desire for both broad theories and practical solutions to important problems, has a name. It is called "Pasteur's quadrant" and it is illustrated in Figure 1.1. Stokes (1997) discussed the evolution of the conventional wisdom that basic (i.e., theory-driven) and applied (i.e., problem-driven) research form the opposite ends of a single continuum. He argued that this convention is wrong and that basic and applied motivations are two independent dimensions, as illustrated in Figure 1.1.

Thus, it is possible, as illustrated by the work of Louis Pasteur, to commit to the ideals of basic science and to the ideals of applied science simultaneously. It is possible to choose research questions based on the potential practical impact for solving important problems (e.g., nuclear or aviation safety) and to approach them with the goal of applying and testing basic theories and with the full scope of empirical and analytical tools of basic research (from field observations to controlled laboratory studies).

### 1.4.1 The Wright Brothers in the Quadrant

We will provide an example that reinforces what it means to be working within Pasteur's quadrant. Our academic institution was named to commemorate the achievements of the Wright brothers. They had a goal that was clearly applied: powered flight. Their efforts to achieve that applied goal were at their nadir toward the end of 1901. Their second visit to Kitty Hawk had been a step backward; modifications to their first glider had produced a new version that was decidedly less air worthy. Wilbur is said to have remarked at this point that "he didn't think man would fly in a thousand years" (Kelly 1996, p. 42).

It turns out that what is sometimes referred to as "mankind's greatest achievement" was in danger of being derailed by basic science. Without getting into too many details (see Jakab, 1990, for a complete description), the design of the glider's wings was based on formulas that included experimentally derived coefficients of air pressure, lift, and drag. Because the second glider produced only about one-third of the lift that was predicted, the Wright brothers suspected that these coefficients were in error.

They conducted a series of basic science experiments to evaluate this possibility. They created a controlled laboratory setting that captured critical elements of the real world (e.g., a wind tunnel and miniature wings) and unique instruments for measurement. The results indicated that the coefficients were indeed wrong, and the rest is history.

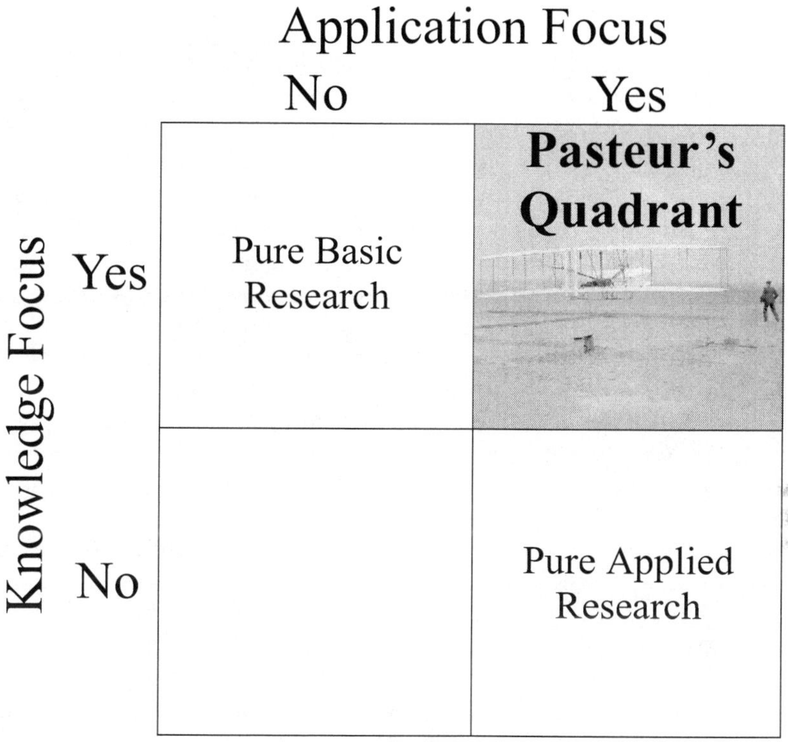


**FIGURE 1.1**
Pasteur's quadrant. (Stokes, D. E. 1997. *Pasteur's Quadrant. Basic Science and Technological Innovation.* Washington, D.C.: The Brookings Institute.) The work of scientists such as Pasteur and the Wright brothers is motivated both by a search for deep understanding (basic theory) and the desire to solve pressing practical problems (application).

Jakab (1990) provides a summary that is particularly relevant to the current discussion:

> They never got bogged down in theoretical matters that were not directly related to the problem at hand. Even though the sophisticated wind tunnel experiments they were about to commence did a great deal to advance the understanding of aerodynamics, the Wrights consciously focused only on those practical questions that would provide them with specific information necessary to building a successful flying machine. They left it to their successors to develop a body of theory that would explain the underlying scientific principles of aerodynamics. (p. 125)

The Wright brothers were clearly working within Pasteur's quadrant.

### 1.4.2 This Book and the Quadrant

Thus, our ambition for this book is to address both basic issues associated with human problem solving and decision making (in Wertheimer's [1959] term: productive thinking) and to address the practical issues of designing graphical interfaces that improve performance in specific work domains. This requires that we consider ontological assumptions and theoretical approaches to human cognition and that we address the specific demands associated with the work domains that we cover. Our ambition is to connect the dots between theory and practice in a way that will help students and designers alike to generalize beyond the specific experiments and display solutions described here, so that they can create new innovations not yet imagined.

Also, it is important that we all realize that the constraints on basic research and applied research are different. Those facing applied challenges cannot always wait for the science to catch up with the challenges. Often they have to rely on heuristics and trial and error to address pressing concerns. If the response of researchers to the requests from designers is always, "Let me do a series of experiments and I will get back to you in a few years," then the designers will eventually stop asking for advice. The time constant for science is typically too long for those who are trying to keep pace with the evolution of technology.

This is where theory becomes most important, as embodied in the form of theoretically based principles of design. These principles will allow scientists to generalize beyond empirical work. In our view, theoretically based principles are the best tool for projecting over the horizon to make guesses about what solutions to try first. Although these principles might not allow scientists to specify the perfect or optimal solution to a new problem, they can often be the basis for ruling out many alternatives—thus greatly simplifying the search process and decreasing the time to solution.

We have seen information and display technologies evolve in ways that we could not have imagined even 10 years ago and the pace of change seems to be accelerating. So, we fully realize that the researchers, students, and designers that we hope to inform today will face challenges and opportunities that we cannot even imagine. We realize that every theory, every principle, and every interface or display solution presented here is a work in progress; each is a step on a continuing journey of discovery. Our goal is not to replace one conventional wisdom with another, but rather to instill an appreciation for the complexity of the problems and for the value of theory for helping to manage this complexity. We hope to instill a skepticism for overly simplistic solutions and an enthusiasm for engaging the challenges to improve human perspicacity and to broaden the sphere of human action in ways that enhance the quality of life.

## 1.5 Overview

The next few chapters are intended to provide a basic theoretical context for studying the topic of display and interface design in terms of work domains and human problem solving and decision making. Chapter 2 lays down an ontological and theoretical context in which we identify the triadic semiotic system, reflecting constraints on awareness, information, and situations as the fundamental unit of analysis. Chapter 3 then expands on the concept of situation constraints. Chapter 4 expands on the concept of awareness constraints. Chapter 5 considers general principles about how these two sources of constraint are coupled in cognitive systems. Chapter 6 introduces the ecological interface design approach, including general principles of design that are applicable across a wide variety of work domains. Chapter 7 provides a general survey of some of the alternative perspectives and approaches to display design that are useful.

Chapters 8 through 11 focus on issues in design for a particular class of interfaces that utilize representations that are primarily analogical in nature. Chapter 8 reviews basic research on visual attention with respect to questions about separability, integrality, and configurality of visual information. Chapter 9 specifically contrasts two sets of display design principles that have been derived from this basic research. Chapters 10 and 11 provide design tutorials that illustrate the principles of ecological interface design applied to the work domains of process control and aviation.

Chapters 12 and 13 focus on issues in design for interfaces that utilize representations that are primarily metaphorical in nature. Chapter 12 explores general issues in the design of metaphorical representations. Chapter 13 provides a design tutorial that considers the iPhone® as an example of an innovative interface that is compatible with the ecological approach. Chapter 14 provides a design tutorial of an ecological interface for military command and control. Chapter 15 describes the principle of visual momentum and associated techniques that can be used to increase it at various levels of an interface (and concrete examples of their use).

Chapters 16 and 17 address issues associated with measurement and evaluation. Chapter 16 contrasts basic assumptions underlying the dyadic and triadic approaches to control and generalizability. Chapter 17 translates these assumptions into a practical guide to the trade-offs involved when evaluating interface solutions.

Finally, Chapter 18 attempts to summarize and reinforce the main themes of the book with the hope of inspiring students to take up the challenge to improve the quality of human experience through interface design.

## 1.6 Summary

> There will be no foolish wand-waving or silly incantations in this class. As such, I don't expect many of you to appreciate the *subtle science and exact art* that is potion-making. However, for those select few who possess

> the pre-disposition ... (Columbus 2001; based on the novel by Rowling 1997, p. 136, emphasis added)

A few words about the title of the book are in order. As the quotes in this chapter indicate, we are not the first to realize that the process of designing displays and interfaces involves elements of both science and art. Rowling's (1997) ability to turn a phrase adds a nice touch. It would be just fantastic if there were an algorithm for display and interface design; we could plug the variables in and produce the most efficient result every time. Unfortunately, to say that there is an algorithm of interface design is a gross exaggeration. At best what we have are theories of cognitive systems and principles of design. As we have emphasized, both the theories and principles are informed by empirical results (so there is a scientific basis); however, they always need to be modified and adapted to the specific circumstances associated with the work at hand—hence the term "subtle science."

The flip side of this coin is that the act of producing effective displays and interfaces is a very creative process. A good display or interface is literally a work of art, especially now that interfaces are highly graphical in nature. It runs deeper than this, however. The design of a display or interface is a creative act, and there may be more than one effective solution (it is certain that there is an infinite number of ineffective solutions). The difference between art in general and art as it manifests itself in display and interface design is that in the latter we are required to convey very specific messages. This involves representing both concrete values (e.g., variables, properties, goals, and the relationships between them) and concepts (e.g., domain entities, potential actions, strategies in execution)—hence the term "exact art."

This book is not designed to provide a catalogue of answers and specific interface solutions. Rather, we hope to illustrate a style of reasoning about interface design. We hope this style of reasoning will contribute to our basic understanding of cognitive systems and improve our ability to make wise generalizations to specific applied problems. We hope to show that the applied interests of the cognitive systems engineering approach are grounded in basic theories of human performance and in basic principles of information and control theory. We hope to show how basic theories of human performance can inform and be informed by the challenges of designing safe human–machine systems. We hope to move the process of ecological display and interface design beyond the realm of magic (wand-waving and incantations) and increase the number of researchers, students, and practitioners who understand it and can apply it beyond the current "select few" who have had the "pre-disposition" to invest years in learning it. Thus, our ultimate goal is to connect the science of cognition with the art of interface design so that the science is applied more subtly and the art is created more exactly.

## References

Bennett, K. B., and J. M. Flach. 2008. Graphical displays: Implications for divided attention, focused attention, and problem solving. In *Best of human factors: Thirty classic contributions to human factors/ergonomics science and engineering,* ed. N. Cooke and E. Salas. Santa Monica, CA: Human Factors and Ergonomics Society.

Burns, C. M., and J. R. Hajdukiewicz. 2004. *Ecological interface design.* Boca Raton, FL: CRC Press.

Columbus, C. 2001. *Harry Potter and the sorcerer's stone.* Warner Bros. Pictures.

Flach, J. M., P. A. Hancock, J. K. Caird, and K. J. Vicente, eds. 1995. *Global perspectives on the ecology of human–machine systems.* Hillsdale, NJ: Lawrence Erlbaum Associates.

Jakab, P. L. 1990. *Visions of a flying machine: The Wright brothers and the process of invention.* Washington, D.C.: The Smithsonian Institution Press.

Kelly, F. C. 1996. *Miracle at Kitty Hawk.* New York: Da Capo Press.

Nadin, M. 1988. Interface design and evaluation—Semiotic implications. In *Advances in human–computer interaction,* ed. H. R. Hartson and D. Hix. Norwood, NJ: Ablex Publishing.

Norman, D. A. 1986. Cognitive engineering. In *User centered system design,* ed. D. A. Norman and S. W. Draper. Hillsdale, NJ: Lawrence Erlbaum Associates.

Rasmussen, J. 1986. Information processing and human–machine interaction: An approach to cognitive engineering. New York: Elsevier.

Rasmussen, J., A. M. Pejtersen, and L. P. Goodstein. 1994. Cognitive systems engineering. New York: John Wiley & Sons.

Rasmussen, J., and K. Vicente. 1989. Coping with human errors through system design: Implications for ecological interface design. *International Journal of Man–Machine Studies* 31:517–534.

———. 1990. Ecological interfaces: A technological imperative in high-tech systems? *International Journal of Human–Computer Interaction* 2 (2): 93–111.

Rowling, J. K. 1997. *Harry Potter and the sorcerer's stone.* New York: Scholastic Press.

Stokes, D. E. 1997. *Pasteur's quadrant. Basic science and technological innovation.* Washington, D.C.: The Brookings Institute.

Suchman, L. A. 1987. *Plans and situated actions: The problem of human–machine communication.* New York: Cambridge University Press.

Tufte, E. R. 2006. *Beautiful evidence.* Cheshire, CT: Graphics Press.

Vicente, K. J. 1999. *Cognitive work analysis: Toward safe, productive, and healthy computer-based work.* Mahwah, NJ: Lawrence Erlbaum Associates.

Vicente, K. J., and J. Rasmussen. 1990. The ecology of human–machine systems II: Mediating "direct perception" in complex work domains. *Ecological Psychology* 2:207–249.

Wertheimer, M. 1959. *Productive thinking.* New York: Harper and Row.

# 2

# *A Meaning Processing Approach*

## 2.1 Introduction

> We may be at the start of a major intellectual adventure: somewhere comparable to the position in which physics stood toward the end of the Renaissance, with lots of discoveries waiting to be made and the beginning of an inkling of an idea of how to go about making them. It turned out, in the case of the early development of modern physics that the advancement of science involved developing new kinds of intellectual sophistication: new mathematics, a new ontology, and a new view of scientific method. My guess is that the same sort of evolution is required in the present case (and by the way, in much the same time scale). Probably now as then it will be an uphill battle against obsolescent intellectual and institutional habits. (Sloan Foundation 1976, p. 10; cited by Gardner 1985, p. 34)

Howard Gardner (1985) suggests that the birth of cognitive science happened in the mid-1950s, when psychology began to move away from behaviorist, stimulus-response views that ignored the construct of mind toward a new view that began to frame questions of mind in the context of developing theoretical and technical achievements related to the processing of information. While it is true that the innovations of the 1950s brought mind back into fashion as a topic for psychology, we believe that, consistent with the predictions in the opening quote, the mathematical, ontological, and methodological changes needed for a revolutionary new scientific approach are only gradually being realized. Thus, cognitive science is still in the middle of an ongoing struggle between multiple paradigms.

The term "paradigm" is one that will be used throughout the book and a short explanation of the meaning of that term is in order. Extending the concepts originally introduced by Kuhn (1962), Lachman, Lachman, and Butterfield (1979) differentiate between the "rational" and "conventional" rules of science. The rational rules refer to general aspects of the scientific method: the formation of hypotheses and the testing of these hypotheses based on observational data. Essentially, this is the type of activity that all scientists must do to qualify as such. In contrast, the conventional rules are the value and belief structures about how this process should proceed. Thus, different groups of scientists studying the same topic can hold different opinions with regard to what should be studied, how it should be studied, and how results should be interpreted. Lachman et al. (1979) emphasize the importance of conventional rules, stating that "a science is shaped as much by paradigmatic judgments [conventional rules] as by the canons of scientific method" (p. 19).

Lachman et al. (1979) define the term paradigm in the following observation: "[W]ithin scientific disciplines, there tend to form subgroups whose members adopt very similar resolutions. When a sufficiently large number of scientists in a field agree to a considerable extent on how such questions are to be resolved, they are said to share a *paradigm*" (p. 6; emphasis original). Note that a paradigm is a far more general term than a theory, an experimental setting, a procedure, or a task (even though the term is often used to describe these entities).

The purpose of this chapter is to provide an overview of two competing paradigmatic stances that have shaped a debate on how the process of interface design should proceed. One paradigm is associated with the classical information processing view of cognition; the other is associated with more ecological approaches to cognition. We will be advocating in favor of this second paradigm, which we will refer to as a meaning processing approach.

## 2.2 Two Alternative Paradigms for Interface Design

The roots for these two paradigmatic approaches to cognition can be traced to a field that has direct relevance for the problem of interface design: *semiotics* (e.g., Nadin 1988). Semiotics is typically referred to as the science of signs or signifying. Its roots can be traced back to early medicine, where the problem of diagnosis was referred to using the Greek word *semeiosis.* This semiotic problem was to identify the nature of a disease based on the patient's symptoms. In a significant sense, the physician's problem is a representative example of issues that are central to any theory of cognition and, by implication, to any theoretically based principles of interface design. Given the available information (e.g., the patient's symptoms, the interface, the specific problem representation), what is the appropriate interpretation? More practically, what is the appropriate action?

### 2.2.1 The Dyadic Paradigm

The roots for the conventional information processing approach to interface design can be traced to the dyadic sign/semiotic model framed by Saussure. Ferdinand Saussure (1857–1913) is considered by many to be the founder of modern linguistics, which in turn had a strong influence on the science of cognitive psychology. He framed the semiotic problem as a relation between the signifier (e.g., symbolic language) and the signified (e.g., mental concepts). The semiotic problem was framed as the study of the nature of the various possible mappings between language and thought or in the context of medicine between symptoms and categories of diseases.

This framework fits well with the computer metaphor for the mind. In this context, thinking is framed as a symbolic computation linking input (e.g., symptoms in the medical context) with an internal classification via rules. In this sense, semiotics is an exploration of the types of mappings or rules that are possible between the signifier (data) and the signified (internal concept).

Saussure's framework seemed to fit well with the goals of linguistics and computer science (i.e., matching symbols to concepts), but with respect to the medical diagnosis problem something is missing. This framework does not seem to include any consideration of the actual state or health of the patient (beyond her symptoms). To what extent do the medical community's categories of disease correspond to the actual states of the patient's health? Further, to what extent do the treatments associated with the categories actually affect that health?

What is missing from Saussure's framework is any basis for connecting the concepts of medicine to the actual health of the patient (e.g., a basis that could help differentiate between the semiotic system of the traveling medicine show charlatan and the well-trained physician). Saussure's semiotics is framed in terms of the relation between the sensory surfaces of an observer and the internal concepts in her mind. This framework fails to close the loop through the object of action: the patient (or, more generally, the ecology). This can be a potential problem for conventional approaches to both interface design and cognition. In the conventional approaches, cognition is almost completely divorced from situations; the only connection is the "scraps of data" at the sensory surfaces. In other words, cognition is disembodied (e.g., Clark 1997). Connections via action tend to be ignored or trivialized.

### 2.2.2 The Triadic Paradigm

A contemporary of Saussure independently framed an alternative triadic model for semiotics. The work of Charles Peirce (1839–1914), which had a strong influence on William James and early functionalist approaches to psychology, has been largely ignored by conventional approaches to cognitive science. However, appreciation for Peirce's work is rapidly growing. Peirce (1931–1935a, 1931–1935b) framed his semiotics in the context of the logical links between the objects of experience and the objects of the world. Peirce's semiotics resulted from his struggles with the processes by which the objects of experience could guide successful interactions with a physical world. How could the physician's knowledge guide successful treatment of disease?

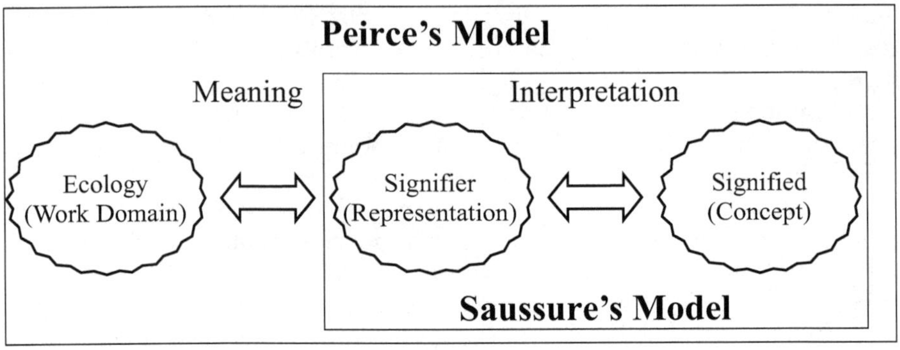


**FIGURE 2.1**
This diagram compares Saussure's dyadic model of semiotics with Peirce's triadic model.

Figure 2.1 compares the dyadic and triadic models of the semiotic problem. We have labeled the third dimension inspired by Peirce's triadic model as the "ecology." In the medical example, this would correspond to the patient's state of health. We use the term "meaning" to refer to the relation between the ecology and the signifier or representation. In the context of medical diagnosis, this would be the relation between the patient's symptoms and the patient's health (i.e., the meaning of the symptoms). We use the term "interpretation" for the relation between the signifier and the signified (concept or belief). This is the relation between the symptoms and the classification by the physician. Treatment is more likely to be successful when the interpretation corresponds with the meaning, although patients can sometimes get well despite the doctors.

### 2.2.3 Implications for Interface Design

For Saussure and for the conventional approaches to cognition and interface design that follow his tradition, the ecology is not part of the sign system. Thus, to the extent that meaning is considered, it is synonymous with interpretation. In this context, it is natural to focus on the relation between the concept (mental model) and the representation (display) and there is little reason to consider anything beyond the representation itself. However, when the sign system is expanded to include the ecology or work domain, meaning can be framed independently from interpretation. In this context, *the role of display designers is to build representations so that the interpretations of the operators using those representations will correspond with the meaning in the ecology.*

For example, the goal is to represent the patient's symptoms so that the doctor's diagnosis will correspond to the actual state of the patient's condition. Thus, understanding the patient's condition and the possible ways that this condition can be represented becomes a significant concern of the interface designer. Ultimately, the goal of any representation is to guide successful action (i.e., treatment).

Thus, the point is not simply to match the mental model of a particular doctor, but rather to consider the extent to which that mental model is aligned with the most up-to-date theories of medicine. We would not want to design an interface that reinforces an antiquated or incorrect mental model. This suggests that it is not sufficient to know how a particular doctor thinks about the problem, but it raises questions about how a doctor *could* and *should* think about the problem. In some respects, *the goal is not to match a particular mental model, but rather to match the best possible models given the collective knowledge of the medical domain.* Note that we used the plural for models. In complex work domains (e.g., medicine or military command and control), there may not be one authoritatively correct model of the domain. In this case, multiple representations or a flexible representation that supports multiple perspectives may be desirable. The ultimate test of any model will be its pragmatic value in guiding successful treatment.

In sum, the field of semiotics offers two frameworks for addressing cognition and meaning. One framework, the dyadic view, focuses exclusively on the relation between a sign (e.g., written word, computer icon, or medical symptom) and the associated mental construct (e.g., interpretation). Within this framework, it is logical to focus theory and research on the mental processes that accomplish the interpretation and it is not surprising that meaning is a disembodied mental construct associated with interpretation. The result is that there are two realities or ontologies: one for mind and another for matter.

An alternative framework is the triadic view. This framework considers interpretation within a larger context that includes the practical value of any associated mental constructs. That is, how do the constructs relate to successful action? In this case, meaning takes on a pragmatic dimension and mind and matter constitute a single reality or ontology.

The result of these two different paradigmatic views is that there is great confusion associated with the construct of meaning. In the dyadic tradition, meaning is a purely mental construct with no grounding outside the mind. It requires a unique ontology from that of the physical world. In the triadic tradition, meaning is grounded in actions and the associated consequences (i.e., the ecology). Thus, the triadic view includes mind and matter in a single ontology that focuses on the relational dynamics between them—on what matters. It is this triadic view that informs our approach to interface design. While it is not necessary that you share our perspective, *it is important that our perspective is clear from the start so that there is no confusion about what we mean when we say that an interface is meaningful. We mean that it is a useful guide for action.* We hope that this will become crystal clear by the end of this chapter.

## 2.3 Two Paths to Meaning

Consider a pilot attempting to land her aircraft. What information is most meaningful?

- The state of the aircraft (e.g., altitude, attitude, airspeed)
- The state of the various instruments (e.g., the altimeter, airspeed indicator, artificial horizon)
- The state of the optical flow field (e.g., the perspective of the runway, the expansion pattern, the flow of texture)
- The state of the pilot (e.g., alertness, skill, knowledge, goals)

To what extent does the meaningfulness of the information depend on the laws of physics (e.g., aerodynamics) and the particulars of the specific situation (e.g., type of aircraft, size of runway, weather)? To what extent does the meaningfulness of the information depend on the state of the pilot (e.g., her previous experience)? In other words, is meaningfulness a property of the situation, is it a property of awareness, or does it somehow depend on both?

### 2.3.1 Conventional Wisdom: Meaning = Interpretation

Conventionally, using a dyadic semiotic system, questions of meaningfulness have been framed in the context of awareness. This is reflected in disproportionate attention to illusions and irrational thinking, rather than to skilled cognition and action. For example, conventional texts on perception tend to focus on optical illusions. There seems to be a working assumption that meaning is constructed based on ambiguous information. For example, Richard Gregory (1974) writes:

> [P]erceptions are constructed, by complex brain processes, from fleeting scraps of data signaled by the senses and drawn from the brain's memory banks—themselves constructions from snippets from the past. On this view, normal everyday perceptions are not part of—or directly related to—the world of external objects, as we believe by common sense. On this view all perceptions are essential fictions: fictions based on past experience selected by present sensory data. (p. xvii)

The large collections of illusions where human judgments tend to be inconsistent with standard measures (e.g., Muller–Lyer illusion) are taken as evidence that the information available to perception is ambiguous. Also, the two-dimensional property of the visual sensory surface is taken as clear evidence that information is missing relative to the three dimensions of the world in which we live.

Similarly, theories of thinking and decision making tend to focus on violations of normative models of rationality (e.g., conjunction fallacy, gambler's fallacy, etc.) and biases and heuristics in judgment and decision making (e.g., anchoring and adjustment, representativeness, hindsight). For example, Wickens (1992) writes:

> Many aspects of decision making are not as accurate as they could be. The limitations of information processing and memory, previously discussed, restrict the accuracy of diagnosis and choice. In addition, limits of attention and cognitive resources lead people to adopt decision-making heuristics, or "mental shortcuts," which produce decisions that are often adequate but not usually as precise as they could be … . Finally, we will sometimes refer to general biases in the decision-making process. These biases are either described as risky—leading to a course of action based on insufficient information—or conservative—leading to the use of less information or less confidence in a decision than is warranted. (p. 261)

In this context, in which meaning is constructed from ambiguous data using limited and biased information processes, it is not surprising that two people can interpret the same situation very differently. For example, two pilots might have different opinions about where the limits of a safe approach are or two drivers may have different ideas about when to initiate braking when approaching a line of traffic. Have you ever instinctively reached for the imaginary brake pedal when you were the passenger of a more aggressive driver?

Thus, conventional wisdom tends to treat meaning as if it were synonymous with interpretation. The meaning is how an individual interprets the situation. This suggests that meaning is a property of awareness and this leads implicitly to the conclusion that the situation is, at best, of secondary interest and, at worst, meaningless or irrelevant. Somehow, meaning is constructed from ambiguous information, based on arbitrary relations to the ecology or situation. Meaning becomes a pure invention or construction of the mind.

In this context, it is not surprising that much of the work in human factors, engineering psychology, and human–computer interaction (HCI) is framed in terms of internal mental processes. In this context, the job of the human factors engineer is to ensure that the designers take into account the limits of these internal computational processes (e.g., perceptual thresholds, memory limitations, decision biases, etc.). In terms of aiding problem solving, the target for design is often framed as a requirement to match the operator's mental model.

There are obvious values of the conventional approach. Certainly, if features of the interface are below perceptual thresholds, if the operator is overwhelmed by an avalanche of data, or if she is constantly surprised by unexpected events, then the value of the interface may be compromised.

However, from the start of our careers thinking about the design of interfaces for safety critical systems such as nuclear power plants and aircraft, a question has nagged us:

What if the mental models of the operators are naive or even wrong?

This possibility was explicitly noted by Norman (1986), who once described mental models as "messy, sloppy, incomplete, and indistinct" (p. 14). It seemed obvious that just any mental model would not be sufficient for operating a nuclear power plant or landing an aircraft safely. Certainly, these systems behave according to physical principles and it seemed obvious that unless the thinking (mental model) at least implicitly takes these principles into account, the control and problem solving might be ineffective and perhaps dangerous. In thinking about how to design interfaces in these contexts, we have found it important to ask, "What is the 'right' mental model?" Or, more conservatively, "Are some models more effective or satisfactory than others? Why?" The point is that the mental model must have some correspondence with the physical process that is being controlled.

### 2.3.2 An Ecological or Situated Perspective: Meaning = Affordance

In thinking about this problem of the right conceptual model, we have come to question the conventional notion of meaning and the conventional notions about the nature of computational processes. We began to look to the situation and to frame questions of meaning in this context. We found ourselves asking how the nuclear power plant and the aircraft actually work. For example, we began to press beyond questions to pilots about how they thought or what they did to ask, "Why was that strategy or that procedure adequate?" In searching for answers to these questions, we had to go beyond the pilots; we had to begin talking with the aeronautical engineers and we had to begin learning about aerodynamics.

For example, in a landing approach to a typical airport, pilots generally begin by setting the throttle to about 70% of full cruising power. They then fly the glide path using their stick (elevators)—not to point the aircraft at the point of touchdown, but rather to keep a constant target airspeed. Most pilots know this strategy. They know that it normally works. However, not all can explain why this strategy works or are aware of the conditions where this might not be the safest strategy (e.g., landing on short fields or aircraft carriers).

In the process of exploring questions about why this works, we learned about the physics of aviation and the relations between total, kinetic, and potential energy. We learned a new way to think about the flight controls. We learned that the throttle's function is to determine the rate of change of energy. Depending on the throttle's setting, total energy will be decreasing (energy-in is less than energy-out due to drag), increasing (energy-in exceeds energy-out), or constant (energy-in is equivalent to energy-out). We learned that the stick's function is to determine the distribution of energy between speed (kinetic energy) and altitude (potential energy).

For example, if energy-in exceeds the energy loss due to drag, then the aircraft must either climb (increase potential energy), accelerate (increase kinetic energy), or both depending on the position of the elevator. Thus, at the correct throttle setting for landing, there will be a roughly constant rate of total energy loss. If the speed (kinetic energy) is constant, then the deficit will result in a constant loss of altitude (potential energy). If the throttle setting is correct and the pilot tracks the correct speed, then the plane will follow the correct constant glide path (e.g., in the range of 2 or 3°). Our collaborations with the aeronautical engineers led to the design of a new landing display that includes an energy path along with a more common flight path display (Amelink et al. 2005).

The work with the aviation display will be discussed in more detail in Chapter 11. The important point for this chapter is the implication for our approach to meaning. As the result of our experiences with displays for safety critical systems, we have found it useful to think about meaning as an attribute of situations. For example, we learned about the functions of the stick and throttle in relation to energy variables (e.g., total, kinetic, and potential energy), rather than simply in terms of the displayed variables (e.g., air speed, altitude, and attitude). In essence, the energy relations (or constraints) provide a context for specifying the functional meaning of throttle and stick actions with regard to achieving the goal of a safe landing. These constraints are meaningful to anyone in the flying situation—no matter what they might believe or know. If someone does not respect these constraints to safe travel, then the consequences will be real and significant! As you will see in Chapter 11, our explorations into the ecology of flight led to some interesting ideas about how to improve landing displays.

An important influence on our movement toward a triadic approach was James Gibson. He was one of the first people to advocate an approach that framed meaning in terms of the situation (or the ecology), rather than as a mental construction. For example, Gibson and Crooks (1982) introduced the concept of the *field of safe travel* in the context of driving. They described this as the "field of possible paths which the car can take unimpeded" (p. 120). This field contains another field that they called the *minimum stopping zone,* which they defined as "the minimum braking distance required to stop the car" (p. 123). They noted that this minimum braking distance depends on the speed of the car, the condition of the road, and the conditions of the brake. They further noted that the field of safe travel is not a "subjective experience of the driver. It exists objectively as the actual field within which the car can safely operate, whether or not the driver is aware of it" (p. 121).

As Gibson's ecological perspective evolved, the insights about the field of safe travel were incorporated into the more general construct of *affordance.* Gibson

(1979) defined the "affordances of the environment" as "what it offers the animal, what it *provides* or *furnishes*, either for good or ill" (p. 127). He goes on:

> The notion of affordance implies a new theory of meaning and a new way of bridging the gap between mind and matter. To say that an affordance is meaningful is not to say that it is "mental." To say that it is "physical" is not to imply that it is meaningless. The dualism of mental vs. physical ceases to be compulsory. One does not have to believe in a separate realm of mind to speak of meaning, and one does not have to embrace materialism to recognize the necessity of physical stimuli for perception. (Gibson 1972/1982, p. 409)

Gibson seems to be using the term "affordance" to address the meaningful aspects of situations. That is, there is a meaningful relation between the animal or human and its environment that is independent of the animal or human's subjective beliefs or opinions; there are objective possibilities and objective consequences. For example, consider the situation of a clear sliding glass door: The surface has objective properties that have consequences for safe and unsafe modes of locomotion. That is, the consequences are independent of the observer's opinion or belief about whether the path is clear or not, though these opinions or beliefs may have obvious consequences for performance. If the person thinks that the path is clear, then she may discover the error of her interpretation when she collides with the solid surface.

As a second example to illustrate the concept of affordance, consider the case of a frozen pond. Will the pond *afford* locomotion? This *affordance* depends on the complementary relation between the thickness (strength) of the ice and the weight of the human. The same pond may afford support for an insect or a small child, but not for a large adult (see Figure 2.2). Thus, it depends on a relation between animal and environmental object; that is, it is impossible to say whether the pond affords support without considering both the ice surface and the animal weight distribution.

Note that it is possible to consider this relation as a property of the situation independently of an interpretation. That is, the pond may afford locomotion for an individual, yet that individual may not be sure and may choose to take a path around the pond. The failure to perceive the possibility does not change the possibility or affordance, although it certainly will constrain the choice of action.

On the other hand, the pond may not afford locomotion, but the individual may believe that it does and proceed across the surface. In this case, the person will soon discover the "objective" value of the affordance as she crashes through the ice. Thus, the person wishing to get to the other side of the pond faces a semiotic problem: Does the pond "afford" support? How do I know? Should I try it and risk the possible consequence of falling through? Or should I expend the additional time and effort required to travel around the pond?

In contrast to Gregory (see previous quote), Gibson believed that a moving eye had access to rich information with respect to the layout of the

**FIGURE 2.2**
The same frozen pond may afford support for one organism, but not another.

three-dimensional world. For Gibson, affordances were the objects of perception—not the products of perception. He hypothesized that, for many situations (such as controlling locomotion), the affordances were directly specified by information (e.g., in an optic array) available to the observer. This means that if the observer were appropriately attuned to this information, she would be able to respond skillfully to those affordances. In Gibson's terms, the well-tuned observer would be able to perceive the affordances directly.

Note that Gibson used the word information in a particular way to refer to the degree of specificity between structure in a medium (e.g., an optic array) and the affordances in the ecology. In the next section, we will explore how this use of the concept of information fits with the technical usage defined by Shannon and Weaver's (1963) information theory.

At this point, however, we would like to introduce the following premises as generalizations of Gibson's ecological approach to the paradigm from which we approach problems of interface design:

- Meaning is a property of situations associated with the possibilities and the consequences of action.

- Skilled action will generally depend on the degree of correspondence between the actual possibilities and consequences of a situation (meaning) and the actor's beliefs about the situation (interpretation).
- Meaning is the input (stimulus, raw material) for cognitive processing, rather than the product of cognitive processing. In other words, meaning is not created by the mind, but it can be discovered by the mind.
- The discovery processes will be more or less difficult depending on the complexity of the situation and the richness of information (degree of specificity) available as feedback from the interactions.

An important implication of these premises is that *a major concern for interface design will be to facilitate the discovery of the meaningfulness of the situation.* That is, the goal of interface design is to develop representations that specify the meaningful properties of a work domain (or problem space) so that operators can discover these meaningful properties and can guide their actions appropriately (increasing the likelihood of positive consequences and minimizing the risk of negative consequences).

In the past, we have framed the perceptual side of these challenges as the semantic mapping principle (Bennett and Flach 1992). Whereas the conventional approaches in human factors and HCI tend to focus exclusively on the relations between the interface and the processes internal to the operator, we are advocating an approach that includes the relations between the interface and meaningful properties of the work domain. We believe that an important concern of the display designer has to be with meaning—not as a property of mind, but rather as a property of the situation or functional problems that operators are trying to solve. The goal is to design representations that are true to the problem—to design representations that support productive thinking (e.g., Wertheimer 1959). Thus, this requires consideration of the constraints of situations as well as the constraints on awareness.

### 2.3.3 Information versus Meaning

The term "meaning processing approach" was chosen to contrast with the more conventional "information processing approach" that has dominated human factors and engineering psychology since its origins during the Second World War (e.g., with the work of Paul Fitts at Wright-Patterson and of Bartlett and colleagues in Cambridge). Ironically, we believe that the meaning processing approach is more consistent with the spirit of information theory (and perhaps with the original spirit of Fitts's and Bartlett's work) than the approach that has since adopted the title information processing approach.

The crux of information theory is that an event must be understood relative to a space of possibilities. For example, the amount of information associated with drawing a particular number from a bin cannot be specified without knowing the numbers contained in the bin. If all the numbers in the bin are identical, then there is no surprise (and thus no information) when that number is drawn. The more alternative numbers in the bin, the greater the uncertainty that is resolved (information that is communicated) when a number is drawn. Thus, *it is impossible to specify the amount of information communicated by an event without some knowledge about the possibilities in the jar.* In this example, the jar (with its associated possibilities) is analogous to the work domain (or ecology). And, for the specific case of a jar of numbered balls, enumerating the alternatives is a good way to characterize this domain.

Research on choice reaction time (Hick 1952) demonstrated that human performance was functionally related to information demands. In these experiments, the task was specifically designed so that the number of alternatives and the probabilities associated with the alternatives was an appropriate way to specify the field of possibilities (analogous to the jar of numbered balls). The results were consistent with the predictions of information theory. For example, in the reaction time task, there was a linear relation between the uncertainty of an alternative (measured in bits) and the time for a correct response.

This was a very important demonstration of the relevance of information theory for modeling human performance. However, we believe that an incorrect generalization resulted. The information processing theory of human performance tended to focus on the number and probability of alternatives as fundamental measures of possibilities since this is how Hick and others (Hyman 1953) manipulated information in their experimental settings. The result was an emphasis on probability theory, rather than on information theory. Thus, any aspect of a stimulus situation that could not be indexed by counting alternatives or computing probabilities tended to be ignored. And even physical processes (e.g., nuclear power plants) that might be characterized in terms of physical laws were reduced to probabilistic descriptions in order to model the human–machine system (e.g., THERP [technique for human error rate prediction]; Swain and Guttman 1983). In this context, it was natural to study human decision making relative to normative models of choice related to probabilistic gambles.

We believe that if one is serious about generalizing the insights of information theory to an understanding of human–machine systems, then one must be serious about modeling the space of possibilities that exist in a work domain. Therefore, one should be interested in the best ways for characterizing these possibilities. We doubt that many aeronautical engineers would be satisfied with probabilistic models of aircraft performance. Nor would many nuclear engineers be satisfied with probabilistic models of the dynamics of feedwater control.

In some cases—for example, the reliability of transistors or the base rates for specific diseases—probability theory will be an important index of possibilities. However, in other cases, physical laws will be more appropriate. Thus, we feel that information theory demands an ecological approach to human–machine systems in general and to interface design in particular. *An important step toward understanding the demands of any problem or work domain is to understand the constraints that bound the space of possibilities.*

Gibson used the term "information" to describe the specificity between variation in the ecology and variation in a medium such as the optical flow field (this is essentially the interface for natural locomotion). Specificity depended on invariance between structure in the optical flow field and structure in the ecology. If the dynamic perspective in the flow field was lawfully related to the motion of the observer, then these laws specified both the layout of the environment and the motion of the observer relative to it. In the classical sense of information, if the flow field is lawfully/invariantly related to properties of the environment and motion relative to it, then it is a good communication channel. That is, the variations in the ecology are well specified by the variations in the representation (i.e., the optical array).

This raises another irony of information theory with respect to display design that is not always appreciated: *For a representation to be a good communication channel, it must be as complex as the thing being represented.* To the extent that the representation is less complex (i.e., variable) than the thing being represented, information is lost. To the extent that the representation is more variable than the thing being represented, information (in this case, noise) is added. Thus, designing a representation to conform to the information processing limits of an operator without considering the consequences in terms of lost information with respect to the domain being controlled can end up trivializing the problem and can lead to a very brittle (unstable) control system. These ideas were originally articulated by Ashby (1956) as the "law of requisite variety."

Fortunately, as early researchers such as Miller (1956) realized, the complexity in terms of information does not place a hard constraint on human processing capacity. This is due to people's capacity to "chunk" or to recode the information presented. Because of people's capacity to organize or recode information, difficulty is not necessarily proportional to complexity (as indexed by amount of information). A key consideration throughout this book will be to consider how designers can utilize the chunking capacity of humans to best advantage. That is, *can we organize information within representations to make it easier for people to chunk data in ways that reflect meaningful properties of the work domain?* For example, the physical relation between total, potential, and kinetic energy can provide a frame for integrating multiple factors into a chunk that reflects structural constraints of the aviation domain.

As Tufte (1990) observed, difficulty of a representation may well be independent of the amount of information that it contains:

> Confusion and clutter are failures of design, not attributes of information. And so the point is to find design strategies that reveal detail and complexity—rather than fault the data for an excess of complication. Or, worse, to fault the viewers for lack of understanding. (p. 53)

Although understanding the space of possibilities is important, it is not sufficient for a meaning processing approach. One must also understand the consequences or values associated with those possibilities in order to address issues associated with meaning. Some possibilities are desirable (e.g., goals) and the system should behave in ways that maximize the probabilities associated with those possibilities. Other possibilities are undesirable (e.g., threats or risks) and the system should behave in ways that minimize the probabilities associated with those possibilities. Information statistics do not consider value. A system that is consistently wrong can be equivalent in terms of information statistics to a system that is consistently right.

However, these two solutions are not equivalent to the designer. *Our goal is to design systems that are consistently right.* Thus, the meaning processing approach considers both the possibilities (as demanded by information theory) and the consequences and values associated with the possibilities. The impact of payoff matrices on performance (e.g., signal detection) demonstrates that value is an additional constraint that must be considered.

For example, in the case of the landing approach, it is important to distinguish between states (e.g., attitudes, positions, and velocities) consistent with the possibility of a soft landing and those that are inconsistent. A good interface should make the distinctions between those states salient to the pilot. In our view, these distinctions are meaningful. Similarly, in diagnosing and treating a patient, it is important that the physician's decision be based on the best possible understanding of the possible states of the patient and the possible consequences of various treatments, including doing nothing. Thus, the point is not to specify all the possibilities uniquely, but rather to make clear the distinctions between possibilities consistent with the functional goals and those that are to be avoided (i.e., the "fields of safe travel" or the "safe or desirable envelopes of performance").

In sum, we would like to make the following assertions about the relation between information, meaning, and the goals of display design:

- Information theory demands that designers attend to the possibilities and constraints of the ecology being represented. Any event must be represented relative to the context of these possibilities.
- Probability is not the only way and generally not the best way to characterize the possibilities in an ecology. Often possibilities or actions will be constrained by known physical laws (e.g., mass and energy balances, laws of motion, etc.).
- In order to be an optimal information channel, a display must be as complex as the problem being represented.
- Difficulty is not necessarily proportional to complexity (in information terms). Processing of complex information can be facilitated through recoding (organizing) information into chunks (preferably reflecting natural properties of the work domain).

- Consideration of the possibilities is necessary but not sufficient to characterize the meanings within an ecology. A second dimension that must be included is the value associated with various possibilities (e.g., consequences, costs, and rewards).
- The ultimate goal of interface design is to help people to distinguish between those possibilities that are desirable and those possibilities that are to be avoided.

## 2.4 The Dynamics of Meaning Processing

The information processing model of human performance is organized as a sequence of effectively independent stages of analysis. Research programs are typically designed to isolate a particular stage, and researchers tend to focus their careers on a particular stage in this process (e.g., one might study perception and another might focus on memory, while others might focus on decision making). Although the stages pass information from one to the others, the dynamics within a stage are at least implicitly treated as if they were isolated from the other stages.

Although the images inspired by these stage models often include feedback, this approach tends to treat the overall system as an open-loop communication channel, rather than as a dynamical closed-loop control system (Jagacinski and Flach 2003). In this respect, this model is not a significant advance beyond the classical stimulus–response models of the behaviorists (essentially the information processing approach simply inserts several dominoes into the chain between stimulus and response).

In our view and those of an increasing number of researchers, this model trivializes the rich, closed-loop coupling of perception and action essential to adaptations to complex ecologies. Those interested in the problems of motor skill and coordination (e.g., Bernstein 1967) have long recognized the significance of these couplings to the overall dynamic. More recently, the field of naturalistic decision making has recognized that, in dynamic environments such as fire fighting, it is impossible to isolate a decision from the processes associated with situation assessment (i.e., recognition as exemplified in Klein 1989; Klein, Orasanu, and Zsambok 1993).

### 2.4.1 The Regulator Paradox

Weinberg and Weinberg (1979) have used the example of driving onto road surfaces that have ambiguous consequences for control (difficult to tell whether they are simply wet or icy—black ice) to characterize the nature of coupling between perception and action that they call the "regulator paradox." Good drivers will typically resolve the ambiguity associated with the state of the road by inserting a test signal with their steering wheel. From the feedback provided by a small pulse to the steering wheel, the driver can sometimes tell whether the surface is simply wet or whether it is icy. If the driver feels a small skid, he may be able to adjust to a more cautious control strategy, to avoid a more catastrophic skid on the black ice. On the other hand, a novice driver may continue driving normally without testing the surface. She may be surprised when her car eventually skids out of control.

The paradox is that, in a changing environment, a regulator must simultaneously function as an observer and as a controller. Error with respect to the control task (the small pulse to the steering wheel) can be information with respect to the observer task. In the process of maintaining control, the driver is both minimizing error and obtaining information about the changing surface that could be valuable to the observer function.

The regulator paradox reflects the intuitions of people who have observed cognition in the wild. In contrast to the conceptualization of a series of information stages (e.g., observation then control), stable performance in a dynamic environment often demands that observer and control processes be accomplished in parallel. Every action is both a means to an end (performatory) and a test of the situation (exploratory). In any dynamic environment, the actor is in the position of learning the rules of the game while simultaneously engaged in playing the game.

### 2.4.2 Perception and Action in Meaning Processing

Figure 2.3 illustrates our conceptualization of meaning processing. This conceptualization is different from more conventional approaches in three important ways. First, this is not framed in terms of processes in the head, but rather in terms of simultaneous dynamics occurring between an actor, information medium, and ecology. Second, this does not reflect a serial sequence of processes; rather, perception and action (or control and observation) are intimately coupled and operating in parallel. In this context every interaction has dual implications. Finally, none of the elements in Figure 2.3 is uniquely associated with either the individual or the environment. That is, the ecology reflects the physical constraints scaled with respect to the organism (i.e., affordances). The medium could include the sensory surfaces as well as more conventional media such as the optical array or the graphical interface. The belief system could include cultural knowledge embedded in cultural, organizational, and physical artifacts (taboos, libraries, manuals, etc.) as well as the internalized experiences of an individual.

Although observation and control function in parallel, in the process of writing we are constrained to a sequential mode of description. We will begin with the dynamic of control since that view is consistent with Weiner's (1948) cybernetic view that has been so influential on conventional images of human performance. For the control dynamic, the medium functions as the comparator. Here an intention (goal) is compared with the current state of the ecology.

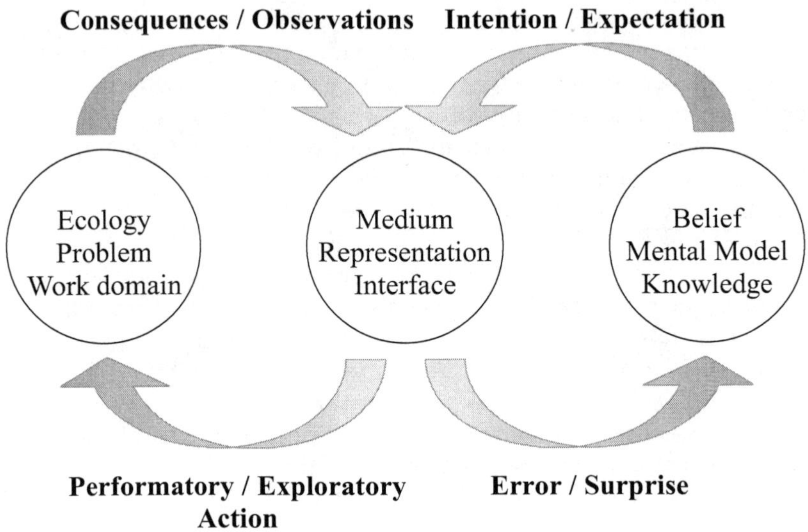


**FIGURE 2.3**
The dynamics of meaning processing. These dynamics involve interactions between a cognitive system and an ecology mediated by an interface (displays and controls). Perception and action are dynamically coupled in parallel so that every interaction has dual implications.

If there is a mismatch, then a performatory action is initiated and the consequences of that action are compared to the intention, leading to further actions. This negative feedback control system acts to reduce the error. In an engineered control system, the intention and feedback are scaled to have common units or currency so that the comparison process is effectively subtraction.

However, in the dynamics of life, this is not the case. A significant challenge to any biological control system is to compare between one currency and another. For example, the pilot must compare patterns in an optical flow field or on her instruments to her intention to land softly and then use that comparison to specify actions on various control surfaces (pedals, stick, and throttle). This is a major challenge for the beginning pilot: How can I tell whether I am on a safe trajectory or not. If I am *not* on a safe trajectory, what actions will most effectively correct the problem?

Conventionally, the intention for the servomechanism is treated as a constant goal and the primary means for reducing error is to act in a way that moves the system toward the goal. The classical example is a thermostatically controlled heating system. However, in Figure 2.3 it should be apparent that there are two ways to reduce error. One can act as a conventional servomechanism or one can change one's intention. For example, a pilot experiencing difficulty maintaining a safe glide path when landing may change her intention from landing to flying a missed approach—recovering altitude and entering back into the traffic pattern. In the engineered system (e.g., the thermostat), the goals are imposed from outside the control system. A person specifies the goal temperature for the heating control system. However, a meaning processing system must choose its own goals. The problem of knowing what goals are feasible leads naturally to the observation problem.

### 2.4.3 Inductive, Deductive, and Abductive Forms of Knowing

The observer problem is most typically encountered in the human performance literature in terms of signal detection theory. Yet, the closed loop dynamics of observers are rarely addressed. The essence of the observer problem was captured well by Peirce's (1931–1935a, 1931–1935b) construct of *abduction.* Peirce presents abduction as a contrast to more conventional forms of logic (i.e., induction and deduction). As forms of knowing, induction provides norms for generalizing from particular observations to a general belief about the world and deduction provides norms for generalizing from a belief about the world to anticipate particular observations.

In contrast, abduction provides a description of how a system learns by doing. That is, an abduction system generalizes from past experiences, consistent with induction, to make hypotheses about the world. However, in an abductive system these hypotheses are tested through action. As long as the actions lead to effects that are consistent with the hypotheses, the hypotheses are maintained. If the observer is surprised by the effects of an action, then hypotheses can be revised to be more consistent with the effects.

As with the controller, for the observer the medium functions as a comparator. However, whereas the comparison between consequence and intention to generate error is most significant for control, the comparison between observations and expectations to generate surprise is most significant to the observer. Whereas the controller operates to eliminate *error,* the observer (i.e., abductive process) operates to eliminate *surprise.* Thus, for the observer an exploratory action is a test of a hypothesis about the ecology. The result of this test is compared with expectations. If the observations are consistent with expectations, then there is no need for further action. However, if there is a mismatch (i.e., surprise), then the observer acts to reduce that mismatch in one of two ways.

First, it can change its beliefs about the ecology so that expectations are better aligned with the results of the test (observation). For example, when starting out in a strange car, your actions will be guided by hypotheses derived from experiences with your own car. However, if the brakes in the strange car are better or worse than those in your car, you may be surprised by the response when you hit the brakes. As a result, you revise your expectations about how this car will behave, based on this observation. With some practice, your expectations about how the new car will behave will become better tuned with the actual behaviors of that car.

Surprise can also be reduced by discounting the observation. This second possibility is clearly seen in person attributions. There is much evidence that once an opinion is formed about a person (e.g., she is a jock), later information (e.g., she has a 4.0 grade point average) can be discounted (e.g., she must have an easy major). There are numerous examples in accident reports where evidence that might have prevented the accident was discounted because it did not match with the expectations of the people involved. Of course, in hindsight, expectations change (i.e., hindsight bias) and it is then hard for investigators to imagine how the information was overlooked (e.g., Woods et al. 1994).

Note that the goal of eliminating surprise is not the same goal as the ideal of truth that motivates formal logic. An abductive system will be perfectly content with a superstitious belief about the ecology, as long as that belief leads to expectations that are consistent with observations. If you expect to win every time you wear your lucky charm, and you do, then the abductive system is happy to retain its faith in the lucky charm and act accordingly. On the other hand, the rules of induction demand that the belief be tested by experiment. Winning does not formally prove the value of your lucky charm. However, *the abductive system is more interested in winning than in proof.* This emphasizes the intimate coupling between observation and control. The meaning processing system is motivated by the instrumentality of its beliefs. Do they lead to success with respect to intentions? In this respect, the meaning processing system is pragmatic, not idealistic. This leads naturally to a rationality that is grounded in the ecology or, rather, in the coupling between perception (observation) and action (control).

Remember that although we discuss control and observation in sequence, these processes are operating simultaneously. When the fire ground commander is faced with a burning building, she is trying to accomplish the goals of rescuing people and saving property while she is simultaneously testing hypotheses about the source and nature of the fire. When she directs her people or her hoses to a specific area of the building, this action is serving both processes. It is a means to her goals and it is a test of her hypotheses. If the fire responds as expected (no surprise), then she continues down the path toward her goals. However, if she is surprised, then she must reevaluate her plan and consider alternative paths to her goals. This is typically a continuous, dynamic process and it is difficult to isolate any choice decision point within the flow. Actions and observations flow together in a continuous stream.

In sum, the meaning processing perspective is chosen to emphasize the dynamic interactions between beliefs, media, and ecologies and the intimate coupling between control (action) and observation (perception/abduction) processes.

## 2.5 Conclusion

This chapter sets the context for the rest of the book. While not everyone will agree with how we use terms like *meaning,* all should now be clear about what we intend. When we refer to *meaning,* we will be talking about the possibilities for action and the associated consequences (value) that a situation offers. It is synonymous with affordance. When we are talking about mental functions, we will use terms such as *belief* or *interpretation.* Our aspirations for this book are to explore effective ways to design interfaces that promote congruence between *meaning, belief,* and *action.* Our goal is to explore effective ways to help people to see the possibilities for action and the consequences of those actions. Our goal is to help people think productively about difficult problems. Our goal is to design more effective meaning processing systems.

We believe that the distinction between the dyadic and triadic semiotic paradigms organizes the literature in a more coherent way. In Table 2.1 we organize some examples of work on cognition and interface design into three categories that reflect different perspectives. Our intention here is not to be exhaustive, but rather to illustrate the implications of the differing perspectives through examples. We picked examples that we hope will be familiar to people who study display design. For those not familiar with this literature, more details about much of this work will be found in later chapters.

The first category in the table, dyadic, includes research and theoretical orientations that fall squarely in the Saussure tradition. That is, these perspectives focus on the relations between representations and interpretations. These approaches tend to be organized around experimental settings and tasks that reflect abstract information processes. There seems to be at least an implicit assumption that the more abstract the experimental context is, the more general or basic the research is. To the extent that a domain is considered, it tends to be treated as a cover story, rather than an essential dimension of the dynamic.

For example, some studies purporting to evaluate cognitive processing in aviation use Sternberg's memory tasks or compensatory tracking tasks in

**TABLE 2.1**

Three Perspectives on Cognition and Interface/Display Design

| Dyadic | Pragmatic | Triadic |
|---|---|---|
| Neisser (1967) | Cockin (1970) | Neisser (1976) |
| Cleveland (1985) | Hollan, Hutchins, and Weitzman (1984); Hollan et al. (1987) | Woods (1984) |
| Card, Moran, and Newell (1983) | Shneiderman (1998) | Rasmussen and Vicente (1989, 1990) |
| Wickens and Carswell (1995) | Tufte (1983, 1990) | Suchman (1987) |
| | Norman and Draper (1986) | Hutchins (1995) |
| | Klein (1989) | Burns and Hajdukiewicz (2004) |

simulators and actual cockpits. In one case, mountain climbers were subjected to standard task batteries implemented on PCs as they climbed Mt. McKinley. Since the data were collected on the mountain, this research was described as ecological. Another example is the development of the Space Fortress game to study skill development. The game was designed as a conglomerate of classical research tasks, rather than being based on the properties of any specific work domain. The point is that the research programs are focused on experimental settings designed to reflect abstract information functions; representativeness with regard to specific human experiences or specific work domains is given little attention.

The idea that experimental settings (and the associated experimental tasks) based on theoretically motivated abstractions might help researchers to identify the long threads with respect to cognitive processing is certainly a reasonable stance. Neisser's (1967) classic book on cognition makes a good case for this. Many of the experimental tasks—for example, the choice reaction time task described earlier—were serious attempts to link theory (e.g., information theory) with natural phenomena (e.g., decision making). However, there is a real danger here. There is a tendency that the research setting or task becomes the phenomenon. That is, researchers can lose sight of both the theoretical motivation and the natural phenomena.

Thus, they begin thinking in terms of probabilities and reaction time, rather than in terms of the motivating phenomena—information and decisions. Further, this perspective tends to dismiss research that takes the details of specific work domains into account as "applied." In this case, applied tends to be used as a code word meaning unscientific.

Note that Neisser's 1967 book is included as an example of the dyadic approach. However, in 1976 Neisser wrote *Cognition and Reality,* in which he introduced the concept of the perceptual cycle (see also the cognitive triad, Woods and Roth 1988). This concept has had a fairly large impact on applied psychology (e.g., Adams, Tenny, and Pew 1991; Smith and Hancock 1995). But, unfortunately, the impact on cognitive science was relatively insignificant. In fact, we have heard the 1976 book being dismissed by people who take the dyadic perspective as the book that Neisser wrote after he gave up science. Even when the people who formulated the conventional wisdom discover a new way of thinking, the legacy of the conventional wisdom can remain behind as a serious constraint to people's understanding.

The second category in Table 2.1 is labeled pragmatic. This category includes research that tends to focus on the practical demands of cognition in natural contexts. For example, Hollan, Hutchins, and Weitzman (e.g., 1987) focused on the problem of designing a graphical interface to train people to control steam engines in ships. Shneiderman's (1998) and Tufte's (1983, 1990) works tend to include many examples of displays explicitly chosen because they work so well in specific contexts. Klein (1989) focused on decision making in natural contexts (e.g., fire commanders). In all of these examples, we feel that there is clear evidence that the demands of work ecologies are taken seriously as a *practically* important consideration for the design of representations or for the understanding of cognition.

However, in this work, the attention to the ecology tends to be implicit. For much of this work, there is no clear attempt at explanation. Rather, the focus is on describing things that work well. To the extent that explanations are attempted, the explanations tend to be framed in terms of the conventional wisdom of the dyadic paradigm. So, although the interfaces are designed to reflect work constraints, the explanations for why a display works tend to be in terms of features of the representation and the dynamics of interpretation processes.

For example, we see early discussions of constructs such as direct manipulation as tending toward a dyadic perspective. That is, the focus tended to be on direct manipulation of objects in the representation, not necessarily of direct manipulation of properties of the work domain. Another example is Klein's (1989) work on naturalistic decision making. We find the descriptions of expertise in natural contexts to be incredibly important data for an understanding of meaning processing that includes all three components of Peirce's semiotic triad. Yet, we find little difference between Klein's *recognition prime decision* model of naturalistic decision making and other models that treat information processing as something that happens exclusively in the head. These approaches tend to focus exclusively on general properties of the interpretation process (e.g., differences between experts and novices), without any explicit consideration of the properties of the specific work domains.

Thus, we have great admiration and respect for the innovations achieved by those who take the pragmatic perspective. But we feel that there is a gap between the theoretical orientation of this group and the triadic perspective. Their methodologies seem to be consistent with the triadic stance, yet they tend to theorize in ways that seem more consistent with the dyadic stance.

The last category, labeled triadic, includes researchers who explicitly acknowledge the work context as an integral part of the cognitive dynamic. These works take a practical and a theoretical stance that is consistent with Peirce's triadic model of semiotics. This stance is reflected in labels such as *situated cognition, distributed cognition,* or *ecological interface design.* This is the perspective that we hope to represent in this book. You will learn much more about the research in this tradition in later chapters.

The point of Table 2.1 is to clarify the different theoretical and methodological orientations of some of the more noted researchers in the field of interface design. There is not always a clear distinction between the dyadic and triadic positions, and some researchers are difficult to classify. This can lead to confusion on what exactly constitutes, for example, an ecological approach. Is it the same as or different from naturalistic decision making or situated cognition? Also, we apologize if we have pigeon-holed anyone where they do not feel they belong. However, this is our attempt to make sense of the field with respect to dyadic and triadic assumptions about the underlying semiotic dynamic of cognition.

## 2.6 Summary

In the end, the major point that we want you to carry forward into the rest of this book is that the work context matters! The effectiveness of an interface ultimately depends on how well it specifies the problems to be solved. And the problems reside not in the head, but rather in the work ecology. The point is to land the plane on the ground or to put out the fire—not to simulate a landing or simulate firefighting in your head. Note that we do not intend to discount the importance of processes in the head (including mental simulations). Rather, the point is that, to be useful, these processes (e.g., mental simulations) must be grounded in the physical, economic, organizational, and social realities of the semiotic dynamic.

The point is not to deny either the situation or awareness dimensions of the problem, but rather to integrate these two dimensions into a single framework (ontology), where both sources of constraint are respected. Thus, the next chapter will suggest some practical ways for thinking about this reality that the head is contained within. This will be followed by a chapter focused on the dynamics of awareness to consider how cognitive processes contribute to the overall dynamic. Then we will try to close the loop to consider the dynamics of situation awareness in the context of a single ontology that considers how situations and awareness jointly interact to shape human experience.

## References

Adams, M. J., Y. J. Tenny, and R. W. Pew. 1991. Strategic workload and the cognitive management of advanced multi-task systems. In *SOAR*. Wright Patterson Air Force Base, OH: Crew Systems Ergonomics Information Analysis Center.

Amelink, M. H. J., M. Mulder, M. M. van Paassen, and J. M. Flach. 2005. Theoretical foundations for a total energy-based perspective flight-path display. *International Journal of Aviation Psychology* 15 (3): 205–231.

Ashby, R. 1956. *An introduction to cybernetics.* New York: Chapman and Hall.

Bennett, K. B., and J. M. Flach. 1992. Graphical displays: Implications for divided attention, focused attention, and problem solving. *Human Factors* 34:513–533.

Bernstein, N. 1967. *The control and regulation of movements.* London: Pergamon Press.

Burns, C. M., and J. R. Hajdukiewicz. 2004. *Ecological interface design.* Boca Raton, FL: CRC Press.

Card, S. K., T. P. Moran, and A. Newell. 1983. *The psychology of human–computer interaction.* Hillsdale, NJ: Lawrence Erlbaum Associates.

Clark, A. 1997. *Being there: Putting brain, body, and world, together again.* Cambridge, MA: The MIT Press.

Cleveland, W. S. 1985. *The elements of graphing data.* Belmont, CA: Wadsworth.

Coekin, J. A. 1970. An oscilloscope polar coordinate display for multidimensional data. *Radio and Electronic Engineer* 40:97–101.

Gardner, H. 1985. *The minds new science.* New York: Basic Books, Inc.

Gibson, J. J. 1972/1982. The affordances of the environment. In *Reasons for realism,* ed. E. Reed and R. Jones. Hillsdale, NJ: Lawrence Erlbaum Associates.

———. 1979. *The ecological approach to visual perception.* Boston, MA: Houghton Mifflin.

Gibson, J. J., and L. Crooks. 1982. A theoretical field-analysis of automobile driving. In *Reasons for realism,* ed. E. J. Reed. Hillsdale, NJ: Lawrence Erlbaum Associates. Original edition, *American Journal of Psychology,* 1938, 51:453–471.

Gregory, R. L. 1974. *Concepts and mechanisms of perception.* New York: Charles Scribner & Sons.

Hick. 1952. On the rate of gain of information. *Quarterly Journal of Experimental Psychology* 4:11–26.

Hollan, J. D., E. L. Hutchins, T. P. McCandless, M. Rosenstein, and L. Weitzman. 1987. Graphical interfaces for simulation. In *Advances in man–machine systems research,* ed. W. B. Rouse. Greenwich, CT: JAI Press.

Hollan, J. D., E. L. Hutchins, and L. Weitzman. 1984. STEAMER: An interactive inspectable simulation-based training system. *The AI Magazine* 5 (2): 15–27.

———. 1987. Steamer: An interactive inspectable simulation-based training system. In *Artificial intelligence and instruction: Applications and methods,* ed. G. Kearsley. Reading, MA: Addison–Wesley.

Hutchins, E. L. 1995. *Cognition in the wild.* Cambridge, MA: MIT Press.

Hyman, R. 1953. Stimulus information as a determinant of reaction time. *Journal of Experimental Psychology* 45:188–196.

Jagacinski, R. J., and J.M. Flach. 2003. *Control theory for humans.* Mahwah, NJ: Lawrence Erlbaum Associates.

Klein, G. A. 1989. Recognition-primed decisions. *Advances in Man–Machine Systems Research* 5:47–92.

Klein, G. A., J. Orasanu, and C. E. Zsambok, eds. 1993. *Decision making in action: Models and methods.* Norwood, NJ: Ablex Publishing Corp.

Kuhn, T. 1962. *The structure of scientific revolutions.* Chicago, IL: University of Chicago Press.

Lachman, R., J. L. Lachman, and E. C. Butterfield. 1979. *Cognitive psychology and information processing: An introduction.* Hillsdale, NJ: Lawrence Erlbaum Associates.

Miller, G. A. 1956. The magical number seven, plus or minus two: Some limits on our capacity for processing information. *Psychological Review* 63:81–97.

Nadin, M. 1988. Interface design and evaluation—semiotic implications. In *Advances in human–computer interaction,* ed. H. R. Hartson and D. Hix. Norwood, NJ: Ablex Publishing.

Neisser, U. 1967. *Cognitive psychology.* New York: Appleton-Century-Crofts.

———. 1976. *Cognition and reality.* San Francisco, CA: W. H. Freeman and Company.

Norman, D. A. 1986. Cognitive engineering. In *User centered system design,* ed. D. A. Norman and S. W. Draper. Hillsdale, NJ: Lawrence Erlbaum Associates.

Norman, D. A., and S. W. Draper, eds. 1986. *User centered system design.* Hillsdale, NJ: Lawrence Erlbaum Associates.

Peirce, C. S. 1931–1935a. *Collected papers of Charles Sanders Peirce,* ed. C. Hartshorne and P. Weiss. Cambridge, MA: Harvard University Press.

———. 1931–1935b. *Collected papers of Charles Sanders Peirce,* ed. A. Burkes. Cambridge, MA: Harvard University Press.

Rasmussen, J., and K. J. Vicente. 1989. Coping with human errors through system design—Implications for ecological interface design. *International Journal of Man–Machine Studies* 31 (5): 517–534.

———. 1990. Ecological interfaces: A technological imperative in high-tech systems? *International Journal of Human–Computer Interaction* 2 (2): 93–111.

Shannon, C. E., and W. Weaver. 1963. *The mathematical theory of communication.* Urbana: University of Illinois Press.

Shneiderman, B. 1998. *Designing the user interface: Strategies for effective human–computer interaction,* 3rd ed. Reading, MA: Addison–Wesley.

Smith, K., and P. A. Hancock. 1995. Situation awareness is adaptive, externally directed consciousness. *Human Factors* 37 (1): 137–148.

Suchman, L. A. 1987. *Plans and situated actions: The problem of human–machine communication.* New York: Cambridge University Press.

Swain, A. D., and H. E. Guttman. 1983. *Handbook of human reliability analysis with emphasis on nuclear power plant applications.* Washington, D.C.: Nuclear Regulatory Commission.

Tufte, E. R. 1983. *The visual display of quantitative information.* Cheshire, CT: Graphics Press.

———. 1990. *Envisioning information.* Cheshire, CT: Graphics Press.

Weinberg, G. M., and D. Weinberg. 1979. *On the design of stable systems.* New York: John Wiley & Sons.

Weiner, N. 1948. *Cybernetics.* Cambridge, MA: MIT Press.

Wertheimer, M. 1959. *Productive thinking.* New York: Harper and Row.

Wickens, C. D. 1992. *Engineering psychology and human performance,* 2nd ed. New York: Harper Collins.

Wickens, C. D., and C. M. Carswell. 1995. The proximity compatibility principle: Its psychological foundation and relevance to display design. *Human Factors* 37 (3): 473–494.

Woods, D. D. 1984. Visual momentum: A concept to improve the cognitive coupling of person and computer. *International Journal of Man–Machine Studies* 21:229–244.

Woods, D. D., L. Johannesen, R. I. Cook, and N. Sarter. 1994. *Behind human error: Cognitive systems, computers and hindsight.* WPAFB, OH: Crew Systems Ergonomic Information and Analysis Center.

Woods, D. D., and E. M. Roth. 1988. Cognitive systems engineering. In *Handbook of human–computer interaction,* ed. M. Helander. Amsterdam: Elsevier Science Publishers B.V. (North-Holland).

# 3

# *The Dynamics of Situations*

## 3.1 Introduction

> The critical point is that understanding computers is different from understanding computations. To understand a computer, one has to study the computer. To understand an information-processing task, one has to study that information-processing task. To understand fully a particular machine carrying out a particular information-processing task, one has to do both things. Neither alone will suffice. (Marr 1982, p. 5)

> It would be perfectly possible for the psychologist to follow the route of the economist: to construct a theory of concept formation that depended on no characteristic of the subject other than his being motivated to perform well. It would be a theory of how perfectly rational man would behave in that task environment—hence, not a psychological theory but a theory of the structure of the task environment … we shall often distinguish two aspects of the theory of problem solving as (1) demands of the task environment and (2) psychology of the subject. These shorthand expressions should never seduce the reader into thinking that as a psychologist he should be interested only in the psychology of the subject. The two aspects are in fact like figure and ground—although which is which depends on the momentary viewpoint. (Newell and Simon 1972, pp. 54–55)

> Perhaps the composition and layout of surfaces constitute what they afford. If so, to perceive them is to perceive what they afford. This is a radical hypothesis, for it implies that the "values" and "meanings" of things in the environment can be directly perceived. Moreover, it would explain the sense in which values and meanings are external to the perceiver. (Gibson 1979, p. 127)

> The emphasis on finding and describing "knowledge structures" that are somewhere "inside" the individual encourages us to overlook the fact that human cognition is always situated in a complex sociocultural world and cannot be unaffected by it. (Hutchins 1995, p. xiii)

As these opening quotes suggest, the pioneers of early work on human information processing and problem solving fully appreciated the significance of the task environment. As suggested in Chapter 2, this follows naturally from Shannon and Weaver's (1963) theory of information that it is essential to situate any action or decision into the larger context of possibilities. In Newell and

Simon's (1972) work, this involved describing the "states" (including the initial and goal states), the "operators" that allowed movement from one state to another, and the "constraints" (i.e., rules) on application of those operators.

The significance of being able to describe the field of possibilities fully led early pioneers to focus much of their attention on well-defined problems (e.g., Tower of Hanoi, Crypto-arithmetic, Tic-Tac-Toe, etc.), where the task environments could be fully specified in terms of the possible states and the rules that limited the "legal" paths through the state space. This was an important first step to allow strong tests of the normative principles of information processing to human performance. This allowed clear specification of optimal solutions and clear comparisons between the solutions of specific algorithms and the solutions of people.

However, an unintended consequence of this choice is that the emphasis placed on problem descriptions by these early researchers faded into the background while the strategies, heuristics, and limitations of the human problem solver became the sole basis for generalizations to the more complex, less well defined problems typical of most natural work domains.

The quotes from Gibson and Hutchins are a bit more radical than the claims of either Marr or Simon and Newell. These quotes suggest that a theory of situations is not simply a complement for a theory of awareness; rather, it provides the foundation for such a theory. They suggest the need for an "ecological" or "situated" approach to cognition. Obviously, these ideas are an important source of inspiration for a meaning processing approach. However, it should be clear from the other quotes that it is not necessary to accept these more radical ideas in order to appreciate the value of a theory of situations for a deep understanding of human performance.

The objective of this chapter is to revive attention to the *situation* side of the *situation awareness* problem by considering task environments (i.e., problem spaces, work domains) and how they can be modeled. We will begin by introducing the general concepts of "state" and "state space" in the context of some classical human performance research on well-defined problems. However, we believe that the second half of the chapter will be more important to interface and display designers. In contrast to early work on human information processing, the emphasis in the second half of this chapter will be on ill-defined problems. In fact, we believe that this is often a fundamental challenge facing interface designers: to discover properties of complex work domains, so that these properties can be integrated into effective problem representations.

## 3.2 The Problem State Space

In essence the term "state" can be a synonym for the properties or dimensions of a problem. One of the goals of situation or work analysis will be to identify properties that help us to visualize significant problem constraints that can be leveraged against the information processing problem. Again, in the context of information processing theory, constraints essentially place boundaries on the field of possibilities. Although constraints do not specify what will happen next, they do help to rule out possibilities—thus reducing the uncertainty about the future.

These bounds can be especially critical for organizing (i.e., chunking) complex information into coherent representations. Thus, this chapter will focus on identifying the problem demands of a work domain independently of any particular strategy or solution algorithm. Chapter 4 will go deeper into the *awareness* side of the equation to explore potential strategies for navigating the problem space. Again, it is important to emphasize that both sides of this equation will be essential to the meaning-processing problem and, ultimately, to the problem of designing effective representations. In the end, a good representation must specify the situational constraints (meaning) in terms that are compatible with the constraints on awareness (interpretation) to solve the triadic semiotic problem. We begin by considering the concept of state in classical approaches to problem solving.

### 3.2.1 The State Space for the Game of Fifteen

Consider the game of Fifteen. This is a game in which two players take turns choosing numbers from the set of numbers from 1 through 9 without replacement. The winner is the first player to have a combination of three numbers that add to 15 (e.g., 1, 9, 5; 2, 8, 5; 1, 8, 6; or 5, 6, 4). The states in this problem can be modeled by placing each number into one of three categories: the pool of available numbers, the choices of player A, and the choices of player B. The initial state has all numbers available and no choices for either player:

| Initial state: | Pool (1, 2, 3, 4, 5, 6, 7, 8, 9) | Player A ( ) | Player B ( ) |
|---|---|---|---|

There are nine possible moves for player A from this initial state:

| | | |
|---|---|---|
| Pool (2, 3, 4, 5, 6, 7, 8, 9) | Player A (1) | Player B ( ) |
| Pool (1, 3, 4, 5, 6, 7, 8, 9) | Player A (2) | Player B ( ) |
| Pool (1, 2, 4, 5, 6, 7, 8, 9) | Player A (3) | Player B ( ) |
| Pool (1, 2, 3, 5, 6, 7, 8, 9) | Player A (4) | Player B ( ) |
| Pool (1, 2, 3, 4, 6, 7, 8, 9) | Player A (5) | Player B ( ) |
| Pool (1, 2, 3, 4, 5, 7, 8, 9) | Player A (6) | Player B ( ) |
| Pool (1, 2, 3, 4, 5, 6, 8, 9) | Player A (7) | Player B ( ) |
| Pool (1, 2, 3, 4, 5, 6, 7, 9) | Player A (8) | Player B ( ) |
| Pool (1, 2, 3, 4, 5, 6, 7, 8) | Player A (9) | Player B ( ) |

A choice by player A produces a new problem state. For example, a choice of "1" results in the following state:

| | | |
|---|---|---|
| Pool (2, 3, 4, 5, 6, 7, 8) | Player A (1) | Player B ( ) |

Player B then has eight possible choices, given this new state:

| | | |
|---|---|---|
| Pool (3, 4, 5, 6, 7, 8, 9) | Player A (1) | Player B (2) |
| Pool (2, 4, 5, 6, 7, 8, 9) | Player A (1) | Player B (3) |
| Pool (2, 3, 5, 6, 7, 8, 9) | Player A (1) | Player B (4) |
| Pool (2, 3, 4, 6, 7, 8, 9) | Player A (1) | Player B (5) |
| Pool (2, 3, 4, 5, 7, 8, 9) | Player A (1) | Player B (6) |
| Pool (2, 3, 4, 5, 6, 8, 9) | Player A (1) | Player B (7) |
| Pool (2, 3, 4, 5, 6, 7, 9) | Player A (1) | Player B (8) |
| Pool (2, 3, 4, 5, 6, 7, 8) | Player A (1) | Player B (9) |

Continuing in this fashion, it would be possible to enumerate all possible states. The possible trajectories through this space of states would all begin with the initial state and would end at either a winning state (one player or the other meets the goal of having three numbers that sum to 15) or a draw state (the pool is empty and neither player has three numbers that sum to 15). For example, here is one possible sequence of play:

| | | |
|---|---|---|
| Move 1 | Player A (5) | Player B ( ) |
| Move 2 | Player A (5) | Player B (3) |
| Move 3 | Player A (5, 4) | Player B (3) |
| Move 4 | Player A (5, 4) | Player B (3, 6) |
| Move 5 | Player A (5, 4, 9) | Player B (3, 6) |
| Move 6 | Player A (5, 4, 9) | Player B (3, 6, 2) |
| Move 7 | Player A (5, 4, 9, 1) | |
| End | Winner because 5 + 9 + 1 = 15 | |

Each move in this sequence could be considered a state. Each of these states has three dimensions or variables: the choices of player A, those of player B, and the numbers remaining in the pool. Since sampling is without replacement, the numbers in the pool can be inferred directly from knowledge of the choices of the two players; thus, it is sufficient to characterize movement through the state variables in terms of the two players alone. The value for each dimension or state variable would be the numbers chosen by each player. In this particular sequence, the state at move 6 is particularly interesting. At that point, player A has guaranteed a victory because, from that state, two moves will lead to winning states. Player B blocks one path to a win by picking 2, thus blocking the path to (5, 4, 9, 2) (4 + 9 + 2 = 15). Therefore, player A chooses "1," which also leads to a winning state.

It is useful to consider the value of specific choices when considering the state space for the Fifteen game with respect to meaning. Was the choice of "5" a good first move? It did lead to a win, but was it the best possible choice? Was the choice of "3" a good first choice for Player B? After all, she did lose. Was there a better choice? A good representation should make it easier to differentiate a good move from a bad move. That is, it should make it easier to see where a good move enhances the chances of winning and a bad move diminishes the chances of winning.

It turns out that the choice of "5" is the best possible first move because "5" allows the most possible combinations of numbers that add to 15: (6, 5, 4; 1, 5, 9; 8, 5, 2; 3, 5, 7). Player B's choice of "3" was not a good choice because it blocked one winning possibility for A and only left one possible win for B (3, 8, 4). Better choices for player B would have been 6, 8, 4, or 2. Each of these choices would eliminate one winning possibility for A while leaving two winning possibilities for player B (e.g., 6, 7, 2 or 6, 1, 8).

The state space for problems such as the Fifteen game is typically represented as a tree diagram where each branch represents a possible sequence of moves. The root of this diagram would be the initial condition. From this root, there would be nine possible first moves; for each of these moves, there would be eight possible responses, in turn followed by seven possible choices, etc. Got the picture? Each state (or node) could be evaluated in terms of the number of remaining branches that lead to wins or losses. The quality of a state might be indexed by the number (or percentage) of its branches that lead to wins, losses, or ties. The challenge for the display designer would be to choose a representation that makes it easy for the human to make this discrimination—to discriminate good choices from bad choices. Note that the value of these discriminations is not mental, but rather reflects the structure of the problem or game independently of any natural or artificial cognitive agent.

It turns out that there is a much better representation for the Fifteen problem than the verbal descriptions we have provided or even the tree diagram that we described. Figure 3.1 shows a spatial representation of the Fifteen problem. It should be readily apparent that the game of Fifteen is isomorphic with the game of Tic-Tac-Toe. The term *isomorphic* means that the states of the two problems can be mapped one to one: Each number choice in the game of Fifteen is identical to the choice of a space in the Tic-Tac-Toe game; the goal state of three numbers adding to 15 is identical to the goal state of three in a row.

Thus, from a logical or computational perspective, the games of Fifteen and Tic-Tac-Toe are identical tasks or situations. Of course, from the perspective of awareness, the games are definitely not the same. The spatial layout reduces the need for memory or mental calculations and makes it much easier for players immediately to see the value of a particular choice. With the

| | | |
|---|---|---|
| 6 | 1 | 8 |
| 7 | 5 | 3 |
| 2 | 9 | 4 |

**FIGURE 3.1**
A spatial representation of the game of Fifteen. Note that this game is isomorphic with the game of Tic-Tac-Toe.

spatial layout, it is much easier to see why "5" is a good first choice, and why "3" is not a good choice for player B.

The discussion of the Fifteen game illustrates the classical information processing approach to analyzing the situation dynamics. For well-defined problems, such as Fifteen (Tic-Tac-Toe), Tower of Hanoi, or Hobbits and Orcs, it is possible to enumerate all possibilities completely. However, when problems get more complex, such as chess or backgammon, enumerating all possibilities becomes prohibitively difficult. It is even more difficult for natural work domains, where the possibilities rival or exceed those of chess and the "rules" can be far more ambiguous or fuzzy. Before we consider the dynamics of ill-defined, natural problems, one other relatively simple problem will be considered that illustrates how the concepts of state, state variable, and state space are used in control theory.

### 3.2.2 The State Space for a Simple Manual Control Task

A convenient task for introducing the control perspective is one of simple manual control: moving a cursor from a start position at the left of the screen to a target on the right side of the screen using a single-axis joystick. The joystick displacement determines the acceleration of the cursor. Thus, a movement of the joystick to the right would cause the cursor to accelerate in that direction. To stop the cursor, the joystick would have to be moved to the left of center, causing a deceleration (actually, acceleration to the left). The cursor would stop (reach zero velocity) when the deceleration exactly cancelled out the acceleration.

Figure 3.2 illustrates this task using a block diagram, the conventional way to illustrate its dynamics. Each block represents an integration process and the output of each block represents the state variables for this dynamic—in

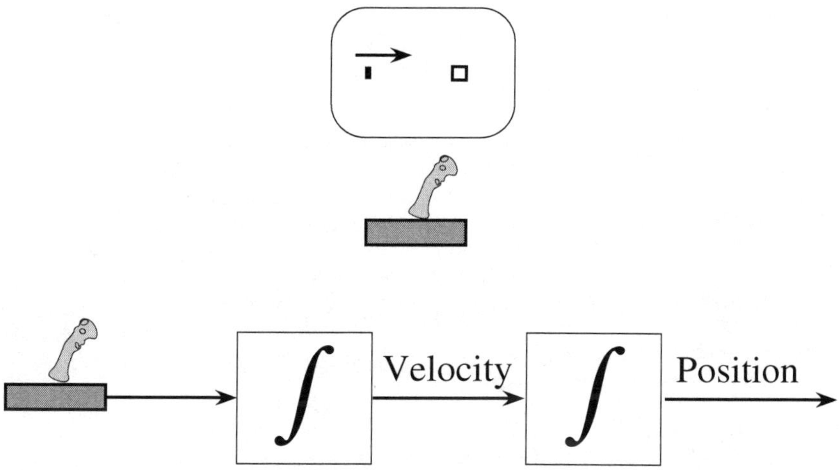


**FIGURE 3.2**
This block diagram illustrates a simple positioning task in which the control dynamics are second order. That is, the control displacement determines the cursor acceleration. Note that the system has two state variables: position and velocity, indicated in the diagram as the outputs of the two integral processes.

this case, the position and velocity of the cursor. This dynamic could also be represented as a second-order differential equation, and the state variables would be the initial conditions. Knowledge of the states (the values of position and velocity) at time $t$ and the control input from $t$ forward is sufficient information to specify the particular trajectory of the cursor from $t$ forward.

Figure 3.3 shows a diagram of the state space for this control system. The target position is represented with the far edge at zero, so the position indicates positive distance from that edge. Velocity is represented as a negative speed, since the distance is being reduced, as the cursor moves toward the target. The two bold, crossing curves represent maximum acceleration (that would result when the control is at full deflection in the direction of the target) and maximum deceleration (that would result when the control is at full deflection away from the target).

These curves divide the space into regions and the labels reflect the significance of each region with regard to the objective of stopping at the target. The curves themselves combine to reflect the minimum-time path from the start to the target. This path would involve a bang-bang style of control. Maximum acceleration would be initiated by full deflection of the stick toward the target; when the state of the target is at the value corresponding to the intersection of the two maxima curves, the stick should be moved to full deflection in the opposite direction (away from the target—maximum deceleration). Finally, the stick should be returned to zero upon entering the target.

The region above the maxima curves (toward the position axis) reflects the states that are reachable from the start position and that allow movement to

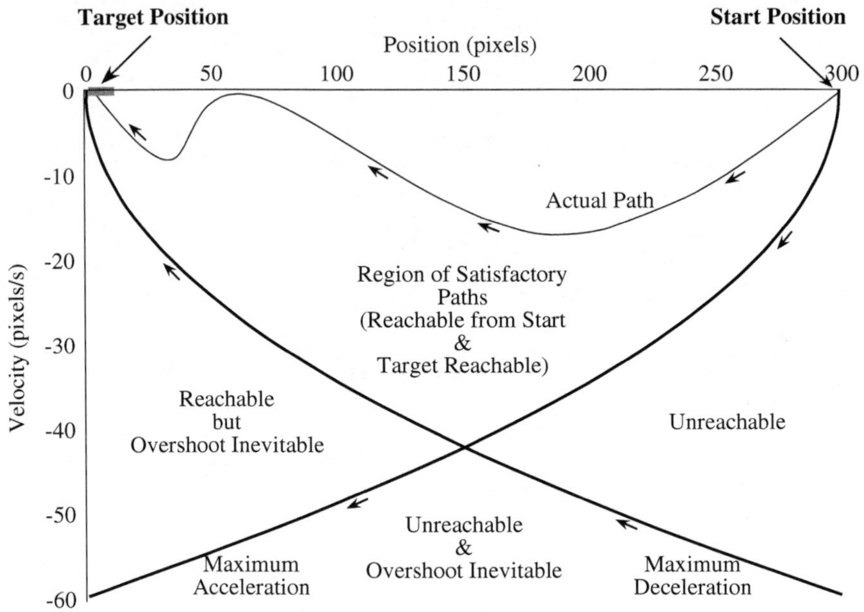


**FIGURE 3.3**
This state–space diagram illustrates significant constraints of the movement control task that result from the task (e.g., target) and motion dynamics (e.g., maximum acceleration).

the target without overshoot. There are an infinite number of satisfactory paths from the start to the target within this region. However, all of them will take more time than the path resulting from the bang-bang input that results in motion along the border of this region.

If the movement reaches a state that is above the maximum-deceleration path to the target (area labeled "reachable but overshoot inevitable"), then there is no action that will avoid overshooting the target. That is, the cursor is too close and the speed is too high to allow the cursor to stop before passing through the target. Also, the states below the maximum-acceleration path are unreachable from the start (i.e., the cursor cannot reach these speeds in a short distance from start).

Thus, the state space diagram in Figure 3.3 provides a good image of the control possibilities. It reflects both the task constraints (i.e., starting position, goal position) and the dynamic constraints (i.e., control limits in terms of the maximal acceleration and deceleration), and it shows a significant performance constraint (minimum-time path). Additionally, it links all the states of motion (position and velocities) to regions that reflect these constraints. The thin line in the diagram illustrates one of many possible trajectories through this space. This trajectory shows two distinct submovements. The first submovement begins at the start position and covers about 80% of the initial distance to the target. It has a peak speed of just under 20 pixels per second.

The second submovement covers the remaining distance to the target with a peak speed under 10 pixels per second.

A display that can greatly enhance performance in a tracking/positioning task such as that illustrated in Figure 3.2 is a "predictive" or "quickened" display. These displays are illustrated in Figure 3.4. With a predictive display, the cursor is replaced with a vector (Figure 3.4b). The tail of the vector represents the current position and the length of the vector is proportional to the velocity of the cursor. Thus, both state dimensions (position and velocity) are explicitly represented in the display. Later, we will call this type of representation a "configural" display since it explicitly represents both individual state values and their joint contribution toward functional goals.

With a quickened display (Figure 3.4c), the cursor is represented as a point (not a vector), but the position of the cursor reflects a weighted sum of position and velocity (rather than only position). In essence, the quickened display would only show the head of the vector used for the predictive display. This type of representation will be referred to as an "integral" display. That is, the two state dimensions are combined in a way that makes the joint functional relation to goals more apparent but masks the individual values of the specific states (i.e., position and velocity).

Research shows that performance in target acquisition can be faster and more accurate with both predictive and quickened representations than with standard representations (where cursor position corresponds to distance from the target and velocity must be perceived based on the cursor's motion). This suggests that the addition of information about the velocity state is useful for this control task.

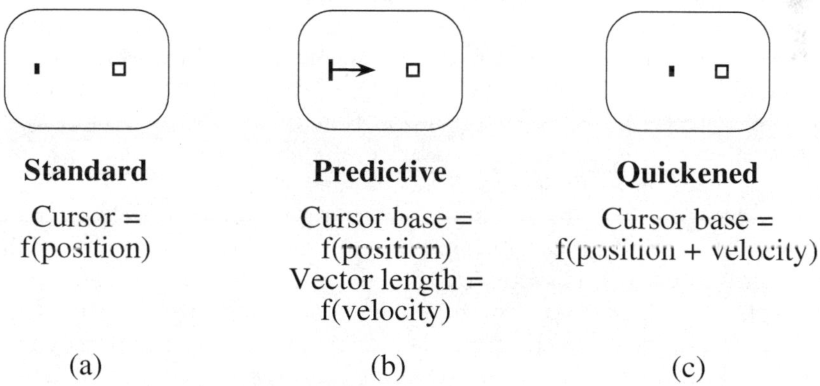


**FIGURE 3.4**
Three alternative mappings of states to displays for a second-order positioning/tracking task are illustrated. The cursor position in the standard format is related to the distance from the target (i.e., the square). The cursor position in the predictive format is also related to the distance from the target; the direction and length of the vector are related to velocity of motion with respect to the target. The cursor position in the quickened format is related to an integral function of the distance and velocity with respect to the target.

### 3.2.3 Implications for Interface Design

Hopefully, these two examples of the state representations for both the Fifteen (Tic-Tac-Toe) problem and the tracking task help to clarify what people mean when they talk about states and state variables. In the case of the Fifteen problem, the state variables would correspond to the positions of the two players (i.e., their choices) and the state of the pool from which items are selected. The value for each state would be the list of items chosen or remaining. For the tracking/positioning task, the state variables are position and velocity and the value at any time would be the measurements for these two variables. Hopefully, it will also be obvious how these representations can help researchers to visualize problem constraints and to relate human performance (e.g., choices in the Fifteen game or motion in the tracking task) to both the goals and the possibilities within the task ecology.

Although the representations for a discrete game and a movement task can be very different, the key idea is to help researchers to understand the behaviors in the context of the task ecology. Note that the space, regions, and boundaries reflect properties of the situation. Behavior can then be represented as a trajectory through the situation. Such representations can be useful to those who are interested in designing "artificial intelligence" systems and those who are trying to understand the meaning processing activities of "naturally intelligent" systems.

However, it is important to appreciate that the Fifteen problem and the tracking task are very simple relative to many of the ecologies that will be of interest to interface designers (e.g., process control, driving, or emergency operations). In these ecologies the state spaces will have many more dimensions than can be easily enumerated in a simple tree diagram or in a two-dimensional graph. Thus, it may not be possible to represent clearly the space of possibilities and the constraints on these possibilities in any single representation. Further, for many natural environments, there may not be a clear consensus with regard to the appropriate or significant dimensions for the state space. Nevertheless, we assert that *efforts toward understanding the ecology as a functional state space will always be worthwhile.*

The challenge is how to go about building a model of problems or work ecologies that are significantly more complicated than the "toy" problems that we presented in this section. This will be the focus of the remainder of this chapter. It is important to reemphasize that this is based on the assumption that a first step to building an effective display representation of a problem or work ecology is to have an understanding of that problem. At the same time, exploring alternative representations may be essential to the development of a deeper understanding.

Thus, it is important to appreciate the iterative nature of learning about work domains and to appreciate that the analysis of situations is not a prerequisite to interface design, but more realistically a co-requisite. The better your understanding is, the better your representations will be; the better your representations are, the better your understanding will be. It is definitely a "chicken and egg" relation where each is essential to the evolution of the other. For this reason, we believe that it is critical that interface designers have a framework to guide their search for a deeper understanding of the ecology. In this respect, the following are critical points for you to appreciate about the concepts of state variable, state, and state space:

- The choice of state variables is a statement about the important or significant dimensions of a situation or problem.
- The state dimensions combine to specify a space of possibilities. Behavior can be visualized as movement through this space.
- The values of the state variables reflect position in the multidimensional state space with respect to significant landmarks (e.g., the initial position, the goal).
- Significant problem constraints can often be represented as boundaries, and qualitative properties of the problem can often be represented as regions within the state space.

## 3.3 Levels of Abstraction

> If one hopes to achieve a full understanding of a system as complicated as a nervous system, a developing embryo, a set of metabolic pathways, a bottle of gas, or even a large computer program, then one must be prepared to contemplate different kinds of explanation at different levels of description that are linked, at least in principle, into a cohesive whole, even if linking the levels in complete detail is impractical. (Marr 1982, p. 20)
>
> For supervisory control decisions, we have found very useful a description of the functional properties of a system according to goals–means or *means–ends relationships* in a functional abstraction hierarchy. Such a hierarchy describes bottom-up what components and functions can be used for, how they may serve higher level purposes, and top-down, how purposes can be implemented by functions and components. Various forms of functional abstraction hierarchies including means–end relationships have been used for design activities. (Rasmussen 1986, p.14; emphasis original)

As we move to consider work domains associated with complex problems, the state dimensions will typically not be as well-defined as in the examples in the previous section. In fact, an important aspect of work analysis will be a search to discover the critical dimensions of the problem (e.g., the state variables). For example, in building flight displays, it is important to consider what the critical state variables are. Is it sufficient to define the states around axes of motion (e.g., the three translational and three rotational axes of motion and the velocities associated with those axes) or is there a value to considering energy states (e.g., kinetic, potential, and or total energy and the associated rates of change)?

Similarly, in developing graphical displays to support tactical decision making for military commanders, it will be necessary to learn about what dimensions are important for assessing and specifying combat strength and effectiveness. In later chapters, specific issues such as these will be addressed in the context of particular work domains. However, in this chapter, our goal is to provide a generic framework to guide the search for a deeper understanding of situations. The goal is to develop descriptions of work domains that can be useful in the design of effective interfaces to complex problems.

### 3.3.1 Two Analytical Frameworks for Modeling Abstraction

The framework that we offer is inspired by the insights of David Marr (1982) and Jens Rasmussen (1986). Both of these researchers were struggling to develop a systems approach to complex cognitive processes. Marr was primarily interested in the design of artificial intelligence systems (specifically, vision systems) and Rasmussen was primarily interested in the design of safe human–machine systems (specifically, nuclear power plants). Marr and Rasmussen recognized that complex systems could be described at multiple levels of abstraction. Further, both realized that each level provided unique insights to specific aspects of performance and that *no* single level was adequate for full understanding. Deep understanding required the consideration of relations within and across levels of abstraction.

Marr proposed three abstract levels for describing perceptual systems, while Rasmussen proposed five levels of abstraction for describing work ecologies. We think that there are many parallel and complementary insights behind both choices. Table 3.1 lists the different levels and illustrates how we believe Rasmussen's five levels nest naturally within Marr's three levels. Each of the following three sections of this chapter will be organized around Marr's three levels, with finer distinctions inspired by Rasmussen nested within the appropriate sections. Thus, we will begin with Marr's top level of description: the computational level.

## 3.4 Computational Level of Abstraction

> The nature of the computations that underlie perception depends more upon the computational problems that have to be solved than upon the particular hardware in which their solutions are implemented. To phrase the matter another way, an algorithm is likely to be

**TABLE 3.1**

Comparison of Levels of Abstraction for Functional Understanding of a Cognitive System

| Marr (1982) | Rasmussen (1986) |
|---|---|
| *Computational theory*<br>What is the goal of the computation? Why is it appropriate? What is the logic of the strategy by which it *can* be carried out? | Functional purpose<br>Abstract function |
| *Representation/algorithmic*<br>How can this computational theory be implemented? In particular, what is the representation for the input and output? What is the algorithm for the transformation? | General functions/activities |
| *Hardware implementation*<br>How can the representation and algorithm be realized physically? | Physical function<br>Physical form |

> understood more readily by understanding the nature of the problem being solved than by examining the mechanism (and the hardware) in which it is embodied.
>
> In a similar vein, trying to understand perception by studying only neurons is like trying to understand bird flight by studying only feathers: It just can't be done. In order to understand bird flight, we have to understand aerodynamics: only then do the structures of feathers and the different shapes of birds' wings make sense. (Marr 1982, p. 27)

The computational level of description is the *most abstract* and the *most fundamental.* It is the most abstract in that it involves a description of a problem that presumes the least in terms of any specific device (hardware) or particular strategy (algorithm) for solving it. It provides the most general description of the problem. For example, in the case of flight, the computational level of description would be one that would apply to both animals and machines. It would apply to both fixed and rotary wing machines. It would apply to both gliders and powered machines. The fundamental laws of aerodynamics would be an important component to a computational level of description. Not only would these laws apply to flight, which reflects a specific goal or aspiration, but such laws also could apply to any body (e.g., a leaf) moving through an air mass.

However, in addition to considering the fundamental laws of flight, to understand problems associated with aviation as a work domain, it is also important to understand the goals for this system. This includes consideration of economic and safety issues and other dimensions associated with the desirable qualities of a successful aviation system.

### 3.4.1 Functional Purpose

For describing work domains, Rasmussen distinguished two levels within the computational level. He associated the top level of description with *functional purpose.* This level reflects the reasons, intentions, and/or values behind why a particular system might exist. As Rasmussen (1986) notes:

> The way in which the functional properties of a system are perceived by a decision maker very much depends upon the goals and intentions of the person. In general, objects in the environment in fact only exist isolated from the background in the mind of the human, and the properties they are allocated depend on the actual intentions. A stone may disappear unrecognized into the general scenery; it may be recognized as a stone, maybe even a geologic specimen; it may be considered an item suitable to scare away a threatening dog; or it may be a useful weight that prevents manuscript sheets from being carried away by the wind—all depending upon the needs or interests of a human subject. Each person has his own world, depending on his immediate needs. (p. 13)

The functional purpose level is considered the highest level of constraint since it addresses the fundamental rationality behind a designed system. What is the functional value of an aviation system? Why was the nuclear power plant built? Why did I buy the iPhone®? Addressing these questions from a rational perspective brings in the ideas of goals and values. And, in this respect, it gets to the essence of quality—that is, what is good or bad, what are the criteria for success or failure, what is desirable or undesirable, what is attractive or what is to be avoided? Questions that might be framed at this level are

- What should be my goal or top priority during an emergency event in a nuclear power plant control room: to maintain normal operations, while figuring out what the alarms mean, or to shut down the plant?
- Should I press on through a potentially dangerous weather system in order to arrive at the airport on schedule or should I divert around the weather system and arrive late?
- Should I risk massive casualties in order to take that position or should I take a defensive stance and wait for the enemy to make a mistake?

Similar questions could be framed for other domains where there is a need to balance multiple goals and potential risks. For the Fifteen problem and the tracking/positioning task discussed in the previous section, the functional purpose level of description would be primarily concerned with the goals and the figures of merit. In the Fifteen problem, there are essentially two goals to consider: the goal of getting three numbers that sum to 15 (or three in a row for Tic-Tac-Toe) and the goal of preventing your opponent from getting there before you. Thus, each move can be considered relative to its offensive and defensive value.

For example, for the first move, "5" is the best choice since it creates four possible sums to 15 and also blocks four. The next best choices are the remaining even numbers (2, 4, 6, 8); each creates and blocks three possible solutions. The worst choices are the remaining odd numbers (1, 3, 7, 9); each creates and blocks only two possible solutions. Perhaps one of the values of the spatial representation is that it uses spatial positions to make the relations of choosing a number (or square) to the offensive (three in a row) and defensive (blocking opponent) values explicit in the interface: The center position (5) is best, the corners are next best (2, 4, 6, 8), and the side edges are worst (1, 3, 7, 9).

For the tracking/positioning task, the target is defined as a particular position–velocity combination (i.e., stopped at the target). In addition, participants are typically instructed to acquire the target as fast as possible, sometimes with the constraint of avoiding overshoot. Thus, minimizing time to acquire the target and minimizing overshoots can also be important criteria with regard to the functional purpose level of abstraction. Considerations about how speed and accuracy trade off become important at this level of abstraction. In normative tracking models based on optimal control theory, minimizing an explicit cost-functional provides an analytic frame for addressing trade-offs between speed and accuracy goals.

These criteria have important implications for the weight given to position and velocity information. If accuracy is important (e.g., overshoots are costly), then more weight (cost) should be given to velocity. That is, the approaches will be more conservative, keeping velocities relatively lower to protect against overshoots. In design of the displays, this would have implications for the proportionality constant for the velocity vector in the predictive display or for the lead constant in the quickened display. Making the velocity component more salient will help the controller to anticipate and avoid overshoots.

### 3.4.2 Abstract Function

Rasmussen's second level, *abstract function,* introduces another important constraint that is fundamental to system performance. This level reflects the ultimate field of possibilities for achieving the system goals. In essence, this level "represents the semantic content of the physical signals and, hence, the overall organizing principle" (Rasmussen 1986, p. 19). This level will typically represent the fundamental dimensions of the state space, which, of course, includes the goal states and also any constraints that limit motion through that space. For many work domains, this will typically involve physical laws and social/regulatory constraints on system performance. This level considers what the possible and/or legal means for achieving the functional purposes are. This is where thermodynamics and the regulatory environment would be considered for nuclear plants and where aerodynamics and the aviation regulatory environment would be considered for aviation systems.

For the Fifteen problem, the key consideration at the abstract functional level would be the rules of the game. That is, the distinction between a legal and an illegal move. This includes the idea of taking turns and the constraint that once a number has been picked, it cannot be used again. For the positioning/tracking task, the abstract function level would include consideration of the motion dynamics (e.g., Newton's second law) and the implications for how the dynamic constraints divide the state space into qualitative regions to reflect the various possible movement paths as illustrated in Figure 3.3.

Note that this does not include any specific paths, but rather the field of possible paths. A good question to ask at this level of analysis is, "What measurement scales are important to the scientists/engineers and regulators who design or manage these systems?" These are not the specific value of the measures (i.e., a particular state) but the dimensions or scales that are used to specify states, possible fields of travel within the state space, and qualitative distinctions such as desirable and undesirable regions within the space of possibilities. The measurement scale or derivatives from these scales are likely to be good choices for dimensions of the state space.

Many people are a bit puzzled that, in Rasmussen's abstraction hierarchy, intentions, goals, and values dominate physical laws. It is important to appreciate that the levels are chosen to reflect means–ends relations, rather than causal relations. Thermodynamic or aerodynamic principles provide potential means for satisfying the goals associated with our energy or transportation needs. The cognitive computation is defined, first and foremost, by the goals or needs to be satisfied. These needs provide the larger context for appreciating the value of specific physical laws and principles. Thus, while goals do not dominate physics with regard to action, they do dominate with regard to the logic or rationale of computations. The goals (*functional purpose constraints*) answer the question why and the physical laws (*abstract functional constraints*) answer the question how.

### 3.4.3 Summary

The computational level of Marr's framework provides the purest description of the nature of the problem and the broadest description of the field of possibilities. The essence of the computational level is quality in terms of both value and feasibility. What are the attributes of a healthy, successful, or well-designed system? It is at this level that normative models of rationality can be framed and where both ideals and minimum criteria for success can be articulated.

For example, it is at this level of abstraction that optimal control principles can be derived and applied to the tracking problem. Such principles can be applied independently from any consideration of the specific type of cognitive or control mechanisms employed. For example, the optimal control principles can identify the minimum time path as the ideal solution to the positioning task. Such an ideal may not be achievable, but it represents a standard or normative benchmark against which the quality of actual solutions can be judged. Similarly, normative moves for the Tic-Tac-Toe game can be specified that will, at least, guarantee a tie.

## 3.5 Algorithm or General Function Level of Abstraction

> At the level of generalized function ... the concepts and language used for description are related to functional relationships that are found for a variety of physical configurations and have general, but typical properties. Examples are feedback loops, which have very characteristic properties, independent of their implementation in the form of mechanical, pneumatic, or electrical systems. (Rasmussen 1986, p. 17)

The second of Marr's three levels and the third level in Rasmussen's means–ends hierarchy frame the problem in terms of algorithms or general functions. Let us begin by considering the Fifteen/Tic-Tac-Toe problem. What general functions would need to be accomplished in order to move from the initial condition to a goal state? You would need decision processes to make a choice (pick a number or select a square), you would need memory processes to keep track of these choices (who has what and what options remain), and you would also need evaluation processes to test whether the goal had been reached (i.e., sum to 15 or three in a row).

Most classical task analyses and task taxonomies, especially cognitive task analyses, are framed at the general function level (e.g., see Fleishman and Quaintance, 1984, for a review of many of these taxonomies). For example, Berliner, Angell, and Shearer (1964) describe a taxonomy of general information processing functions that are divided into perceptual, mediational, communication, and motor processes. Each can be divided further into subfunctions (activities) and sub-subfunctions (behaviors). But, at all levels, the terms reflect general functions that could be implemented in many different devices (e.g., a brain, a digital computer, or a social organization). Similarly, a power plant could be described in general terms that would include heating, cooling, and regulating flow of materials and/or energy. An aviation system could be described in terms of piloting, navigating, and communication functions, which could be further subdivided. Many of these processes would be required independently of the choice of representations.

A typical way to represent a system at the general functional level is in terms of block diagrams. Each box in such a diagram would indicate a particular process and the links between the boxes would reflect the flow of information and/or materials through the system. This form of diagram is particularly good for illustrating general organizational constraints such as precedence relations, bottlenecks, parallel processes, and feedback loops. Thus, this level is particularly good for visualizing the flow of information; in other words, it reflects the algorithm or the steps of a complex process or program. It is this type of analysis that Rasmussen (1986) envisioned for the general functional level.

However, Rasmussen realized that considerations about the potential optimality or stability of a particular organization (e.g., whether control is hierarchical or distributed) needed to be based on more abstract considerations (e.g., information or control theory) that would typically be framed at the computational level. Again, this would involve consideration of intentions, values, and laws (including physical, social/economic, and systems). Typically, many different general functional organizations can at least satisfy the computational level constraints.

For example, consider the current consideration of different ways to manage air traffic (the current highly centralized process vs. more distributed control scenarios, such as free flight). In the current system, many feedback loops with regard to the flow of traffic are closed only through the air traffic control centers. In the free flight design, many of these loops would be closed through individual pilots. Either system could work, but the different organizations would have clear implications for the types of information display that pilots would require.

An important motivation for the abstraction hierarchy was Rasmussen's concern that human factors considerations were being framed based only on a horizontal analysis at the general functional level (in which functions were reduced to subfunctions and sub-subfunctions) without consideration of the importance for vertical analysis across levels of abstraction to answer questions about why a specific organization is satisfactory.

## 3.6 Hardware Implementation

The lowest level of description involves consideration of the actual physical implementation. Here, the differences between a bird, a fixed-wing aircraft, or a rotary-wing aircraft become significant. Here, one must consider details such as the size and shape of a particular wing. In the context of the Fifteen problem, we can begin to consider things such as whether the memory function will be carried out via human working memory or via pen and paper. Further, if it is via pen and paper, will we use lists of numbers, tree diagrams, or a spatial grid arrangement? In the tracking problem, consideration of the type of control device (e.g., spring-centered joystick, force stick, or mouse) and of the physical details of the motions of the stick in relation to the motions on the screen (e.g., stimulus–response compatibility) might be important considerations.

### 3.6.1 Physical Function

At the hardware level, Rasmussen's abstraction hierarchy includes two levels: physical function and physical form. At the level of physical function, the focus is on the type of system that will accomplish a given function. For example, for the Fifteen problem, one might consider what type of physical system will be used to accomplish the memory function: human working memory, paper and pen, or a computer. Each system brings in its own constraints (e.g., the limits of human working memory).

### 3.6.2 Physical Form

At the level of physical form, one considers the physical details—for example, specifically how information will be arranged on a piece of paper (as lists of numbers or as a spatial grid). For the tracking/positioning task, the general physical function might consider the type of control device (spring-loaded joy stick, pressure stick, or mouse). At the physical form level, the compatibility relations and the proportionality constants (gains) between stick motion and cursor states might be important details to consider.

Consider these dimensions from the point of view of the operator in a power plant control room. Knowledge about the various types of physical functions (e.g., electrical vs. hydraulic) could have important implications for thinking about the possible ways that a device might fail. For example, a hydraulic system with a leak might be expected to fail gradually as the fluid levels drop. However, an electrical system with a short would be expected to fail in a discrete, all-or-none fashion. Thus, if a hydraulic system fails suddenly without any previous indication of loss of fluid, one might suspect that the failure might actually be a problem with the electrical sensor. On the other hand, knowledge about the specific physical form (e.g., the collocation of hydraulic and electrical systems in a common conduit) may suggest the possibility of interactions (e.g., a hydraulic leak shorting out an electrical sensor) that might be difficult to understand unless one knows the physical details of the plant layout.

Thus, at the hardware implementation level, a situation is described in terms of the physical details and the constraints associated with those details. This level provides the most concrete answers to "how" something works. This level is often best represented in terms of wiring diagrams or maps that show the spatial arrangement of the distinct physical components (represented as icons) of the system.

## 3.7 Putting It All Together

> One way of coping with the complexity of the real-life environment is to change the resolution of one's considerations when the span of attention is increased, i.e., to aggregate physical items into higher-level objects or to integrate patterns of actions into routines controlled by higher level intentions. Complexity is not an objective feature of a system. (Rasmussen and Lind 1981)

> The complexity perceived depends upon the resolution applied for the information search. (Rasmussen 1986, p. 118)

It is perhaps an irony that in order to see the big picture, it is necessary to partition the larger picture into smaller, more manageable chunks. Yet, at the end of the day, success in managing the system can often depend on the ability to put it all back together into a coherent big picture. In a real sense, it is this ability to construct a coherent picture from multiple chunks that motivates the construct of *situation awareness*. In the next chapter, we will delve much deeper into the aspects of human cognition that constrain awareness.

However, before we leave the problem of situations, it is essential to consider representation of situations in relation to expert performance. This is necessary because we are not interested in situations as an ultimate truth, but rather as a guide to building representations that help people to approach problems productively. Thus, the goal of our situation analysis is to help inspire the design of effective representations. And as a first guess about what an effective representation might look like, it may be useful to think about how experts represent complex problems.

### 3.7.1 Abstraction, Aggregation, and Progressive Deepening

Figure 3.5 shows a two-dimensional matrix that crosses Rasmussen's five levels of abstraction with five levels of decomposition. Decomposition reflects the whole-part aggregation of the work domain, or the parsing of the work domain into finer details. Rasmussen (1986) states that "the whole–part dimension is necessary for control of the span of attention and continuously changes the definition of 'objects' to consider" (p. 14). Rasmussen used the combined matrix to illustrate the relation between abstraction and decomposition as two strategies for managing complexity. In observing expert troubleshooting, Rasmussen noted that these two strategies (while conceptually independent) tend to be at least loosely coupled. That is, experts tend to shift levels of abstraction and levels of detail together. As they consider lower levels of abstraction, they tend to bring in finer levels of detail. Thus, if you plot the reasoning process of an expert during troubleshooting onto this matrix, you will see that much of the time is spent along the diagonal. This illustrates a process of progressive deepening that was described by Duncker (1945).

In this progressive deepening (Duncker 1945), the higher levels of abstraction provide a context for framing the problem at the next level. For example, consideration of the general goal or purpose provides the context for focusing on specific dimensions (e.g., constraints or possibilities) at an abstract level (which, in turn, directs attention to specific processes or functions related to those dimensions). This level directs attention to specific components related to those functions that draw attention to certain locations containing those components. Even when the troubleshooting is initiated by a detail—for

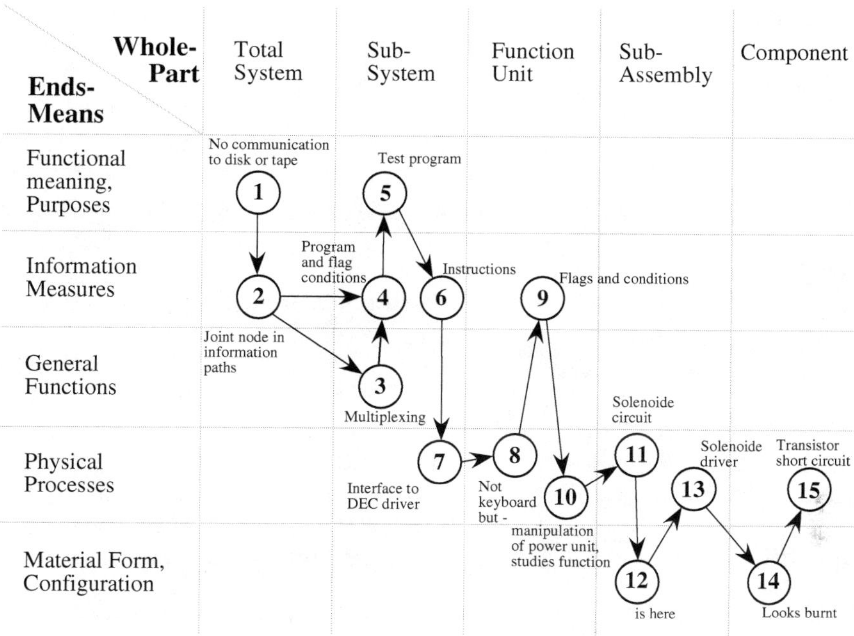


**FIGURE 3.5**
This diagram illustrates the varying span of attention of a maintenance technician's search for the location of a faulty component in a computer-based instrumentation system and the different levels of abstraction in representation that he applies. (Adapted with permission from Rasmussen, J. 1986. *Information Processing and Human–Machine Interaction: An Approach to Cognitive Engineering.* New York: Elsevier. All rights reserved.)

example, a piece that is left over after reassembly of a complex machine—it may be very difficult to understand where the piece fits into the machine without considering the functional purpose or the functional organization of the larger system first.

### 3.7.2 Implications for Work Domain Analyses

In a search for meaning or in a search for understanding of complex situations, it seems that one needs to be able to navigate through levels of abstraction and levels of decomposition. Yet, many of the classical approaches to task, work, and behavior analyses have tended to emphasize decomposition, with little regard for the benefits of abstraction. For example, in early approaches to behavioral modeling, researchers broke down global behaviors (e.g., maze running) into a chain of microgoals. In retrospect, such analyses did little to improve our understanding of behavior. In fact, little progress was made until researchers began to consider more abstract functional aspects of performance (e.g., stages of information processing) and abstract constructs like mental maps.

Similarly, classical task analyses seem to focus on decomposing behaviors into finer components at the general functional level (cognitive functions, subfunctions, and sub-subfunctions). The classical approaches to task analysis tend to be framed exclusively at the algorithmic level of abstraction, with little consideration of the more abstract computational constraints (e.g., values and process dynamic constraints) that shape performance. A risk of such representations is that fundamental computational dimensions of the problem can be overlooked.

Another problem of classical task analysis is that it tends to focus exclusively on behavior or activities in existing systems. This perspective will always be biased because the activities are organized around the current tools and the current representations. It is very unlikely that a description of the activities in the Fifteen problem would inspire designers to consider the spatial representation of Tic-Tac-Toe. Rather, they are more likely to consider ways to support verbal memory and mental computations (i.e., the activities demanded by the verbal representation). Innovation in interface design will often require consideration of the problem to be solved independently from the activities associated with conventional tools, processes, or interfaces.

This illustrates a potential danger of user-centered approaches to work design. The skills of users are often organized around conventional interfaces; thus, it is difficult for them to envision alternative approaches and they typically will initially resist change from the interfaces that they have become adapted to even when there is clear evidence in favor of more effective interfaces. Although we are not recommending that the user's preferences be ignored, we are suggesting that it is important also to consider a "use-centered" or "problem-centered" perspective (Flach and Dominguez 1995) so that the user's preferences can be balanced against the demands of the tasks and the opportunities offered by new technologies.

Figure 3.6 illustrates how a process of work analysis might address multiple levels of abstraction in a way that reflects the productive thinking of experts (à la Figure 3.5). To understand the rationale about why something is done, one moves to higher (more abstract) levels of analysis. To explore how something can be accomplished, one moves to lower (more concrete) levels of analysis. Exploration along the diagonal of this abstraction–decomposition space may be essential for understanding the meaningful aspects of a situation.

### 3.7.3 Summary

Thus, we would like to offer the following claims:

- Productive thinking about situations involves a kind of progressive deepening in which concrete details are illuminated in the context of more abstract organizational, functional, and purposeful properties of the larger situation.

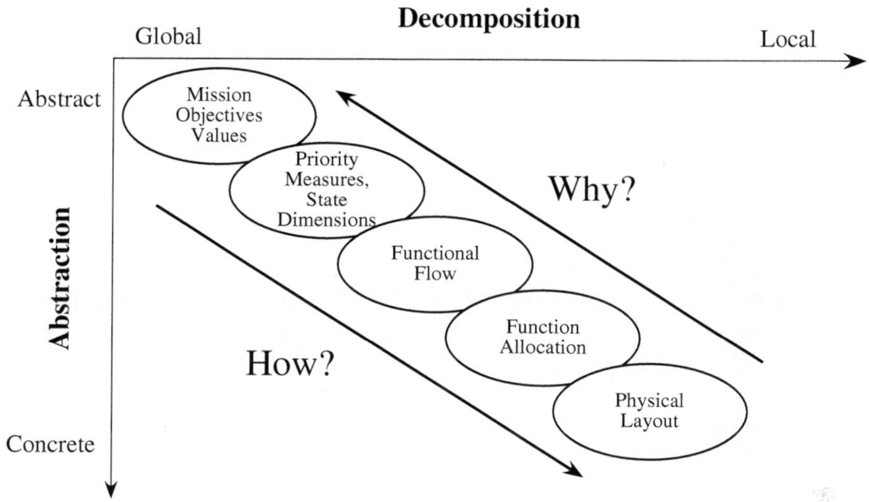


**FIGURE 3.6**
This figure illustrates work analysis as a process of exploring across levels of abstraction and decomposition. The diagonal pattern suggests that more detail becomes important at more concrete levels of analysis.

- A prerequisite for the design of good representations is a work analysis process that provides insights into the couplings across levels of abstraction and decomposition.
- Good representations leverage the loose coupling between levels of abstraction and decomposition to organize detailed information in ways that reflect constraints at higher levels of abstraction.

An excellent example of an interface that explicitly represents the couplings across levels of abstraction is the P + F interface that Vicente (1991, 1992, 1999) designed for DURESS II. In this feedwater control task, the P + F interface explicitly links output goals through abstract functional constraints (i.e., mass and energy balance) and general functional constraints (e.g., valves and reservoirs) to help operators to see the situation in light of the computational constraints. This representation has been shown to support productive thinking with regard to both process control and fault diagnosis (Pawlak and Vicente 1996).

Another example where there was an explicit attempt to integrate computational constraints into a display representation is Beltracchi's (1987, 1989) use of a Rankine cycle diagram to show the pressure–temperature–entropy relations in a power plant control system. Both Vicente's and Beltracchi's displays combine aspects of a spatial flow diagram to show input–output relations with more abstract configural geometries to highlight relations across levels of abstraction. In a later chapter, we will illustrate how abstract energy relations can be integrated into a landing display to help explicate specific control functions to enhance situation awareness for landing.

## 3.8 Conclusion

The goal of this chapter was to draw attention to the challenge of modeling situations. Without a good model of a situation, it is unlikely that it will be possible to develop representations that will enhance awareness or performance relative to that situation. In order to enhance performance, a representation must have a correspondence with meaningful dimensions of the situation. This is the goal of cognitive work analysis: to model work situations.

Vicente (1999) provides a comprehensive approach to cognitive work analysis that is most compatible with our approach to display design. A key component of this approach is Rasmussen's (1986) abstraction hierarchy—an analytical tool to model important categories of information, as well as the inherent relationships between these categories. This model then provides us with a description of the informational content (and relations) that need to be made apparent in the interface. The ultimate challenge represented in the aggregation hierarchy is to provide a description of information in terms of parts/wholes (or chunks) in a way that will allow the agent flexibility to control the grain of resolution and the span of attention in ways that lead to productive thinking about complex problems.

The bottom line is that our ability to design effective representations (i.e., representations that support good situation awareness) depends both on our theories of awareness and our theories of situations. While a good theory of the situation is necessary to ensure correspondence with the work or problem situation, it is not sufficient to ensure good situation awareness. In addition, the representation must be coherent or comprehendible. This brings us to questions of awareness that will be addressed in the next chapter.

To conclude, it is important to note that the abstraction hierarchy is not the answer to this situation or work analysis problem. Rather, it is a challenge for researchers to take the complexity of situations seriously and a way to frame questions about situations. It is a guideline about how to search for meaning or for the deep structure of a problem and it is a caution that no single level of description is likely to capture the full complexity of most work domains. Once again, we want to emphasize that this search will be iterative with our attempts to design more effective interfaces. At the same time that our work analysis, guided by the abstraction hierarchy, will inspire the design of more effective interfaces, the design of more effective interfaces will offer new insights into the work domain.

Finally, let us reconsider the two forms of the Fifteen/Tic-Tac-Toe problem presented in this chapter. The verbal representation presented first is not as effective as the spatial representation presented second. Why not? To answer this question, it is necessary to consider both the problem (e.g., the goals and rules of the game) and the capabilities of the person (or other agent) that must solve the problem. But it is important to understand that these are not independent questions because, as we argued in Chapter 2, we are dealing with a single ontology of experience rather than two independent ontologies—one of mind and one of matter.

In this chapter we have focused on the situation constraints, but we have considered these constraints in light of the demands on productive thinking. In the next chapter, we will consider the constraints on awareness, but again, it will be in light of demands for productive thinking about situations. Our ultimate goal is to find ways to partition problems that make the dimensions that are significant in relation to the situation demands salient in relation to the agent's awareness. Our ultimate goal is to build representations that enhance situation awareness and that enhance the quality of experience.

## References

Beltracchi, L. 1987. A direct manipulation interface for heat engines based upon the Rankine cycle. *IEEE Transactions on Systems, Man, and Cybernetics* SMC-17:478–487.

———. 1989. Energy, mass, model-based displays, and memory recall. *IEEE Transactions on Nuclear Science* 36:1367–1382.

Berliner, D. C., D. Angell, and J. Shearer. 1964. Behaviors, measures, and instruments for performance evaluation in simulated environments. In *Symposium and Workshop on the Quantification of Human Performance.* Albuquerque, NM.

Duncker, K. 1945. On problem-solving. *Psychological Monographs* 58 (5): 1–113.

Flach, J. M., and C. O. Dominguez. 1995. Use-centered design. *Ergonomics in Design* July:19–24.

Fleishman, E. A., and M. K. Quaintance. 1984. *Taxonomies of human performance.* Orlando, FL: Academic Press, Inc.

Gibson, J. J. 1979. *The ecological approach to visual perception.* Boston, MA: Houghton Mifflin.

Hutchins, E. L. 1995. *Cognition in the wild.* Cambridge, MA: MIT Press.

Marr, D. 1982. *Vision.* San Francisco, CA: W. H. Freeman and Company.

Newell, A., and H. A. Simon. 1972. *Human problem solving.* Englewood Cliffs, NJ: Prentice Hall.

Pawlak, W. S., and K. J. Vicente. 1996. Inducing effective operator control through ecological interface design. *International Journal of Human-Computer Studies* 44 (5): 653–688.

Rasmussen, J. 1986. *Information processing and human–machine interaction: An approach to cognitive engineering*. New York: Elsevier.

Rasmussen, J., and M. Lind. 1981. Coping with complexity. In *European Annual Conference on Human Decision Making and Manual Control*. Delft, the Netherlands.

Shannon, C. E., and W. Weaver. 1963. *The mathematical theory of communication*. Urbana, IL: University of Illinois Press.

Vicente, K. J. 1991. Supporting knowledge-based behavior through ecological interface design. Urbana-Champaign, IL: Engineering Psychology Research Laboratory, Department of Mechanical Engineering, University of Illinois.

———. 1992. Memory recall in a process control system: A measure of expertise and display effectiveness. *Memory & Cognition* 20:356–373.

———. 1999. *Cognitive work analysis: Toward safe, productive, and healthy computer-based work*. Mahwah, NJ: Lawrence Erlbaum Associates.

# 4

# *The Dynamics of Awareness*

## 4.1 Introduction

> It is a little dramatic to watch a person get 40 binary digits in a row and then repeat them back without error. However, if you think of this merely as a mnemonic trick for extending the memory span, you will miss the more important point that is implicit in nearly all such mnemonic devices. The point is that recoding is an extremely powerful weapon for increasing the amount of information that we can deal with. In one form or another we use recoding constantly in our daily behavior. (Miller 1956, pp. 94–95)
>
> Acquired skill can allow experts to circumvent basic capacity limits of short-term memory and of the speed of basic reactions, making potential basic limits irrelevant ... the critical mechanisms [that mediate expert performance] reflect complex, domain-specific cognitive structures and skills that performers have acquired over extended periods of time. Hence, individuals do not achieve expert performance by gradually refining and extrapolating the performance they exhibited before starting to practice but instead by restructuring the performance and acquiring new methods and skills. (Ericsson and Charness 1994, p. 731)
>
> To reach a proper human–computer cooperation it will be necessary to study ... strategies that are actually used by operators in different situations ... Furthermore, it is very important to analyze the subjective preferences and performance criteria that guide an operator's choice of strategy in a specific situation. Unless these criteria are known, it will not be possible to predict the strategy that an operator will choose, faced with a specific interface design. (Rasmussen 1986, p. 25)

The opening quote is from George Miller's classic paper, "The Magical Number Seven, Plus or Minus Two: Some Limits on Our Capacity for Processing Information." Miller's work is typically cited as key evidence about the limits of human information processing (in particular, working memory). However, if you read the original paper carefully, a major insight is that due to the human's capacity to recode information, there is *no* effective information limit to working memory.

Yes, working memory is limited to about seven chunks, but "the span of immediate memory seems to be almost independent of the number of bits per chunk, at least over the range that has been examined to date" (Miller 1956, p. 93). He continues that "by organizing the stimulus input simultaneously into several dimensions and successively into a sequence of chunks, we manage to break (or at least stretch) this informational bottleneck" (p. 95). The performance he is describing is the performance of Sidney Smith, who trained himself to recode five binary digits into single chunks (e.g., 10100 = 20). With this encoding, strings of about 40 binary digits were within the five- to nine-chunk capacity of working memory.

### 4.1.1 Zeroing In as a Form of Abduction

The second quote from Ericsson and Charness (1994) amplifies Miller's insight based on an extensive review of the literature on expertise. They observe that the ability of experts to organize information in ways that reflect the deep structure of their particular domains of expertise allows them to effectively bypass the information processing limits that are typically seen with novices or with context-free experimental tasks.

The work of de Groot (1965) and Chase and Simon (1973a, 1973b) in the domain of chess has set the stage for much of this research. For example, the principal difference between stronger and weaker chess players that de Groot observed was the ability of experts to grasp the situation quickly so that the first options that they considered turn out to be among the best choices. De Groot (1965) writes that "within the very first five to ten seconds, the master subject is apt to have more relevant information about the position available to him than the lesser player can accumulate in, say, a quarter of an hour of analysis" (p. 324). Dreyfus (1992) uses the term "zeroing in" to characterize this ability of experts to assess complex situations quickly in order to make smart decisions. He contrasts this with more deliberate search processes that work systematically through the possibilities—"counting out." He writes:

> The human player whose protocol we are examining is not aware of having explicitly considered or explicitly excluded from consideration any of the hundreds of possibilities that would have had to have been enumerated in order to arrive at a particular relevant area of the board by counting out. Nonetheless, the specific portion of the board which finally attracts the subject's attention depends on the overall position. To understand how this is possible, we must consider what William James has called "the fringes of consciousness"; the ticking of a clock which we notice only if it stops provides a simple example of this sort of marginal awareness. Our vague awareness of the faces in a crowd when we search for a friend is another, more complex and more nearly appropriate, case. (p. 103)

Reynolds (1982) provides some insights into the zeroing-in process based on studies of the eye movements of chess players (Tikhomirov and Poznyanskaya 1966) and a reanalysis of some of de Groot's verbal protocols. These studies show that weaker players tend to focus attention based on the distribution of pieces, but that stronger players tend to focus attention based on the distribution of spaces affected by those pieces. Reynolds (1982) concludes:

> The reanalysis of de Groot's protocols indicates that master and grandmaster chess players direct their attention to a different area of the board from that of players of lesser expertise. While the beginning tournament player is captivated by configurations of black and white pieces of wood, the masters and grandmasters center their attention on those squares affected by the pieces. (p. 391)

It is tempting to hypothesize, given our discussion of the levels of abstraction in Chapter 3, that the attention of the experts is guided by more abstract constraints (e.g., a kind of functional center of mass), whereas the attention of less skilled players tends to be captured by more concrete properties of the situation. We suggest that this may be an important hint about how to integrate our analyses of situations together with intuitions about awareness into a more comprehensive understanding of situation awareness. This idea will be expanded in the ensuing discussion.

The opening quote from Rasmussen is a challenge to generalize the insights from basic research on expertise to the problems encountered in work domains. Klein's (1993) work on naturalistic decision making is an important step toward meeting this challenge. In describing the evolution of his recognition-primed decision (RPD) model, Klein's observations are very similar to de Groot's observations regarding experts and their ability to zero in quickly on the best option:

> The RPD model was developed to describe how people can make good decisions without ever comparing options. The initial studies were done with fireground commanders. We expected that they would use their experience to cut down the number of options they compared, maybe just looking at two. We were wrong—they insisted that they hardly ever compared options. (p. 32)

Klein's RPD model includes two major components (see also Chapter 7). The first component, situation assessment, reflects the ability of practitioners to zero in quickly on the situation and to choose a single course of action that often leads to a satisfactory solution. The second component involves an evaluation of the course of action. However, this evaluation does not involve comparisons with other options, but rather a kind of mental simulation in which the course of action is mentally played out in an attempt to discover any potential difficulties. This is a kind of counting out and a similar process can be seen in the protocols of chess players (e.g., Dreyfus 1992). But the deliberate evaluation is concentrated on a single option, rather than across a full set of options, greatly reducing the computational load.

In our view, this process matches well with Peirce's (1931–1935) intuitions about the nature of human rationality as a closed-loop, abductive process. In Peirce's language, the first component is the generation of a good hypothesis by experts that is based on a quick assessment of the situation. The second component is a test of the hypothesis to close the loop and provide feedback. In a situation where there is high risk, the loop might first be closed through mental simulations in an attempt to anticipate potential problems that might possibly lead to rejection of this option. In less risky situations, we posit that the evaluative component may not involve mental simulation, but may involve acting on the choice to directly test the hypothesis. We further suggest that this closed-loop, abductive process is fundamental to the dynamics of human rationality.

### 4.1.2 Chunking of Domain Structure

The clearest empirical evidence about the qualitative differences across levels of expertise in chess are de Groot's (1965) observations, confirmed by Chase and Simon (1973b), that expert chess players are able to remember more information when presented with a meaningful configuration taken from a chess game for a brief period (on the order of a few seconds). These observations bring us naturally back to Miller's observations about recoding or chunking. It seems apparent that more experienced chess players are better able to integrate multiple bits of information (or chess pieces) into larger coherent patterns or chunks, thus effectively extending the limits of working memory.

Chase and Simon included an important control condition. They showed that the advantage in remembering the placement of pieces was reduced (if not completely eliminated) when the configurations were random arrangements of pieces. Thus, it seems clear that the ability to extend the limits to working memory depends in part on the coherent structure of the game of chess that is eliminated in the random configurations.

This inference is strengthened by an experiment reported by Reynolds (1982), who created three random arrangements that were systematically constrained with respect to the proximity of the pieces to a functional center of mass. The differences across skill levels increased as the configurations became more constrained relative to this functional center, with no differences at the low-proximity condition and a more than two to one advantage for the highest class players over the others in the high-proximity condition. Again, this suggests that the improved performance is somehow linked to functional properties of the chess problem.

One hypothesis to explain the improved expert performance in these experiments is that they have a large bank of stored patterns that they can use to help recognize situations and chunk information. For example, it has been estimated that a master chess player would require storage of some 50,000 chess configurations (Simon and Gilmartin 1973). However, Vicente and Wang (1998) offer an alternative explanation. They suggest that the ability of experts is based on their experience with the constraints of the problem, as might be characterized in an abstraction hierarchy. The idea is that the knowledge of the expert is not a collection of memorized patterns, but rather an implicit model of the domain constraints. These might include recognizing the intention of an opponent (functional purpose level); identifying offensive and defensive strengths and weaknesses related to functional centers of mass (abstract function); identifying patterns related to distinct processing stages, such as a generic opening (general function); etc.

The general hypothesis of Vicente and Wang (1998), which we share, is that experts are more effectively tuned to the higher, more abstract levels of constraint than less experienced players are. It is their knowledge of these constraints that allows experts to chunk information into functionally relevant patterns, and thus to zero in quickly on good options. And it is the violation of these constraints in the random conditions that evens the playing field between experts and novices. Thus, the ability to bypass inherent information processing limitations is tied to the structure of the situation. In this sense, expertise is the ability to utilize the natural constraints of situations to recode or chunk the stimuli in ways that enhance situation awareness and that effectively reduce the computational complexity related to the search for good solutions.

The hypothesis that experts are tuning in to constraints at higher levels of abstraction than those available to people with less experience is consistent with the literature on expertise. For example Chi, Feltovich, and Glasser (1981) conclude from their observations of differences in how novices and experts approach physics problems that novices respond to the surface features of a problem while experts respond to its deep structure. Rasmussen's abstraction hierarchy provides an explicit, a priori way to differentiate "deep" structure (e.g., more abstract constraints related to intentions, physical laws, and legal properties of a situation) from "surface" structure (e.g., more concrete constraints associated with physical properties of a situation). Clearly, it takes experience with a domain to discover the more abstract constraints. But once they are discovered, dividends can be reaped with regard to computational efficiencies.

### 4.1.3 Implications for Interface Design

The challenge is how to integrate these intuitions about the qualitative differences between novices and experts into an explicit model of human information processing or awareness. This will be addressed in the next section of this chapter. However, before moving on to a model for awareness, let us take a few moments to highlight what we believe are important implications for the major theme of this book—interface design. In our view, the ultimate goal of interface design is to support or enhance human performance. In other words, a good interface is one that helps the people using it to perform more like experts.

Extrapolating further, then, a good interface is one that helps people to see the deep structure and to utilize that structure in ways that improve the efficiency of processing information. *Thus, a good interface is one that makes the domain or situation constraints (at all levels of abstraction) salient. A good interface is one that helps to externalize or make explicit the implicit representations that experts use to make fast, frugal, smart decisions. The coherence in these internal representations of experts has its ultimate source in the inherent correspondence with constraints of the situations themselves.*

## 4.2 A Model of Information Processing

Figure 4.1 shows a conventional model of human information processing. This model includes a set of logical information processing stages through which an input stimulus is successively transformed into a percept, a concept, a decision, and, eventually, a behavior. This model may or may not include a feedback loop. However, even when the feedback loop is included, there has been a tendency to treat each stage in this process as an independent component of an open-loop communication channel. Many researchers in the information processing tradition have tended to focus on a specific stage of this process and have tended to study that stage in isolation from other stages. Thus, one researcher studies perception, another studies memory, another studies decision making, and still another studies motor control. For each stage, one might ask:

- How is information coded?
- How much information can be stored?
- How long does processing take?

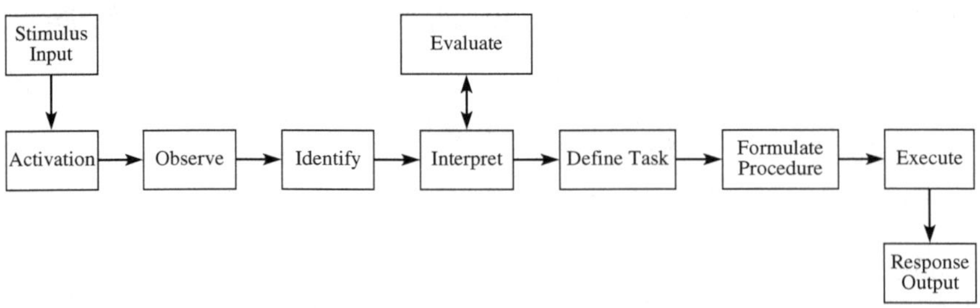


**FIGURE 4.1**
This figure shows the conventional model of cognition as an information communication channel.

Researchers who study attention have tried to localize the "bottleneck" or "resource limits" to specific stages in the process. Is the constraint localized early or late in the process? Do specific stages draw on a common resource or on multiple resources? With regard to the overall dynamic, there is at least an implicit assumption that each stage is a necessary component of the whole process, although the contributions (e.g., processing time) may depend on the specific nature of the input to the specific stage. Further, there is the assumption that the dynamics of the individual stages will add up to provide insight into the overall dynamic.

### 4.2.1 Decision Ladder: Shortcuts in Information Processing

To make sense of his observations of expert electronic troubleshooting, Rasmussen (1986) took the conventional model of information processing stages (Figure 4.1) and folded it in a way that suggests how the information process itself can be parsed in terms of depth of processing (or, alternatively, levels of abstraction). Figure 4.2 illustrates the relation between Rasmussen's decision ladder and the more conventional communication channel model.

Figure 4.3 provides a more complete representation of Rasmussen's decision ladder. In the complete model, Rasmussen shows both information processes (the rectangles in Figure 4.3) and the states of knowledge that are the product of these processes (the circles in Figure 4.3). A key insight that suggested this decision ladder was Rasmussen's discovery that the thought processes of the experts were characterized by associative leaps from one state of knowledge to another (the dashed lines in Figure 4.3). Rasmussen (1986) wrote:

> Immediate associations may lead directly from one state of knowledge to the next. Such direct associations between states of knowledge is the typical process in familiar situations and leads to very efficient bypassing of low-capacity, higher-level processes. Such associations do not follow logical rules; they are based on prior experience and can connect all categories of "states of knowledge." (p. 69)

By folding the model, it was easier for Rasmussen to illustrate these associative leaps as shortcuts from one knowledge state to another. For example, an observation (e.g., a printer will not turn on) might not lead to a detailed causal analysis, but it might stimulate a memory from this morning when you forgot to plug in the coffee pot, thus, triggering a simple task (e.g., check the power source to see if it is plugged in). Note that the task was not triggered by an analytic process of interpreting the situation, evaluating multiple possible causes, and systematically eliminating those possibilities.

Rather, in this particular case it was driven by an automatic association with a rather serendipitous event. If the morning had been different, the first hypothesis to come to mind might have been different. (For example, the

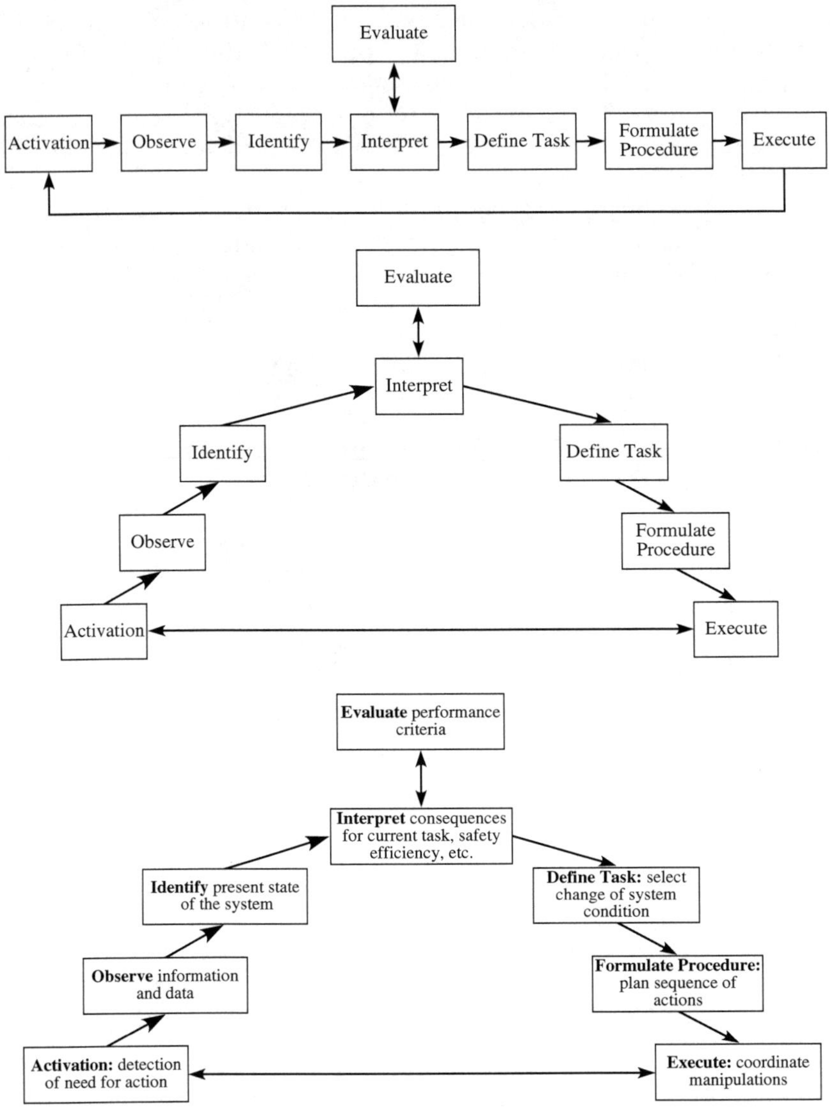


**FIGURE 4.2**
The evolution from the classical information channel model toward Rasmussen's decision ladder model.

printer is broken. Have you ever called the technician to fix a broken piece of equipment, only to have him come in and discover that it is unplugged?) But even if this were the first hypothesis, it might have resulted from a different process—perhaps based on a more logical evaluation of possible causes. The key point is that there are many paths through the cognitive system.

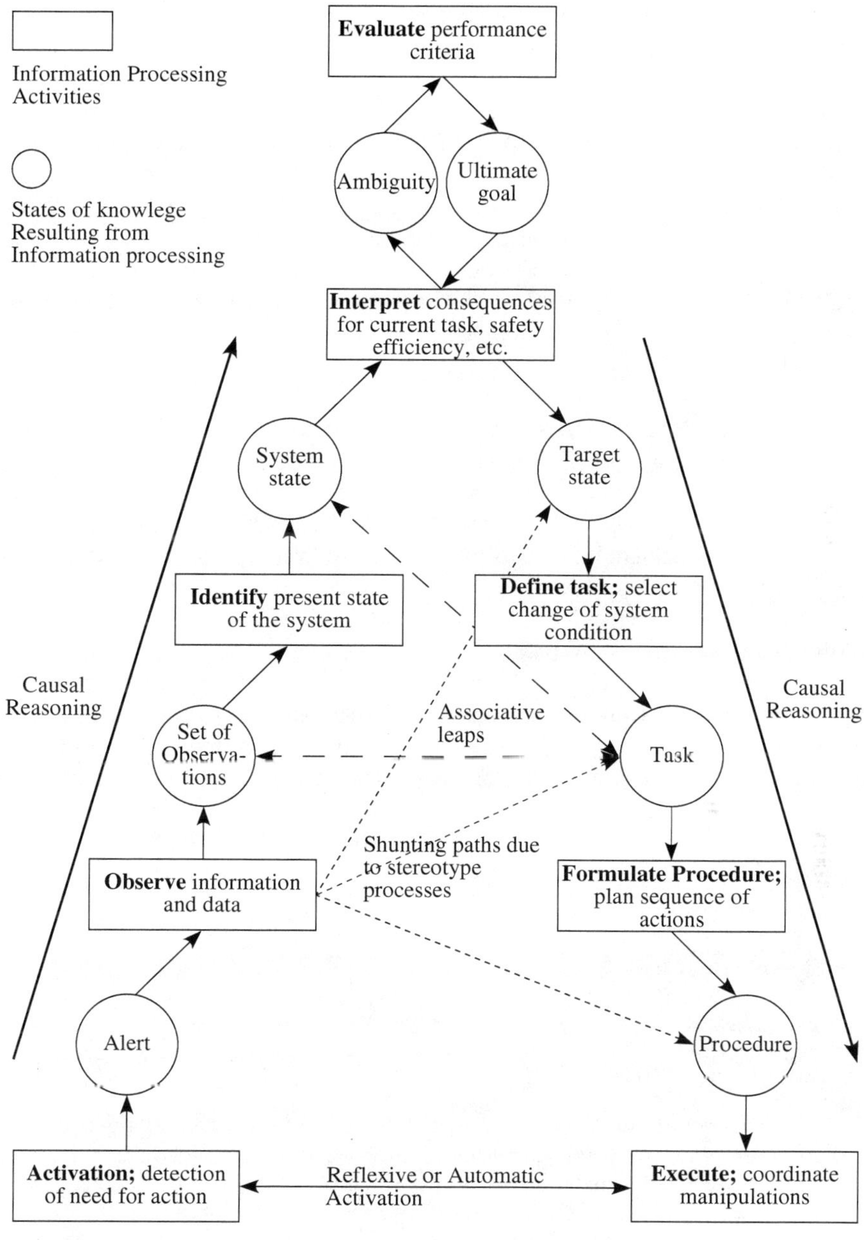


**FIGURE 4.3**
Rasmussen's (1980) decision ladder. (Adapted with permission from Rasmussen, J. 1980. In *Human Interaction with Computers*, ed. H. T. Smith and T. R. G. Green. London: Academic Press. )

The shunts or shortcuts shown in Figure 4.3 illustrate just some of the many possible associations between knowledge states. In principle, Rasmussen realized that any knowledge state can be associated with any other and, further, that these associations can go in any direction (e.g., in the process of formulating or mentally simulating a procedure, an observation may become more salient or a hypothesis about the system state might be triggered). Thus, the decision ladder models the cognitive process as a richly connected associative network, rather than as a multistage communication channel.

Another important aspect of the decision ladder is that the higher processes (e.g., interpreting and evaluating) are the most demanding from a computational point of view. In other words, the higher processes represent the processing bottleneck. Thus, the shortcuts across the decision ladder represent specific hypotheses about how experts can bypass the limitations typically associated with human information processing. The general hypothesis is that, due to their extensive experience, experts have developed many associations that provide them with opportunities to bypass the intensive counting-out processes associated with the higher processes in the decision ladder. These associations allow experts to "recognize" solutions rather than analytically interpreting the situation and evaluating options (activities that are necessary for novices, who have not learned the shortcuts yet).

This idea that human cognition involves the use of shortcuts is not unique to the expertise literature or to Rasmussen. In fact, this intuition is quite pervasive. However, it may not be easy to appreciate since the terms used to frame this intuition can be different depending on the research context. In the following sections, we will consider three different constructs not typically associated with the decision ladder that we believe serve to confirm Rasmussen's intuitions and the value of the decision ladder as a general model for human awareness. These constructs are (1) automatic processing, (2) direct perception, and (3) heuristic decision making.

## 4.3 Automatic Processing

> One contemporary composer has said that before a professional musician feels ready to give a solo performance, there must be sufficient training that one can enter the "mindless state." That is, it is necessary to practice so much, to know the entire situation so well, that the actions rattle off effortlessly without conscious awareness. During a public performance, it is essential that if the mind panics or wanders, perhaps burdened with concern about the reaction of the audience, "the fingers will go on in their mindless way, playing their part until finally the mind can come back to the piece." The "mindless" state of performance is what could be called "automatization"—the automatic, unconscious performance of a skill. (Norman 1969, p. 201)

Probably everyone has experienced the mindless state described in this quote from Norman at one time or another. For example, when you react to an emergency situation when driving, you feel that your feet are already reaching for the brake before you have had a chance to think or decide what to do. In these situations, it seems as though the typical information processing stages are bypassed and the correct response is evoked reflexively or automatically without any conscious assessment of the situation. In some circumstances, this automatic response is necessary for success—that is, there is no time to think; any significant delay would result in a catastrophe.

Schneider and Shiffrin (1977) were able to create an experimental situation that produced this type of "mindless" or automatic behavior in the laboratory. They used a combination of visual and memory search. Participants were given a set of target characters (e.g., the letters R, T, U, S) to store in short-term memory. They were then presented a display that contained another set of letters (e.g., X, V, U, B). The participants were then to respond as quickly as possible to indicate whether one of the target characters was present in the visual display. Thus, in this instance, the correct response would be *yes.* In addition, they manipulated the relation between the target characters (potential memory set items) and the distracter characters (characters in the display that were not targets).

### 4.3.1 Varied and Consistent Mappings

In a *varied mapping condition,* targets on one trial could be distracters on another trial. That is, targets and distracters were drawn from a common set of characters. Thus, referring back to our example, X, V, or B, which are distracters on the sample trial, might be targets on a later trial, and R, T, U, or S, which are targets on this trial, may appear as distracters on a later trial. In a *consistent mapping condition,* targets and distracters were drawn from disjoint sets; that is, a target element would never be a distracter. Again, in terms of our example, X, V, and B could never be used as a target on future trials and R, T, U, and S could never be used as distracters on future trials. The number of items in the memory set and the number of items in the display were also varied.

### 4.3.2 Two Modes of Processing

Early in practice in this task, regardless of whether the mapping was consistent or varied, the time to detect a target was an increasing function of the number of comparisons between items in memory (M) and items in the display (D) ($RT \propto M \times D$). This suggests a kind of counting-out process in which each of the memory items is exhaustively compared to each of the display items to test for a match. Schneider and Shiffrin (1977) called this *controlled processing* and noted that it "requires attention, uses up short-term capacity, and is often serial in nature … . In search, attention, and detection tasks, controlled processing usually takes the form of a serial comparison process at a limited rate" (pp. 51–52). In the varied mapping condition, this pattern persisted relatively unchanged despite extensive practice.

However, a very different pattern of performance emerged in the consistent mapping condition with extensive practice. Detection time eventually became independent of the number of items in memory or the display, so search for one of four memory items among four display items did not take significantly longer than search for one target item in a display of one item. The participants reported that after extensive practice, the consistent targets would "pop out" of the display, so it became unnecessary to search through either memory or the display. Schneider and Shiffrin (1977) hypothesized that this performance was evidence for a distinctly different process that they called *automatic processing:*

> [Automatic processing] is triggered by appropriate inputs, and then operates independently of the subject's control … [they] do not require attention, though they may attract it if training is appropriate, and they do not use up short-term memory capacity … . In search, detection, and attention tasks, automatic detection develops when stimuli are consistently mapped to responses; then the targets develop the ability to attract attention and initiate responses automatically, immediately, and regardless of other inputs or memory load. (p. 51)

### 4.3.3 Relationship to Decision Ladder

Let us consider these two distinct styles of processing (controlled vs. automatic) in light of Rasmussen's decision ladder. We propose that controlled processing reflects a process that involves the higher levels in this model. The conscious, deliberate comparison process involves higher stages such as identification, interpretation, and evaluation. On the other hand, automatic processing reflects a direct link between lower stages in the process—perhaps even a direct link from activation to response—thus bypassing the more demanding higher stages of processing.

However, the important question is what characteristic of the consistent mapping condition allows this processing efficiency (i.e., this associative leap, this shortcut) to occur. We believe this reflects our inherent ability to utilize structure or constraints in situations. We hope it is obvious that the consistent mapping condition imposes a significantly different constraint on performance relative to the varied mapping condition. We believe that automatic processing reflects the ability of the participants to use this constraint to minimize processing demands.

Note that this constraint is discovered (at least implicitly) by the participants as a result of experience over many trials. This is why it takes time for the automatic processing to develop. It is also very likely that conscious knowledge of the consistency is neither necessary nor sufficient to enable the associative leap. It seems that the only way to Carnegie Hall is practice,

practice, practice. It takes many repetitions to develop the associations that underlie the "mindless" state characteristic of automatic processes.

This seems to be at the heart of Zen approaches to skilled behaviors such as swordsmanship and archery. Practitioners are subject to endless repetitions until they can achieve a "no mind" state where they do not concentrate to *make* the arrow hit the target, but rather empty their mind in order to *let* the arrow hit the target. In terms of the decision ladder, they practice until the higher aspects of information processing (the mind?) get out of the way of success.

It is also important to note that under the varied mapping condition, despite extensive practice, automatic processing never develops. In some sense, this may be the most surprising result from the Schneider and Shiffrin (1977) experiments. In everyday life, we are familiar with the qualitative shifts that occur as the result of extensive practice. It seems that in the varied mapping condition, a perverse situation has been created where experience offers little benefit. It is tempting to conclude that some degree of consistency is essential for skill to develop.

### 4.3.4 Implications for Interface Design

What is the lesson for those interested in designing interfaces to facilitate the development of expert performance or to enhance situation awareness? One simple lesson is that consistency will play an important role in helping people to bypass information processing limits. When possible, displays should be designed to achieve a consistent mapping. That is, those aspects of the work situation that are targets or signals should be represented in a way that is consistently distinct from aspects of the work situation that are distracters or noise. Further, *the mapping between targets, displays states, and the desirable responses should be consistent.* However, take heed of Emerson's (1841) caution: "A foolish consistency is the hobgoblin of little minds." The challenge is to discern what consistencies will serve the goal of expert performance and what consistencies will lead us down the garden path to foolishness. We will revisit this point later.

Finally, along with Emerson's caution, it is important to be aware that automatic processes have a cost. For example, some of the participants in the Schneider and Shiffrin experiment reported that, after a time, the consistently mapped target letters would sometimes leap out at them from pages of books—outside the experimental context. That is, the automatic processes could interfere in contexts where the consistent mapping was not relevant to the performance goals.

Norman (1981) describes a class of action slips that occur when "schemas not part of a current action sequence become activated for extraneous reasons, then become triggered and lead to slips" (p. 6, Table 1). Two classes of these slips are "capture" errors and unintended activation. In both situations, an association with a well-learned motor sequence (or habit) or with an external stimulus captures attention away from the intended goal. For example, a driver who normally drives a car with manual transmission may automatically reach for the clutch when driving a rental car with an automatic transmission. In this instance, the well-learned habit automatically captures behavior.

Norman offers the classic Stroop task to illustrate unintended activation. In one version of this task, the participant is presented strings of characters (e.g., 2, 11, 444, 3333) and is asked to respond with the number of items in the string. This requires that the participant ignore the characters, which are numbers that are different from the string counts. This can be surprisingly difficult because the associations with the character names will tend to capture attention automatically. Thus, another consideration in the design of interfaces is to avoid consistencies that may evoke counterproductive action sequences or that may divert attention to attributes that are unrelated to the system goals.

## 4.4 Direct Perception

Consistency is the key issue in generalizing the insights of Schneider and Shiffrin (1977) to performance in tasks other than the visual and memory search tasks. That is, when we see performance in natural contexts that seems to bypass information limitations and that seems automatic, we should expect to find consistent mappings. So, what are the consistent mappings for the pianist that allow her to achieve the mindless state needed for a flawless performance under what can be stressful conditions? There is the consistent layout of the keyboard. But there are also more subtle consistencies associated with music. Music is not a random sequence of notes; rather, it typically reflects constrained choices that are associated with particular scales. Practicing these scales so that they become automatic can be a significant part of music training. Note that it is the constraints associated with these scales (e.g., the blues scale) that allow musicians to jam and to improvise without getting totally lost in the possibilities.

This is where music theory may be important: It describes the constraints that underlie different forms of music. Musical performance skill may not require that the musician have a conscious awareness of music theory. However, we hypothesize that the music theory may be critical to the human performance researcher who wants to explain the musician's skill because the music theory will help her to understand the constraints (consistent mappings) that make the skill possible.

### 4.4.1 Invariance as a Form of Consistent Mapping

What about more mundane situations where humans and animals seem to do the right thing (e.g., walking through a cluttered environment without bumping into things) automatically (i.e., with little conscious awareness)? Are there consistencies that support these smooth, coordinated (one might say, expert) actions? Gibson's (1979) theory of direct perception is based on the assumption that there are indeed consistent mappings (in his terms, invariance) that underlie the abilities of animals to move skillfully through the environment. In particular, Gibson was interested in how people and animals use visual information to move through their environments skillfully (Gibson 1958/1982).

As Gibson studied this problem, he began to discover consistent mappings associated with visual optics. All of us have been introduced to the laws of perspective in grade school art classes. We learned that parallel lines in a three-dimensional world will project as lines that converge as they approach the optical horizon when depicted on a two-dimensional picture plane. Perhaps you also learned that the angle of convergence will depend in part on the height of the observation point. If the observation point is lower to the ground, the inside angle will be wider (Figure 4.4a) than if the observation point is higher above the ground (Figure 4.4b).

Now imagine what would happen if you moved an eye (or a camera) smoothly from a higher point to a lower point (e.g., from Figure 4.4b to Figure 4.4a): You would see a specific optical transformation in which the angles would splay out at a rate that increased as the distance to the surface was closed. This optical transformation is lawful. That is, there is a consistent mapping between motion over the surface and the optical transformations associated with those motions—mappings that can be described using differential geometry (e.g., see Gibson, Olum, and Rosenblatt 1955; Flach and Warren 1995). In fact, these lawful properties are what allow computer game and virtual reality designers to simulate motion on computer screens. The transformations on the screen are a two-dimensional projection that corresponds invariantly with specific motions over a specific three-dimensional surface.

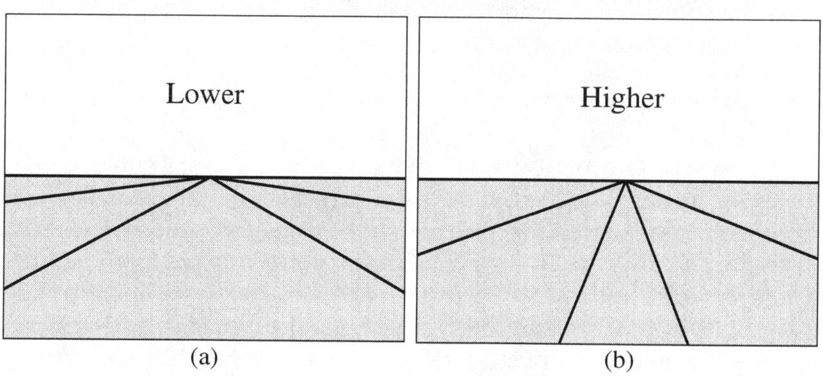


**FIGURE 4.4**
Parallel lines on the ground will be seen as converging lines in a picture plane. The angles of projection will become more acute as the point of observation is moved to a higher point.

### 4.4.2 Perception and the Role of Inferential Processes

The radical aspect of Gibson's theory was the postulate of *direct perception*. Conventionally, perception has been conceived to involve higher mental processes that must compute an image of the world from a collection of largely ambiguous cues together with knowledge (e.g., assumption of size constancy) about the world. Thus, perception is considered to be an inferential process, which would imply involvement of the higher levels in the decision ladder.

However, Gibson proposed that animals could take advantage of the invariant optical relations to effectively bypass the higher levels of processing. They could form direct associations between the optical invariants and the actions without the need for mediation via an internal inferential model of the world. The development of these direct associations would in many ways be similar to the development of automatic processing described by Schneider and Shiffrin (1977).

Gibson (1958/1982) described how the invariant optical relations could allow direct control of motion through simple control processes that operated directly on properties of the two-dimensional projection, without the need of an inverse model to reconstruct the three-dimensional world. Here is Gibson's description of steering and aiming:

> The center of the flow pattern during forward movement of the animal is the direction of movement. More exactly, the part of the structure of the array from which the flow radiates corresponds to that part of the solid environment toward which he is moving. If the direction of his movement changes, the center of flow shifts across the array, that is, the flow becomes centered on another element of the array corresponding to another part of the solid environment. The animal can thus, as we would say, "see where he is going." The act of turning or steering is, therefore, a visual as well as a muscular event. To turn in a certain direction is to shift the center of flow in that direction relative to the fixed structure of the optic array. The amount of turn is exactly correlated with the angular degree of shift. The behavior of aiming at a goal object can now be specified … . To aim locomotion at an object is to keep the center of flow of the optic array as close as possible to the form which the object projects. (p. 155)

At the time that Gibson formulated his theory of direct perception, it was very difficult to manipulate optical structure dynamically and to measure the behavioral consequences in order to test his hypotheses empirically. However, with advances in graphical displays and virtual reality, we now have a fairly large body of empirical research to support Gibson's hypotheses (e.g., Flach and Warren 1995). It seems clear that people are able to utilize the geometric properties of optical arrays to minimize computational demands and to enhance skilled interactions with their environment. When we consider this work in light of the decision ladder, what Gibson called *direct perception* would reflect people's ability to discover associations between optical invariants and movement control. This would allow direct associations between the visual stimulation (e.g., the point of radial outflow) and the appropriate action (e.g., turning to follow moving prey).

Note that in the academic debate between conventional perceptual psychologists and Gibsonians, people have tended toward two extreme positions. The conventional view is that higher computations are a necessary part of every perception; the extreme Gibsonian view is that higher computations are never involved in perception. The decision ladder offers a middle ground. This model allows for both direct associations (for those situations where there are consistent mappings/invariants relevant to the functional goals) and the involvement of higher computational/inferential processes (for those situations where the mappings are less consistent or where the consistencies have not yet been learned).

The idea of global invariants (specifying relations) as the basis for skilled action was a very important heuristic to help free behavioral researchers from an almost exclusive focus on illusions and failures of perception that provided little insight into human skill. However, there is increasing evidence that people are very opportunistic in their ability to use either global or local constraints (nonspecifying variables). Jacobs, Runeson, and Michaels (2001) write:

> Specifying relations granted by global constraints (e.g., natural laws) might seem more useful for perceiver-actors than those granted by local constraints .... Because they allow accurate performance in a wider range of task ecologies. One might expect observers always to take advantage of the existence of global constraints. However, the present results suggest that observers merely search for variables that are useful in the ecology encountered in practice.
>
> The variables observers come to detect after practice often appear to be the more useful nonspecifying variables. Whether observers ultimately move on to the use of specifying variables seems to depend on particular characteristics of the stimulus set. To the extent that nonspecifying variables happen to correlate highly with the property to be perceived, perceivers seem to become trapped in a local minimum. Thus, great care must be taken in the selection of a stimulus set; otherwise, what may appear to be global cognitive principles can, in fact, be local solutions to local problems. (pp. 1034–1035)

There seems to be a natural efficiency (or laziness) in which people gravitate toward the simplest solution that satisfies their local functional goals. They utilize any associations that work; some of the illusions and errors in the perceptual literature might best be understood as being similar to Norman's (1981) action slips (unintended activation), where a local association is generalized to a situation that violates a consistency that is typical in other contexts. In the language of Jacobs et al. (2001), people become "trapped in a local minimum."

In terms of the decision ladder, people will tend to utilize associations at lower levels in the ladder to avoid the computational demands of higher levels whenever possible.

### 4.4.3 Implications for Interface Design

In sum, Gibson proposed that skilled control of movement reflected an attunement to optical structure (optical invariants) that directly specified both properties of the environment (e.g., layout of surfaces) and the motion of the observer relative to that layout. We consider this to reflect a kind of automatic processing very similar to that described by Schneider and Shiffrin (1977). We believe that a very effective way to visualize this automatic processing or direct perception is as a shortcut within the decision ladder that allows a direct link between observation and action that bypasses higher stages of information processing.

In terms of a question raised in the previous section about distinguishing between mindless consistency and valuable consistency, we think that a key lesson from Gibson's ecological optics is that it rests on the mapping of optical structure to functional behaviors (e.g., controlling locomotion). *Thus, for interface design, we suggest that the key is not consistency in and of itself. Rather, the key will be to create consistent mappings between the interface representations and functional properties of the task situation.*

This is where the constraints identified in the abstraction hierarchy become important. A central hypothesis guiding an ecological approach to display design will be that expert human performance results when people utilize the constraints inherent in the situation (as reflected in an abstraction hierarchy) to minimize computational demands and to facilitate skillful control. A goal for interface design is to build representations that consistently map to these constraints. That is, to build displays where properties of the representation (e.g., a configural geometry) are invariantly linked to properties of the work domain.

For example, in Coekin's (1970) polar display (also, Woods, Wise, and Hanes 1981), the geometrical symmetry of an octagon is invariantly mapped to the normal state of a nuclear power plant, and different distortions of this symmetry specify different types of failure conditions (e.g., loss of coolant accident). With such a configural display, it is likely that operators will be able to detect failures automatically and quickly recognize the type of failure and the associated actions that could be expected to bring the system back under control.

Gibson's intuitions about the significance of the links between perception (e.g., optical invariants) and action (e.g., affordances) for cognitive systems anticipates the constructs of *direct manipulation* and *direct perception* that will be discussed in later chapters of the book. The key point is that it is possible (in the context of optical control of locomotion) and desirable (in context of graphical interface design) that the interface directly specifies the possibilities for action and the associated consequences.

## 4.5 Heuristic Decision Making

In 1978 Herbert Simon received the Nobel Prize in economics for his work on organizational decision making. A central intuition of Simon's work was the recognition that human rationality is bounded. That is, humans do not seem to follow the prescriptions of normative models of economic decision making or rationality. Rather, they consider only a subset of the potentially relevant information and they tend to seek satisfactory, rather than optimal, solutions (i.e., humans satisfice). In other words, people use heuristics (take shortcuts), rather than carrying out the computationally difficult processes associated with most normative models of rationality. Sound familiar?

Todd and Gigerenzer (2003) have considered three different ways to examine the bounds on human rationality that Simon described:

- Bounded rationality as optimization under constraints
- Bounded rationality as cognitive illusions
- Bounded rationality as ecological rationality

### 4.5.1 Optimization under Constraints

The first approach to bounded rationality, optimization under constraints, focuses on the practical limits of problem solving. It is clear that, in many naturalistic contexts, there can be both time pressure and ambiguities that make it difficult for people to carry out the full analysis of the situation that would be required by many of the normative models of rationality.

This approach tends to assume that satisficing and the use of heuristics is an adaptation (in this case, a compromise) to the dynamics of the decision situation. There is an implicit assumption that, given the time and information, human rationality would tend to reflect the normative models (e.g., subjective utility theory). For example, this approach would explain Klein's (1993) observation that firefighters make a decision based on a quick assessment of the situation because the situation demands quick action. Every moment lost while analytically comparing options could increase the threats to property and human life. Klein (1993) writes:

> One of the features of naturalistic settings is the presence of acute stressors that are sudden, unexpected, and of short duration. Acute stressors include:
>
> - time pressure
> - ambiguity
> - noise
> - threat
> - impending failure

> • public scrutiny of performance
> • high workload
>
> These conditions make it difficult to carry out many analytical strategies, and may also disrupt decision making in general. (p. 67)
>
> Even for tasks where quantitative, analytical decision strategies are appropriate, these stress effects will make it difficult if not impossible to carry out such strategies. It seems reasonable for people to use the simpler, singular evaluation strategies, and to use experience in sizing up situations to avoid option comparison altogether.
>
> Moreover, the simpler, naturalistic strategies such as recognition primed decisions do not necessarily result in poor decisions. In many studies, high levels of acute stress did not disrupt decision quality and sometimes even increased decision quality. (p. 70)

In short, the "optimization under constraints" view of bounded rationality sees the use of satisficing and heuristics as a kind of speed–accuracy–efficiency trade-off reflecting dynamic constraints of the situation.

### 4.5.2 Cognitive Illusions

The second approach to bounded rationality—as cognitive illusions—focuses on the information processing limitations of humans (e.g., limited working memory). In this view, heuristics (e.g., representativeness) tend to take on a negative connotation associated with cognitive biases in which humans make choices that violate the prescriptions of normative theory (e.g., gambler's fallacy). This research on heuristics as biases tends to parallel early research on perception, producing a collection of examples to show how people violate normative models of rationality and probability (Kahneman, Slovic, and Tversky 1982). For example, Wickens and others (2004) list five "heuristics and biases in receiving and using cues," six "heuristics and biases in hypothesis generation, evaluation, and selection," and four "heuristics and biases in action selection."

The "availability heuristic" is a good example of the cognitive illusions approach to bounded rationality. Suppose that you were asked to estimate the likelihood or risk of people dying from various causes (e.g., cancer, diabetes, automobile accident, aircraft accident). According to probability theory, this should involve comparing the relative frequencies of the various causes. However, often people will judge likelihood based on how easy it is to bring to mind or remember instances of a specific cause. Thus, a person's estimate of probability will be biased by factors, such as *salience*, that influence memory but that are not logically related to probability. This may lead them to judge (erroneously) cancer as a more likely cause of death than diabetes or to assume (incorrectly) that flying is more dangerous than driving. Cancer deaths and aircraft accidents tend to be more newsworthy—thus more salient, thus more memorable, and thus judged more likely.

Although many factors could influence the salience of a memory (e.g., recency, emotional intensity, personal relevance, etc.), one of these factors is the number of repetitions in past experience, which in many situations may be correlated with the actual frequency of an event. Thus, although salience is not *logically* related to probability judgments, in many situations it may be *circumstantially* related to frequency and therefore to probability. Thus, there are at least some situations and perhaps many where the availability heuristic will lead to sound judgments. This leads naturally to the third sense of bounded rationality: ecological rationality.

### 4.5.3 Ecological Rationality

The focus of ecological rationality considers both the limitations and abilities of humans and the constraints of problems to discover how it might be possible for people to be accurate, fast, and frugal. The key intuition, as you might now be able to guess, is that people can make

> … good (enough) decisions by exploiting the structure of the environment … Heuristics that are matched to particular environments allow agents to be ecologically rational, making adaptive decisions that combine accuracy with speed and frugality. (We call the heuristics "fast and frugal" because they process information in a relatively simple way, and they search for little information.) … . By letting the world do some of the work—by relying on the presence of particular useful information patterns—the decision mechanisms themselves can be simpler; hence our focus on simple, fast, and frugal heuristics. (Todd and Gigerenzer 2003, p. 148)

Suppose that you were given pairs of German cities and were asked to decide which of each pair had the larger population?

- Berlin versus Munich
- Hamburg versus Cologne
- Stuttgart versus Düsseldorf

#### *4.5.3.1 Less Is More*

Would additional information be useful? Suppose you knew that Berlin was the capitol of Germany, or that Berlin, Hamburg, and Munich were state capitols, or that Hamburg, Munich, Cologne, and Stuttgart had teams in the top professional league. Would this information be helpful? It turns out that U.S. students do slightly better at this task when German cities are used and that German students do better with American cities (Goldstein and Gigerenzer 1999). This is an example of the *less is more* effect.

One explanation for this effect is that people use a recognition heuristic. That is, if only one of the two objects is recognized, then the recognized object is chosen. The hypothesis is that this heuristic works better for the U.S. students for German cities (and vice versa) because, for those students from a different country, recognition will be highly correlated with population. Thus, this simple heuristic works well due to the relation between *salience* in memory and the task variable (population).

#### 4.5.3.2 Take the Best

Gigerenzer and Goldstein (1996) describe a slightly more complete heuristic (or satisficing algorithm) called *take the best*. The first step involves the recognition heuristic:

Step 1: "If only one of the two objects is recognized, then choose the recognized object. If neither of the two objects is recognized then choose randomly between them. If both of the objects are recognized, then proceed to Step 2" (p. 653).

Step 2: "For the two [recognized] objects, retrieve the cue values of the highest ranking cue from memory" (p. 653). For example, this might be whether the cities have teams in the top professional soccer league.

Step 3: Does the cue discriminate? "The cue is said to discriminate between two objects if one has a positive cue value and the other does not" (p. 653).

Step 4: "If the cue discriminates, then stop searching for cue values. If the cue does not discriminate, go back to Step 2 and continue with the next cue until a cue that discriminates is found" (p. 653).

Step 5: "Choose the object with the positive cue value. If no cue discriminates, then choose randomly" (p. 653).

In the context of the decision ladder, the first step might be visualized as a shortcut based on a memory association. The second step, if required, might involve higher cognitive processing in order to evaluate what the most discriminating cue would be. However, Gigerenzer and Goldstein (1996) have suggested other satisficing algorithms, such as *take the last*, that allow other shortcuts. Rather than using the most discriminating cue, *take the last* uses the last cue that worked—thus again using an association in memory to bypass a need to evaluate the situation to identify the most diagnostic cues.

#### 4.5.3.3 One-Reason Decision Making

Gigerenzer and Goldstein (1996) suggest that satisficing often takes a form that they call *one-reason decision making:* "The inference or decision is based on a single good reason. There is no compensation between cues" (p. 662).

They continue that although the choice is made exclusively on one reason or cue, "this reason may be different from decision to decision. This allows for highly context-sensitive modeling of choice" (p. 662).

One of the significant aspects of Gigerenzer and Goldstein's (1996) approach to satisficing is that not only does it fit well with empirical observations of human choice performance, but they have also been able to demonstrate that, given incomplete data, the one-reason algorithms result in performance that is comparable to that of many of the more complex normative models. Thus, these algorithms not only are fast and frugal with regard to demands on cognitive resources, but also lead to generally smart decisions. Therefore, these shortcuts are not necessarily weaknesses, but can actually reflect the strengths of experience. In fact, Todd and Gigerenzer (2003) go so far as to argue that the limitations themselves are adaptive. For example, rather than seeing working memory as a bottleneck, it may be viewed as a filter that is tuned to meaningful aspects of the environment. Thus, it is designed to block out the noise and to focus attention on the relevant aspects of the environment (e.g., consistent mappings, optical invariants).

## 4.6 Summary and Conclusions

What is the long thread that links research on expertise to research on naturalistic decision making, to research on automatic attention, to research on visual control of locomotion, or to ecological rationality? And what are the implications for interface design? To sum up the dynamics of awareness in a single word, we would describe awareness as *opportunistic*. In particular, humans use experience opportunistically—such that their assessments and choices are strongly influenced by information that is salient in memory. In novel contexts (including decontextualized laboratory tasks), this opportunism appears as a *bias* that often results in visual illusions and deviations from the prescriptions of normative models of rationality. However, in familiar situations, salience in memory will be closely linked to consistencies and invariant properties of the situations. Thus, this opportunism is more likely to appear as *expertise, skill,* or *situated cognition.*

The implication for interface design is that an important goal is to *bias* the human in ways that lead to productive thinking in a particular domain. That is, the goal is to make sure that the information that is *salient* to the human is also *important or significant* to the work domain. The goal is to make the significant consistent mappings, invariants, laws, or constraints salient to the human. This shifts the focus of the designer from being *user centered* to being *use centered* (Flach and Dominguez 1995).

In a user-centered focus, the mantra is to honor thy user, and the focus is on understanding the information processing limits and biases and protecting the system against the negative consequences of these limitations. In a use-centered focus, the mantra is honor the problem/work domain, and the focus is on understanding the situation constraints so that important or meaningful aspects of the problem can be made salient in the representation (either in the external display or in the internal experience of the user). The use-centered focus assumes that, by making meaningful information salient, it is possible to facilitate the development of expertise, to leverage the ability of humans to zero in on essential properties, and to make fast, frugal, good decisions.

In Chapter 2 we introduced the argument for a single ontology that spans both mind and matter in order to appreciate how both shape human experience. In Chapters 3 and 4 we have parsed this single ontology into a perspective of situations (Chapter 3) and a perspective of awareness (this chapter). However, it is important to note that our discussion of situations was *not* independent from considerations of awareness and that our discussion of awareness was *not* independent from considerations of situations. The emphasis was different, but in both cases we were considering a single ontology for a science of human experience. In the next chapter we will try to close the loop by talking about the dynamics of situation awareness in a way that explicitly links the situation constraints and the awareness constraints into a unified framework for thinking about human experience and, ultimately, for guiding the design of interfaces.

## References

Chase, W., and H. A. Simon. 1973a. The mind's eye in chess. In *Visual information processing,* ed. W. G. Chase. New York: Academic Press.

———. 1973b. Perception in chess. *Cognitive Psychology* 4:55–81.

Chi, M. T. H., P. J Feltovich, and R. Glaser. 1981. Categorization and representation of physics problems by experts and novices. *Cognitive Science* 5 (2): 121–152.

Coekin, J. A. 1970. An oscilloscope polar coordinate display for multidimensional data. *Radio and Electronic Engineer* 40:97–101.

de Groot, A. D. 1965. *Thought and choice in chess.* The Hague, the Netherlands: Mouton.

Dreyfus, H. L. 1992. *What computers still can't do.* Cambridge, MA: MIT Press.

Emerson. 1841. Self reliance. In *Essays.* New York: Harper Perennial, 1981.

Ericsson, K. A., and N. Charness. 1994. Expert performance: Its structure and acquisition. *American Psychologist* 48:725–747.

Flach, J. M., and C. O. Dominguez. 1995. Use-centered design. *Ergonomics in Design* July: 19–24.

Flach, J. M., and R. Warren. 1995. Low-altitude flight. In *Local application of the ecological approach to human–machine systems,* ed. P. Hancock, J. M. Flach, J. Caird, and K. J. Vicente. Hillsdale, NJ: Lawrence Erlbaum Associates.

Gibson, J. J. 1958/1982. Visually controlled locomotion and visual orientation in animals. In *British journal of psychology*, ed. E. J. Reed. Hillsdale, NJ: Lawrence Erlbaum Associates. Original edition, *British Journal of Psychology*, 49:182–194.

———. 1979. *The ecological approach to visual perception*. Boston, MA: Houghton Mifflin.

Gibson, J. J., P. Olum, and F. Rosenblatt. 1955. Parallax and perspective during aircraft landings. *American Journal of Psychology* 68:372–385.

Gigerenzer, G., and D. G. Goldstein. 1996. Reasoning the fast and frugal way: Models of bounded rationality. *Psychological Review* 103:650–669.

Goldstein, D. G., and G. Gigerenzer. 1999. The recognition heuristic: How ignorance makes us smart. In *Simple heuristics that make us smart*, ed. G. Gigerenzer, P. M. Todd, and A. R. Grou. New York: Oxford University Press.

Jacobs, D., S. Runeson, and C. F. Michaels. 2001. Learning to visually perceive the relative mass of colliding balls in globally and locally constrained task environments. *Journal of Experimental Psychology: Human Perception and Performance* 27:1019–1038.

Kahneman, D., P. Slovic, and A. Tversky, eds. 1982. *Judgments under uncertainty: Heuristics and biases*. Cambridge, England: Cambridge University Press.

Klein, G. 1993. Naturalistic decision making: Implications for design. Wright Patterson Air Force Base, OH: CSERIAC.

Miller, G. A. 1956. The magical number seven, plus or minus two: Some limits on our capacity for processing information. *Psychological Review* 63:81–97.

Norman, D.A. 1969. *Memory and attention*. New York: John Wiley & Sons, Inc.

———. 1981. Catergorization of action slips. *Psychological Review* 88:1–15.

Peirce, C. S. 1931–1935. *Collected papers of Charles Sanders Peirce*, ed. C. Hartshorne and P. Weiss. Cambridge, MA: Harvard University Press.

Rasmussen, J. 1980. The human as a system component. In *Human interaction with computers*, ed. H. T. Smith and T. R. G. Green. London: Academic Press.

———. 1986. *Information processing and human–machine interaction: An approach to cognitive engineering*. New York: Elsevier.

Reynolds, R. I. 1982. Search heuristics of chess players of different calibers. *American Journal of Psychology* 95 (3): 383–392.

Schneider, W., and R. M. Shiffrin. 1977. Controlled and automatic human information processing: I Detection, search, and attention. *Psychological Review* 84:1–66.

Simon, H. A., and K. A. Gilmartin. 1973. A simulation of memory for chess positions. *Cognitive Psychology* 5:29–46.

Tikhomirov, O. K., and E. D. Poznyanskaya. 1966. An investigation of visual search as a means of analyzing heuristics. *Soviet Psychology* 5:3–15.

Todd, P. M., and G. Gigerenzer. 2003. Bounding rationality to the world. *Journal of Economic Psychology* 24:143–165.

Vicente, K. J., and J. H. Wang. 1998. An ecological theory of expertise effects in memory recall. *Psychological Review* 105 (11): 33–57.

Wickens, C. D., J. Lee, Y. D. Liu, and S. Gordon-Becker. 2004. *Introduction to human factors engineering*, 2nd ed. Upper Saddle River, NJ: Prentice Hall.

Woods, D. D., J. A. Wise, and L. F. Hanes. 1981. An evaluation of nuclear power plant safety parameter display systems. Paper read at the Human Factors Society 25th Annual Meeting, at Santa Monica, CA.

# 5

# *The Dynamics of Situation Awareness*

## 5.1 Introduction

> Human rational behavior … is shaped by a scissors whose two blades are the structure of the task environments and the computational capabilities of the actor. (Simon 1990, p. 7)
>
> These two blades—the two sources of bounds on our rationality—must fit together closely for rationality to cut. While the external bounds may be more or less immutable from the actor's standpoint, the internal bounds compromising the capacities of the cognitive system can be shaped, for instance by evolution or development, to take advantage of structure of the external environment. (Todd and Gigerenzer 2003, p. 144)

Simon's scissors metaphor provides a great image to help us wrap up the theoretical context for our approach to interface design. Situation awareness is sometimes treated as if it were a single entity that resides completely within the head of a domain practitioner. As the previous two chapters suggest, situation awareness is more properly considered as a joint function of both the constraints of the work domain (Chapter 3) and the constraints of the human (Chapter 4). The scissors metaphor emphasizes the need for a single ontology that encompasses both blades: one blade that reflects constraints of the task situation and a second blade that reflects constraints on human ability or awareness.

To expand the metaphor further, one might compare the interface to the scissors' hinge since it provides a crucial link (see Figure 5.1). Thus, sharpening the blade on the ecology side involves the creation of associations that are closely linked with the task constraints (e.g., to build in consistencies such that the domain constraints are salient features of the representation). Sharpening the blade on the agent side involves the design of controls and displays with features that are consistent with the formidable perception, action, and reasoning capabilities of the human. The interface is the hinge that allows these two sharp blades to fit together closely, which allows rationality to cut (i.e., to support fast, frugal, and accurate performance).

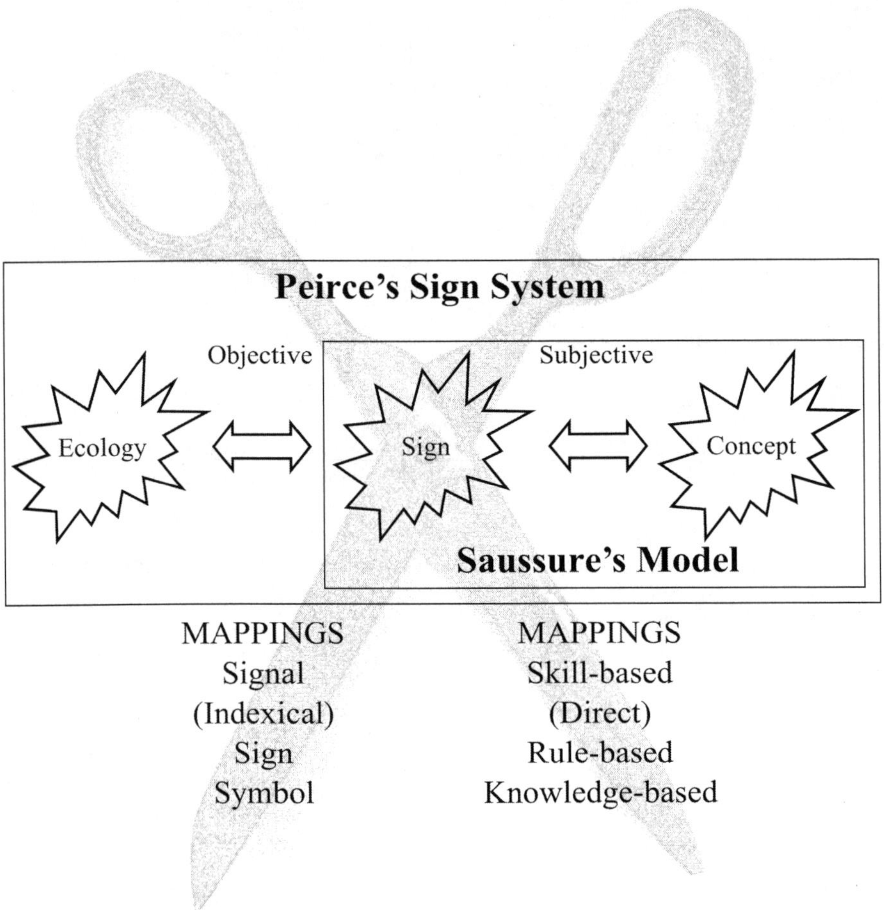


**FIGURE 5.1**
This illustrates how Rasmussen's framework maps into Peirce's semiotic system. Skill-based mappings are consistent with Gibson's concept of direct perception and Peirce's concept of indexical specification. In essence the signal is an analog of the ecology. Sign and symbol relations are effectively cues that can require associative or inferential mediation between concept and ecology.

## 5.2 Representation Systems and Modes of Behavior

In many respects, the scissors described by Simon is equivalent to the semiotic system described by Peirce, with the hinge being the sign that connects awareness to a situation. It is not surprising that Rasmussen (1986) drew heavily from semiotics in order to link the two blades of the scissors conceptually. As was described in Chapter 3, Rasmussen's decision ladder suggests that there are many different possible paths through the information processing system. He organized these paths into three general modes of processing: skill based, rule based, and knowledge based.

Furthermore, these modes depend on different types of representation systems: signals, signs, and symbols. Note that these representation systems reflect different types of interactions over the ecology–observer–representation triad. They are *not* properties of the representation that are independent of the ecology and observer (i.e., the exact same display can serve as a different representation system depending upon the agent and/or the state of the ecology). Figure 5.1 shows how these distinctions might be illustrated in the context of Peirce's (1931–1935) model for semiotics.

### 5.2.1 Signal Representations/Skill-Based Behavior

One type of representation system described by Rasmussen is the *signal* relation. In this relation, the ecology and the observer are linked through continuous space–time properties of the representation. Gibson's optical invariants are examples of this category. In driving, the relation between the car and the road can function as a signal relation for a skilled driver. The space–time relations between the vehicle and the road can be specified by space–time properties of an optical flow field. If the expansion pattern in the flow is centered on the driving lane, then the motion of the vehicle will be down the center of the lane. Similarly, a safe following distance behind a lead vehicle might be specified by the angular extent and the expansion rate of the optical contour of the lead vehicle.

When a functional mapping exists between the space–time properties of the ecology and the representation, then it is possible for the observer to respond *directly* to the ecology; that is, there is no need for inferential leaps to close gaps between the ecology as it is and the ecology as represented. The properties of the representation (e.g., optic flow) *specify* the state of the world. For example, when the image of the car in front of you expands in a particular fashion, the imminence of collision is specified; no inference is involved. Schiff (1965) has shown that many animals will reflexively brace or move to avoid collision when presented with this optical pattern of looming. And they will do this at the earliest ages at which they can be tested.

The representation is a direct analog of the ecology when signal relations exist. In essence, the representation is functionally transparent and it is possible for the observer to see the ecology directly as it is. Peirce (1931–1935) used the term *indexical* to refer to situations where there was a direct link between the representation and the object being represented; examples are the relation between temperature and the height of a column of mercury or the relation between the wind direction and the orientation of a weather vane.

This situation, in which the observer can coordinate directly with the ecology through the space–time properties of the representation, is the situation that Rasmussen called *skill-based* behavior. It is skilled in two senses. In one sense, it is skilled because discovering the space–time relations and learning to coordinate actions through them requires practice through doing; these skills cannot be developed through discussion and/or lecture. For example, skilled pilots must learn what a safe approach looks like (see Langewiesche, 1944, for a very interesting discussion of the optical patterns that specify a safe landing).

In the other sense, it is skilled because, once the coordination is mastered, the result is typically very smooth, fluid interactions between observer and environment that generally require minimal mental effort. The skilled athlete makes things look so easy! The skilled driver can weave through highway traffic safely with little mental effort. Also, many people who have mastered the skill may not be able to describe or articulate how the fluid behavior is accomplished; the link between seeing and doing is often *tacit.* Many skilled drivers will not be skilled driving instructors. Can you explain how you walk, how you know how much force to use when hopping from one rock surface to another when crossing a stream, or how you know when to brake when approaching a line of stopped traffic?

A phenomenological consequence of skill-based interactions is that the interface tends to disappear so that the person feels as if he or she is directly interacting with the work processes. In essence, the interface technology becomes an extension of the person, just as the cane becomes an extension of the arm so that, when touching an object with the cane, the cane wielder feels the texture of the surface being touched, not the cane (Hoff and Overgard 2008).

### 5.2.2 Sign Representations/Rule-Based Behavior

The second type of representation system described by Rasmussen is the *sign* relation. In this relation, the ecology and the observer are linked through consistency (high degrees of correlation). For example, in driving, the relation between the braking lights or turn signals of the lead vehicle and its behavior can function as a sign system. That is, there is a correlation between the brake lights coming on and the behavior of the lead vehicle. An observer who has experienced this relation over time will learn to be alert (and be prepared to brake) when the brake lights of the lead vehicle come on.

Note that the link between brake lights and stopping is not through space–time properties, but rather through convention. The color of brake lights and the link between the onset of the lights and brake activation are conventions established in the design of automobiles and highway systems. It is possible for the observer to develop fixed response sequences to these signs once these conventions are learned. The responses to these *consistent* properties of representations can be as smooth and effortless as the skilled response to signals.

The importance of *consistency* (consistent mapping) for smooth or "automatic" processing of information was clearly established by Schneider and Shiffrin (1977; Shiffrin and Schneider 1977) and this is true for both signals and signs. The key differentiator between signals and signs is whether the consistency is inherent in the coupling of perception and action (signals) or whether it is mediated by a constraint (e.g., a cultural or design convention) outside the perception–action dynamic (sign).

Rasmussen described interactions mediated by sign relations as *rule-based* behavior. They are rule based in the sense that the consistent relation between the representation and the environment or between the representation and the appropriate response can be described in terms of a simple rule: If brake lights come on, then the lead vehicle will be decelerating, or if brake lights come on, be prepared to decelerate. It seems that, as with skill-based relations, extensive practice by doing is necessary for rule-based relations to function in an automatic way.

However, the fact that rules mediate the gap between the representation and the ecology makes it much more likely that people will be able to articulate the rules (whether or not they have mastered them). So, rule-based behaviors are less likely to be *tacit* than skill-based behaviors. Rule-based behaviors will typically be associated with heuristics (i.e., shortcuts), where perception and action are mediated by the learned rules and higher levels of information processing will not be necessary. Rule-based behavior is less direct than skill-based behavior in the sense that the relation between representation and interpretation is *mediated* by the rule. However, both rule- and skill-based behaviors can become functionally *automatic* or *reflexive* as a result of leveraging consistency (taking advantage of constraints) in the semiotic relations.

### 5.2.3 Symbol Representations/Knowledge-Based Behaviors

The third type of representation system described by Rasmussen is the *symbol* relation. In this relation, the link between observer and ecology is ambiguous or hypothetical. That is, the relation between the form of the representation and either the objective or subjective meaning is not immediately apparent without some degree of analysis or problem solving. The best that the observer can do is to guess or hypothesize about the relation. Note that both patterns in space–time and consistencies only exist over time. So, whenever a person is presented with a novel situation, the relation between the form of the representation and the *meaning* (in both senses) will be ambiguous to some degree until the observer has had the time necessary to pick up and utilize the patterns or consistencies.

Natural language, once learned, will function as a sign relation in which the words will be automatically linked to meaning through a history of consistent associations and learned rules. However, a foreign language will at first be a symbol system in that the rules and consistent links between the sounds and the meanings will be ambiguous. The foreign language is not arbitrary, but it is ambiguous and requires analysis in order to figure out the meaning. Note that as Rasmussen uses the terms, signal, sign, and symbol refer to relations across the semiotic system. Thus, whether a representation (e.g., a word, brake lights, or graphical icon) functions as a sign, symbol, or signal depends on the experiential history of the semiotic dynamic.

In the driving context, the four-way stop often functions as a symbol relation for drivers unaccustomed to driving in the United States. That is, the conventions of behavior in this context are often ambiguous or mysterious to such drivers. For example, when one tries to explain that it is a simple rule that order of arrival at the intersection determines the order of passing through the intersection, many foreign drivers are still mystified because it is not clear to them what "arrival at the intersection" means. That is, they often guess that this refers to arrival at the end of the queue for the intersection and they find it impossible to tell how many of the people in the other lanes were there before them. It is not obvious to them, as it is to drivers experienced in driving in the United States, that first to arrive refers only to those vehicles at the threshold of the intersection and does not depend on arrival at the back of the queue. Similarly, round-about or traffic circles that are familiar sign systems for European drivers can be symbol systems for U.S. drivers who have not yet discovered the conventions.

Rasmussen characterized interactions mediated by symbol relations as *knowledge-based behavior.* With symbol relations, the observer must act as a problem solver (e.g., scientist or detective). That is, the observer must evaluate and act on hypotheses or guesses about the relation between the ecology as represented and the ecology as it is. The problem of diagnosis faced by the physician or nuclear plant operator confronted with a novel set of symptoms is an example of knowledge-based behavior. The drivers unaccustomed to driving in the United States are challenged by a knowledge-based situation with respect to four-way stops. They test their hypothesis about what "first to arrive" means through acting in a certain way; the reaction of other drivers provides empirical evidence about whether their hypothesis is right or wrong.

Knowledge-based behavior tends to require deep levels of processing or mindfulness. This type of situation places the greatest demands on the information processing system. Knowledge-based behavior requires the systematic reasoning that Piaget (1973) called formal operations. It requires high degrees of abstraction and integration. In essence, the person must do problem solving—generating and evaluating hypotheses.

It is an irony of language that knowledge-based behaviors are typically demanded in situations where people have the least experience or skill. As people get more experience with a representation system, they will naturally become more aware and more able to utilize the time–space patterns and the consistencies that are available. The first time you try to drive, the car and the highways are full of ambiguities that you have to resolve through knowledge-based interactions. With practice (trial and error), you discover the patterns and consistencies and come to rely more and more on skill- and rule-based modes of control to the point where now you can sometimes find yourself arriving at your workplace or home without any conscious awareness of the events that transpired during the drive. You managed the drive on autopilot while higher knowledge-based levels of information processing were engaged by more challenging puzzles, such as your latest design problem. The irony of the language is that the more experience or knowledge you have about a particular situation the less dependent you will be on what Rasmussen labeled as the knowledge-based mode of interaction.

## 5.3 Representations, Modes, and the Decision Ladder

Conventional texts on cognition typically have a lot to say about rule- and knowledge-based interactions. However, one will usually find very little discussion of skill-based interaction. Again, this reflects a dyadic model of the cognition dynamic where all the action is in the relation between the observer and the representation; this action is reflected in the rules that link the representation to responses and in the rational processes associated with hypothesis testing or problem solving. This meshes very nicely with the computer metaphor in which the computer is a disembodied symbolic engine or rule-based device. It is disembodied in the sense that its relation to a physical ecology is mundane compared to its relation to a virtual ecology of rules and symbols.

The dyadic model tends to ignore the ecological dimension of the triad. In the context of Three Mile Island, this model has much to say about the operator and the display representations, but has almost nothing to say about the nuclear power plant that the displays are representing. Similarly, this model has lots to say about how a driver might think about driving, but has difficulty explaining the fluidity of a skilled driver—particularly, those aspects of the skill that the driver cannot clearly articulate. How does our image of cognition change when we consider the possibility of signal relations and skill-based interactions? How does our image of cognition change when we bring in the ecology as an essential dimension of the meaning dynamic (the semiotic system)? How can we address experience in everyday life without considering common situations associated with navigating through the environment?

### 5.3.1 Skill-Based Synchronization

Figure 5.2 illustrates how the different semiotic systems map onto the decision ladder. Skill-based processing tends to utilize lower regions of the decision ladder—automatic or reflexive links between activation and execution. This is called skill based because the ability to utilize these automatic paths seems to depend on high levels of practice in consistent environments. This type of

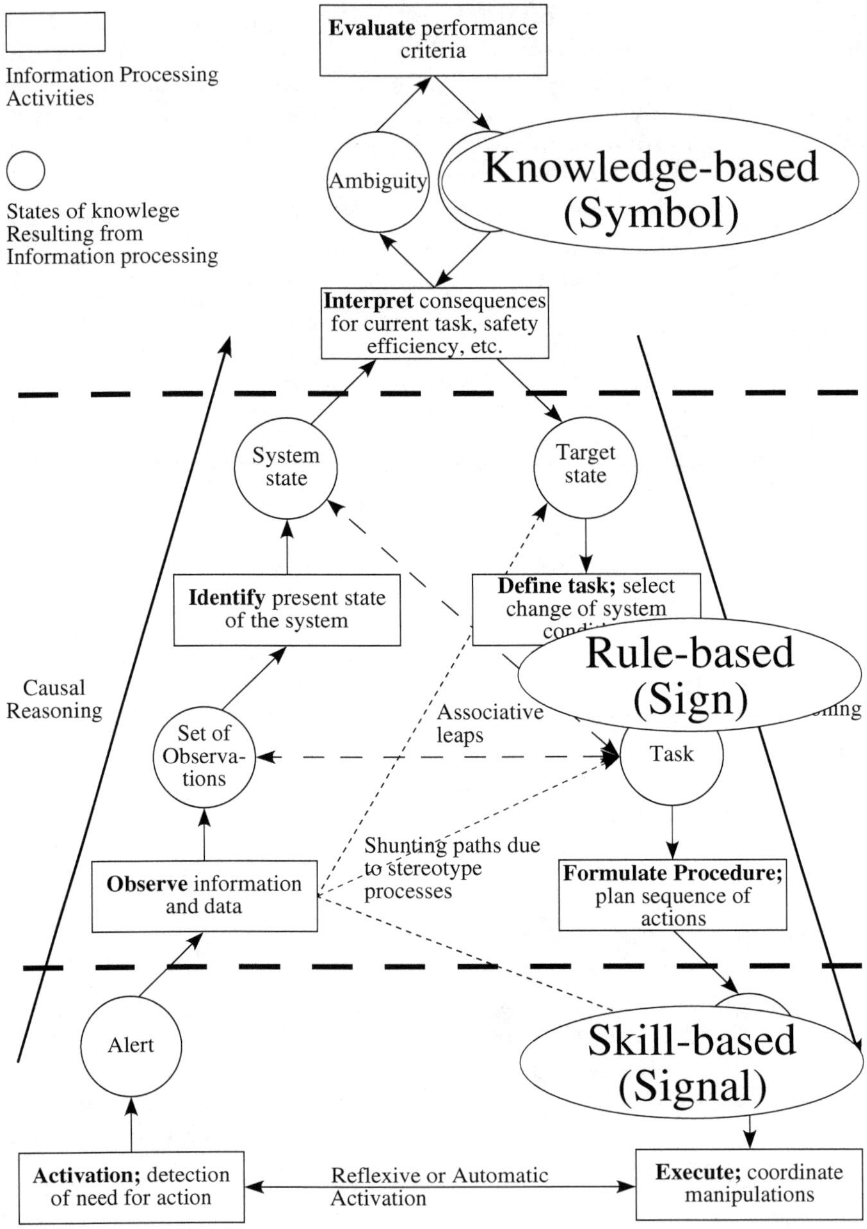


**FIGURE 5.2**
This diagram illustrates how qualitatively different semiotic systems (signal, sign, and symbol) enable qualitatively different types of information processing (skill-, rule-, and knowledge-based). (Adapted with permission from Rasmussen, J. 1986. *Information Processing and Human–Machine Interaction: An Approach to Cognitive Engineering.* New York: Elsevier. All rights reserved.)

processing takes advantage of invariant relations between the form of the activation and the form of the response (signals). The automaticity and the dependence on a consistent environment are most apparent when the consistency is broken; then we see the action slips and illusions that reflect the generalization of an association beyond the boundaries of the associated constraint.

### 5.3.2 Rule-Based Shortcuts

Rule-based processing tends to utilize different shortcuts across the middle regions of the decision ladder. This type of processing tends to take advantage of stereotypical aspects of situations (signs) to bypass the need for deeper analysis. Whereas the skill-based level takes advantage of consistency in space–time patterns (e.g., optical expansions associated with collision), the rule-based level takes advantage of more conventional or situational consistencies.

For example, when we drive a rental car, we may automatically reach for the turn signal and find that we have activated the windshield wipers. The actions were guided by an expectation created by our experience in our own car. Similarly, when the car does not start in the morning, the assumption that the battery is dead could be stimulated by the memory of a similar situation last week. Checking the battery in this case does not reflect a deep analysis of how cars work, but simply an association with a recent event.

### 5.3.3 Knowledge-Based Reasoning

Knowledge-based processing refers to situations where the full range of information processing activities must be engaged to make sense of a situation. This is required when the form of the problem representation (symbolic) does not have any obvious or apparent relation to the form of the response required. Under these conditions, the human must function as a scientist or detective who must piece together a solution through systematic analysis.

For example, when your usual route across town is jammed due to an accident, you may engage in knowledge-based processing to decide whether you should wait in the slowed traffic or whether you should try to get off the highway and try an alternate route. Or when you discover that your battery is fine, then you have to think about why your car will not start: "What are other potential explanations?" "How can I rule out alternative explanations to discover the solution?"

### 5.3.4 Summary

Each of the qualitative modes of processing identified by Rasmussen (skill-, rule-, or knowledge-based) utilizes different sources of constraint (signals, signs, or symbols, respectively) that open different paths through the decision ladder. In the context of Simon's scissors metaphor, signal, sign, or symbol refers to qualitatively different types of consistency or constraint associated with a situation or work domain; skill-, rule-, and knowledge-based refer to differential abilities. Note that, in this context, every link in the decision ladder can be evaluated from the perspective of awareness (is this an association that the person has learned to use?) and from the perspective of situations (is this association correlated with structure in the task context?). In this respect, the decision ladder can be used as a descriptive model to illustrate the shortcuts that a particular expert actually uses in his or her problem solving.

For example, Rasmussen (1986) originally used the decision ladder to illustrate the troubleshooting behavior of electronic technicians. However, the decision ladder can also be used as a prescriptive model to illustrate associations and shortcuts that might be (or should be) made available due to consistent mappings and invariant properties of task situations. In this context, the decision ladder reflects an analysis of the problem constraints (i.e., cognitive work analysis), as opposed to the experience constraints of a particular cognitive agent or of a particular strategy.

## 5.4 Ecological Interface Design (EID)

> The principal goal behind EID is to design an interface that will not force cognitive control to a level higher than that required by the demands of the task, and yet that provides the appropriate support for each of the three levels. In order to design such an "ecological interface," the following factors must be taken into consideration. First, it is necessary to merge the observation and action surfaces so that the time–space loop is maintained, thereby taking advantage of the efficiency of the human sensorimotor system. In addition, it is also necessary to develop a consistent one-to-one mapping between the abstract properties of the internal process to be controlled and the cues provided by the manipulation/observation surface … the goal is to make visible the invisible, abstract properties of the process (those that should be taken into account for deep control of the process) visible to the operator. In semiotic terms, this means that the cues provided by the interface have a consistent mapping onto the symbolic process properties. In this way, the same conceptual model may act as a symbolic representation when considered in relation to the elements of the environment and the laws controlling their relationships, and as a system of prescriptive signs when considered in relation to the rules for model actions of the system. (Rasmussen and Vicente 1989)

Rasmussen and Vicente's (1989) construct of ecological interface has been an important inspiration for this book. In the last few chapters, we have been trying to outline a broader conceptual framework to help elucidate and motivate the idea of designing ecological interfaces. However, although we fully resonate to Rasmussen and Vicente's (Rasmussen and Vicente 1989, 1990; Vicente and Rasmussen 1990, 1992) ideas, from the very start, we have been worried that the term "ecological" might lead to misunderstanding.

This is because, although Rasmussen and Vicente (1989) clearly articulate the links to ecological approaches to human performance, the ecological approach tends to remain an enigma to many people. In part, we feel that this may be because people are trying to understand this approach from a dualistic ontology or from a dyadic semiotic perspective. Thus, people are more inclined to link the term "ecological" with the term "natural" and the implication is that an ecological interface should be simple and natural to use. In this context, natural reflects ease of use, rather than the mapping to the task ecology that was the intended emphasis.

### 5.4.1 Complementary Perspectives on EID

We hope that the intention behind the term "ecological interface" is becoming clearer. The ideal for an ecological interface is to design a representation that is faithful to the task ecology. When viewed through the lens of human problem solving, the goal for an ecological interface is to make the deep structure of a problem salient. When viewed through the lens of the abstraction hierarchy described in Chapter 3, the goal for an ecological interface is to make the constraints at all levels of the abstraction hierarchy visible in the representation.

For example, the operator should be able to see the state of the work domain in relation to the goals, the costs, and the fields of possibilities associated with physical and regulatory laws and organizational layout. This ideal is illustrated most clearly by Vicente's work with the DURESS system (Dinadis and Vicente 1999; Pawlak and Vicente 1996; Vicente 1991, 1992, 1999; Vicente, Christoffersen, and Pereklita 1995). Further illustrations will be provided in later chapters.

When viewed through the lens of the decision ladder (Chapter 4 and Figure 5.2), the goal of an ecological interface is to support both direct perception and direct manipulation. Direct perception means that it is possible to perceive the state of the system through consistent patterns in the display (analogous to optical invariants or consistent mappings). In other words, the representation should provide signals and signs that map directly onto the states/constraints of the work processes to support productive thinking (e.g., chunking, automatic processing, and recognition primed decisions). Similarly, direct manipulation means that it is possible to act on the system via controls that provide space–time signals and signs that map directly onto required inputs.

Direct manipulation also becomes important in those inevitable situations where a complex system gets into a state that was not and probably could not have been anticipated by the designers (e.g., faults that violate local design assumptions). In this case, knowledge-based processing will be required. The goal of direct manipulation in the context of knowledge-based processing is to design a representation that allows the operator to test hypotheses directly through actions on the representation whose consequences will be visible. In this case, the display should support reasoning in terms of Piaget's formal operations. It should help the operator to manage converging operations toward solution of the puzzle.

In sum, the goal of the ecological interface is to create direct links between perception and action that support all three levels of information processing—skill-, rule-, and knowledge based. The goal is that thinking be externalized in the perception–action loop, reducing as much as possible the dependence on memory-intensive, logical computations.

### 5.4.2 Qualifications and Potential Misunderstandings

Will an ecological interface be easy to use? Hopefully, the answer to this question will be yes. However, it is a very qualified yes. The first qualification involves Ashby's (1968) "law of requisite variety." In essence, this law requires that, for a representation to be a good interface for control, it must be at least as complex as the phenomenon that is being controlled. This law suggests that representations can overly simplify a problem. Such representations may suffice in restricted regions (e.g., under normal operations), but these representations risk trivializing the problem and potentially lead to a naive view of the work domain and a restricted solution space for solving problems. The representation of a complex problem domain will itself need to be complex! There is no escaping this general systems constraint.

The second qualification is that the ease of use for complex domains will depend heavily on learning. Automatic processes do not come for free. They depend on practice. It takes time to discover the patterns and to build the associations that link those patterns to correct inferences and actions. Tanabe (personal communication) has argued that ecological interfaces require us also to reconsider approaches to training. This includes reconsideration of how knowledge content is organized and structured and increases the importance of simulation so that people have opportunities to explore and learn by doing.

It is not unusual that people are surprised and disappointed when they first see an interface that is intended to be ecological. Often, it is not at all easy to use initially. In fact, it can seem unnecessarily complicated relative to more traditional displays. For example, compare the mimic display with the ecological (P + F) interface for DURESS (e.g., Vicente 1999). The mimic display seems much simpler. It typically takes time and practice to appreciate and take advantage of the benefits of an ecological interface.

Thus, the goal of an ecological interface is not to take a complex domain and make it seem naturally simple to a naive user. Rather, the goal is to allow the user to discover and learn the consistencies and eventually to develop a rich network of associations that will support good situation awareness and fast, frugal decision making. In essence, the ecological interface allows one to get maximal benefit from experience (i.e., from trial and error). The consistencies and patterns in the ecological interface should reveal the constraints in the work domain. Experience with the ecological representation should lead to a deeper appreciation of the possibilities in the work domain, making the critical properties of the domain salient.

Thus, with the appropriate experience and training, the ecological interface should become easier to use because there will be many opportunities to bypass the computationally intensive aspects of information processing through the use of direct associations with patterns in the representation. However, the ecological interface definitely does not make things easier for the designer. In fact, the ecological interface approach is a challenge to the designer to go beyond simple display prescriptions based on generic models of human information processing to learn about the problem/work space and to discover the constraints that shape the possibilities for moving through that space. This creates a demand for thorough work analysis that goes beyond simply talking to a few domain experts.

The more diligent that the designer is in identifying properties of the problem and the more creative she is in building display geometries that map invariantly to these properties, the more valuable will the representation be to the opportunistic human information processors. In effect, the goal of the ecological interface is to bias the operator so that she will choose the heuristics that will leverage the structure of the problem most effectively. Remember that humans are essentially opportunistic agents that will use whatever associations are available to ease their computational burden. The goal for the designer is to help ensure that the associations available in the representation lead to smart situation assessments, satisfactory decisions, and appropriate control input. The goal is to ensure that the associations that are most salient are the ones that support productive thinking and situated action.

## 5.5 Summary

In many respects, the hinge that determines whether the two blades of Simon's scissors will fit together well enough to cut is the interface representation. This hinge must be designed so that the two blades are aligned well, one to the other. We do not believe that it is possible to do this if the blades are designed from two separate, disconnected ontologies. It is often convenient to parse the world into the objective task constraints that are independent of any potential computational agent or device and the subjective constraints that are inherent to a specific agent or computational device, independently of any particular problem.

However, if we want to support human problem solving, we must not get trapped in this parsing. We believe the abstraction hierarchy provides a context for thinking about the situation constraints in a way that respects the properties of human agents and that the decision ladder provides a context for thinking about awareness constraints in a way that respects the properties of work domains. Together, these conceptual tools help us to bridge the gap between awareness (mind) and situations (matter) to address the question of what matters and to address questions of meaning.

To reiterate, humans are essentially opportunistic information processing agents. They are going to be strongly biased by saliency. The goal of both training and interface design is to help ensure that the associations that are salient are the ones that are relevant to the problem. The goal of training and interface design is to help ensure that the first choice that comes to mind is a very good choice. The goal is not to help people to conform to the prescriptions of logic and rational decision models. Rather, the goal is to shape experience so that humans can form the associations that let them automatically zero in on relevant aspects of the problem, minimizing as much as possible the need for laborious logical analysis or computation.

## References

Ashby, R. W. 1968. Variety, constraint, and the law of requisite variety. In *Modern systems research for the behavioral scientist,* ed. W. Buckley. Chicago, IL: Aldine.

Dinadis, N., and K. J. Vicente. 1999. Designing functional visualizations for aircraft systems status displays. *International Journal of Aviation Psychology* 9 (3): 241–269.

Hoff, T., and K. I. Overgard. 2008. Ecological interaction properties. In *Embodied minds—Technical environments,* ed. T. Hoff and C. A. Bjorkli. Trondheim, Norway: Tapir Academic Press.

Langewiesche, W. 1944. *Stick and rudder.* New York: McGraw–Hill.

Pawlak, W. S., and K. J. Vicente. 1996. Inducing effective operator control through ecological interface design. *International Journal of Human–Computer Studies* 44 (5): 653–688.

Peirce, C. S. 1931–1935. *Collected papers of Charles Sanders Peirce,* ed. C. Hartshorne and P. Weiss. Cambridge, MA: Harvard University Press.

Piaget, J. 1973. *The child and reality.* Translated by R. Arnold. New York: Grossman. Original edition: *Problemes de psychologie genetique.*

Rasmussen, J. 1986. *Information processing and human–machine interaction: An approach to cognitive engineering.* New York: Elsevier.

Rasmussen, J., and K. Vicente. 1989. Coping with human errors through system design: Implications for ecological interface design. *International Journal of Man-Machine Studies* 31:517–534.

———. 1990. Ecological interfaces: A technological imperative in high tech systems? *International Journal of Human-Computer Interaction* 2 (2): 93–111.

Schiff, W. 1965. Perception of impending collision: A study of visually directed avoidant behavior. *Psychological Monographs* 79 (11): 1–26.

Schneider, W., and R. M. Shiffrin. 1977. Controlled and automatic human information processing: I. Detection, search, and attention. *Psychological Review* 84:1–66.

Shiffrin, R. M., and W. Schneider. 1977. Controlled and automatic human information processing: II. Perceptual learning, automatic attending, and a general theory. *Psychological Review* 84:127–190.

Simon, H. A. (1990) Invariants of human behavior. *Annual Review of Psychology*, 41, 1–19.

Todd, P. M., and G. Gigerenzer. 2003. Bounding rationality to the world. *Journal of Economic Psychology* 24:143–165.

Vicente, K. J. 1991. Supporting knowledge-based behavior through ecological interface design. Urbana-Champaign: Engineering Psychology Research Laboratory, Department of Mechanical Engineering, University of Illinois.

———. 1992. Memory recall in a process control system: A measure of expertise and display effectiveness. *Memory & Cognition* 20:356–373.

———. 1999. *Cognitive work analysis: Toward safe, productive, and healthy computer-based work.* Mahwah, NJ: Lawrence Erlbaum Associates.

Vicente, K. J., K. Christoffersen, and A. Pereklita. 1995. Supporting operator problem solving through ecological interface design. *IEEE Transactions on Systems, Man and Cybernetics* 25 (4): 529–545.

Vicente, K. J., and J. Rasmussen. 1990. The ecology of human–machine systems II: Mediating "direct perception" in complex work domains. *Ecological Psychology* 2:207–249.

———. 1992. Ecological interface design—Theoretical foundations. *IEEE Transactions on Systems, Man and Cybernetics* 22 (4): 589–606.

# 6

# *A Framework for Ecological Interface Design (EID)*

## 6.1 Introduction

> Marshall McLuhan observed that "the medium is the message." Designers send a message to the users by the design of interactive systems. In the past, the message was often an unfriendly and unpleasant one. I believe, however, that it is possible to send a more positive message that conveys the genuine concern a designer has for users. If the users feel competent in using the system, can easily correct errors, and can accomplish their tasks, then they will pass on the message of quality to the people they serve, to their colleagues, and to their friends and families. In this way, each designer has the possibility of making the world a little warmer, wiser, safer, and more compassionate. (Shneiderman 1992, p. iv)

The growing attention to and interest in interface design has been fueled by advances in both hardware (particularly graphics) and software. This technology provides opportunities for design innovation that make Shneiderman's lofty goals to make "the world a little warmer, wiser, safer, and more compassionate" far more realistic than they were a generation ago. To make this point explicit, consider the hardware and software of the first interactive interfaces. The teletype interface (see Figure 6.1) was essentially a glorified typewriter that included a keyboard for input and a roll of paper for output. The user typed in alpha-numeric input one line at a time via a command line interface; the computer typed out lines of alpha-numeric symbols.

The text editors were also based on a command line convention. One of the more (in)famous of these editors was TECO (standing for either *t*ape *e*ditor and *co*rrector, or *t*ext *e*ditor and *co*mposer). Cryptic alpha-numeric commands were required to reposition a cursor and manipulate the text, etc. This was quite difficult since neither the text of a document or the cursor could be seen, at least without special instructions. For example, to view 10 lines of text around the current location required the command "-5l10t" to be typed in, followed by a press of the return key. This command instructed TECO

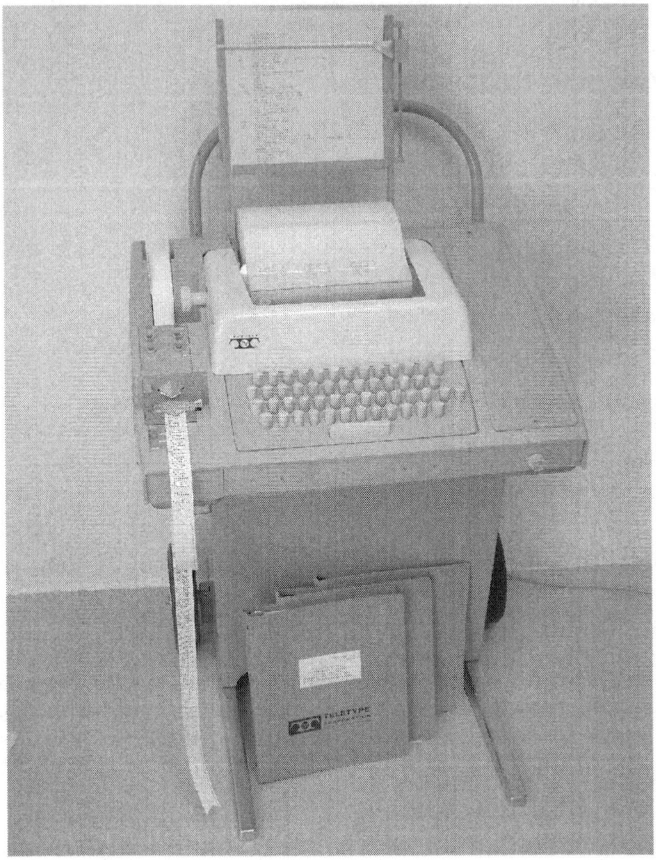

**FIGURE 6.1**
The ASR 33 Teletype interface. (Photo courtesy of David Gesswein, www.pdp8online.com.)

to move back five lines of text ("-5l") and then type the next 10 lines of text ("10t"). With tongue in cheek, Post (1983, p.264) observed:

> Real Programmers consider "what you see is what you get" a bad concept in text editors. The Real Programmer wants a "you asked for it, you got it" text editor; one that is complicated, cryptic, powerful, unforgiving, and dangerous. TECO, to be precise.
>
> It has been observed that a TECO command sequence more closely resembles transmission line noise than readable text. One of the more entertaining games to play with TECO is to type your name in as a command line and try to guess what it does.

In the 1980s we referred to this game as "TECO roulette" after its Russian counterpart!

Obtaining a well-formatted final document was also an adventure. Special instructions for formatting the document needed to be embedded within the text. For example, the characters ".i5" were instructions to indent the ensuing text by five characters. The original document then was submitted to a formatting program to produce a final version. The intrepid writer could then print the document out on paper to evaluate the ultimate success (or, more often, failure) of these formatting commands.

Obviously, these interfaces put a much different (and a much more difficult) set of constraints on performance than the interfaces of today. Current interfaces are much more powerful and accommodating and offer a far greater potential to provide effective decision making and problem solving support. However, despite a plethora of books designed to take advantage of these advances (e.g., Eberts 1994; Preece et al. 1994; Sharp, Rogers, and Preece 2007; Shneiderman 1998), this potential fails to be realized with alarming regularity.

## 6.2 Fundamental Principles

> The aim of ecological interface design can be described as trying to make the interface transparent, that is, to support direct perception directly at the level of the user's discretionary choice, and to support the level of cognitive control at which the user chooses to perform. The mapping across interfaces, which will support the user's dynamic switching among levels of focus and control, must support control of movements, acts, and plans simultaneously. To do this, the designer must create a virtual ecological environment that maps the intended invariants of the functional system design onto the interface. (Rasmussen and Vicente 1990, p. 102)

This chapter will provide a broad framework for ecological interface design with the goal of realizing this potential on a more consistent basis. This framework builds upon insights drawn in previous chapters and provides principles of design, strategies of design, and domain/agent constraints that are relevant to their use. The three triadic components of a sociotechnical system, first introduced in Chapter 2 (Figure 2.3), have been re-represented in Figure 6.2. Note that the dynamic interactions between components (the labeled arrows) have been removed and some terms associated with each component have been added.

The ultimate goal of ecological interface design is to provide effective decision-making and problem-solving support. The key point from the triadic perspective is that informed decisions about interface design can only be made within the context of both the domain (i.e., the work to be done; situations) and the cognitive agents (i.e., humans and machines; awareness) responsible for doing the work. The fundamental challenge for the design

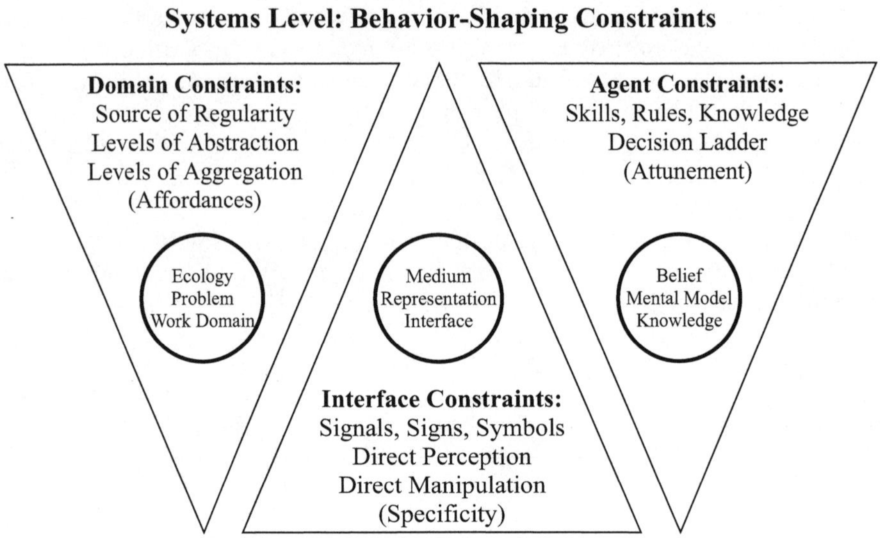


**FIGURE 6.2**
The three sets of behavioral-shaping constraints in a sociotechnical system.

of interfaces is to make sure that the constraints contributed by the interface are tailored (i.e., well matched) to the constraints of both the work domain and the cognitive agents. As suggested in Figure 6.2, this might be conceptualized as the degree of fit between pieces of a puzzle. In this case, the complementary shapes of the triangular wedges suggest a good fit. We begin our discussion of the design framework with the consideration of two general principles that are the foundation of effective interface design: direct manipulation and direct perception.

### 6.2.1 Direct Manipulation/Direct Perception

> In a system built on the model world metaphor, the interface is itself a world where the user can act, and that changes state in response to user actions. The world of interest is explicitly represented and there is no intermediary between user and world. Appropriate use of the model world metaphor can create the sensation in the user of acting upon the objects of the task domain themselves. We call this aspect "direct engagement." (Hutchins, Hollan, and Norman 1986, p. 94)
>
> Design, evaluation and research in HCI and HICI (human–intelligent computer interaction) should be seen in terms of representational issues. The display/interface is a referential medium where visual (and other

> elements) are signs or tokens that function within a symbol system. (Woods 1991, p. 174)

> In Gibson's terms, the designer must create a virtual ecology, which maps the relational invariants of the work system onto the interface in such a way that the user can read the relevant affordances for actions. (Rasmussen, Pejtersen, and Goodstein 1994, p. 129)

The term *direct manipulation* was coined by Shneiderman (1983) and analyzed extensively by Hutchins et al. (1986). The term *direct perception* was first applied in the context of interface design by Flach and Vicente (1989) as a way to link the insights from Gibson's theories of direct perception to the challenge of interface design. This was in conjunction with the evolution of the construct of ecological interface design. The key to appreciating both terms is the meaning of *direct* in relation to the construct of *psychological distance* or ease of use.

Hutchins et al. (1986) write:

> There are two separate and distinct aspects of the feeling of directness. One involves a notion of the distance between one's thoughts and the physical requirements of the system under use. A short distance means that the translation is simple and straightforward, that thoughts are readily translated into the physical actions required by the system and that the system output is in a form readily interpreted in terms of the goals of interest to the user. We call this aspect "distance" to emphasize the fact that directness is never a property of the interface alone, but involves a relationship between the task the user has in mind and the way that task can be accomplished via the interface. Here the critical issues involve minimizing the effort required to bridge the gulf between the user's goals and the way they must be specified to the system. (pp. 93–94)

In essence, Hutchins et al. use psychological distance as a metaphor for ease of use. The smaller the distance is (i.e., the more direct the interaction is) the less demands will be placed on cognition. In simple terms, an interface that is direct is one that makes it easier for humans to do their tasks or to accomplish their goals. Again, in Hutchins and others' words: *"At the root of our approach is the assumption that the feeling of directness results from the commitment of fewer cognitive resources"* (p. 93). A second aspect of directness identified by Hutchins et al. was the *"qualitative felling of engagement, the feeling that one is directly manipulating the objects of interest"* (p. 94). Another way to say this is the degree to which the controls and displays of an interface are transparent relative to the problem being represented or, alternatively, the degree of specificity or correspondence between the behavior of the interface and behavior of the domain. In essence, as implied in the opening quotes, a *direct* representation is a high-fidelity simulation of the domain.

Shneiderman (1982, p. 251) identified some of the properties of direct manipulation interfaces that tend to reduce or ease cognitive demands:

continuous representation of the object of interest

physical actions or labeled button presses instead of complex syntax

rapid incremental reversible operations whose impact on the object of interest is immediately visible

Note that Shneiderman's recommendations address both perception (representation of) and action (physical actions), so the original construct of direct manipulation interfaces is one that spans both perception and action. This is reinforced by Hutchins et al., who associate direct manipulation with both the gulf of evaluation (perception) and the gulf of execution (manipulation). Thus, in some senses, the two terms are redundant. That is, direct manipulation and direct perception both refer to the direct coupling between perception and action, which is exactly what Gibson was emphasizing as a fundamental aspect of many of our natural interactions with the world. Thus, in our view these terms are not distinct constructs, but rather simply two aspects of bridging the psychological distance between a human and a problem.

Despite this observation, we find that the ability to refer to one facet of this overall dynamic serves to facilitate discussion. Throughout the remainder of the book, we will be using both direct perception and direct manipulation to refer to facets of interaction and principles of interface design. Moreover, we see potential confusion between our use of these terms and the precedence that was set by Shneiderman and Hutchins et al. Therefore, we begin at the outset by being perfectly clear. When we use the term "direct perception," we will be referring to objects in the interface that directly specify domain or ecological constraints and that are available to be perceived by an agent (similar to what Hutchins et al. referred to as the gulf of evaluation). Conversely, when we use the term "direct manipulation," we will be referring to objects in the interface that can be acted upon by an agent to control input to the work domain (similar to the gulf of execution).

In this chapter we would like to present ecological interface design as a natural extension of the intuitions of Shneiderman and Hutchins et al. The common goal is to reduce the cognitive demands of work. And the common intuition is that one powerful way to do this is to enrich the coupling between perception and action. However, the challenge is how to achieve this goal. The added value of the ecological interface design construct is to ground the shared goals in the triadic framework. This means we need an explicit way to analyze work domains in order to specify the objects of interest. We also need an explicit model of cognition in order to specify what is easy and what is hard. In other words, an ecological interface design approach begins with the proposition that unless you can specify the task demands on the one hand and the cognitive abilities and limitations on the other hand, you cannot begin to bridge the gulf between the two.

The point of ecological interface design is that reducing the psychological distance between humans and complex problems is not simply a matter of replacing text with graphics or of replacing typing with point-and-click interactions. It is not simply a matter of changing the syntax in the interface; rather, it demands attention to semantic issues (Bennett and Flach 1992). It is not simply a matter of changing the mode of interaction (e.g., from menus to icons; see associated discussion in Chapter 13). Rather, it is a matter of organizing or chunking information in ways that respect the constraints of the work domain and the cognitive constraints of the people doing the work. Reducing psychological distance means improving the match between structure in the ecology (e.g., the patterns or regularities that limit possibilities in the problem space) and the belief structure of the people involved (e.g., associative networks that link awareness and action).

## 6.3 General Domain Constraints

An important distinction has been made with respect to the nature of the ways in which work domains can be coupled with their users. At one extreme, there are safety critical systems such as aviation and process control, where the work is tightly constrained by physical laws on the work domain side of the triad and by operators who are carefully selected and trained on the cognitive agent side of the triad. At the other extreme are general purpose domains such as libraries or office productivity tools, where the work domain side of the triad is primarily constrained by functional purposes that may differ from individual to individual and by operators who will be extremely diverse in both their experiences and skill levels.

This continuum is reflected in the following observation:

> The weight of the *intentional constraints* compared with the *functional, causal constraints* can be used to characterize the regularity of different work domains. In this respect, the properties of different work domains represent a continuum. At the one end are located tightly coupled, technical systems the regularity of which have their origins in stable *laws of nature.* At the other extreme are the systems in which the entire *intentional* structure depends on an actual user's own subjective preferences and values. In the middle are a wide variety of systems which owe their regularity to influences from formal, legal constraints as well as institutional and social practices .... Thus the relationship between the causal and intentional structuring of a work system and the degree to which the intentionality of the system is embedded in the system or brought to

play by the individual actor is an important characteristic. (Rasmussen et al. 1994, p. 49; emphasis added)

Figure 6.3 illustrates this continuum. At one end, on the right-hand side, are domains in which the primary determinants of the unfolding events that occur are "law-driven" constraints (i.e., the causal, highly coupled, and regular constraints of the system itself). Vicente (1999) called this end of the continuum "correspondence-driven" domains. At the opposite end of the continuum, on the left-hand side, are domains in which the primary determinants of the unfolding events are "intent-driven" constraints (i.e., the human agent's needs, goals, and intentions). Vicente (1999) called this end of the continuum "coherence-driven" domains. These fundamental distinctions will be considered in greater detail.

### 6.3.1 Source of Regularity: Correspondence-Driven Domains

The relational invariants (or, alternatively, the behavior-shaping constraints) of correspondence-driven domains arise primarily from immutable physical laws of nature (e.g., thermodynamics, conservation of mass and energy) as opposed to the intentions of the agents who are controlling them. Process control (e.g., power generation) provides one category of this class of domains. A process control system is designed to accomplish very specific goals (e.g.,

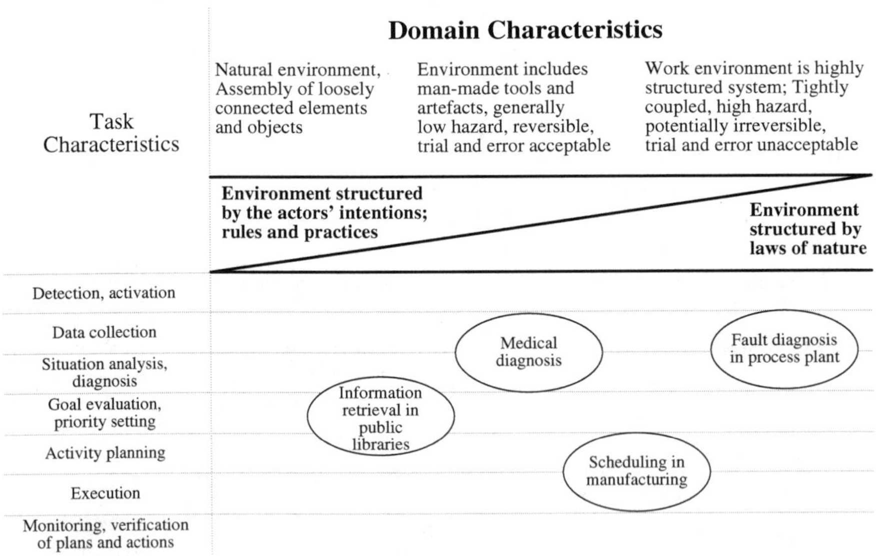


**FIGURE 6.3**
Sources of regularity in work domains. (Adapted from Rasmussen, J., A. M. Pejtersen, and L. P. Goodstein. 1994. *Cognitive Systems Engineering*. New York: John Wiley & Sons, Inc. With permission.)

produce energy) and to do so in a reasonably specific fashion. The various parts of the system are both physically and functionally interconnected. That is, the relational invariants of the domain are tightly coupled. As a result, there is a high degree of causality in the system. In essence, the law-driven constraints must be the fundamental consideration in interface design; the goal is to design representations that accurately reflect them.

Of course, the information processing and strategic constraints of the operators cannot be ignored. Even in a domain that is primarily driven by the laws of nature there will still be some level of discretion on the part of controlling agents. Often, more than one resource and/or more than one strategy can be used to accomplish a particular goal, and an agent may make a choice based on subjective intentional constraints (e.g., personal preference). However, some strategies may be more effective than others, and one consideration in design might be to bias the cognitive agents intentionally toward better choices. In summary, the causal constraints in law-driven domains take precedence over the needs, goals, and intentions of the agents, but neither can be ignored

The ultimate goal for correspondence-driven domains is typically to shape the awareness of the human operators so that it is possible to leverage the physical constraints fully to reduce the information processing demands. For example, to ensure that the pilots' understanding of aircraft or the nuclear power plant operators' understanding of the thermodynamic processes is based on the deep structure (physical laws) that actually determines the behavior of the system. The assumption is that the better the pilots or plant operators understand the process being controlled, the easier it will be for them to achieve their objectives and the easier it will be for them to diagnose problems and adapt to unexpected events.

### 6.3.2 Source of Regularity: Coherence-Driven Domains

At the opposite end of the continuum are intent-driven domains such as general information search and retrieval (e.g., the Internet) and consumer electronic devices (e.g., mobile phones). This category of domains stands in sharp contrast to law-driven domains: The behavior-shaping constraints are loosely coupled to any physical laws. The unfolding events in these domains depend more tightly upon the needs, goals, and intentions of the agents and are therefore far less constrained by the physics of the work domain. In fact, each agent may have different functional goals and even for the same agent the functional goals may vary from one situation to another. For example, it will be difficult, perhaps even impossible, to predict the Internet sites to which an individual user will "surf" on any particular day or which book of fiction a patron will find appealing on a particular visit to the library.

This category of domains also includes office productivity tools such as calendars, e-mail programs, word processors, and spreadsheets. In these cases, there is no single work domain guiding the design. It is more realistic to think of these as general purpose tools that can be used to support work in many different domains. Thus, there is no single work ecology to be specified, but rather a diverse range of ecologies that reflect the different functional goals of a wide range of potential users.

Thus, for coherence-driven domains, the strategic constraints (e.g., knowledge, values, motivations, beliefs, and strategies) take precedence as the primary inspiration for the design of effective representations. Rather than conforming to hard physical constraints, representations for coherence-driven domains might be designed to support specific strategies that have been shown to be generally effective. For example, the design of the BookHouse interface was inspired by analysis of the different strategies used by librarians to help people find interesting books (Pejtersen 1980, 1992).

The other important principle for coherence-driven domains will be consistency. In the context of Rasmussen's decision ladder discussed in Chapter 4, the goal is to take advantage of previous associations or heuristics that people bring to the interaction (i.e., semiotic system). The point is that a representation that is in line with people's experiences (i.e., beliefs and expectations) will be more coherent than one that is inconsistent with those expectations. Thus, in designing general productivity tools, it may be very important to consider the skills that people bring from experience with other similar products or from legacy systems. Again, it is important to avoid mindless consistency; however, it is also important to appreciate the skills and expectations that people bring to the interface as a result of a history of interaction with information technologies. The experience resulting from this history should be respected and used to advantage when possible.

### 6.3.3 Summary

In summary, it is important to note that both types of constraints (law driven and intent driven) will exist in all domains. What varies between domains is the relative importance of each type of constraint in determining the unfolding events that occur (i.e., the behavior of the overall system). This is represented graphically by the diagonal line that separates the two labels on Figure 6.3: Placement of a work domain farther to the right-hand side indicates an increasing role of law-driven constraints (and vice versa). The relative importance of these two general categories of constraints can be equal in some domains (e.g., medical diagnosis and tactical operations in the military are located in the middle of the continuum).

This continuum proves to be quite useful in categorizing various types of work domains and in determining where a designer should look for inspiration about how to structure a representation that will provide the best coupling for a given semiotic system. For correspondence-driven domains, the domain analysis to identify the nested means–ends constraints (typically illustrated using an abstraction hierarchy) will be an important source for inspiration about how to structure information. For coherence-driven domains, the analysis of the knowledge and heuristics that cognitive agents are likely to employ (typically illustrated using the decision ladder) will often be a more important source of inspiration about how to structure information.

Of course, it is dangerous to ignore either side of the triad when designing the interface; however, depending on the nature of the work (law driven or intent driven), it may be more productive to begin your search for meaning on one side or the other of the semiotic system. However, it will also be important to be open to discovering key sources of regularity on both sides of the semiotic system as a particular problem or design opportunity is iteratively explored. The ultimate goal is that the representation *corresponds* with the deep structure of the problems to be solved and that it is *coherent* with respect to the expectations of the people working on the problem.

## 6.4 General Interface Constraints

Having established fundamental distinctions between domains, we will now consider fundamental distinctions between interfaces. Three fundamentally different types of representational formats are typically used in the interface: analogical, metaphorical, and propositional (e.g., Woods 1997). These three formats constitute alternative interface design strategies. All three formats have a long history in the human–computer interaction (HCI) and human factors literatures. The first two formats have often been (and still are) used interchangeably; there was (and still is) considerable confusion regarding the difference between them. We will clearly differentiate all three interface design strategies and describe the factors that are relevant in choosing between them.

### 6.4.1 Propositional Representations

The first representational form to be discussed is what Woods (1997) has referred to as propositional. Klatzky (1980) provides a definition and an example from the field of cognitive science:

> In theories of memory, the principal alternative to a spacelike representation is called a *propositional* code. Such a code is like a digital clock in that it adequately represents the corresponding stimulus, but it does not do so by being analogous to it. It is more abstract than a picture, and it is not continuous, but discrete. For example, the propositional representation of two points in space might convey, "There is a point 1 at location $x_1$, $y_1$; there is a point 2 at location $x_2$, $y_2$." It would indicate the locations of the points, but it would not represent the space in between. (p. 174)

In terms of interface design, propositional representations refer to the use of digital values (i.e., numbers), the alphabet (i.e., words and language), and other forms of alpha-numeric labels. Propositional representations are compact and precise. They capitalize on an extensive knowledge base. Chapanis (1967) once referred to propositional representations as "a very large and important area of human factors … that is almost entirely neglected" (p. 1). They provide the opportunity for the most detailed and precise representation of an ecology.

However, there is also a downside to the use of propositional representations. The relative merits of digital versus analog representations have a long history (e.g., Hansen 1995; Bennett and Walters 2001). Note that in contrast to graphical representations, the mapping between symbol and referent is an arbitrary one for propositional representations. In general, due to the symbolic nature of propositional representations, they will generally not provide good support for either rule- or skill-based interactions. This form of representation is generally the most computationally expensive in terms of placing demands on knowledge-based processes. These processes are needed to *remember,* to *decode,* to *infer logically,* or to *compute* the relations (e.g., distance in the Klatzky example) that are being represented by the symbols.

Clearly, the overall trend in interface innovations is to move away from propositional types of representations (e.g., command line interfaces) toward graphical representations that exploit the powers of computer graphics and the powers of human perception (graphical interfaces). For example, compare the instruction manuals (which are, in fact, a propositional interface) associated with early computer interfaces like the ASR 33 with the manuals for modern computers or other information technology like the iPhone®. Innovations in graphical interfaces are clearly reducing the need for extensive propositional representation in the form of large manuals (although propositional help menu systems remain components of most software). However, propositional representations still have their place as an effective interface design strategy.

### 6.4.2 Metaphors

The second representational form to be discussed is metaphor. When used in a literary sense, the term metaphor refers to the comparison of one idea or concept to another. When used in the interface, the term refers to an interface design strategy that uses graphical images to represent various objects or actions in a domain. When the images are small and can be acted upon, they have typically been referred to as icons. Thus, the perceptual component of an icon is a spatial metaphor (e.g., the representation of a wastebasket on desktop operating systems). These images can range from pictorial to abstract in nature; they are typically static representations.

Although icons can be dynamic through animation (painfully so on the World Wide Web), the dynamic variations rarely represent meaningful changes in the underlying domain. We will often use the term "local spatial metaphor" in referring to an interface icon. The term "spatial metaphor" will also be used to refer to collections of icons and other graphical images that provide higher order structure in the interface (but have no action component).

Halasz and Moran (1982) provide some insight regarding the fundamental purpose of metaphors:

> Consider the nature of literary metaphor. When we say that "Turks fight like tigers," we mean only to convey that they are fierce and cunning fighters, not that we should think about the Turks in terms of tigers. We mean only to convey a point, *not a whole system of thought*—the tiger is only a vehicle for expressing the concepts of ferociousness and cunningness. Literary metaphor is simply a communication device meant to make a point in passing. Once the point is made, the metaphor can be discarded. (p. 385; emphasis original)

Additional insights are provided by Alty et al. (2000):

> Literary theory characterizes the role of metaphor as the presentation of one idea in terms of another … . Critical to the power of metaphor is that the convocation of ideas must involve some transformation, otherwise there is simply analogy or juxtaposition and not the idea of metaphor. *Metaphors draw incomplete parallels between unlike things* … . Thus, in the design of the Apple Macintosh interface, the real-world desktop acts as a metaphor … metaphors do not make explicit the relationship between metaphor and functionality. Users actively construct the relationships that comprise the metaphor during interaction with the system. (p. 303; emphasis added)

These comments speak to the strengths and utility of the metaphor as an interface design strategy. Metaphors are used to relate new and unknown domains of application (e.g., deleting files from a computer) to more familiar and well-known domains (e.g., interactions with a physical trash basket). The overall purpose is to allow individuals who are relatively untrained and naive to understand the new work domain by thinking about it in terms of concepts with which they are already familiar. For example, Alty et al. (2000) state that an interface with metaphors "seeds the constructive process through which existing knowledge is transformed and applied to the novel situation" (p. 303). Furthermore, metaphors are expressly designed to convey limited similarities between a familiar and an unfamiliar domain (i.e., to make a point in passing). They are not intended to support more complicated forms of reasoning that rely upon detailed functional similarities between the two domains.

### 6.4.3 Analogies

The third representational form to be discussed is analogy. First consider the traditional meaning of analogical representations in the interface. In this sense, analogies, like metaphors, are used to relate unfamiliar aspects of a new domain to more familiar domains. Halasz and Moran (1982) provide a concrete example of analogy in the interface drawn from the domain of personal computing:

> Given the analogy, the new user can draw upon his knowledge about the familiar situation in order to reason about the workings of the mysterious new computer system. For example, if the new user wants to understand about how the computer file system works, he need only think about how an office filing cabinet works and then carry over this same way of thinking to the computer file system. (p. 383)

Gentner and her colleagues provide a comprehensive analysis of analogy in their "structure-mapping theory of analogical thinking" (Gentner and Gentner 1983; Gentner 1983; Gentner and Markman 1997). Gentner (1983) provides a description of the basic aspects of this theory:

> The analogical models used in science can be characterized as structure-mappings between complex systems. Such an analogy conveys that like relational systems hold within two different domains. The predicates of the base domain (the known domain)—particularly the relations that hold among the objects—can be applied in the target domain (the domain of inquiry). Thus, a structure mapping analogy asserts that identical operations and relationships hold among nonidentical things. The relational structure is preserved, but not the objects. (p. 102)

Thus, much like metaphors, analogies are used to relate familiar and unfamiliar domains. However, there are important differences. Specifically, an analogy provides more than a simple point of similarity (or an "incomplete parallel"); it provides clues regarding functional similarities between the two domains (i.e., a system of thought). Thus, analogies tend to contain more structure and can have the potential to support more complicated forms of reasoning. In terms of Gentner and her colleagues, the objects in the unfamiliar domain will be expected to behave in the same way as the objects in the familiar domain. Thus, knowledge about objects and relations in the familiar domain can be used to draw inferences and to make predictions about the behavior of the new domain.

### 6.4.4 Metaphor versus Analogy

As the previous sections indicate, there is a considerable amount of overlap between metaphors and analogies. Not surprisingly, the distinction between metaphors and analogies and their role in interface design has been the source of both confusion and debate over the years. For example, Carroll and Mack (1999) simply dispense with the concept of analogy and subsume it under metaphor:

> In an effort to resolve issues regarding the underlying representational mechanism of metaphor, the notion has been given definition in relatively more structural terms .... A typical analysis is that of Gentner (1980, 1983), who has developed a "structure-mapping" analysis of metaphor. This view interprets metaphor as a mapping between two (graph theoretically expressed) domains. (p. 387)

Alty et al. (and probably Gentner as well) would clearly not agree (2000):

> Although many designers believe they are using metaphor in their designs, many current so-called "metaphor-based" interface systems actually adopt *analogical or model-based approaches. These techniques employ direct mappings.* (p. 303; emphasis added)

To summarize the literature, the human–computer interface design community has adopted metaphors as a fundamental design construct and has abandoned analogy as a construct. The basic problem appears to be the inherent difficulties in "reasoning by analogy." The utility of an analogy will depend upon the quality of the structural mapping between the familiar domain and the new domain. If there are sufficient similarities between domains, then the analogy is likely to be helpful in understanding the new domain. However, even when the quality of this mapping is high, the analogy will eventually break down. There are at least two ways in which this may occur. First, the familiar domain may suggest relations or properties that do not hold true in the new domain. Conversely, relations or properties that are important in the new domain may not naturally be suggested by the analogy. Halasz and Moran (1982) summarized these concerns early on:

> Computer systems are unique. The tasks they carry out may often be familiar, but their underlying conceptual structures are not. The basic problem with analogical models is that they attempt to represent this conceptual structure with familiar concepts that are fundamentally inappropriate for representing computer systems ... analogy is dangerous when used for detailed reasoning about computer systems. (pp. 384–386)

### 6.4.5 Analog (versus Digital)

When we use the term "analogical" to describe a type of representational form, we are emphasizing a slightly different meaning of the word than suggested in the previous sections. We use it to refer to representations that literally have analog properties: A continuous incremental change in a domain variable or property is reflected by a corresponding continuous and incremental change in its graphical representation. For example, a watch with hands constitutes an analog representation (continuous, incremental changes). In contrast, a watch with numbers provides a digital representation (a propositional representation with discrete changes). Our use of the term analogical is consistent with Woods's observation (1991) that "*in analogical representation the structure and behavior of the representation … is related to the structure and behavior of what is being represented.* This means that perceptions about the form of representation correspond to judgments about the underlying semantics" (p. 185).

Note that these analog representations also qualify as analogies in the sense suggested by Gentner and the HCI design community. The objects in the interface (i.e., dynamic, abstract geometrical forms like a bar graph) are clearly distinct from the actual objects in the domain. Therefore, reasoning about the domain on the basis of these representations constitutes a form of reasoning by analogy. Unlike metaphors, these analogical representations are specifically designed to support detailed reasoning about the work domain (i.e., to provide a system of thought). The success of an interface designed with analogical representations will, in fact, depend upon the quality of the mapping (i.e., structural mapping, specificity) between it and the domain.

## 6.5 Interface Design Strategies

In this section we integrate the previous discussions on general types of domains and general types of representations into a broad framework for ecological interface design. The practical differences between metaphor and analogy make more sense in the context of the triadic semiotic system. With an interface design based on metaphor, the focus tends to be on the general strategies of the cognitive agent. That is, the goal is to help the agent to transfer skills and strategies associated with the metaphor to the target domain of application. Thus, the desktop metaphor is designed to help people to generalize knowledge about manipulations of files and folders in an office to the domain of managing computer files.

With an interface design based on analogy, the focus tends to be on the structure within the problem space or work domain. In this sense, the analogy is an explicit model of the work domain so that relations and changes in the model correspond to relations and changes in the underlying work domain. Typically, the analogy provides a "concrete" visualization of domain constraints that would not otherwise be apparent.

Figure 6.4 illustrates how the concepts of metaphor and analogy map onto the relative constraints associated with the work domain (problem structure)

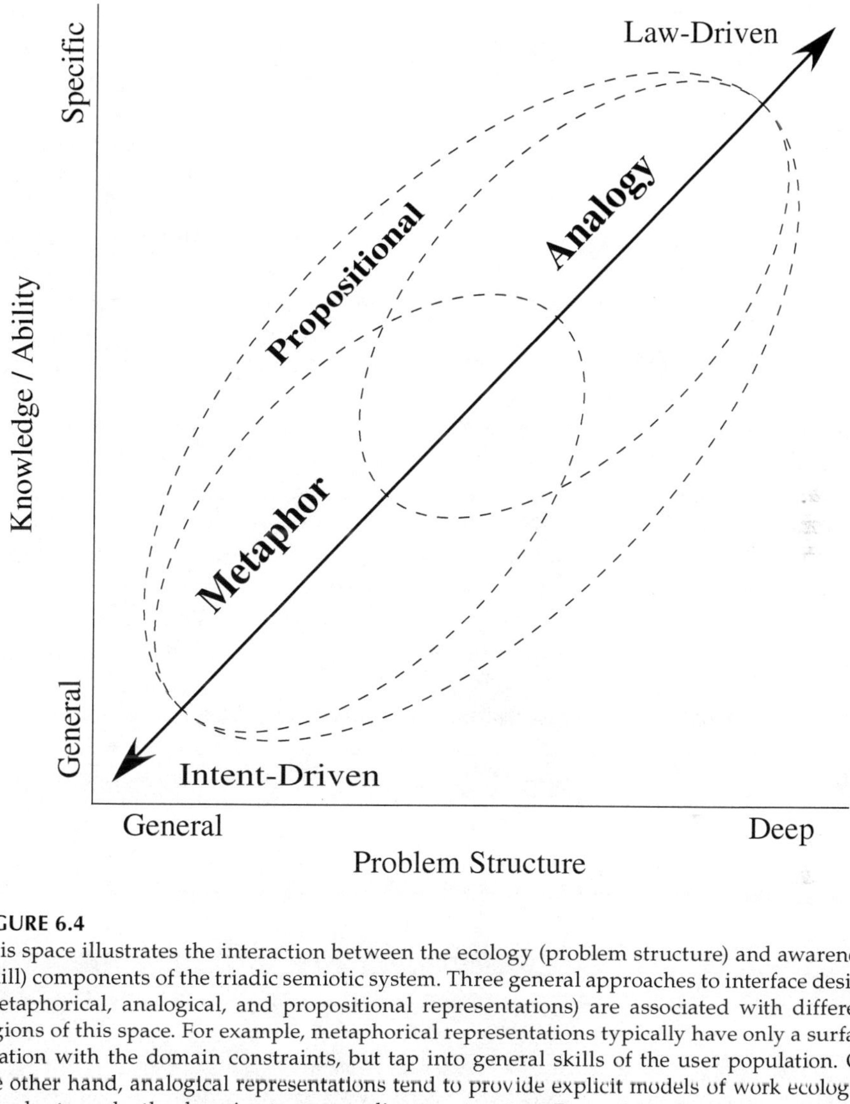


**FIGURE 6.4**
This space illustrates the interaction between the ecology (problem structure) and awareness (skill) components of the triadic semiotic system. Three general approaches to interface design (metaphorical, analogical, and propositional representations) are associated with different regions of this space. For example, metaphorical representations typically have only a surface relation with the domain constraints, but tap into general skills of the user population. On the other hand, analogical representations tend to provide explicit models of work ecologies in order to make the domain structure salient to operators. Propositional representations tap general abilities (i.e., reading), but do not leverage more specialized skills tuned to specific structures of a work domain.

and the cognitive agents (ability). Metaphors tend to connect to the domain at a superficial or surface level; however, they typically tap into very general strategies of operators. Thus, metaphors will typically be most useful for intent-driven domains to the extent that they can tap into general skills shared by the user population. In terms of Rasmussen's (1986) semiotic distinctions, metaphors will typically function as signs and thus will generally support rule-based interactions.

Analogies, on the other hand, tend to be better for tapping into the deep structure of a domain in a way that can attune the skills of operators to the demands of that domain. Particularly, in complex law-driven domains, significant training may be required so that operators can appreciate and leverage the power of the analogy against the problem domain. If the structural mapping is good, then analogies can function as signals. Thus, analogies offer the possibility of supporting skill-based interactions.

Propositional representations are also illustrated in Figure 6.4. These representations tap into general language abilities (assumed of most operators) and the associated representation in Figure 6.4 suggests a general utility. Command-line interfaces use propositional representations exclusively, and they are still in use today (e.g., UNIX). These interfaces offer some advantages relative to graphical interfaces: "There is power in the abstractions that language provides" (Hutchins et al. 1986, p. 96). For example, it is easier to delete a group of files with a particular extension interspersed with other types of files using a UNIX command as opposed to direct manipulation. But it takes a lot of time and training to harness that power.

With today's graphical interfaces, propositional representations are becoming much like the figurative fingers in the dike: to plug in holes or cracks not filled by the metaphorical or analogical graphic displays. Thus, as will be discussed in later chapters, propositional representations can be an important source of information when they are configured within metaphorical and analogical forms of representations. The following sections will explore the two primary interface design strategies, analogy and metaphor, in greater detail.

## 6.6 Ecological Interface Design: Correspondence-Driven Domains

The constraints in correspondence-driven domains have a high degree of regularity (i.e., they are tightly coupled and law driven). The agent must consider the causal relations that exist within the inherent structure of the domain (e.g., goals, functions, physical makeup) if it is to be controlled properly. In such domains, agents will typically be trained extensively to have relatively similar levels of skills, rules, and knowledge, as mandated by the complexities and inherent risks often associated with these domains. It follows, then, that the interface must incorporate representations of this structure if it is to provide effective decision-making and problem-solving support. What is needed in the interface are detailed analogies of the domain constraints; this is typically done using abstract geometric forms that are situated within performance boundaries and that dynamically change in concert with changes in the underlying domain.

Consider a simple example from process control. Temperature may be an important state variable. However, a digital display of temperature alone would not support direct perception since it is not related to critical action boundaries. This context would have to be supplied by some cognitive agent (e.g., remembering that a particular temperature is critical). On the other hand, an analog display that maintained a compatible relationship with a control device and included explicit representations of critical boundaries (e.g., normal operating regions) would allow direct perception in which the functional constraints could be represented explicitly. Thus, the operator does not need to see temperature; rather, she needs to see where temperature is with regard to the critical boundaries. Is the system in the intended operating regions? Is it approaching instability? This can be further improved if the temperature graphic were explicitly linked to the controls for regulating it. The interface would then also support direct manipulation—for example, by click-and-drag operations performed directly on the temperature icon to specify commands to reduce or increase the temperature.

If the critical boundaries are a function of a single state variable, then building direct perception interfaces can be a trivial exercise. In many cases, the problem reduces to a one-dimensional compensatory or pursuit-tracking task. However, in complex systems, critical action boundaries will be determined by the interaction of multiple state variables. The stability of the process will depend on temperature, pressure, concentrations of chemicals, and the rate of change of these concentrations. Further complicating the problem is the fact that, as noted earlier, the constraints of the system will be defined at various levels of abstraction and aggregation. In other words, there will be a hierarchical nesting of affordances that are defined by the domain. Each of the different levels will have its own critical action boundaries and its own time scale. How can these complex interactions and multiple intention levels be integrated in a way that allows direct perception?

### 6.6.1 Nested Hierarchies

Natural environments are also characterized by multiple interacting variables and levels of abstraction. In describing the terrestrial environment, Gibson (1979) noted that "there are forms within forms both up and down the scale of size. Units are nested within larger units. Things are components of other things. They would constitute a hierarchy except that this hierarchy is not categorical but full of transitions and overlaps" (p. 9). Direct perception is possible because this nesting of structure in the environment is revealed in an ambient optic array that preserves this nested structure. "If a surface is composed of units nested within larger units, its optic array is composed of solid angles nested within larger solid angles" (Gibson 1979, p. 108). In fact, a hierarchy of structure can be identified within the optic array. Local invariants can be nested within higher order invariants, which can in turn be nested within global invariants.

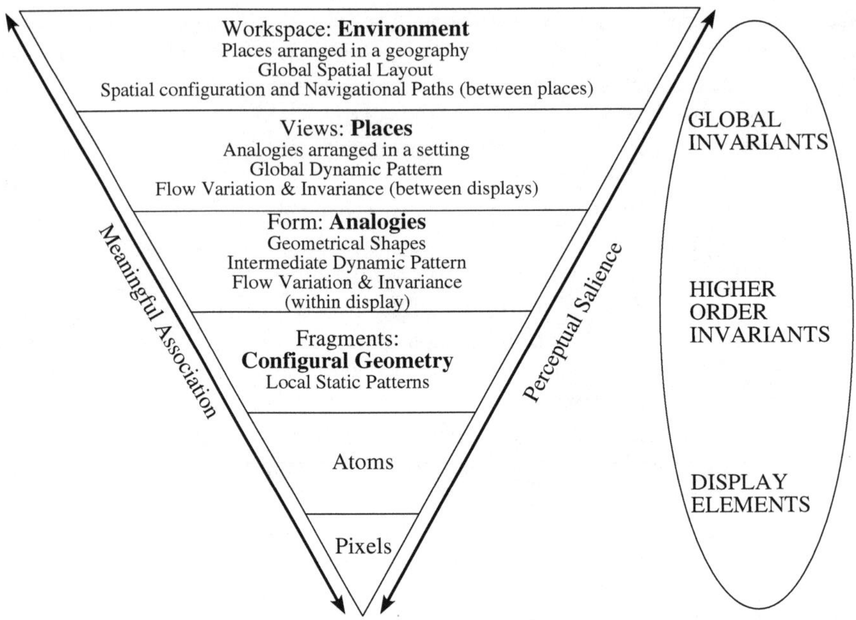


**FIGURE 6.5**
The hierarchically nested, analogical display geometries required to support direct perception by a cognitive agent in complex law-driven domains.

Displays and interfaces can also be characterized by multiple interacting variables and nested hierarchies. Figure 6.5 illustrates the nested hierarchies of graphical information (adapted from Woods 1997) in the interface of a law-driven domain. Pixels will configure to produce meaningful, low-level patterns such as lines and curves (i.e., graphical atoms). Graphical atoms (e.g., two lines) will configure to produce graphical fragments (i.e., emergent features) with higher order emergent properties (e.g., angles). Collections of graphical fragments will produce analog, geometric forms such as Coekin's (1970) polar star display. Several configural displays may be required to provide views that testify with regard to higher level aspects of the domain (i.e., functions, subsystems, modes). Finally, collections of views constitute the work space that specifies the system in its entirety. Each level will be nested within adjacent levels, producing local invariants, higher order invariants, and global invariants up and down the complete hierarchy.

The concepts of higher order and global invariants are similar to the concepts of configural and emergent features from research on perceptual organization (e.g., Pomerantz 1986). Treisman (1986) wrote a fairly comprehensive review of perceptual organization and cognition. In this review, she observed that "if an object is complex, the perceptual description we form may be hierarchically structured, with global entities defined by subordinate elements and subordinate elements related to each other by the global description" (p. 35.54). This description of hierarchical organization in perception mirrors Rasmussen's abstraction hierarchy. These two hierarchies (perceptual and functional) may provide a basis for a new theory of compatibility for display design. A compatible interface for complex system control is one where the hierarchical organization engendered by the representation matches the hierarchical organization of function (i.e., the abstraction hierarchy).

### 6.6.2 Nested Hierarchies in the Interface: Analogical Representations

Direct perception in complex human–machine systems will require visual representations that contain global invariants (configural properties) that map to high levels in the abstraction hierarchy and elements that map to lower levels of the abstraction hierarchy. This will provide a most fluent and effective means for communication with a controller. It will provide a basis for the operator "seeing" how actions relate to higher level functions (why) and to lower level physical instantiations (how).

This principle for mapping organizational structure within a representation to functional structure within a work domain is illustrated in Figure 6.6. In this scheme, global invariants are used to provide the global perspective required of high levels in the abstraction hierarchy. This structure should reveal critical action boundaries with regard to the functional purpose of the system.

The polar coordinate display (Woods, Wise, and Hanes 1981) designed for nuclear power plant control rooms provides an illustration of how this might be accomplished (see Figure 6.7). In this display, over 100 state variables are represented as the eight vertices of an octagon. The display is scaled so that, when the plant is in a normal state, the configuration of variables produces a symmetrical geometric form. A break in the domain constraints (i.e., a failure) produces a break in the geometrical form. The polar coordinate display

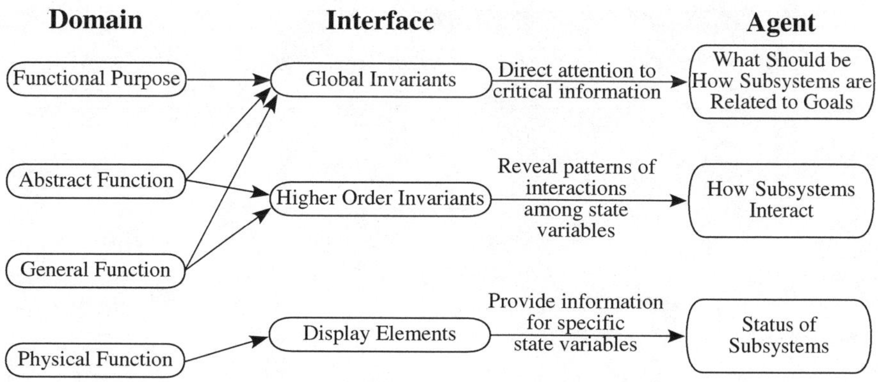


**FIGURE 6.6**
The mapping of the semantics of a work domain, as characterized by levels within the abstraction hierarchy, onto nested display geometries to allow direct perception by a cognitive agent.

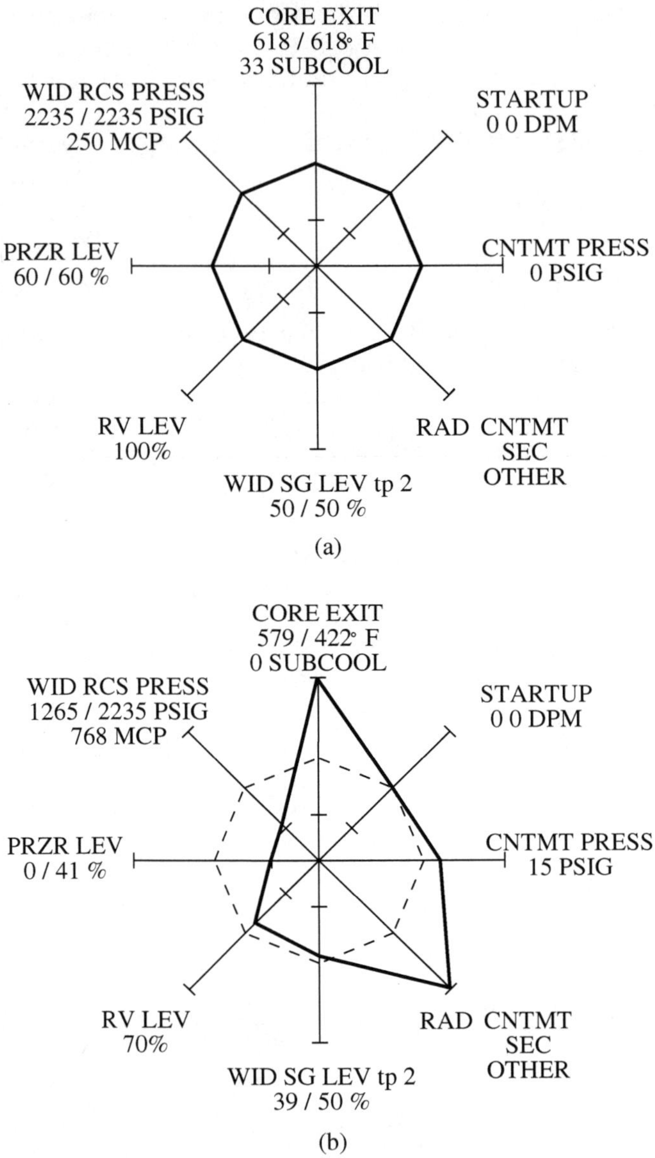


**FIGURE 6.7**
The polar graphic display format. (Adapted from Schaefer, W. F. et al. June 23, 1987. Generating an integrated graphic display of the safety status of a complex process plant, United States Patent 4,675,147.)

was designed so that specific types of asymmetries specify specific types of failures. We have referred to this general type of representational format as a configural display (Bennett and Flach 1992), borrowing terms from the perceptual organization literature.

Higher order invariants in configural displays should be chosen so that patterns of interactions among state variables are revealed. For example, state variables that are tightly coupled should be perceptually linked and functionally related variables might be positioned on adjacent or perhaps opposite vertices in the polar coordinate display. Thus, the interactions among variables can be graphically revealed. However, note that great care must be taken to ensure that the graphical interactions produced by the display are meaningful in terms of the underlying domain semantics (see the associated discussions in Chapters 8 through 11). Finally, local elements should provide information for lower level data, thereby directing attention to critical information at lower levels of abstraction. In the polar coordinate display, position of a vertex provides this information.

The key to direct perception is the perceptual coordination across the levels of global, higher order, and local invariants of both the configural display and the underlying domain. The nested hierarchy of invariants in the domain (e.g., goals, properties, functions, and physical components) must be reflected in the nested hierarchy of invariants in the graphical representations (e.g., overall shape, contours, line orientations, and individual vertices). The quality of this mapping (i.e., the specificity of the display with regard to the underlying domain) will determine the effectiveness of the display. In the polar coordinate display, some, but not all, of this coordination is accomplished through nested structure within a single geometric form.

As suggested by this analysis, a single geometric display (i.e., a form) or even collections of graphical forms in a single window (i.e., a view) will not be sufficient for monitoring and control of a complex system; several or many windows of information will be required. This will be particularly true when the functional purposes of the system are diverse. However, the prescription for design remains the same. Direct perception will be possible to the extent that a global invariant can be designed that directs attention and links functions across the various views.

Woods's (1984) discussion of "visual momentum" addresses the problem of maintaining a global perspective over multiple views or windows to avoid "keyhole effects." For complex work domains, it may be impossible to include a complete representation of the work domain in a single window. For example, in a complex military command and control setting, it can be impossible to include all the important entities in a single representation: One view might be optimized with respect to time constraints (e.g., a time line or Gantt chart representation), while other views might reflect spatial constraints (e.g., situation map displays). For large operations, it may be impossible to show all the different assets (e.g., air, land, and sea forces) on a single map without that map becoming a cluttered jumble of noise.

Thus, in designing the interface, consideration must be given to transitions between windows and to the challenge of navigating across multiple perspectives on the work space. Woods (1984) suggests the metaphor of a well-cut versus poorly-cut film to illustrate this challenge. The goal is that a multiwindow interface be designed like a well-cut film so that the transitions from one view to another provide a logical flow, where each window provides a meaningful context for the others. The main point here is to recognize that when an interface is configured, it is important not to think exclusively in terms of spatial relations within a window. In many cases, it will be important to think about the dynamic relations (i.e., over time) across multiple windows (and perhaps across multiple modalities as well). Chapter 15 delves more deeply into the construct of visual momentum.

## 6.7 Ecological Interface Design: Coherence-Driven Domains

As described previously, the behavior-shaping constraints of coherence-driven domains are fundamentally different from those for correspondence-driven domains. In these work domains, the goals and intentions of the user play a far more predominant role in the events that unfold (hence, the use of the term intent-driven). The element of physical causality is missing and these constraints are much more loosely coupled. For example, the exact book of fiction that a particular library patron will choose to read on a particular day is far less predictable (i.e., far less regular) than the level of a reservoir in a process control system. Furthermore, the skills, rules, and knowledge of the agents interacting with these intent-driven domains will *not* be similar. The agents will typically be untrained and infrequent users of the system who are of all ages and from all walks of life. The interface will need to support a wide variety of knowledge about the particular domain and about interfaces and computers in general.

The appropriate interface design strategy for intent-driven domains is the use of metaphors. As described earlier, the purpose of metaphors is to support novice or infrequent users by relating the requirements for interaction to concepts or activities that are already familiar. For example, variations of the lock metaphor are often used to indicate that the value of some parameter or object in the interface (e.g., security settings, properties of graphics, etc.) either can (open lock) or cannot (closed lock) be changed. Although the specific techniques used in interface design are dissimilar (i.e., metaphors vs. analogy), the overall approach and goals for designing effective virtual ecologies for intent-driven domains are very similar to those for law-driven domains.

### 6.7.1 Objective Properties: Effectivities

In principle, the starting point, once again, is to understand the objective properties of the work domain. However, when designing for intent-driven domains, there generally is not a well-established physics of the problem space. In these cases, the best guess about the physics is typically the judgment of domain experts. Thus, the innovations in these domains are often based on an analysis of the strategies that domain experts use.

For example, Pejtersen (1992) developed a computerized database and retrieval system (the BookHouse) to help library patrons find books of fiction to read (a clearly intent-driven domain). Much of the work analysis in the design of the BookHouse interface focused on the strategies that expert librarians used to help identify interesting books. The ecology of this work domain ultimately boils down to the fundamental ways in which books of fiction can vary (i.e., the meaningful distinctions between them). It is an abstract categorical structure, as opposed to the physical causal structure that typifies law-driven domains. Nonetheless, it is the landscape upon which a search for any book of fiction can be conducted. These strategies became the basis for the AMP classification scheme for fiction developed by Pejtersen (e.g., 1980).

Subjectively, the AMP classification scheme specifies the differences between works of fiction that the librarians judged would be meaningful to a reader (i.e., reflect differences that are important with respect to preferences or choices among books). The basic dimensions of the classification scheme are illustrated in the middle column of Figure 6.8. These dimensions include why the author wrote the book (to educate, to scare), the way in which it was written (literary style), its setting (the context), what happens (the course of events), and its readability (font size, reading level). The relationship between these objective dimensions and the needs and intentions of the reader are made explicit in the right column of Figure 6.8.

Finally, each of these dimensions possesses a level of abstraction that corresponds to a category in the abstraction hierarchy, as illustrated in the left column of Figure 6.8. Thus, the domain semantics also reflect the nested hierarchies (local, higher order, and global invariants) that were described in the previous section for law-driven domains. Note that this is a model of the work domain, but not a model from the perspective of a classical physical analysis: It is a perspective derived from the strategies of expert librarians. In essence, for intent-driven domains, the judgments of experts may provide our best guesses about the nature of the underlying domain constraints.

This classification scheme, based on domain semantics, is a key component in assisting library patrons in finding a book of interest. Each book in a library's stock is indexed in terms of each classification dimension and a record is incorporated into a database in the BookHouse system. A patron locates a book of interest by conducting a database search defined by the specification of a particular descriptor within one or more of these dimensions. This

| Abstraction Hierarchy | Document: Content AMP Classification | User Needs/Intentions |
|---|---|---|
| Functional meaning, Purposes | Why: Author Intention<br><br>Information; Education; Emotional Experience | Goals: Why ?<br><br>User's Ultimate Task and Goal |
| Information Measures | Why: Professional Paradigm<br><br>Style Literary or Professional Quality; or School | Value Criteria: Why ?<br><br>Value Criteria Related to Reading Process and/or Product |
| General Functions | What: General Frame of Content<br><br>Cultural Environment, Historical Period, Professional Context | Content Frame: What ?<br><br>General Topical Interest of Historical, Geographical or Social Setting |
| Physical Processes | What: Specific, Factual Content<br><br>Episodic, Course of Events; Factual Descriptions | Subject Matter: What ?<br><br>Topical Interest in Specific Content |
| Material Form, Configuration | How: Level of Readability<br><br>Physical Characteristics of Document; Form, Size, Color, Typography, Source, Year of Print | Accessibility: How ?<br><br>User's Reading Ability, Perceptual and Cognitive Capability |

**FIGURE 6.8**
The AMP classification scheme developed by Pejtersen to describe the ecology of fictional books. Placed in the context of the abstraction hierarchy and corresponding user needs/intentions. (Adapted from Rasmussen, J., A. M. Pejtersen, and L. P. Goodstein. 1994. *Cognitive Systems Engineering*. New York: John Wiley & Sons, Inc. With permission.)

type of search is based on meaning as opposed to convenience. Compare it to the traditional computerized search based on relatively meaningless bibliographical data. The classification scheme enables "requests for specific subjects in fiction like 'exciting books about everyday life of children on farms in Guatemala' or 'critical books about physical demands in modern sports'" (Rasmussen et al. 1994, p. 239).

### 6.7.2 Nested Hierarchies in the Interface: Metaphorical Representations

The design of interfaces for coherence-driven domains poses a different set of problems from those of the equivalent process for correspondence-driven domains. One does not have the luxury of well-trained, knowledgeable, and homogeneous agents interacting with the system. As a result, it is

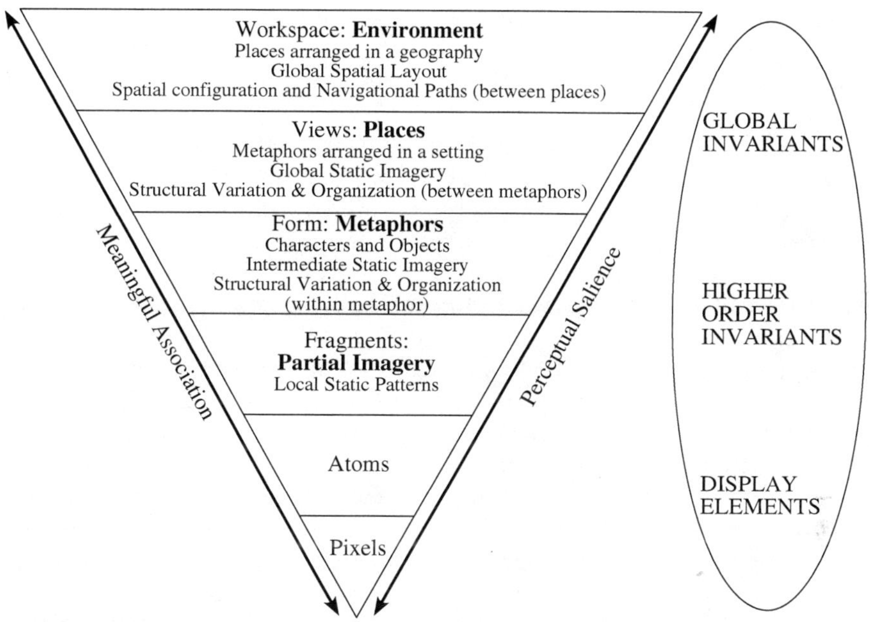


**FIGURE 6.9**
The hierarchically nested, metaphorical display imagery required to support direct perception by a cognitive agent in intent-driven domains.

absolutely essential that the system be designed so that it is both intuitive to learn and easy to use. The interface is the primary means to achieve that goal. Thus, a fundamental challenge in interface design for intent-driven domains is to leverage existing conceptual knowledge and skills common to the diverse set of humans who will use the system. Specifically, the objects, activities, and sequences of events that are required as control input in the work domain should be related to commonplace, normal objects and activities encountered and executed in everyday life to allow maximum transfer of skills from everyday experience to the task.

Figure 6.9 illustrates the hierarchical nesting of visual information in interfaces for intent-driven domains. This figure replicates Figure 6.5 (law-driven domains); the major difference is that metaphorical representations have been substituted for analogical representations. This basic representational form is used to represent higher order invariants. In addition, coordinated sets of spatial metaphors at multiple levels of granularity will be required. Collections of metaphors will be arranged in places. These collections will represent the various objects and actions that are required to complete activities in the domain. Finally, collections of places will be required to represent the modes, functions, or subsystems of complex work domains (the work space). This introduces the requirement to navigate between different screens in the interface. A global metaphor will be required to organize the overall work space and to relate the various places (thereby facilitating navigation or visual momentum).

The exact form that the metaphors, places, and environments take in the interface of an intent-driven domain will vary. Once again, the BookHouse system will be used to provide one concrete example. The interface consists of hierarchically nested metaphorical representations. The global metaphor of a virtual library was chosen to organize the overall interface at the work space level (i.e., the virtual environment). The act of finding a book of fiction is equated to the process of navigating through this virtual space. The various activities that are required to find a book (e.g., narrow down a search; implement a search strategy) are related to substructures within the overall virtual space. Thus, a patron chooses a subset of books or a different search strategy by navigating through different rooms (places) in the virtual library (environment). Finally, the details required to execute a specific search are related to items within a substructure.

Thus, a patron specifies a search parameter (e.g., time frame) by manipulating an object (a metaphor) in a room (a place). Note that we are not necessarily recommending virtual three-dimensional spaces like the library metaphor used in the BookHouse. Rather, we are recommending a nested organization of functions within the interface space, which in many contexts may be accomplished with a simple two-dimensional space.

Thus, these hierarchically nested metaphors capitalize upon familiar conceptual knowledge and skills by relating the structure and function of the new library retrieval system to objects and activities with which almost all patrons will be familiar. The lower level metaphors (e.g., a globe to represent geographical setting) are designed to provide an associative link between the objects or actions that are needed in the work domain (unfamiliar) and preexisting concepts in semantic memory (familiar). At a higher level, the global metaphors map the actions required to interact with the system onto a set of common, natural activities (i.e., locomotion through a spatial ecology) that people have become naturally skilled at accomplishing through the ages. Almost all library patrons will have a wealth of experience and conceptual knowledge about navigating through buildings or other spatial structures. Thus, the sequence of events and activities required to conduct searches of the database are crafted to be similar to commonplace, normal activities carried out constantly in everyday life.

## 6.8 Summary

At the end of Chapter 2 we outlined general approaches to cognition and interface design and briefly discussed some of the confusion regarding ecological research (e.g., the "ecological" research conducted on Mount McKinley).

Exactly what constitutes an ecological interface has also been a source of some confusion. In Chapters 2 through 6 we have provided more details regarding cognitive systems engineering and the ecological approach to interface design and are now in a position to address this issue more completely.

One common misconception is that the definition of an ecological interface is one that relates interactional requirements to activities that are more "natural." From this perspective, the BookHouse interface is ecological because the process of finding a book has been translated into the act of navigating through a virtual ecology. This aspect is certainly consistent with Gibson's approach to ecological psychology and is clearly a contributing factor. However, we believe that the defining characteristic of an ecological interface is that it has been explicitly designed on the basis of a detailed understanding of the work ecology. In the case of the BookHouse interface, this translates into the fact that its foundation lies in the sources of regularity in the domain of fiction (i.e., the dimensions of the AMP classification framework).

Thus, it is not the graphical metaphor that makes the BookHouse ecological, but rather the detailed work analysis to identify the meaningful ways to distinguish among works of fiction. In other words, the detailed knowledge elicitation with librarians to identify what differences make a difference in terms of the problem of searching for interesting books is the key to the ecological nature of the BookHouse. The virtual library metaphor is one way to make those distinctions salient to people, but we posit that such a graphical interface would be impotent without the structure derived from the work analysis.

In summary, we have conceptualized interface design as the process of constraint mapping between three high-level system components (domain, agent, interface). The general interface design strategy that is most effective (i.e., metaphors vs. analogies) will depend upon the nature of the constraints that are contributed by the domain and the agent. However, the overall goals of interface design are central to both strategies: to support direct perception and manipulation and to maintain an intact perception–action loop—in other words, to support skill-based and rule-based forms of interaction. To do so, the objective properties of the domain must be represented in a way that makes them salient to the agents. Direct perception must be supported by building a representation that reflects the nested structure across levels of abstraction and aggregation that characterize the work environment. That is, *the interface must provide global and local structure (or invariants) in the representation that correspond to the inherent structure in the domain and that specify the potential for action in relation to goals and values.*

The nuances of design associated with these overall goals will be explored in subsequent chapters. Several caveats to the topics introduced in this chapter are in order. Although propositional formats were described as a fundamental representation type, they are not listed as a primary interface design strategy because propositional representations are used most effectively as a complement to both analogies and metaphors, rather than as the primary choice of representation (although this often happens, unfortunately).

Furthermore, analogies and metaphors have been presented in an either–or fashion. In practice, the distinction between them can blur; effective interfaces will often be constructed from all three representational types. For us, *the practically most important distinction between analogy and metaphor is best seen in the context of the semiotic triad; with metaphor, the emphasis is on tapping into general skills of the agent that can be leveraged against a problem and, with analogy, the emphasis is on making the deep structure of the domain salient.* The common goal is to enhance coordination within the overall semiotic system.

Finally, this chapter has emphasized the need to support skill-based behaviors in interface design. This is rightly so because the leveraging of powerful perception–action skills of the human agent lies at the heart of effective interface design. However, it is still very important to support rule- and knowledge-based behaviors.

Thus, we want to close by considering ecological interface design (EID) in relation to knowledge-based processing. In describing the semiotic dynamic of meaning processing, we noted the two conjoint perspectives on the dynamic (1) as a control system and/or (2) as an observer/abductive system. This raises two functional roles for direct perception/manipulation. In the context of control, the function of manipulation is to get to the goal. This has been the focus of most of the discussion in this chapter and is typically the focus of discussions of direct manipulation. The key here is direct feedback relative to progress toward the goal. However, with respect to knowledge-based processing, the context of observation or abduction becomes very important. This also reflects Shneiderman's third attribute of direct manipulation (i.e., reversible operations). In this context, the role of manipulation is exploratory. That is, manipulations are experiments on the interface. Thus, the function of the manipulation is to test hypotheses about the system or to learn about the system; this is a knowledge-based activity.

Skill- and rule-based interactions will typically evolve as the result of interactions with structured environments. However, no matter how structured the environment is, some initial learning or experience will generally be required. Thus, early interactions will demand some degree of knowledge-based processing. In this context, an ecological or a direct manipulation interface is one that supports trial-and-error learning. There are two aspects that will be important to this learning by doing.

First, there should be salient associations between the actions and the consequences of those actions (explicit feedback). Second, actions that lead to undesirable consequences should be reversible. Note that especially in high-risk, correspondence-driven domains, this will not always be possible. In these cases, not only will feedback be important, but feed forward will also be important so that the crossing of irreversible boundaries can be specified well in advance, when there is still an opportunity for correction. In domains where there is little risk, interfaces that encourage exploration and trial-and-error experiments may be more engaging and more inclined to support human creativity and discovery. In domains where there is great risk, extensive training (perhaps using high-fidelity simulators) will be required as an important complement to ecological interface design.

## References

Alty, J. L., R. P. Knott, B. Anderson, and M. Smyth. 2000. A framework for engineering metaphor at the user interface. *Interacting with Computers* 13:301–322.

Bennett, K. B., and J. M. Flach. 1992. Graphical displays: Implications for divided attention, focused attention, and problem solving. *Human Factors* 34:513–533.

Bennett, K. B., and B. Walters. 2001. Configural display design techniques considered at multiple levels of evaluation. *Human Factors* 43 (3): 415–434.

Carroll, J. M., and R. L. Mack. 1999. Metaphor, computing systems, and active learning. *International Journal of Human–Computer Studies* 51:385–403.

Chapanis, A. 1967. Words words words. *Human Factors* 7 (1): 1–17.

Coekin, J. A. 1970. An oscilloscope polar coordinate display for multidimensional data. *Radio and Electronic Engineer* 40:97–101.

Eberts, R. E. 1994. *User interface design*. Englewood Cliffs, NJ: Prentice Hall.

Flach, J. M., and K. J. Vicente. 1989. Complexity, difficulty, direct manipulation and direct perception. Urbana-Champaign: Engineering Psychology Research Laboratory, University of Illinois.

Gentner, D. 1983. Structural mapping: A theoretical framework for analogy. *Cognitive Science* 7:155–170.

Gentner, D., and D. Gentner. 1983. Flowing waters or teeming crowds: Mental models of electricity. In *Mental models,* ed. D. Gentner and A. L. Stevens. Hillsdale, NJ: Lawrence Erlbaum Associates.

Gentner, D., and A. B. Markman. 1997. Structure mapping in analogy and similarity. *American Psychologist* 52 (1): 45–56.

Gibson, J. J. 1979. *The ecological approach to visual perception.* Boston, MA: Houghton Mifflin.

Halasz, F., and T. Moran. 1982. Analogy considered harmful. Paper read at CHI '82 Human Factors in Computing Systems, New York.

Hansen, J. P. 1995. An experimental investigation of configural, digital, and temporal information on process displays. *Human Factors* 37:539–552.

Hutchins, E. L., J. D. Hollan, and D. A. Norman. 1986. Direct manipulation interfaces. In *User centered system design,* ed. D. A. Norman and S. W. Draper. Hillsdale, NJ: Lawrence Erlbaum Associates.

Klatzky, R. L. 1980. *Human memory: Structures and processes.* San Francisco, CA: W. H. Freeman and Company.

Pejtersen, A. M. 1980. Design of a classification scheme for fiction based on an analysis of actual user–librarian communication, and use of the scheme for control of librarians' search strategies. In *Theory and application of information research,* ed. O. Harbo and L. Kajberg. London, England: Mansell.

———. 1992. The BookHouse: An icon based database system for fiction retrieval in public libraries. In *The marketing of library and information services 2*, ed. B. Cronin. London, England: ASLIB.

Pomerantz, J. R. 1986. Visual form perception: An overview. In *Pattern recognition by humans and machines,* ed. H. C. Nusbaum and E. C. Schwab. Orlando, FL: Academic Press.

Post, E. 1983. Real programmers don't use Pascal. *Datamation* 29 (7): 263–265.

Preece, J., Y. Rogers, H. Sharp, D. Benyon, S. Holland, and T. Carey. 1994. *Human–computer interaction.* Harlow, England: Addison–Wesley.

Rasmussen, J. 1986. *Information processing and human–machine interaction: An approach to cognitive engineering.* New York: Elsevier.

Rasmussen, J., A. M. Pejtersen, and L. P. Goodstein. 1994. *Cognitive systems engineering.* New York: John Wiley & Sons.

Rasmussen, J., and K. J. Vicente. 1990. Ecological interfaces: A technological imperative in high-tech systems? *International Journal of Human–Computer Interaction* 2 (2): 93–111.

Schaefer, W. F. et al. June 23, 1987. Generating an integrated graphic display of the safety status of a complex process plant, United States Patent 4,675,147.

Sharp, H., Y. Rogers, and J. Preece. 2007. *Interaction design: Beyond human–computer interaction,* 2nd ed. Chichester, England: John Wiley & Sons.

Shneiderman, B. 1982. The future of interactive systems and the emergence of direct manipulation. *Behavior and Information Technology* 1:237–256.

———. 1983. Direct manipulation—A step beyond programming languages. *Computer* 16 (8): 57–69.

———. 1992. *Designing the user interface: Strategies for effective human–computer interaction,* 2nd ed. Reading, MA: Addison–Wesley.

———. 1998. *Designing the user interface: Strategies for effective human–computer interaction,* 3rd ed. Reading, MA: Addison–Wesley.

Treisman, A. M. 1986. Properties, parts, and objects. In *Handbook of perception and human performance,* ed. K. Boff, L. Kaufmann, and J. Thomas. New York: John Wiley & Sons.

Vicente, K. J. 1999. *Cognitive work analysis: Toward safe, productive, and healthy computer-based work.* Mahwah, NJ: Lawrence Erlbaum Associates.

Woods, D. D. 1984. Visual momentum: A concept to improve the cognitive coupling of person and computer. *International Journal of Man–Machine Studies* 21:229–244.

———. 1991. The cognitive engineering of problem representations. In *Human–computer interaction and complex systems,* ed. G. R. S. Weir and J. L. Alty. London: Academic Press.

———. 1997. The theory and practice of representational design in the computer medium. Columbus: Ohio State University, Cognitive Systems Engineering Laboratory.

Woods, D. D., J. A. Wise, and L. F. Hanes. 1981. An evaluation of nuclear power plant safety parameter display systems. Paper read at the Human Factors Society 25th Annual Meeting, Santa Monica, CA.

# 7

# Display Design: Building a Conceptual Base

## 7.1 Introduction

> Another feature of the Wright brothers' creative thought process that figured prominently in their advance toward powered flight was the great extent to which they used *graphic mental imagery* to conceptualize basic structures and mechanisms, even aerodynamic theory. Wilbur and Orville's facility for *nonverbal thought* was among the most prevalent and salient aspects of their *inventive method.* (Jakab 1990, pp. 4–5; emphasis added)

The process of building human–computer interfaces is one that involves the construction of visualization to support productive thinking, as described in Chapter 6. The fundamental challenge lies in the design of graphical representations (i.e., visual displays). This is a very complicated endeavor and it is not surprising that researchers have considered it from a variety of conceptual perspectives. The purpose of this chapter is to review and critique some of these perspectives. Each was chosen because it emphasizes a particular facet of effective display design and provides valuable insights. It is also true, at least in our opinion, that each perspective has some limitations, which will also be described briefly. The focus of the chapter is somewhat skewed toward issues in the design of analogical representations. However, overarching issues that are relevant for all representations are also considered. In summary, the goal of this chapter is to provide a broad introduction to and a foundation for effective interface design.

Several of the graphical displays in the following sections use variables from a simple process control system that is illustrated in Figure 7.1 This system has a reservoir for storing liquid, two input streams, and an output stream. The measured variables associated with this system are the valve settings ($V_1$, $V_2$, and $V_3$) and flow rates ($I_1$, $I_2$, and O) associated with each stream and the reservoir level (R). The system also has one goal for reservoir volume ($G_1$) and another for mass output ($G_2$). A more detailed description is provided in Chapter 10.

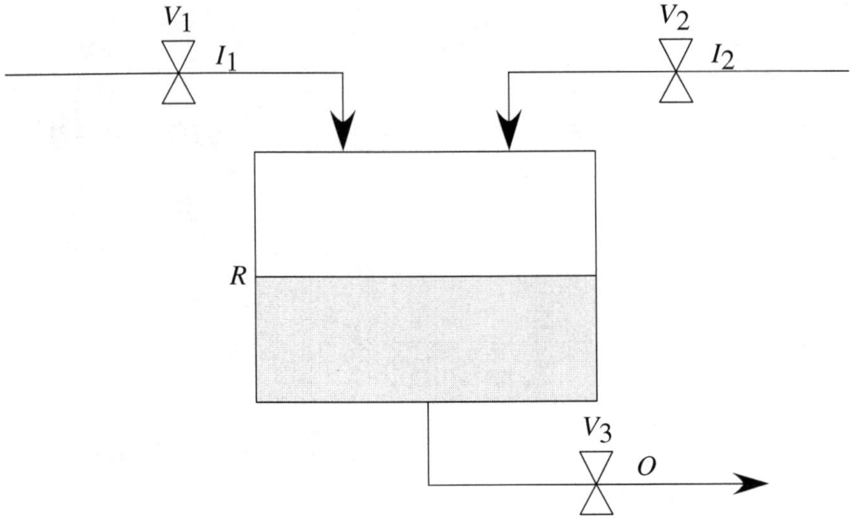


**FIGURE 7.1**
A simple process control system and its measured variables. (Adapted with permission from Bennett, K. B., A. L. Nagy, and J. M. Flach. 1997. In *Handbook of Human Factors and Ergonomics,* ed. G. Salvendy. New York: John Wiley & Sons. All rights reserved.)

## 7.2 Psychophysical Approach

The first perspective on display design to be considered is that of Cleveland and his colleagues (Cleveland and McGill 1984, 1985; Cleveland 1985). Their goal was to develop a "scientific basis for graphing data" (Cleveland 1985, p. 229). This approach has its roots in psychophysics, drawing upon that discipline's empirical data, principles (e.g., Weber's law, Stevens's law), and experimental methodologies. In terms of the global conceptual distinctions outlined in Chapter 2, this approach is clearly a dyadic one: The primary consideration is the relationship between the physical properties of representations and the perceptual capabilities of the observers (i.e., their ability to extract quantitative information from those representations). Consider the following quotation (Cleveland 1985):

> When a graph is constructed, quantitative and categorical information is encoded by symbols, geometry, and color. Graphical perception is the visual decoding of this encoded information. Graphical perception is the vital link, the raison d'etre, of the graph. No matter how intelligent the choice of information, no matter how ingenious the encoding of the

> information, and no matter how technologically impressive the production, a graph is a failure if the visual decoding fails. To have a scientific basis for graphing data, graphical perception must be understood. Informed decisions about how to encode data must be based on knowledge of the visual decoding process. (p. 229)

Issues in the design of graphical representations were studied using psychophysical methodologies. For example, Cleveland and McGill (1984) encoded the same quantitative information into alternative graphical representations that required discriminations involving alternative "elementary graphical perception" tasks, as illustrated in Figure 7.2. The participants were given a "standard" display (see the examples labeled "1" in Figure 7.2a through 7.2f) that graphically illustrated a value of 100%. Participants were then provided with "reference" displays (see the examples labeled "2," "3," and "4" in Figure 7.2a through 7.2f) that graphically illustrated some portion of the standard (e.g., examples 2, 3, and 4 portray 25, 75, and 50% of the standard, respectively). The participants' task was to provide a quantitative estimate of the percentage.

### 7.2.1 Elementary Graphical Perception Tasks

The results of these and similar evaluations were used to develop principles of display design. Performance on the elementary graphical perception tasks was ranked in the following order (from best to worst): position along a common scale (Figure 7.2a); position along identical, nonaligned scales (Figure 7.2b); length (Figure 7.2c); slope (Figure 7.2d)/angle (Figure 7.2e); area (Figure 7.2f); volume; and color hue color saturation/density (Cleveland 1985, p. 254).

The resulting principles of design are relatively straightforward: The designer should choose the highest ranking visual features available when a graphical representation is built. For example, consider the choice between a bar chart (Figure 7.3) and a pie chart (Figure 7.4). The bar chart would be the preferred representational format because it involves judging position along a common scale (i.e., the vertical extent of the various bar graphs), which is ranked at the top of the list.

In summary, the work of Cleveland and his colleagues provides designers with useful information about the dyadic relationships between display and observer (i.e., factors that influence the quality of format mapping). Their principles of design can be used to make informed decisions with regard to basic representational formats that should be used and those that should be avoided. For example, the displays that we have developed for process control (Figure 7.5 and Chapter 10) and for military command and control (Chapter 14) make extensive use of bar graph displays to represent the value of individual variables.

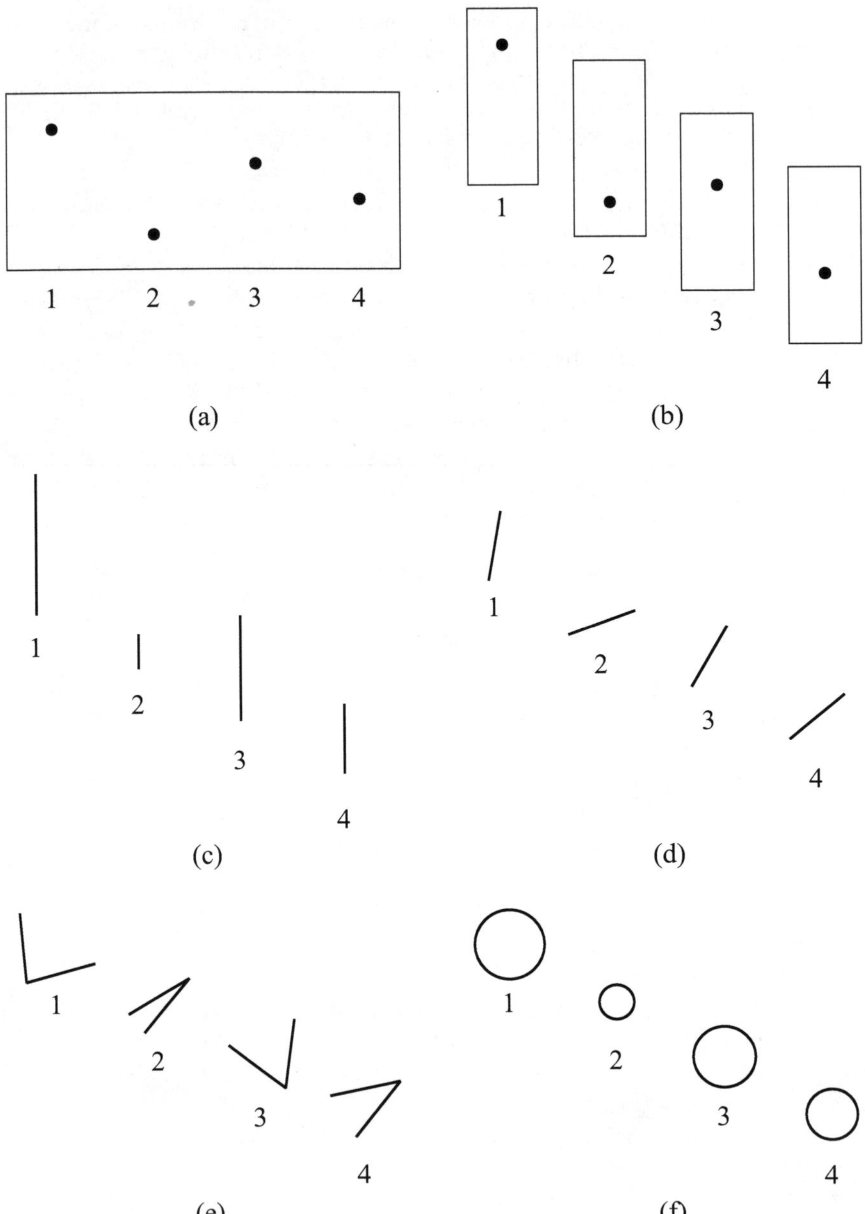


**FIGURE 7.2**
Stimuli used to evaluate performance at various elementary graphical perception tasks. (Adapted with permission from Cleveland, W. S. 1985. *The Elements of Graphing Data*. Belmont, CA: Wadsworth. All rights reserved.)

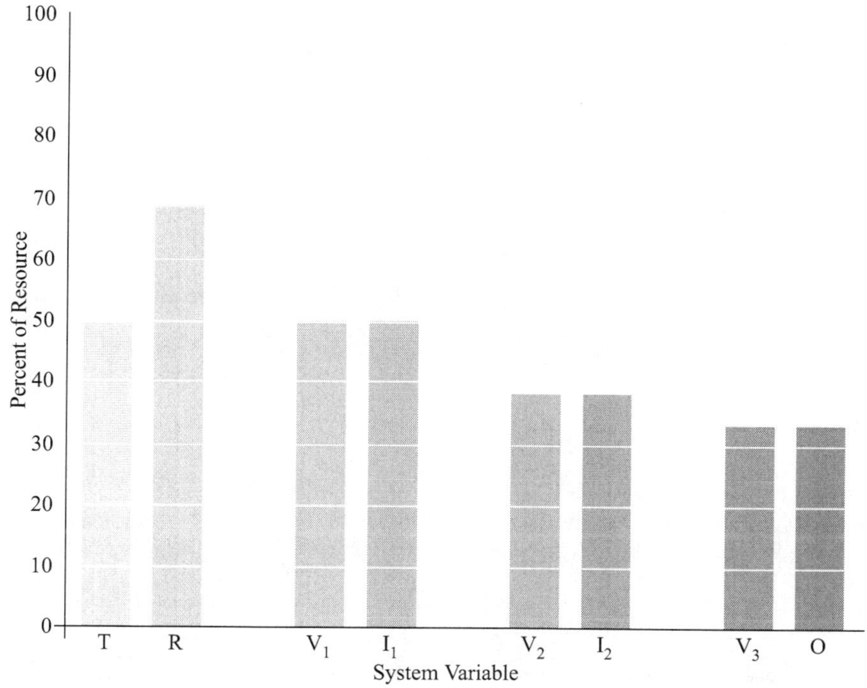


**FIGURE 7.3**
A bar graph display. It is a well-designed graphic because it involves the elementary graphical perception task that Cleveland identified (position along a common scale) and is also consistent with some of the aesthetic design principles suggested by Tufte. (Adapted with permission from Bennett, K. B., A. L. Nagy, and J. M. Flach. 1997. In *Handbook of Human Factors and Ergonomics,* ed. G. Salvendy. New York: John Wiley & Sons. All rights reserved.)

## 7.2.2 Limitations

Although this approach provides some useful insights, there are some limitations. Cleveland's work is valuable to consider when choosing elements of a display, but it fails to provide a framework for scaling up these elemental considerations to the design of more complex analogical representations. For example, the configural display in Figure 7.5 (see Chapter 10) conveys information through six elementary graphical perception tasks. The nested and hierarchical properties of complex work domains require correspondingly higher order visual properties in the virtual ecology (e.g., symmetry, parallelism, closure, good form, perpendicularity). These emergent properties are critical (see Chapters 8 through 11), but were not considered. Furthermore, the exclusive focus on static displays ignores changes that occur when displays are updated dynamically (see Chapter 8). Finally, this approach is explicitly dyadic (human <-> representation) and ignores the critical aspects of mapping with regard to

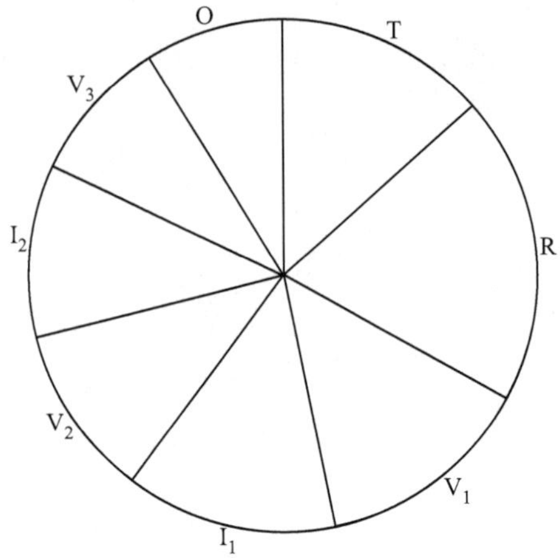


**FIGURE 7.4**
An example of a graph using a pie chart display format. It requires discriminations involving a less effective elementary graphical perception task than the bar graph. (Adapted with permission from Bennett, K. B., A. L. Nagy, and J. M. Flach. 1997. In *Handbook of Human Factors and Ergonomics,* ed. G. Salvendy. New York: John Wiley & Sons. All rights reserved.)

the underlying work domain, a shortcoming that is characteristic of the dyadic approach.

## 7.3 Aesthetic, Graphic Arts Approach

Tufte (e.g., 1983, 1990) approaches representation design from an aesthetic, graphic arts perspective. In direct contrast to the psychophysical approach just described, Tufte eschews empirical data in developing principles of design. Instead, Tufte relies heavily upon intuitive judgments about what constitutes effective (and ineffective) display design. Numerous examples of both good and bad display design are presented, compared, contrasted, and otherwise dissected to illustrate the associated design principles.

### 7.3.1 Quantitative Graphs

An early focus of this work (Tufte 1983) was on principles of design for graphs of quantitative data. One principle Tufte discusses is "data–ink ratio": a measurement technique to assess the relative amount of ink used in presenting data, as opposed to nondata elements, of a graph (a higher proportion of data ink is viewed as more effective). A second principle is "data density": the number of data points in the graphic divided by the total area (a higher data

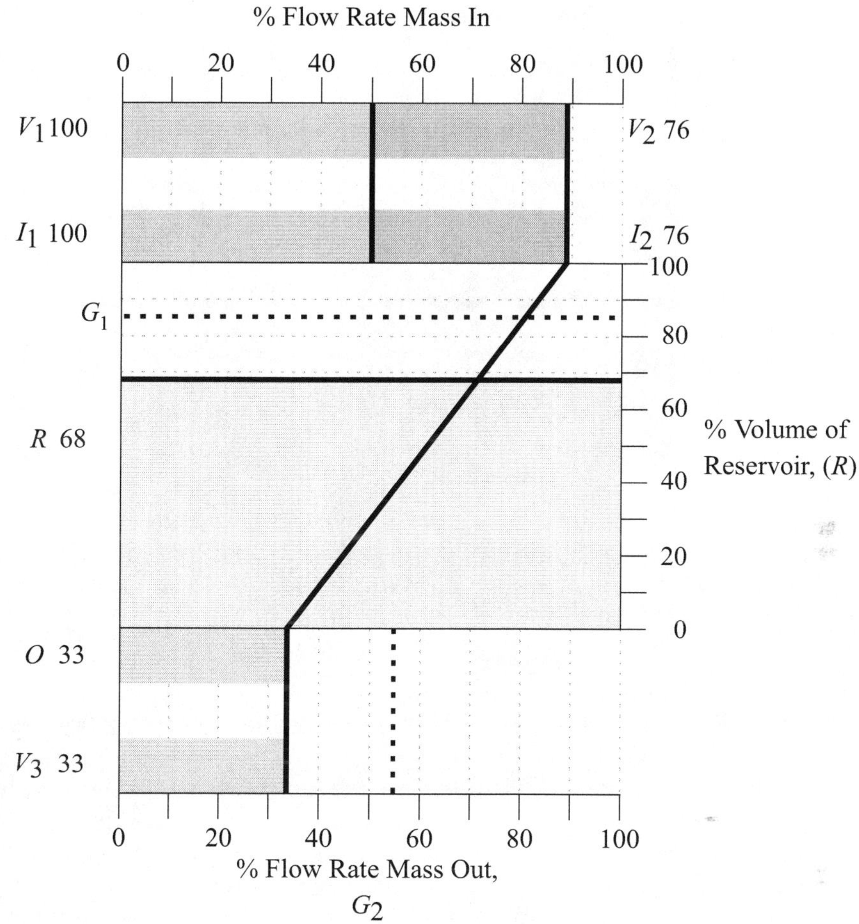


**FIGURE 7.5**
A display with multiple graphical perception tasks and multiple levels of layering and separation. (Adapted with permission from Bennett, K. B., A. L. Nagy, and J. M. Flach. 1997. In *Handbook of Human Factors and Ergonomics*, ed. G. Salvendy. New York: John Wiley & Sons. )

density is more effective). Other principles include eliminating graphical elements that interact (e.g., moire vibration), eliminating irrelevant graphical structures (e.g., containers and decorations), and other aesthetics (e.g., effective labels, proportion and scale).

Compare Figure 7.3 (a well-designed bar graph incorporating Tufte's principles) to Figure 7.6 (a poorly designed graph illustrating numerous violations of Tufte's principles). First, the striped patterns used on the bar graphs interact to produce an unsettling moire vibration. Second, a tremendous amount

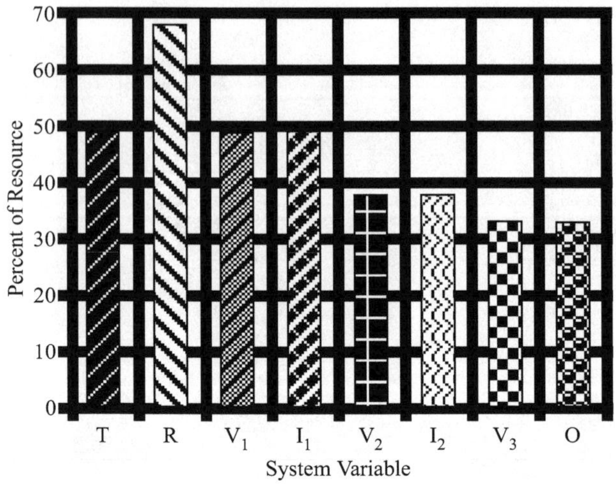


**FIGURE 7.6**
A poorly designed bar graph display. (Adapted with permission from Bennett, K. B., A. L. Nagy, and J. M. Flach. 1997. In *Handbook of Human Factors and Ergonomics*, ed. G. Salvendy. New York: John Wiley & Sons. All rights reserved.)

of ink is devoted to non-data graphical structures: An irrelevant data container (the box) imprisons the graph and the grid lines are heavy, bold, and conspicuous. Together, these two factors produce a very low data–ink ratio.

The bar graph illustrated in Figure 7.7 exacerbates these poor design choices through the incorporation of a third spatial dimension. Perspective geometry is used to introduce a number of visual features that are quite salient and difficult to ignore. Unfortunately, they are also essentially irrelevant, serving only to complicate visual comparisons. For example, the representations for variables that are plotted deeper in the depth plane are physically different from the representations positioned at the front. Thus, the physical area of the three-dimensional bar graph representing the variable O is approximately six times as large as the bar graph representing T, even though the percentage of resources for the latter (50%) is nearly twice the size of the former (33%).

The violations of Tufte's principles that are evident in Figure 7.7 relative to the well-designed version in Figure 7.3 will be made more explicit. The vacuous third dimension produces visual artifacts due to perspective geometry. This difference alone tremendously degrades the graph. Several visual properties serve to produce a lower data–ink ratio in the graph. First, the box around the graph is unnecessary and distracting. Second, the lines forming the X and Y axes of the display and the grid lines are unnecessarily bold. The location of the grid lines is still represented in Figure 7.3; however, they are made conspicuous by their absence.

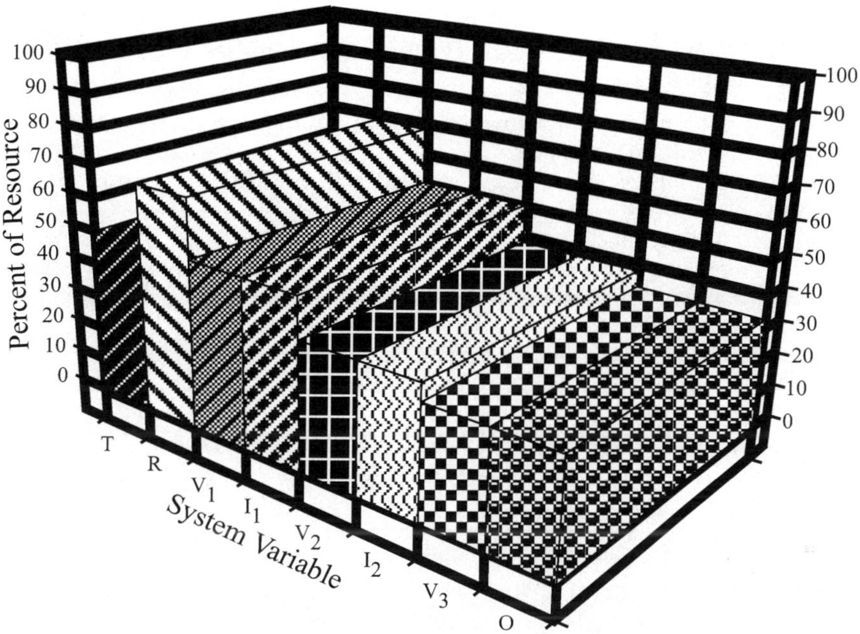


**FIGURE 7.7**
A bar graph display that has been made even less effective through the incorporation of a three-dimensional perspective.

There were several additional aesthetic and meaningful design violations. The patterns used to "paint" the bar graphs in Figures 7.6 and 7.7 are vibrant and distracting; those in Figure 7.3 are uniform shades of gray. This manipulation is far less exotic (or technologically sophisticated), but far more effective in conveying information. The bar graphs in Figure 7.3 have been visually segregated by nonuniform spacing between bar graphs. Thus, the functional relationship between pairs of related variables (e.g., $V_1$ and $I_1$) is graphically reinforced by placing the corresponding bar graphs close together. Finally, labels have been added to the axes in Figure 7.3 to assist the viewer in interpretation. The end product is a much more effective design.

### 7.3.2 Visualizing Information

Tufte (1990) broadens the scope of these principles and techniques by considering nonquantitative displays as well. Topics that are discussed include micro- and macrodesigns (the integration of global and local visual information), layering and separation (the visual stratification of different categories of information), small multiples (repetitive graphs that show the relationship between variables across time, or across a series of variables), color (appropriate and inappropriate use), and narratives of space and time (graphics that preserve or illustrate spatial relations or relationships over time). The following quotations summarize some of the key principles and observations:

> It is not how much information there is, but rather, how effectively it is arranged. (p. 50)
>
> Clutter and confusion are failures of design, not attributes of information. (p. 51)
>
> Detail cumulates into larger coherent structures .... Simplicity of reading derives from the context of detailed and complex information, properly arranged. A most unconventional design strategy is revealed: to clarify, add detail. (p. 37)
>
> Micro/macro designs enforce both local and global comparisons and, at the same time, avoid the disruption of context switching. All told, exactly what is needed for reasoning about information. (p. 50)
>
> Among the most powerful devices for reducing noise and enriching the content of displays is the technique of layering and separation, visually stratifying various aspects of the data .... What matters—inevitably, unrelentingly—is the proper relationship among information layers. These visual relationships must be in relevant proportion and in harmony to the substance of the ideas, evidence, and data displayed. (pp. 53–54)

This final principle, layering and separation, is graphically illustrated by the differences between Figures 7.5 and 7.8. These two figures present exactly the same underlying information; however, they vary widely in terms of the visual stratification of that information. In Figure 7.8 all of the graphical elements are at the same level of visual prominence; in Figure 7.5 there are several levels of visual prominence. Collections of mats are used to group visual elements together perceptually that are related to the functional structure of the underlying domain (at the level of views described in Chapter 6).

The lowest level of visual prominence in the display is associated with the nondata elements. The various display grids have thinner, dotted lines and their labels have been reduced in size and made thinner.

The individual variables in the display are represented with a higher level of perceptual salience. The graphical forms that represent each variable have been gray-scale coded, which contributes to separating these data from the nondata elements. Similarly, the lines representing the system goals ($G_1$ and $G_2$) have been made bolder and dashed. In addition, the labels and digital values that correspond to the individual variables are larger and bolder than their nondata counterparts.

The highest level of visual prominence has been reserved for graphical elements that represent higher level system properties (e.g., the bold lines that connect the bar graphs). The visual stratification could have been further enhanced through the use of color (e.g., for individual variables). The techniques of layering and separation will facilitate an observer's ability to locate and extract information.

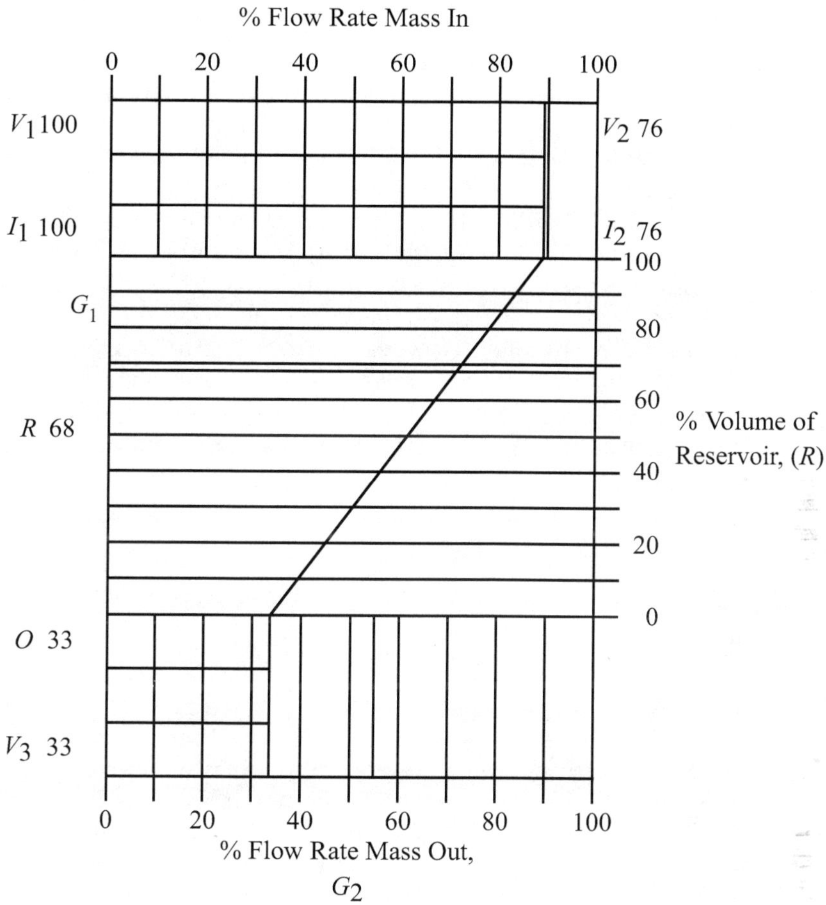


**FIGURE 7.8**
A display that has no layering and separation. (Adapted with permission from Bennett, K. B., A. L. Nagy, and J. M. Flach. 1997. In *Handbook of Human Factors and Ergonomics*, ed. G. Salvendy. New York: John Wiley & Sons. All rights reserved.)

Tufte's work is widely acclaimed, and rightfully so. His work illustrates the benefits of viewing issues in display design from a graphics arts, or aesthetic, perspective. Many of the design principles developed by Tufte relate to areas of research in the display and interface design literatures. The general principles and lucid examples provide important insights that can be applied in the design of displays for complex, dynamic domains. In many ways, Tuft's aesthetic approach to the design of printed materials represents a direct parallel to the efforts of Mullet and Sano (1995), who have applied a graphic arts perspective to interface design. Both of these perspectives are particularly relevant to the design of metaphorical representations for intent-driven domains (see Chapters 12 and 13).

### 7.3.3 Limitations

There are, however, some limitations. The work focuses on the design of static displays and therefore suffers the same set of limitations that were outlined earlier in the critique of Cleveland's work. For example, Tufte only briefly discusses the principle of "1 + 1 = 3" in the context of static representations. This is essentially the concept of emergent features, which turns out to be a critical consideration for the design of dynamic analogical displays (see Chapters 8 through 11). Perhaps the primary limitation of Tufte's work lies in the fact that it is descriptive, but not prescriptive. Thus, the focus is on analyzing how well a particular graph has conveyed meaningful aspects of a domain in an effective manner. However, very little guidance is provided with regard to what aspects of the domain semantics should be present in a representation in the first place. As described in Chapter 6 and illustrated throughout the remaining chapters in the book, this is the foundation of ecological interface design.

## 7.4 Visual Attention

A third perspective on display design is to consider the problem in terms of visual attention and perception. These concerns (e.g., dimensional structure of stimuli, emergent features, and form perception) play a critical role in our approach to analogical display design. Chapter 8 summarizes the basic literature and Chapter 9 compares and contrasts two approaches to display design that are grounded in this literature. Chapter 10 illustrates how these principles can be applied to the design of analogical displays for a simple process control system.

## 7.5 Naturalistic Decision Making

One of the primary purposes of displays in the interface is to provide a form of decision-making support. Therefore, another important perspective on display design comes from the literature on decision making. Recently, there has been an increased appreciation for the creativity and insight that experts bring to human–machine systems. Under normal operating conditions, an individual is perhaps best characterized as a decision maker. Depending on the perceived outcomes associated with different courses of action, the amount of evidence that a decision maker requires to choose a particular option will vary. In models of decision making, this is called a decision criterion.

Research in decision making has undergone a dramatic change in recent years. Historically, decision research has focused on developing models that describe how experts generated optimal decisions. Decision making was considered to be an extremely analytical process. Experts were viewed as considering all relevant dimensions of a problem, assigning differential weights to these dimensions, generating multiple alternatives (potentially all alternatives), evaluating (ranking) these alternatives, and selecting the optimal solution. Thus, cognitive processes were heavily emphasized while perceptual processes were essentially ignored.

### 7.5.1 Recognition-Primed Decisions

Recent developments in decision research, stimulated by research on naturalistic decision making (e.g., Klein, Orasanu, and Zsambok 1993), portray a very different picture of decision making. As suggested by their name, these approaches to decision making were developed from insights obtained from observing experts making decisions in natural work domains. One example of naturalistic decision making is the recognition-primed decisions (RPDs) described by Klein and his colleagues. RPD is described in the following manner (Klein 1989):

> We were surprised to find that the Fire Ground Commanders (FGCs) argued that they were not making any choices, considering alternatives, assessing probabilities, or performing trade-offs between the utilities of outcomes ....
>
> Instead the FGCs relied on their abilities to recognize and appropriately classify a situation. Once they knew it was "that" type of case, they usually also knew the typical way of reacting to it. They would use the available time to evaluate an option's feasibility before implementing it. *Imagery might be used to "watch" the option being implemented, to discover if anything important might go wrong.* (p. 49; emphasis added)

#### *7.5.1.1 Stages in RPD*

As suggested by the preceding quote, there are several stages in RPD (see Figure 7.9). The first stage is to recognize the problem (i.e., to categorize it as one of several different types of problems that have been encountered in the past). This is followed by a "situational assessment" phase, which involves establishing goals, looking for critical perceptual cues, developing expectancies about how upcoming events should unfold, and identifying typical actions that have proved successful for similar situations in the past. Before implementing a potential solution, experts will normally engage in a form of mental simulation where each step in the potential solution is checked for its potential to succeed or fail.

Naturalistic decision making emphasizes dynamic action constraints in decision making (see Figure 2.3 and the associated discussion). Contrary to the classical view of decision making, RPD views experts as satisficing, not optimizing.

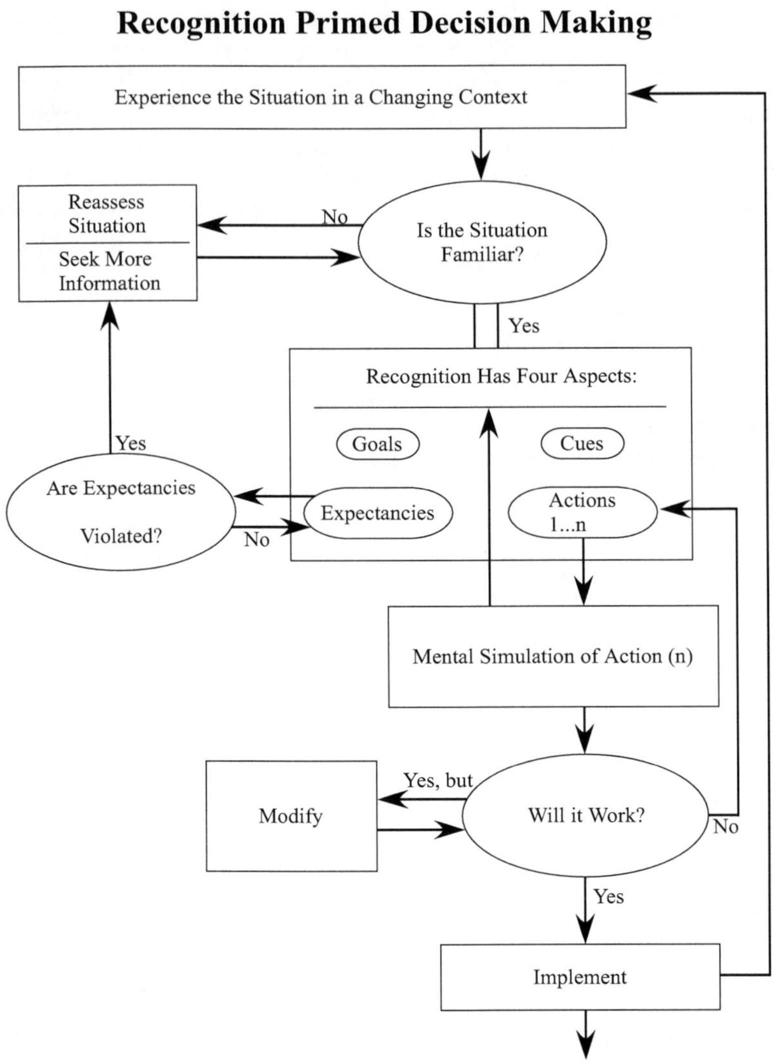


**FIGURE 7.9**
The recognition-primed decision making model. (Adapted with permission from Klein, G. A. 1989. *Advances in Man–Machine Systems Research* 5:47–92. JAI Press. All rights reserved.)

Their studies (e.g., Klein 1989) revealed that experts do not generate and evaluate all possible solutions; they generate and evaluate only a few good alternatives. Essentially, the experts are looking for the first solution that has a good chance of working and viable alternatives are normally considered in a serial fashion until one is found. Thus, this class of decision-making theories gives more consideration to the generation of alternatives in the context of dynamic demands for action.

#### *7.5.1.2 Implications for Computer-Mediated Decision Support*

A second change of emphasis is in the increased awareness of perceptual processes and the critical role that they play in decision making (as the term "recognition-primed decision" implies). Initially, experts utilize perceptual cues, in conjunction with their prior experience, to determine the prototypicality of a particular case (e.g., how is this case similar or dissimilar to those that I have encountered before?). Later, in the evaluation stage, visual or spatial reasoning often plays an important role in the form of mental imagery. For example, Klein (1989) observed that "deliberate imagery is found when new procedures are considered .... One fireground commander reported 'seeing' ... the effects if an exposure was to begin burning ... the FGC imagined flames coming out of a particular window and evaluated whether the flames would endanger certain equipment" (p. 60).

This new naturalistic perspective on decision making emphasizes the critical role that the interface will play when decision making must be computer mediated (i.e., when the constraints of a domain cannot normally be seen directly). Under these circumstances, designers must build interfaces that are rich with visual cues that provide external representations of the constraints in the domain. This will provide a set of perceptual cues that support the domain practitioner in (1) recognizing the type of situation currently encountered, (2) assessing details of the current situation that have implications for action, and (3) even suggesting the appropriate action to take (see Chapter 10 for an expanded discussion of these points).

Thus, domain practitioners will literally be able to see current system states and appropriate actions, rather than reasoning about them. In fact, this is the express goal of ecological interface design: to transform decision making from a cognitive activity (requiring limited capacity resources such as working memory) to a perceptual activity (using powerful and virtually unlimited perceptual resources). This ability is illustrated by the shortcuts in Rasmussen's decision ladder at the rule- and skill-based levels. These shortcuts reflect the smart, frugal heuristics that allow experts to "see" good options without exhaustive searches or intense cognitive computations (see Figures 4.3 and 5.2 and associated discussion).

## 7.6 Problem Solving

Another primary purpose of displays in the interface is to provide a form of problem-solving support. Therefore, another extremely important literature for display designers to consider is problem solving. We differentiate problem solving from decision making by virtue of the fact that problem solving involves novel situations (i.e., unanticipated variability) that domain practitioners have not encountered previously and for which preplanned guidance (training, procedures) has not been developed. Under these abnormal or unanticipated operating conditions, an individual is most appropriately characterized as a creative problem solver. The cause of the abnormality must be diagnosed, and steps must be taken to correct the abnormality (i.e., an appropriate course of action must be determined). This involves monitoring and controlling system resources, selecting between alternatives, revising diagnoses and goals, determining the validity of data, overriding automatic processes, and coordinating the activities of other individuals.

While it is certainly important to support decision-making performance with effectively designed graphical displays, it is perhaps even more important to support problem solving. Experience has shown that the stakes are higher during accidents: Poor performance during abnormal conditions can have far-reaching implications that go beyond the money, equipment, and lives that are lost. For example, consider the decades-long impact that the Three Mile Island and Chernoble accidents have had on the nuclear power industry.

The bottom line is that, during abnormal conditions, even domain experts will be placed in the role of a problem solver, the stakes will be high, and the interface should be designed to provide support under these circumstances. In essence, the interface provides a potential window into the deep structure of the problem that Rasmussen (1986) models using the abstraction hierarchy. The vast literature on problem solving provides some valuable insights, ranging from the seminal work of the Gestalt psychologists (e.g., Wertheimer 1959) and the paradigmatic contributions of Newell and Simon (1972) to contemporary approaches. We will concentrate on the contributions of the Gestalt psychologists.

### 7.6.1 Gestalt Perspective

The Gestalt psychologists produced an impressive body of work on problem solving. Much like theories of naturalistic decision making, this approach emphasized the importance of spatial reasoning in problem solving. Perception and cognition were believed to be intimately intertwined: to think is to perceive and to perceive is to think. For example, Wertheimer (1959) states that "thinking consists in envisaging, realizing structural features and structural requirements" (p. 235). Although this runs counter to the normal tendency of information processing psychologists to compartmentalize each nuance of cognitive activity (i.e., to add another box inside the head), it is a possibility that is actually reflected in common language (e.g., "I see what you mean"; "a light-bulb went off inside my head"; "that was a very illuminating talk").

#### *7.6.1.1 Gestalts and Problem Solving*

The interrelatedness of perception and problem solving and the critical role of representation were perhaps most evident in the title with which the

Gestalt psychologists chose to refer to themselves. The term *gestalt* translates roughly into form, shape, or configuration. However, this term, essentially perceptual in nature, was used to describe the end result of both thinking and perceiving. The Gestalt psychologists viewed the act of problem solving as the process of achieving a proper gestalt. When faced with a novel problem, the problem solver is likely to have a poor understanding of it (i.e., an incomplete or bad gestalt, or representation).

The key to successful problem solving was viewed as the transformation of this bad gestalt into an appropriate or good gestalt. This good gestalt would reveal the structural truths of a problem and allow the invention of a novel solution or the application of a previously successful solution. Thus, problem solving was viewed very much in perceptual terms: It required seeing a problem from a different perspective or representing the problem in a different way.

Wertheimer's work on geometry (1959) will be used to illustrate some of these ideas more concretely. Consider a student who is learning how to calculate the area of various geometrical forms. The student has learned that multiplying the length of one side of a rectangle ($S_1 = 6$) by the length of the second side ($S_2 = 3$) provides the area of a rectangle (see Figure 7.10a).

The student is then presented with a new problem: to find the area of a parallelogram (Figure 7.10b). Wertheimer (1959) outlined two different approaches or types of responses that could be applied by a student faced with a new problem. The student could blindly apply the rules for a rectangle to this new problem. This rote application of inappropriate rules in this case would, of course, result in the calculation of an area that is too large for the parallelogram, as illustrated in Figure 7.10b.

The alternative, and more appropriate, response is to modify the procedures in a sensible fashion. In this case the student must consider the similarities and differences between the constraints associated with the old and new problems and then modify the procedures accordingly. The student begins with a good gestalt for finding the area of a rectangle and understands that finding the area of a rectangle essentially involves finding the number of square units that fit inside that particular form (see Figure 7.10c). The gestalt is a good one, primarily because the shape of the units used to do the measuring is consistent with the general shape of the form to be measured (i.e., the squares fit nicely into the rectangle, with no partial or overlapping squares). This, however, is not a particularly good gestalt for the parallelogram because there are partial squares at the corners that the student most likely has no way of dealing with (see Figure 7.10d).

#### *7.6.1.2 Problem Solving as Transformation of Gestalt*

As mentioned previously, the Gestalt psychologists believed that the sensible modification of procedures often involved the transformation of a bad gestalt into a good gestalt. In this case the student seeks a way to transform

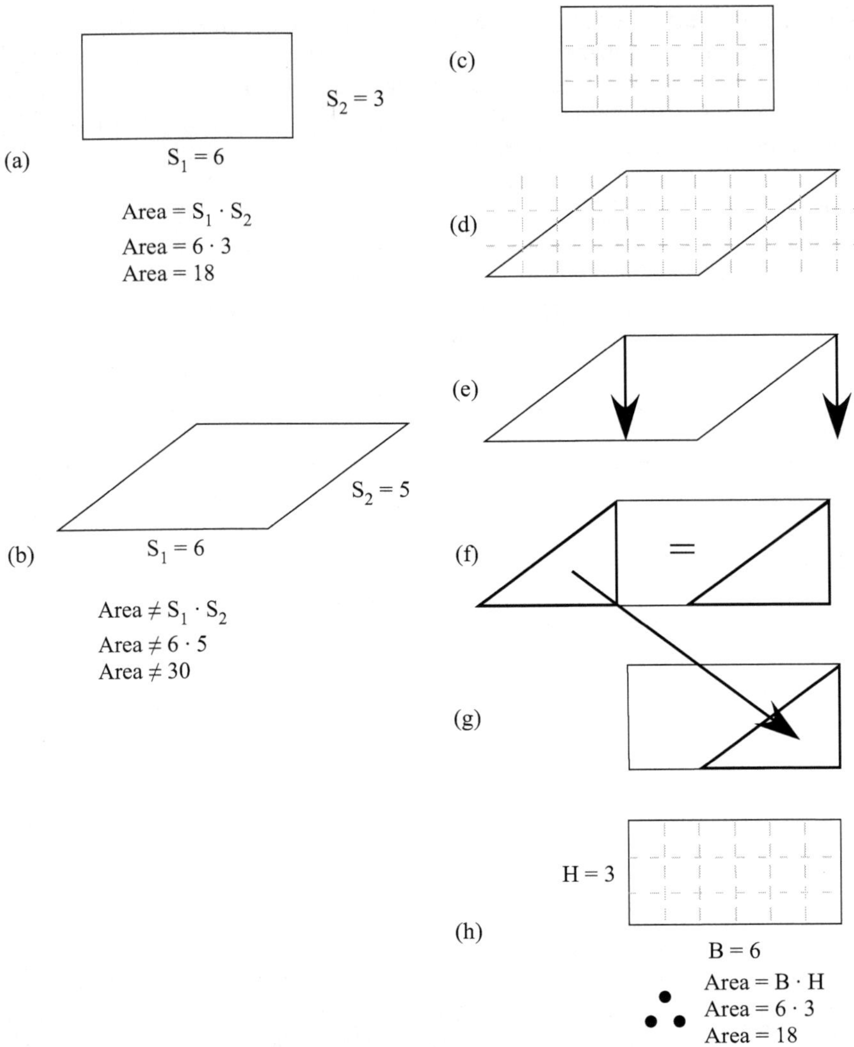


**FIGURE 7.10**
A graphic illustration of the sensible adaptation of a problem-solving procedure from the Gestalt perspective. The learner solved the problem (finding the area of a parallelogram) by transforming a bad gestalt into a good gestalt (i.e., transforming the parallelogram's initial form into an equivalent rectangular form).

the new form (the parallelogram) into one to which he or she can apply the successful procedures that have already been learned (a rectangle). The student might consider alternative ways to achieve this goal; one of them is to drop perpendicular lines from the top-most edges of the parallelogram (see Figure 7.10e). Because the opposite sides of a parallelogram are parallel, the sizes of the two triangles thus formed are equal (see Figure 7.10f). A rectangle form with equivalent area to the original parallelogram is completed by replacing the triangle on the right with the triangle on the left (see Figure 7.10g).

The student has now transformed the bad gestalt (a parallelogram) into a good gestalt (a rectangle) and can apply the previously learned procedures to arrive at the area. This student now understands why the area of a parallelogram must be calculated using the formula of base times height, as opposed to multiplying the two sides (see Figure 7.10h).

### 7.6.2 The Power of Representations

This example suggests that the way in which a problem solver represents the problem plays a key role in its solution. In fact, this is perhaps the primary lesson to be learned from the problem-solving literature: The representation of a problem can have a profound influence on the ease or difficulty of its solution. To illustrate this point clearly, consider the following problem (Duncker 1945):

> On a mountain trip, on which descent was by the same path as had been the ascent of the previous day, I asked myself whether there must be a spot en route at which I must find myself at exactly the same time on the descent as on the ascent. It was of course assumed that ascent and descent took place at about the same time of day, say from five to twelve o'clock. But without further probing, I could arrive at no conclusive insight. Since then, I have put this question to dozens of naive persons as well as of intellectuals, and have observed with great satisfaction that others had the same difficulty. Let the reader himself ponder a bit. (p. 56)

Through the years, this basic problem has been refined and presented in a way that makes the critical elements clearer (e.g., see Halpern 1996) *and it is now generally referred to as the problem of the "monk's pilgrimage":*

> Once there was a monk who lived in a monastery at the foot of a mountain. Every year the monk made a pilgrimage to the top of the mountain to fast and pray. He would start out on the mountain path at 6 a.m., climbing and resting as the spirit struck him, but making sure that he reached the shrine at exactly 6 o'clock that evening. He then prayed and fasted all night. At exactly 6 a.m. the next day, he began to descend the mountain path, resting here and there along the way, but making sure that he reached his monastery again by 6 p.m. of that day.
>
> The problem is to prove, or disprove, the following statement: "Every time the monk makes his pilgrimage there is always some point on the mountain path, perhaps different on each trip, that the monk passes at the same time when he is climbing up as when he is climbing down." (Halpern, p. 329)

The first step in solving a problem from the Gestalt perspective is to find the structural truths of a problem. Consistent with the cognitive systems engineering approach, we will refer to these as the constraints of a problem. In this case the constraints are related to both time and space. The spatial constraints are those associated with the mountain path and the fact that it (not any other path) is traversed not once, but twice. These constraints could be considered from a birds-eye perspective that illustrates the twists and turns of the path. However, it is more appropriate to think about these spatial constraints in terms of the elevation of the path (i.e., the number of feet from the base of the mountain to the top of the mountain). There are also temporal constraints associated with the problem: Each of the two trips (i.e., one up and one down) began at the same time (6 a.m.) and was completed at the same time (6 p.m.); they were completed on successive days.

Once the constraints of a problem are understood, the next step is to translate them into a proper gestalt (i.e., graphical representation). Consider Figure 7.11. The spatial and temporal constraints are represented on the two axes of the graph. Each of the two lines represents a trip taken by the monk (dashed line = day 1; solid line= day 2). This graph makes it absolutely, positively clear that the assertion has been proved. The vagaries of "climbing and resting here and there along the way" have absolutely no consequence for the final answer: There must be an exact time at which the monk is at the exact same point in the path. Note that the time and place cannot be predicted beforehand, but there will be one. The time and place can be identified after the fact, given the representation that is used in Figure 7.11 (i.e., the point where the two lines intersect). Thus, finding a solution to a problem can be a trivial exercise if the representation portrays the constraints effectively.

It should be pointed out that this may not be the most direct or the preferred solution to the problem. Duncker describes his preferred solution in the following fashion (1945):

> Certainly there exist several approaches to an evident solution. But probably none is, I might almost say, more drastically, evident than the following view of the situation: let ascent and descent be divided between two persons on the same day. They must meet. Ergo ... with this, from an unclear, dim condition not easily surveyable, the situation has suddenly been brought into full daylight. The answer becomes evident, inspection sufficiently exact. (p. 56)

### 7.6.3 Functional Fixedness

The previous section describes how appropriate representations (i.e., those that accurately reflect the constraints of a problem domain) can improve problem-solving performance. We now consider a variation of this theme: Duncker's original work on the problem of "functional fixation." Five different versions

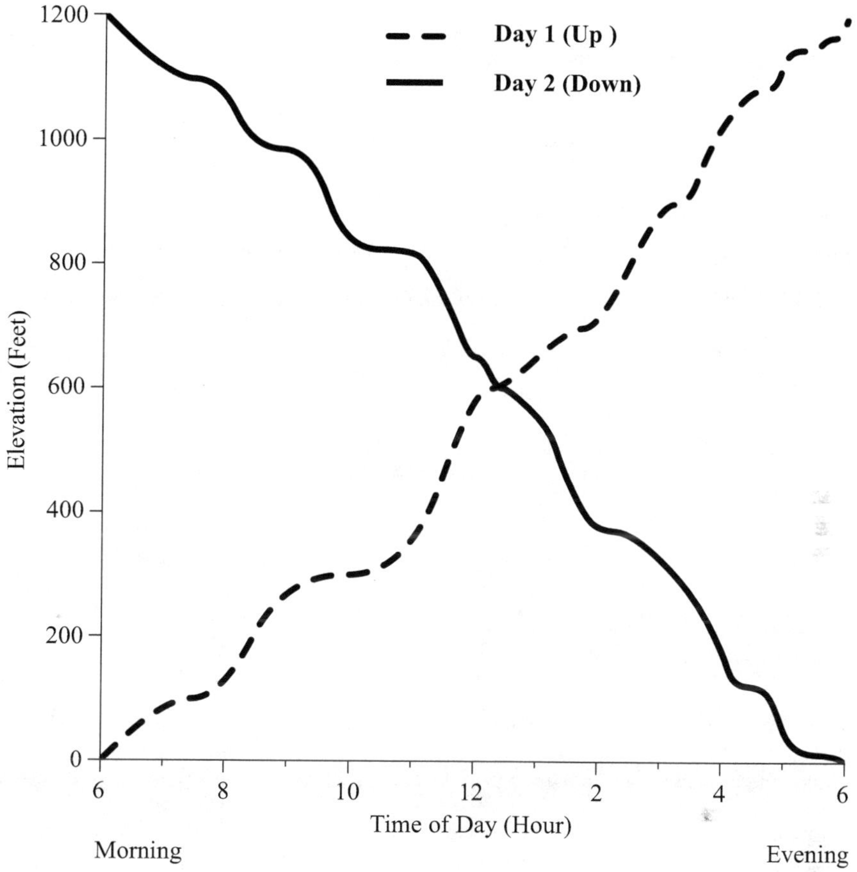


**FIGURE 7.11**
A visual solution of the monk's pilgrimage problem (see text for details of this problem).

of the basic problem were investigated: gimlet, box, pliers, weight, and paper clip. The details of perhaps the most famous version, the box problem, will be provided to make the discussion concrete (Duncker 1945):

> The "box problem": On the door, at the height of the eyes, three small candles are to be put side by side ("for visual experiments"). On the table lie, among many other objects, a few tacks and the crucial objects: three little pasteboard boxes (about the size of an ordinary matchbox, differing somewhat in form and color and put in different places). (p. 86)

A description of the problem (e.g., to attach three candles to a door at eye height) is given to the problem solver and a variety of items for potential use in its solution are placed on a table (see Figure 7.12). Some of these items

**FIGURE 7.12**
The box problem in the "w.p." (without pre-utilization) mode of presentation. The critical items for the novel solution (i.e., the matchboxes) are not presented in a way that reflects their normal functionality (i.e., the boxes are not being used as containers) and functional fixedness does not occur. (Copyright 2009, Dalton Bennett.)

are irrelevant to the solution (e.g., paper clips, pencils, tinfoil, ashtrays) while others are relevant (e.g., matches, matchboxes, candles, and tacks). The solution requires that the problem solver discover a novel functionality for a critical item (i.e., a use that goes beyond the normal functionality of the item). Duncker referred to the normal functionality of an item as "$F_1$" and a novel functionality as "$F_2$". Thus, in the case of the box version, the problem solver must see beyond the normal functionality of boxes as containers ($F_1$) and discover their potential as a mounting platform ($F_2$): The solution involves tacking the boxes to the wall and mounting the candles inside the boxes.

The key experimental manipulation lies in the way in which the problem was presented. There were two different types of presentation that varied on a dimension referred to as "pre-utilization." In one presentation type, the critical items are presented to the problem solver in ways that are consistent with their normal functionality. For example, the tacks, candles, and matches are initially located inside the boxes. This functionality (i.e.,

$F_1$, container) is consistent with normal use and is referred to as the "a.p." (after pre-utilization) condition. In the second type of presentation mode (i.e., the presentation mode depicted in Figure 7.12), the critical items are not initially shown fulfilling their normal functionality (i.e., the boxes are empty). Duncker refers to this as the "w.p." (without pre-utilization) condition (Duncker 1945):

> Solution: with a tack apiece, the three boxes are fastened to the door, each to serve as platform for a candle. In the setting a.p., the three boxes were filled with experimental material: in one there were several thin little candles, tacks in another, and matches in the third. In w.p., the three boxes were empty. Thus $F_1$: "container"; $F_2$: "platform" (on which to set things). (p. 86)

The results indicate that presentation mode had a major impact on performance. Participants were able to solve problems effectively in the w.p. condition (see Figure 7.12), when critical items were not initially depicted in terms of their prototypical use (e.g., the boxes were initially shown as empty). All of the participants (seven out of seven) were able to solve the box problem; overall completion rate was 97% across all five problems. Thus, problem solvers were very effective in discovering novel functionalities for the critical items when they were presented in this mode.

In sharp contrast, initially presenting critical items in a way that was consistent with their normal usage (e.g., boxes filled with tacks, candles, and matches) dramatically lowered performance. For example, less than half of the box problems in the a.p. condition (three out of seven) were solved successfully. Across all five problems, only slightly more than half (58%) of the problems were completed successfully. Duncker's interpretation of these results was that problem solvers became fixated on the normal functionality of critical items ($F_1$, e.g., a container to hold items), which made it more difficult to envision or discover the alternative functionalities needed for successful problem solution (e.g., $F_2$, a platform to set things on).

The Gestalt psychologists broadened the concept of functional fixedness substantially. In a general sense, functional fixedness is the tendency of a problem solver to view a problem in one particular way, to the exclusion of others. This general tendency was referred to as "perceptual set." The Gestalt psychologists also investigated a reciprocal tendency—that of responding in a particular way (at the expense of others). Perhaps the most well-known empirical demonstration of this tendency is the water jar problems of Luchins (1942). These studies demonstrate that problem solvers will tend to respond in ways that have proven to be successful in the past, even when there are alternative solutions that are more efficient. This tendency has been referred to as "response set."

These findings clearly generalize beyond the confines of the experimental laboratory. Various forms of functional fixedness have been observed in today's sociotechnical systems. The very nature of the problem to be solved can change over time, and domain practitioners often ignore disconfirming evidence and fail to revise their original hypotheses (Woods 1988). This is closely related to "tunnel vision" (Moray 1981); these forms of perceptual set are usually associated with unfortunate consequences. Response set occurs when domain practitioners respond in stereotypical ways that are at best inefficient and at worst inappropriate. We will explore the possibility that effective graphical displays will serve to decrease the potential for functional fixedness in today's sociotechnical systems in later chapters.

### 7.6.4 The Double-Edged Sword of Representations

> [T]he notion that one can capture some aspect of reality by making a description of it using a symbol and that to do so can be useful seems to me a fascinating and powerful idea. But ... there is a trade-off; any particular representation makes certain information explicit at the expense of information that is pushed into the background and may be quite hard to recover. (Marr 1982, p. 21)

Earlier we made the assertion that the fundamental lesson from the problem-solving literature is that the representation of a problem can have a profound influence on the ease or difficulty of its solution. The graphical solution to the monk problem demonstrates the positive potential for graphical displays. However, as Marr's quote emphasizes, graphical displays are a double-edged sword. For the same reasons that a good representation can make a problem easy to solve, a poor representation can make a problem much more difficult to solve. For example, Duncker's demonstrations of functional fixedness provide an example illustrating the negative impact of inappropriate representations: the primary experimental manipulation (i.e., presentation of full or empty boxes) is ultimately a change in the representation of the problem.

Numerous examples in the problem-solving and display design literature could be used to illustrate the potential downside associated with poor representations. For example, Tufte (1990) provides many examples of poor static displays. The power of today's graphics technology provides system designers with the capability to build displays that fail in ways that are even more spectacular. Mullet and Sano (1995) provide numerous examples of how good intentions can go quite bad; several more subtle examples will be examined in greater detail in subsequent chapters. The difficulty in providing appropriate representations arises from the fact that we are dealing with semiotics, as discussed previously.

This is an important point of emphasis. Graphic displays only provide representations of a domain; they are not the objects, properties, or constraints themselves. An infinite number of different representations could be devised for the same referent in the domain. The utility of each representation will depend upon specific sets of mappings between its visual features, the observer's capability to perceive them, and the tasks to be performed. Woods (1991) echoes this important point: "There are no a priori neutral representations" (p. 175). In closing, consider the interesting possibility that poor choices in representation could have impeded the intellectual progress of an entire culture (Marr 1982):

> [I]t is easy to add, to subtract, and even to multiply if the Arabic or binary representations are used, but it is not at all easy to do these things—especially multiplication—with Roman numerals. This is a key reason why the Roman culture failed to develop mathematics in the way the earlier Arabic cultures had. (p. 21)

In summary, the Gestalt psychologists and researchers in naturalistic decision making (e.g., RPD) have highlighted the critical role of visual reasoning and problem representation in decision making and problem solving. These trends have, either directly or indirectly, led researchers in interface design to focus on the representation problem. Perhaps the first explicit realization of the power of graphic displays to facilitate understanding was the STEAMER project (Hollan, Hutchins, and Weitzman 1984, 1987; Hollan et al. 1987), an interactive, inspectable training system. STEAMER provided alternative conceptual perspectives—"conceptual fidelity" of a propulsion engineering system through the use of analogical representations. In addition, earlier approaches to the design of human–computer interfaces (Hutchins, Hollan, and Norman 1986; Shneiderman 1983) can be viewed as an outgrowth of this general approach that has adopted metaphorical representations.

More recently, scientific visualization (the role of diagrams and representation in discovery and invention) is being vigorously investigated (e.g., Brodie et al. 1992; Earnshaw and Wiseman 1992). We view cognitive systems engineering and ecological interface design as the culmination of these trends. The goal is to provide practitioners with the perceptual cues about the domain that are required to support these experts in their decision-making and problem-solving activities.

## 7.7 Summary

This chapter introduced the topic of display design, the wide variety of complex issues that need to be addressed, and many key concepts (both applied and theoretical). The focus, however, was to describe several alternative approaches to display design that have been developed. These approaches were chosen because they contribute valuable insights to effective display design. However, each of these approaches is incomplete. Some fail to consider the additional complexities that become relevant when a display is animated (e.g., visual interactions between graphical elements). Some focus on basic elements of display design that are difficult to translate into the complex graphics that are required for complex dynamic domains. Almost all of these approaches are not triadic: There is a lack of concern for the meaning behind the display. The tasks to be performed during evaluation are usually defined by the visual characteristics of the display itself, rather than the semantics of a domain that the display has been designed to represent.

However, the mapping between the visual structure in a representation and the constraints of the problem domain is fundamental to a triadic perspective, as will be shown in subsequent chapters. While Gestalt concepts like "deep structure" or "structural understanding" suggest the importance of links to the problem domain or ecology, there is little discussion of the problem independently from its representation. For ecological interface design, a primary function of the abstraction hierarchy is to help guide the search to uncover the deep structure in the problem domain that in turn might be graphically illustrated using some of the techniques described in this chapter.

## References

Bennett, K. B., A. L. Nagy, and J. M. Flach. 1997. Visual displays. In *Handbook of human factors and ergonomics,* ed. G. Salvendy. New York: John Wiley & Sons.

Brodie, K. W., L. A. Carpenter, R. A. Earnshaw, J. R. Gallop, R. J. Hubbold, A. M. Mumford, C. D. Osland, and P. Quarendon, eds. 1992. *Scientific visualization: Techniques and applications.* Berlin: Springer–Verlag.

Cleveland, W. S. 1985. *The elements of graphing data.* Belmont, CA: Wadsworth.

Cleveland, W. S., and R. McGill. 1984. Graphical perception: Theory, experimentation, and application to the development of graphical methods. *Journal of the American Statistical Association* 79:531–554.

———. 1985. Graphical perception and graphical methods for analyzing scientific data. *Science* 229:828–833.

Duncker, K. 1945. On problem-solving. *Psychological Monographs* 58 (5): 1–113.

Earnshaw, R. A., and N. Wiseman, eds. 1992. *An introductory guide to scientific visualization.* Berlin: Springer–Verlag.

Halpern, D. F. 1996. *Thought and knowledge: An introduction to critical thinking.* Mahwah, NJ: Lawrence Erlbaum Associates.

Hollan, J. D., E. L. Hutchins, T. P. McCandless, M. Rosenstein, and L. Weitzman. 1987. Graphical interfaces for simulation. In *Advances in man–machine systems research,* ed. W. B. Rouse. Greenwich, CT: JAI Press.

Hollan, J. D., E. L. Hutchins, and L. Weitzman. 1984. STEAMER: An interactive inspectable simulation-based training system. *The AI Magazine* 5 (2): 15–27.

———. 1987. Steamer: An interactive inspectable simulation-based training system. In *Artificial intelligence and instruction: Applications and methods,* ed. G. Kearsley. Reading, MA: Addison–Wesley.

Hutchins, E. L., J. D. Hollan, and D. A. Norman. 1986. Direct manipulation interfaces. In *User centered system design,* ed. D. A. Norman and S. W. Draper. Hillsdale, NJ: Lawrence Erlbaum Associates.

Jakab, P. L. 1990. *Visions of a flying machine: The Wright brothers and the process of invention.* Washington, D.C.: The Smithsonian Institution Press.

Klein, G. A. 1989. Recognition-primed decisions. *Advances in Man–Machine Systems Research* 5:47–92.

Klein, G. A., J. Orasanu, and C. E. Zsambok, eds. 1993. *Decision making in action: Models and methods.* Norwood, NJ: Ablex Publishing Corp.

Luchins, A. S. 1942. Mechanization in problem solving: The effect of Einstellung. *Psychological Monographs* 54 (248): 1–95.

Marr, D. 1982. *Vision.* San Francisco, CA: W. H. Freeman and Company.

Moray, N. 1981. The role of attention in the detection of errors and the diagnosis of failures in man–machine systems. In *Human detection and diagnosis of system failures,* ed. J. Rasmussen and W. B. Rouse. New York: Plenum Press.

Mullet, K., and D. Sano. 1995. *Designing visual interfaces: Communication oriented techniques.* Englewood Cliffs, NJ: SunSoft Press.

Newell, A., and H. A. Simon. 1972. *Human problem solving.* Englewood Cliffs, NJ: Prentice Hall.

Rasmussen, J. 1986. *Information processing and human–machine interaction: An approach to cognitive engineering.* New York: Elsevier.

Shneiderman, B. 1983. Direct manipulation—A step beyond programming languages. *Computer* 16 (8): 57–69.

Tufte, E. R. 1983. *The visual display of quantitative information.* Cheshire, CT: Graphics Press.

———. 1990. *Envisioning information.* Cheshire, CT: Graphics Press.

Wertheimer, M. 1959. *Productive thinking.* New York: Harper and Row.

Woods, D.D. 1988. Coping with complexity: The psychology of human behavior in complex systems. In *Mental models, tasks and errors: A collection of essays to celebrate Jens Rasmussen's 60th birthday,* ed. L. P. Goodstein, H. B. Andersen, and S. E. Olsen. New York: Taylor & Francis.

———. 1991. The cognitive engineering of problem representations. In *Human–computer interaction and complex systems,* ed. G. R. S. Weir and J. L. Alty. London: Academic Press.

# 8

# *Visual Attention and Form Perception*

## 8.1 Introduction

> My thesis today is simply that too many of the experiments, interpretations, and concepts that have been used in studies of information processing have emphasized the processing part of the problem to the neglect of the information part of it .... In preparing this talk, I reread, with the usual pleasure, James Gibson's (1960) presidential address to the Eastern Psychological Association on the concept of the stimulus in psychology ... we need experiments and concepts pertinent to the input justas we need them pertinent to the processing .... We would never consider drawing conclusions from an experiment based on a single subject ... Why, then, are we apparently so happy drawing sweeping conclusions about how information is processed when we have used only one kind of stimulus? I ... want to argue that we become equally concerned about the nature of the information input. (Garner 1970, p. 350)

In Chapter 6 we described the process of building virtual ecologies and outlined general design strategies that are based on two fundamental representational formats (analogies, metaphors) that constitute the building blocks of interface design. In Chapter 7 we described a variety of alternative perspectives that are relevant to the design of these representational forms. In this chapter we narrow our focus to the consideration of one perspective that we have found particularly useful in developing principles of design for the analog geometric forms needed for law-driven domains. That perspective is visual attention and form perception; we will review the concepts that have been developed, the methodologies used to investigate them, the results of some pivotal experimental studies, and alternative interpretations of those results.

We have previously described the hierarchically nested analogical display geometries required to support direct perception by a cognitive agent in law-driven domains (see Figure 6.5 in Chapter 6 and the associated discussion). The literature on visual attention and form perception has provided insights into the basic issues that are relevant to the perception of these nested geometries. What are the fundamental units of perception? What are the basic types of visual information available? What are the relationships between these types of information? How do parts group into wholes? Is the perception of the parts of a form secondary to the perception of the whole or vice-versa? What constitutes visual attention and how can it be distributed over parts and wholes? The answers to these questions are relevant to principles of display design because visual displays need to provide different types of information and to support a range of activities.

One of the fundamental issues in display design can be conceptualized in terms of a continuum of demands that are placed on an observer. At one end of the continuum are tasks that require selective responses to specific elements in the visual field. This type of task will be referred to as a "focused" task. An observer in a visual attention study might be required to attend selectively to a part of the visual array; an observer in a display design study might be required to consider the value of a single sensor. An example of a focused task is to check the speedometer (a single measured variable) when a state trooper is detected on the highway.

At the opposite end of this continuum are tasks that require a consideration of the relationships among several variables. An observer in a visual attention study might be required to divide attention across more than one element of the visual array. An observer in a display design study might need to consider higher order properties that are derived from the relationships between variables. An example of a divided attention task is the decision to brake or to change lanes when the car in front of you slows down—a decision that requires the consideration of a number of variables.

It is clear that domain practitioners who use visual displays will need to complete successfully a range of tasks requiring the consideration of both parts and wholes of visual forms. The question for visual display design is whether or not different specialized display formats are required for each category of tasks (i.e., one for focused tasks, another for divided tasks) or whether a single display format has the capability to support both types of tasks. The present chapter will organize the concepts and findings from the visual attention and form perception literature that are relevant to this critical issue in design.

## 8.2 Experimental Tasks and Representative Results

Concrete examples from experiments that have been conducted in the literature on visual attention and form perception are described in this introductory section. This description includes representative examples of the visual stimuli that were evaluated, the experimental methodologies that were employed, and the experimental tasks that observers were required to perform. We describe the results of a representative study at the end of the section. These concrete examples provide a context in which to consider alternative perspectives on

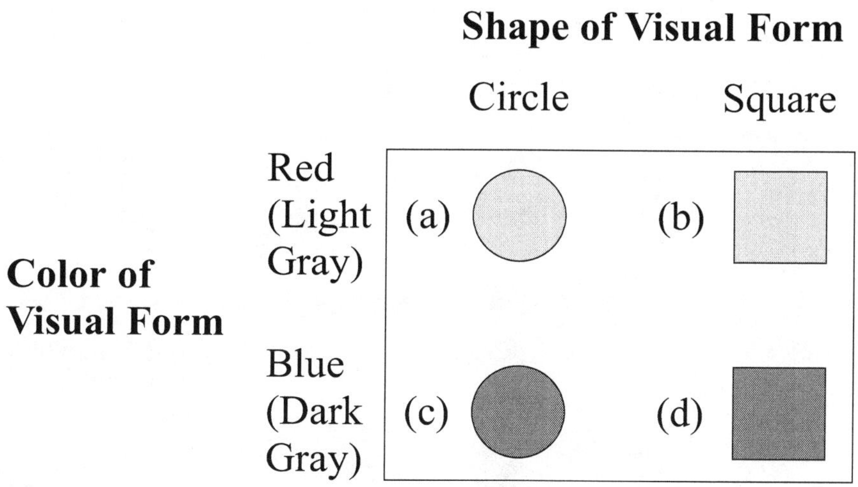


**FIGURE 8.1**
A typical set of four stimuli for speeded classification in a visual attention experiment. The $2 \times 2$ matrix is formed by a factorial combination of two stimulus dimensions (shape and color) and two values for each dimension (i.e., circle vs. square; red [light gray] vs. blue [dark gray]).

visual attention and form perception. This is useful since these different perspectives have provided the basis for alternative principles of display design and because some of these principles are conflicting.

The "speeded classification" task is an important and representative methodology that has a long history in the literature. This methodology typically involves the presentation of four different visual stimuli. These four stimuli are produced by a factorial combination of two perceptual dimensions (e.g., color and form) and two values on each dimension (e.g., square vs. circle; red vs. blue), as illustrated in Figure 8.1. (Note that we have used shades of gray instead of colors in the examples to facilitate discussion.)

Participants are presented with just one of these four visual forms in an individual trial. They are required to consider a visual property of the stimulus (e.g., its color) and to produce one of two responses (i.e., "1" or "2"). For example, a "1" response might be required when the color of the stimulus was light gray (and a "2" response for the dark gray color). The primary measure of performance is response time. The specific visual property that must be considered (or ignored) and the type of task that must be performed are systematically varied across four different experimental conditions. Each of these conditions will be examined in turn.

### 8.2.1 Control Condition

The control condition assesses the capability of an observer to discriminate changes in one perceptual dimension. There are four versions of the

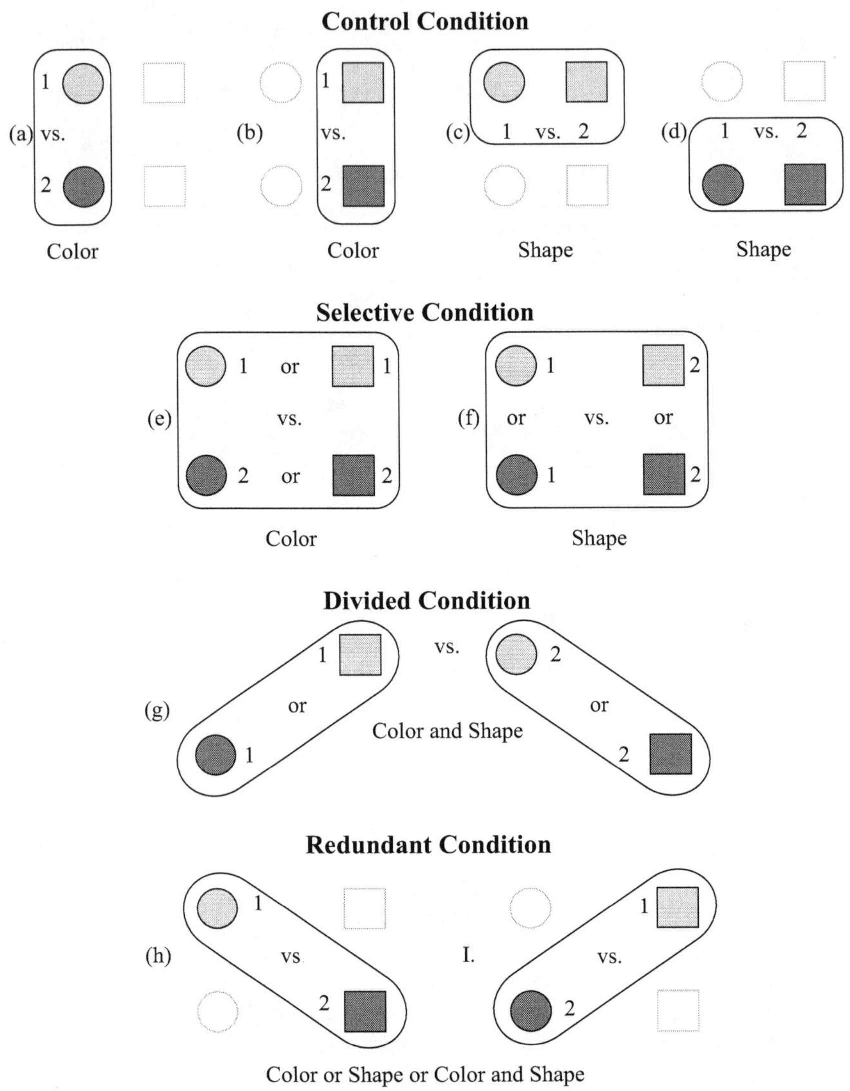


**FIGURE 8.2**
Four different experimental conditions that are used to produce different information extraction demands in a typical visual attention study.

control condition, as illustrated at the top of Figure 8.2 (labeled A, B, C, and D). Consider condition A. An observer would see only two of the four visual forms (in this case, the light gray circle and dark gray circle) during a block of trials in a condition. As illustrated in Figure 8.2a, these two forms are filled and are enclosed by the outlined form (the two unused forms are only outlined). The observer is presented with an individual visual form during a trial and is asked to make a discrimination on the relevant perceptual dimension. Thus, in condition A the correct response for a light gray circle is "1" and the correct response for a dark gray circle is "2" (the correct response appears to the left of the visual form). Similarly, the color of the square must be considered in condition B while shape is critical for conditions C and D.

In summary, in the control condition, decisions are contingent on variations in one perceptual dimension (e.g., color), while variations in the other perceptual dimension (e.g., shape) are held constant. This control condition qualifies as a focused task since only one dimension is relevant to the decision. An important point to note is that the value of the unattended perceptual dimension remains constant within a block of trials for each control condition. For example, an observer must discriminate shape in control condition C, but the color of the two stimuli is always the same.

### 8.2.2 Selective Attention

The second condition to be considered is selective attention. This condition also qualifies as a focused task and is very similar to the control condition. Consider selective condition E (see Figure 8.2e). The observer must discriminate on the perceptual dimension of color: A light gray circle or a light gray square requires a "1" response; a dark gray circle or a dark gray square requires a response of "2." The difference, relative to the control condition, is that all four stimuli will be seen in the same block of trials during selective attention. Therefore, unlike the control condition, the unattended perceptual dimension (e.g., shape in condition E) is free to vary (i.e., both circles and squares can appear in a block of trials). Thus, the observer must focus on one perceptual dimension (like the control condition) and is required to ignore the variation that occurs in the irrelevant perceptual dimension (unlike the control condition).

### 8.2.3 Divided Attention

The third condition is referred to as divided attention. All four stimuli are seen during a block of trials and the response (1 or 2) is contingent on the conjunction of both perceptual dimensions (e.g., color and shape). Therefore, unlike both the control and selective conditions, attention must be divided across both dimensions to complete the task successfully since consideration of either dimension alone is insufficient.

Note that there is only one possible version for this task (condition G in Figure 8.2g). One can conceptualize a correct response as falling on a diagonal of the 2 × 2 matrix. This is graphically illustrated in Figure 8.2g; note that the two diagonals have been pulled apart and separated in space for clarity of presentation. For example, consider the correct response when a circle is presented. If the color of the circle is dark gray, then the correct response is "1"; if the color is light gray, then the correct response is "2." Thus, an appropriate response depends upon a conjunction of values over both perceptual dimensions, and observers must divide their visual attention to respond correctly.

### 8.2.4 Redundant Condition

The final condition is the redundant condition and there are two versions (Figure 8.2h and 8.2i). Only two of the four stimuli are seen in each version. Each pair of stimuli was carefully chosen so that there is simultaneous variation in both dimensions. Consider the redundant condition H (Figure 8.2h). The observer is presented with either the light gray circle or the dark gray square in an individual trial. The observer can provide a correct response by considering (1) the color of a visual form (light gray or dark gray), (2) the shape of a visual form (square or circle), or (3) the conjunction of both its color and shape. This task can be done successfully as either a focused (basing responses on a single attribute) or a divided (basing the responses on the conjunction of both attributes) attention task.

### 8.2.5 A Representative Experiment

A representative study will be described to provide a concrete set of results that can be used to illustrate critical issues in visual attention and form perception. Pomerantz and Schwaitzberg (1975) conducted an experiment using the speeded classification methodology and three of the four conditions described in the previous section: control, selective, and divided attention. These three conditions are particularly critical for visual display design since they examine potential trade-offs between the need to focus visual attention on a particular element within a visual form (i.e., consider an individual variable) or, alternatively, the need to divide attention across these visual elements (i.e., consider relationships between variables).

Pomerantz and Schwaitzberg (1975) used normally oriented pairs of parentheses as their visual stimuli, as illustrated in Figure 8.3. This figure maintains the underlying structure of the matrix presented in Figure 8.1; there are two perceptual dimensions (spatial location and orientation), which can assume one of two values (location of parenthesis = right or left; orientation of parenthesis = right facing or left facing). They also systematically varied the spatial distance between the two parentheses in a stimulus pair, ranging from 0.25 to 8° of visual angle.

Figure 8.4 is fashioned after Figure 8.2 to illustrate the visual stimuli and the critical visual features of the three relevant experimental conditions. As in Figure 8.2, the outlined enclosure and "filled" visual forms signify stimuli that were present in an experimental condition (i.e., "((" and "()" in control condition A). However, the light and dark fills of parentheses in Figure 8.4 do not denote differences in color. Rather, they are used to differentiate between critical and noncritical visual features in an experimental condition. The

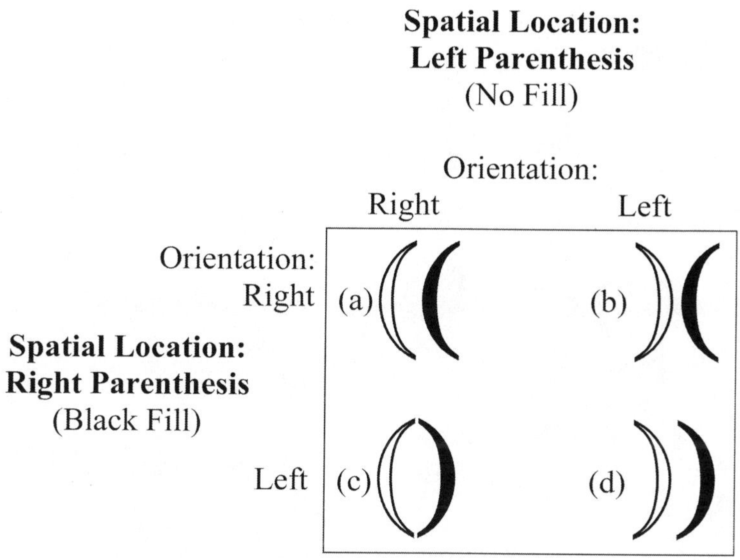


**FIGURE 8.3**
A set of four stimuli formed from a factorial combination of two stimulus dimensions (i.e., spatial location of parenthesis, left vs. right) and two values for each dimension (i.e., right facing vs. left facing). (Adapted with permission from Pomerantz, J. R., and W. R. Garner. 1973. *Perception & Psychophysics,* 14:565–569. Copyright 1973 by the Psychonomic Society. All rights reserved.)

critical visual feature is represented by a black fill; the appropriate response is listed beside this feature.

For example, in control condition A, the critical visual feature is the right parenthesis. When it faces to the right, a response of "1" is required and, when it faces left, a response of "2" is required.

The results obtained by Pomerantz and Schwaitzberg (1975) are illustrated in Figure 8.5. It is clear that there are substantial differences between conditions and that these differences depend heavily upon the spacing between parentheses. Both the divided attention and the control condition showed a general trend of performance degradation as the spacing between parentheses was increased. The selective attention condition produced the opposite trend: Performance improved as the spacing between parentheses was increased.

## 8.3 An Interpretation Based on Perceptual Objects and Perceptual Glue

An interpretation of these results from one conceptual perspective will be explored in this section. This will be referred to as the "object-based"

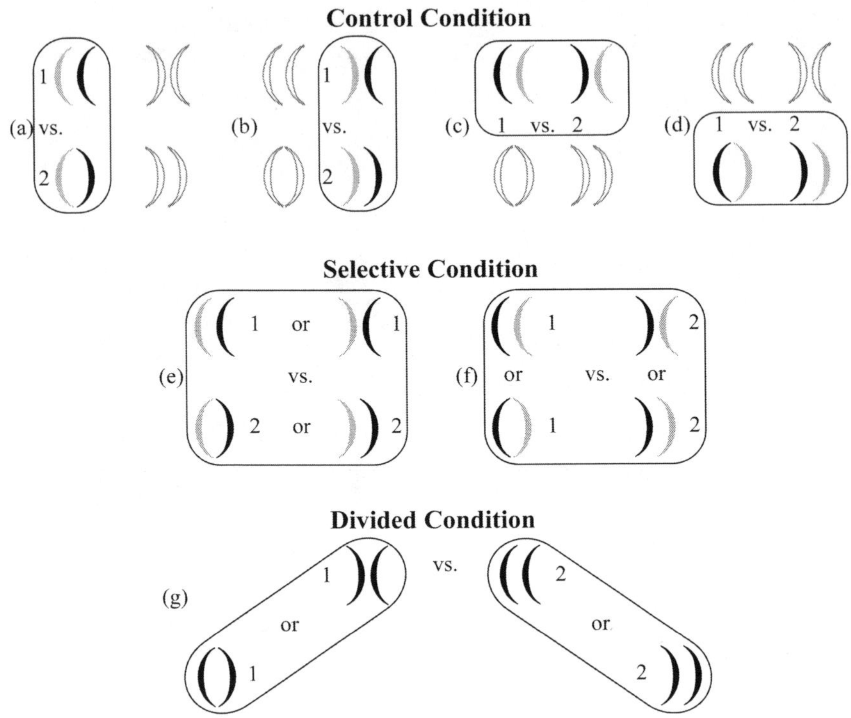


**FIGURE 8.4**
The three experimental conditions of the Pomerantz and Schwaitzberg (1975) study, including the subsets of stimuli that were available during a block of trials outlined enclosures (and filled parentheses), the critical visual features (black fill), and the appropriate response (numbered labels adjacent to parentheses).

interpretation since it is predicated on the idea that perceptual objects play an important role in visual attention and form perception. From this perspective, the parts of a visual array (e.g., graphical fragments; see Figure 6.5) may be organized into perceptual objects (e.g., graphical forms). This process is also sometimes referred to as perceptual "grouping" and there are associated implications for visual attention.

### 8.3.1 Gestalt Laws of Grouping

The Gestalt psychologists were perhaps the most well-known advocates of this position. At least five Gestalt "laws" (see Figure 8.6) are related to perceptual grouping: proximity, similarity, closure, good continuation, and symmetry (Pomerantz and Kubovy 1986).

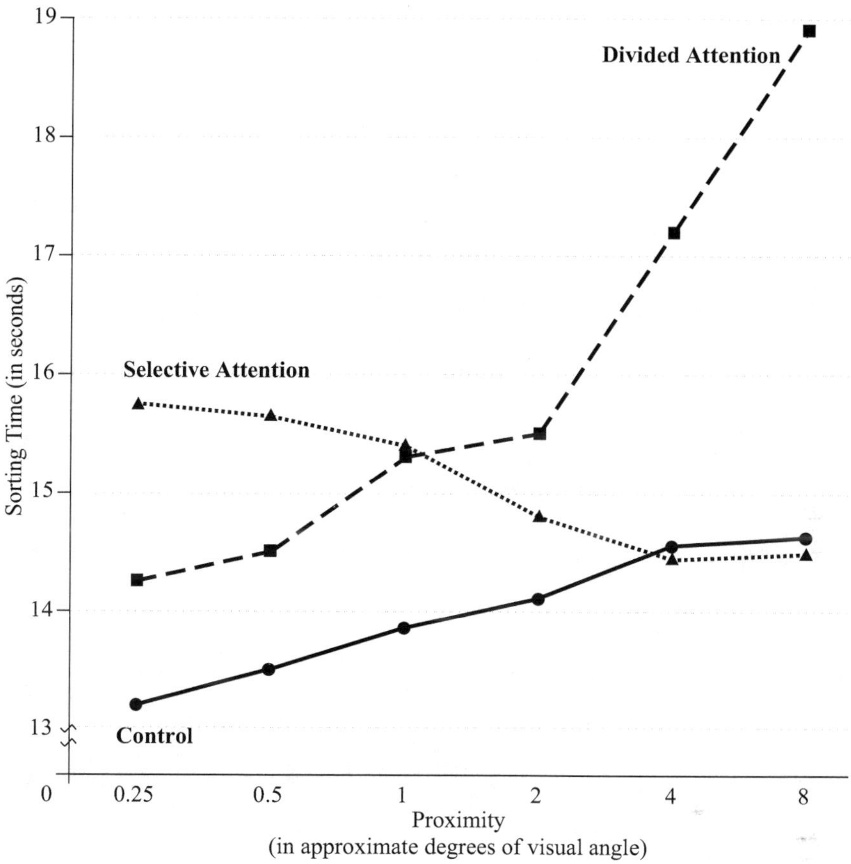


**FIGURE 8.5**
Levels of performance across three experimental conditions for pairs of normally oriented parentheses (Pomerantz and Schwaitzberg 1975). (Adapted with permission from Pomerantz, J. R. 1986. In *Pattern Recognition by Humans and Machines,* ed. H. C. Nusbaum and E. C. Schwab. 

The law of proximity refers to the fact that parts that are physically closer will tend to be grouped together. Thus, the spacing between visual elements in Figure 8.6a results in the parts being grouped into columns on the left and rows on the right.

The law of similarity refers to the fact that physically similar parts will tend to be grouped together. Thus, the parts in Figure 8.6b are organized into homogeneous rows of circles or squares, as opposed to heterogeneous columns containing both circles and squares (note that the physical spacing between parts is exactly the same).

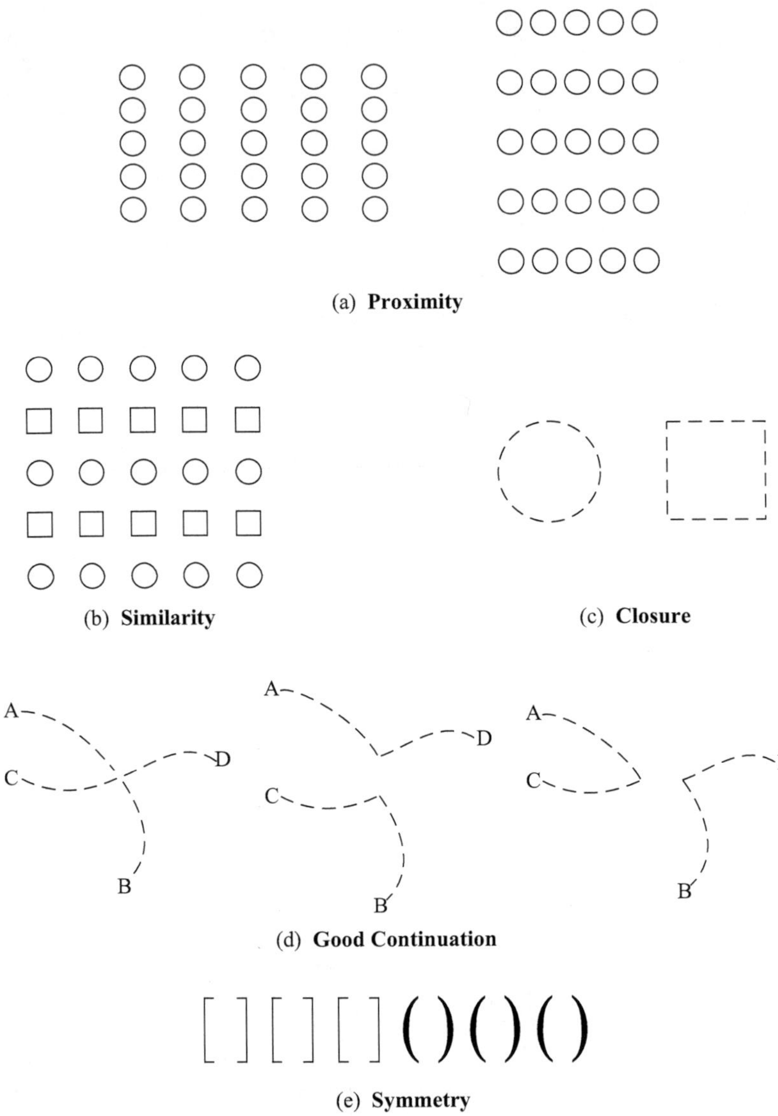


**FIGURE 8.6**
Gestalt laws of perceptual grouping. (Adapted with permission from Pomerantz, J. R., and M. Kubovy. 1986. In *Handbook of Perception and Human Performance,* ed. K. Boff, L. Kaufmann, and J. Thomas. Copyright 1986, New York: John Wiley & Sons.)

The law of closure refers to the fact that "when elements are arranged so they define a closed region, they will group together to form perceptually unified shapes" (Pomerantz and Kubovy 1986, p. 36-4). Thus, the parts in Figure 8.6c (i.e., the dashed lines) are grouped into a circle and a square even though these parts are not physically connected.

The law of good continuation refers to the fact that graphical "trajectories" can have an influence on perceptual grouping. Thus, one sees two intersecting lines in Figure 8.6d (e.g., a line with end points A and B) rather than two lines that meet and then reflect backward (e.g., a line with end points A and C).

The law of symmetry refers to the fact that, when parts have a symmetrical visual relationship, they are more likely to be grouped together. Thus, one sees three pairs of brackets on the left of Figure 8.6d as opposed to two "pillars" enclosed by two brackets or six individual parts.

### 8.3.2 Benefits and Costs for Perceptual Objects

According to the object-based interpretation, automatic benefits and costs are incurred on the performance of focused and divided attention tasks when graphical fragments (parts) are organized into perceptual objects (whole forms). The benefits to divided attention tasks are reflected directly in one of the more celebrated Gestalt adages: "The whole is greater than the sum of its parts." The Gestalt psychologists believed that the perception of the whole took precedence over the perception of the parts, arguing "that the brain is structured to deal directly with the holistic properties of the stimulus, such as the configuration, symmetry, and closure of a visual form" (Pomerantz and Kubovy 1986, p. 36-7). The modern-day version of this perspective is that the global properties of a perceptual object are processed automatically and in parallel.

The inherent cost to focused attention tasks is the result of a hypothetical substance referred to as "perceptual glue" (Pomerantz 1986). Once perceptual grouping occurs, it produces a strong coherence of form that binds individual graphic elements into a whole (i.e., a perceptual object), thereby making any attempt to focus attention on the component parts difficult to achieve. Boring (1942) summarizes the Gestalt position on perceptual glue: "A strong form coheres and resists disintegration by analysis into parts or by fusion with another form" (p. 253). Similarly, Pomerantz (1986) observes that "although conceding that the component parts of a stimulus can be attended to, the Gestalt psychologists argued that dismantling a stimulus into its parts is not the norm in perception and that such analysis can be achieved only through deliberate acts of scrutiny" (pp. 36-7–36-8).

These concepts provide the basis for predictions and interpretations from the object-based perspective. First consider the results obtained by Pomerantz and Schwaitzberg (1975) for divided attention. The parts were perceptually grouped together into a unitary whole when the distance between the parentheses was small (the laws of proximity, closure, and symmetry seem particularly relevant). The perceptual "object" that resulted from this grouping possessed global perceptual qualities (i.e., a gestalt) above and beyond the simple summation of the perceptual qualities of the parts themselves.

Thus, observers were not really processing two parts to perform the divided attention task when the parentheses were closely spaced: They were processing a unitary whole—one that was likely to be very salient and distinctive. Therefore, performance was good (see Figure 8.5). As the parentheses were placed farther and farther apart, however, the process of perceptual grouping became progressively more difficult. Under these conditions, observers were required to process two parts independently and performance suffered as a result.

The object-based interpretation also provides a ready interpretation of the costs in performance obtained in the selective attention condition: perceptual glue. A small spatial distance between parentheses facilitated perceptual grouping and resulted in the formation of a strong perceptual object. The parts of this object are glued together, thereby hindering the observers in their attempts to focus on a part of this object. This produces the cost for the selective attention condition obtained by Pomerantz and Schwaitzberg (1975; see Figure 8.5) when the parentheses were closely spaced. The perceptual grouping is weakened or eliminated when the spacing between the parts is increased; the cost to focused attention disappears. Thus, the object-based interpretation provides reasonable interpretations of the performance obtained in both the divided and selective conditions.

However, the results obtained for the control condition are much more difficult to interpret. This condition produced a pattern of performance that is essentially a mirror image of the pattern obtained in the selective attention condition: Performance was best when the parentheses were in close spatial proximity and became progressively worse as the distance between parentheses was increased (see Figure 8.5). Both of the focused conditions require observers to focus on a part to complete the experimental task. According to the law of proximity, the cost of "ungluing" this part from the perceptual object should have been highest at close spatial distances and should have systematically decreased as the parts were separated in space (and the form became less "object like"). Yet, the opposite pattern of results was obtained for the control condition. This diametrically opposed pattern of results is difficult to reconcile with the principles of perceptual objects and perceptual glue.

## 8.4 Configural Stimulus Dimensions and Emergent Features

In this section we will describe the theoretical foundations of an alternative perspective on visual attention and form perception. As will be shown, the concept of a perceptual object plays no role in this alternative perspective. Instead, Garner and his colleagues (Garner 1970, 1974; Garner and Felfoldy 1970; Pomerantz and Garner 1973) focused on the dimensional structure of stimuli (i.e., the fundamental types of information that might be present in a visual stimulus such as color or form) and the implications of that structure for processing (i.e., potential relationships or interactions that could occur between these perceptual dimensions). This work was continued and expanded considerably by Pomerantz and his colleagues (Pomerantz and Schwaitzberg 1975; Pomerantz, Sager, and Stoever 1977; Pomerantz 1986; Pomerantz and Kubovy 1986; Pomerantz and Pristach 1989), particularly with regard to issues in form perception.

### 8.4.1 The Dimensional Structure of Perceptual Input

Although it may come as a surprise to many researchers working within the confines of traditional information processing, Garner's work was heavily influenced by Gibson's ecological approach (see the quote at the beginning of the chapter). Much like Gibson, Garner was very concerned with describing the nature of the information that was out there to be perceived. He proposed three qualitatively different types of stimulus dimensions: "separable," "integral," and "configural" (Garner 1974). The focus was on the relationships between stimulus dimensions. Were the perceptual dimensions processed independently or together? Did variations in one perceptual dimension influence the perception of the other? The perceptual dimensions were defined by different patterns of performance across the four task conditions described in Section 8.2.

#### *8.4.1.1 Separable Dimensions*

A separable relationship occurs when perceptual dimensions are processed independently. Each dimension retains its unique perceptual identity within the context of the other dimension, resulting in a lack of interaction among stimulus dimensions. For example, the two perceptual dimensions illustrated in Figure 8.1 are separable dimensions: The perception of color does not influence the perception of shape and vice versa. Observers can focus on an individual dimension and ignore variations in the irrelevant dimension.

Figure 8.7a provides a hypothetical but representative set of results for separable perceptual dimensions across each of the four conditions in the speeded classification task. As illustrated by this graph, performance with separable dimensions will be better when the experimental task requires the observer to consider variations within a single perceptual dimension (e.g., shape alone or color alone). The control, selective, and redundant conditions can all be performed on this basis and performance should be relatively good, as illustrated in Figure 8.7a.

On the other hand, performance on divided attention tasks should be poor. Separable stimulus dimensions are processed independently and do not interact to produce new higher order visual properties. As a result, variations in not one but two perceptual dimensions need to be considered; rules governing the various conjunctions and correct responses then need to be applied. An implication is that mental processing or effort is required to combine

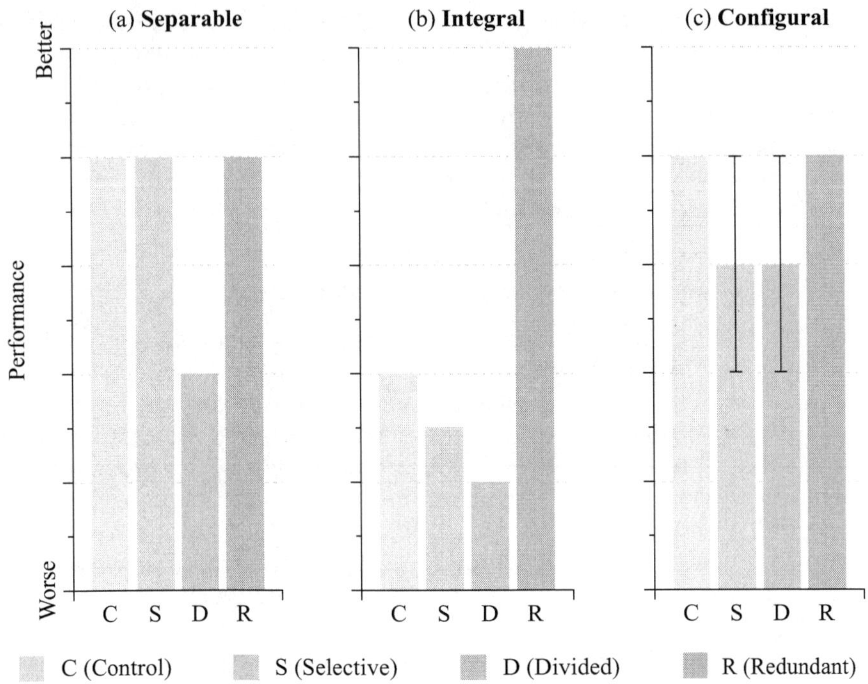


**FIGURE 8.7**
Proposed categories of stimulus dimensions (i.e., separable, integral, and configural) and prototypical patterns of results for each of the speeded classification conditions (i.e., control, selective, divided, and redundant).

the independent dimensions. As illustrated in Figure 8.7a, performance with separable perceptual dimensions should therefore suffer in divided attention tasks, when both dimensions must be considered.

### *8.4.1.2 Integral Dimensions*

In contrast to separable dimensions, an integral relationship is defined by a strong interaction between stimulus dimensions. Garner (1970) states that "two dimensions are integral if in order for a level on one dimension to be realized, there must be a dimensional level specified for the other" (p. 354). For example, hue and brightness are often proposed as integral perceptual dimensions; perceived color is a function of both. Integral stimulus dimensions are processed in a highly interdependent fashion; a change in one dimension necessarily produces a change in the second dimension.

As a result of this highly interdependent processing, a redundancy gain occurs: "These stimuli are perceived not dimensionally but in terms of their overall similarity, and an additional, redundant dimension makes two stimuli more discriminable" (Pomerantz 1986, p. 17). Thus, performance in the redundant condition is particularly good (see Figure 8.7b). However, the unique perceptual identities of the individual dimensions are lost and performance suffers when attention to one (control and selective attention conditions) or both (divided attention condition) dimensions is required. An implication is that the integral stimulus dimensions are naturally processed as a single entity and that additional processing or effort is required to dissect this single entity. This pattern of results is illustrated in Figure 8.7b.

#### *8.4.1.3 Configural Dimensions*

A configural relationship refers to an intermediate level of interaction between perceptual dimensions. Each dimension maintains its unique perceptual identity, but the two can also interact or "configure" to produce new higher order properties. Information about the individual stimulus dimensions (e.g., the orientation of each parenthesis) remains available with configural dimensions. However, additional configural properties are created through their interaction. For example, the parenthesis pairs in Figure 8.3a and 8.3d ["((" and ")) "] demonstrate the higher order configural property of vertical parallelism: The two parts are aligned at their tops and separated by the same amount of space continuously throughout their vertical axis. Conversely, the parenthesis pairs in Figure 8.3b and 8.3c demonstrate another higher order visual property: vertical symmetry [")(" and "()"]. These two parts are aligned at their tops and form mirror images of each other along the vertical axis (i.e., at each point of vertical, extent the distance from the vertical axis to each mark is exactly the same).

The general pattern of results typically obtained with configural dimensions is presented in Figure 8.7c. Information regarding individual stimulus dimensions is still available. Thus, levels of performance for conditions that require each dimension to be considered independently (i.e., control, selective, divided conditions) is substantially better with configural dimensions than with integral stimulus dimensions. On the other hand, these stimulus dimensions are not processed in the totally independent manner that is seen with separable dimensions. Thus, there appears to be a cost in focusing attention on one perceptual dimension (this apparent cost will be examined more closely later in the chapter).

The configural properties produced by the interaction between dimensions (e.g., the parallelism and symmetry of the parentheses in Figure 8.3) can facilitate performance when variations on both dimensions must be considered simultaneously. The error bars on the bar graphs for the selective and divided attention conditions indicate a range of possible outcomes, a point that will be discussed later. The implication is that these stimuli require little or no extra mental processing or effort" to combine or to dissect the dimensional components.

It is no coincidence that the terms "separable," "integral," and "configural" have been used to describe visual displays. In fact, these findings have provided a conceptual foundation for several different approaches to display design (as described in Chapter 9). Our approach to the design of analog geometric displays has been heavily influenced by the concept of configural stimulus dimensions. Therefore, this concept will now be examined in greater detail.

### 8.4.2 Emergent Features; Perceptual Salience; Nested Hierarchies

The higher order visual properties produced by the interaction of configural stimulus dimensions (e.g., the symmetry and parallelism in Figure 8.3) have historically been referred to as "emergent features." Pomerantz and Pristach (1989) describe emergent features in the following fashion: "Basically, emergent features are relations between more elementary line segments, relations that can be more salient to human perception than are the line segments themselves" (p. 636).

The term "salience" is synonymous with visual prominence (i.e., conspicuity in a psychophysical sense). Thus, it refers to how well a particular visual feature stands out relative to other features that are present (i.e., how salient or discriminable or distinctive that visual feature is, independently of any semantic associations). The definition offered by Pomerantz and Pristach is overly restrictive, however, since graphical elements other than line segments can produce emergent features. A sampling of other emergent features that can be produced by configural dimensions includes colinearity, equality, closure, area, angle, horizontal extent, vertical extent, and good form.

In Chapter 6 we described the nested, hierarchical geometries that characterize both natural and virtual ecologies (Section 6.6.2). These visual properties can now be reconsidered in terms of emergent features. For example, consider the pie chart illustrated in Figure 8.8; this graphical form contains a

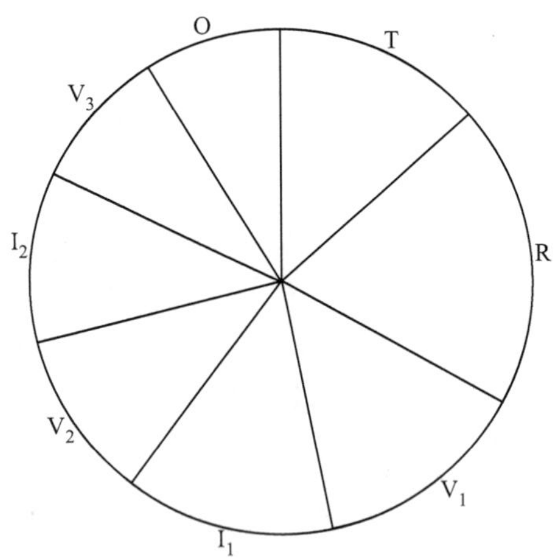


**FIGURE 8.8**
A simple pie chart (see text for a description of the nested, hierarchical emergent features it contains). (Adapted with permission from Bennett, K. B., A. L. Nagy, and J. M. Flach. 1997. In *Handbook of Human Factors and Ergonomics*, ed. G. Salvendy. Copyright 1997, New York: John Wiley & Sons. All rights reserved.)

number of emergent features. The most salient emergent feature is its overall shape; the global outline is that of a closed geometrical form that is perfectly symmetrical around any axis (i.e., its form is a circle). The area of this circle is also a global emergent feature. The parts of the circle provide emergent features at an intermediate level of salience; each slice of the pie has a shape, an orientation, and an area.

At a lower level of salience are the elementary emergent features (i.e., the graphical fragments) that define the pie slice: the size of the angle at the center and the intersections formed by the radius lines and the perimeter arc. At the lowest level of salience are the graphical fragments that ultimately define the boundaries of the wedge: the line segments and arc. Thus, one way to consider visual forms is as a set of nested hierarchical features (including elemental, configural, and global features) that vary in their relative salience.

### 8.4.3 Configural Superiority Effect

The configural superiority effect (Pomerantz et al. 1977) is a particularly compelling example that demonstrates several important points about emergent features. The stimuli in the baseline condition consisted of an array of four widely spaced and normally oriented parentheses (see Figure 8.9a1 and Figure 8.9b1). The observer's task was to identify the one element of the array that had a different orientation (i.e., left facing or right facing) than the other three. In Figure 8.9a1 and 8.9b1 the disparate stimulus is the left-facing parenthesis located in the lower right quadrant. The observer's task is to indicate which quadrant of the display (e.g., lower right) contains the anomalous element. Thus, this experimental task qualifies as a divided attention task: The orientation of each parenthesis in the array must be considered to determine an appropriate response. The average reaction time obtained for the baseline condition (i.e., individual parentheses) was 2400 ms, as illustrated in the left-hand side of Figure 8.9c.

#### *8.4.3.1 Salient Emergent Features*

Two additional experimental conditions were also evaluated. In the "salient" experimental condition, a constant visual context was added to the individual parentheses. This context consisted of four identical, normally oriented parentheses (see Figure 8.9a2). This produces the set of four stimuli illustrated in Figure 8.9a3. Participants performed the same experimental task (to indicate the quadrant containing the parenthesis with a different orientation).

One might reasonably predict that the addition of this constant visual context would degrade performance relative to the baseline condition (i.e., individual parentheses); the task would be more difficult because the observer is now required to sort through (or process) twice as many visual elements than before (in Tufte's terms, the data–ink ratio has decreased). These

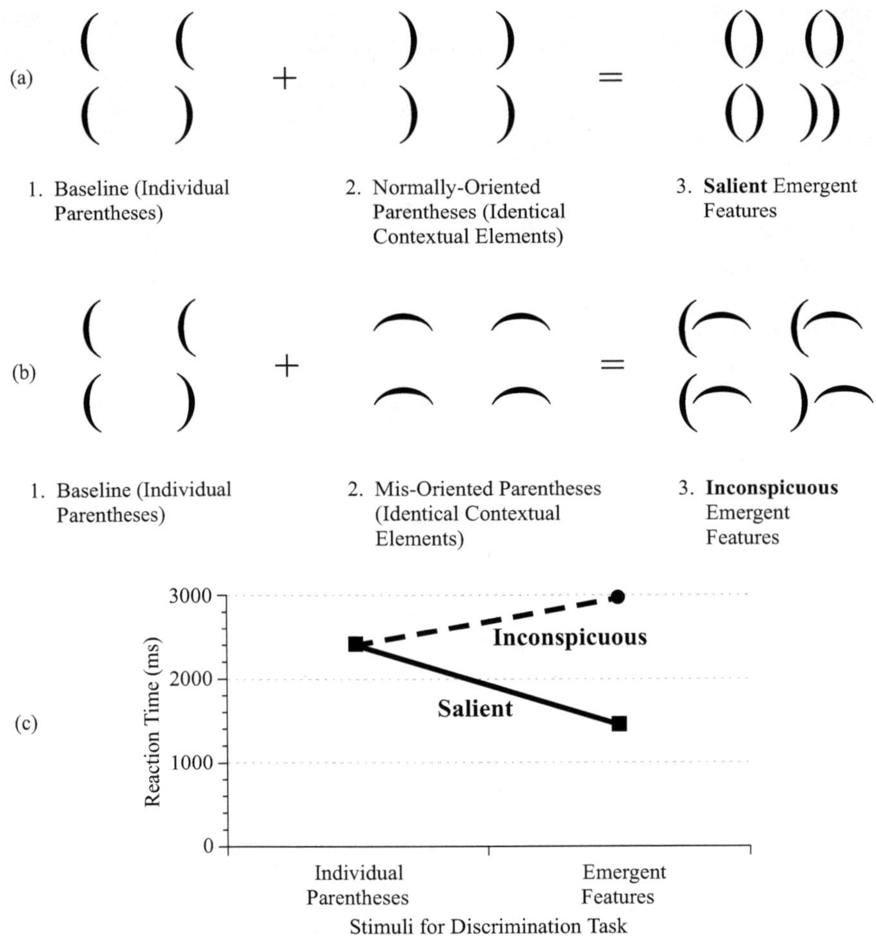


**FIGURE 8.9**
A graphical depiction of the configural superiority effect observed by Pomerantz, Sager, and Stoever (1977). Two identical visual contexts (except for rotation of the elements) were added to individual parentheses and produced dramatic differences in performance. (Adapted with permission from Pomerantz, J. R. 1986. In *Pattern Recognition by Humans and Machines,* ed. H. C. Nusbaum and E. C. Schwab. Copyright 1986, Orlando, FL: Academic Press. All rights reserved.)

additional graphical elements would appear to have no intrinsic value in and of themselves; the possibility that they could clutter the visual display and therefore degrade performance is a very real one. Instead, Pomerantz et al. (1977) found that the addition of this constant visual context resulted in performance that was substantially better than performance with individual parentheses; reaction time was almost cut in half (40% reduction; see the results labeled "salient" on the right-hand side of Figure 8.9c).

The interpretation of these results requires the explicit consideration of the emergent features that were present, their salience (both in general and relative to each other) and their mapping to the constraints of the task. As described earlier, the emergent features produced by closely spaced pairs of normally oriented parentheses are symmetry and parallelism. These emergent features are (1) highly salient in their own right and (2) easily discriminated from each other.

Furthermore, these emergent features are consistently mapped to the constraints of the task. Consider the case when a left-facing parenthesis was the target (see Figure 8.9a3). The emergent feature of parallelism ["))"] will appear in the visual quadrant containing the target (lower right), while the emergent feature of symmetry ["()"] will appear in all other quadrants. Conversely, when a right-facing parenthesis is the target (not illustrated in Figure 8.9), the emergent feature of symmetry will appear in the target quadrant (and parallelism will appear in all other quadrants). Thus, the location of the target parenthesis was consistently and directly specified by the presence of a highly salient and highly discriminate feature.

These observations, in combination with the drastically reduced reaction time, would strongly suggest that the observers were performing qualitatively different activities in these two conditions. The widely spaced parentheses in the baseline condition required that the task be completed by considering and remembering the direction of each of the individual parentheses in a serial fashion. The addition of the identical context produced higher order, more salient, and more discriminate visual properties (i.e., emergent features) that were well mapped to the task constraints. The disparate emergent feature "popped out" from the visual array and specified the location of the disparate parenthesis directly. The observer did not have to search for, remember, and compare the directions of individual parentheses; the appropriate response was directly specified by the presence of the salient emergent feature.

#### *8.4.3.2 Inconspicuous Emergent Features*

A second experimental condition (the "inconspicuous" condition) added a visual context that contained the same four parentheses. However, these parentheses were rotated 90° in a counterclockwise direction (see Figure 8.9b2). These "misoriented" parentheses produced a substantial decrement in performance (see the right-hand side of Figure 8.9c). The amount of time required to perform the task with the misoriented parentheses (Figure 8.9b3) was far worse than that required for the baseline condition (an increase of 23%). The performance differences relative to the salient emergent features condition (Figure 8.9a3) were even more dramatic: Observers took more than twice as long to identify the appropriate quadrant.

The interpretation of these results requires the consideration of the same factors described earlier: emergent features, their salience, and their mapping to the constraints of the task. It is not particularly clear whether the addition of the four misoriented parentheses actually produced emergent features or not. Pomerantz (1986) debates this issue and concludes that "these misoriented parenthesis pairs appear neither to group nor to contain any emergent features .... However, it is possible that they do group and produce emergent features but that all four stimuli possess the *same* emergent feature" (p. 16; emphasis original). What is abundantly clear is that the alternative emergent features produced by the misoriented parentheses were not easily discriminated from each other (i.e., the perceptual differences between emergent features were not salient or conspicuous).

The end result is that the correct quadrant to report was *not* specified by a salient emergent feature that "pops out" from the visual field automatically and in parallel. The observers were forced to search the target parentheses element by element (i.e., serially) to locate the disparate stimulus. In fact, the increased latencies relative to the baseline condition would strongly suggest that the additional visual context served to clutter or perhaps camouflage the location of the anomalous parenthesis (consistent with Tufte's data–ink principle).

### 8.4.4 The Importance of Being Dynamic

These findings emphasize critical differences between the design of static displays (i.e., those that appear on the printed pages of a book) and the design of dynamic displays (i.e., those that appear on the interface of a sociotechnical system). Pomerantz (1986) astutely observed that emergent features are "dependent on the identity and arrangement of the parts" (p. 8). This fact is clearly illustrated in the results of the configural superiority effect. The identity of the parts that were added in the well-mapped and the poorly mapped conditions was exactly the same (i.e., both conditions contained four parentheses of exactly the same size and curvature). The drastic differences in performance were obtained solely on the basis of differences in the arrangement of these parts (i.e., the counterclockwise rotation of the parentheses by 45°).

These results underscore the increased difficulty in the design of dynamic displays (e.g., a display that represents the changing conditions in a power plant) relative to the design of static displays (e.g., a graph on a printed page). The arrangement of parts in dynamic displays will constantly be in flux because they are directly coupled to the changing variables and properties of the domain. The configurality superiority effect highlights the implications of this fact: Different spatial arrangements of exactly the same parts can produce salient emergent features in one case but not in another. Thus, it is critical that dynamic displays are developed and evaluated under experimental contexts that allow the range of their dynamic behavior to be observed (i.e., dynamic simulations of the work domain). More than once we have designed displays that looked great on paper, but turned out to be confusing or ineffective when they were integrated into a simulated work domain.

## 8.5 An Interpretation Based on Configurality and Emergent Features

The previous section described a theoretical perspective on visual attention and form perception based upon configural stimulus dimensions and emergent features. This provides a theoretical foundation for an alternative interpretation of the results obtained by Pomerantz and Schwaitzberg (1975). As we will see, this interpretation appears to be more robust and comprehensive than that provided by perceptual objects and perceptual glue. Figure 8.10 will be used to clarify this interpretation. It replicates Figure 8.4 but provides textual labels (located beneath the parentheses) that explicitly describe the mapping between emergent features and the correct response.

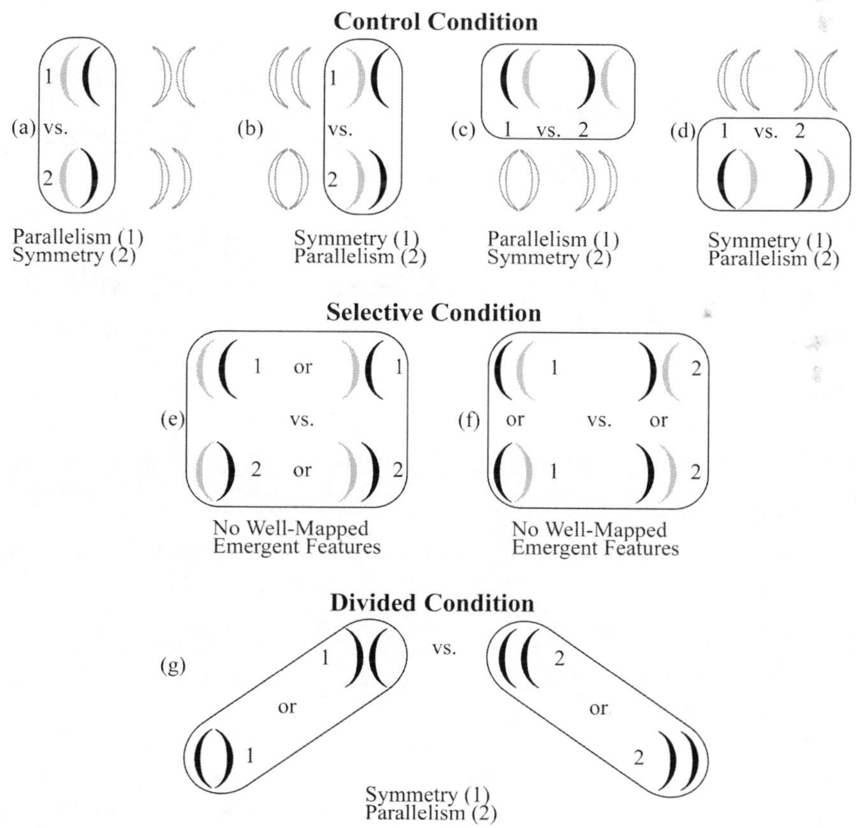


**FIGURE 8.10**
This figure replicates Figure 8.4 with additional textual descriptions of the mapping between emergent features and the required responses.

### 8.5.1 Divided Attention

As described previously, the divided attention condition is one that requires the consideration of a conjunction of stimulus values across both perceptual dimensions. In terms of Pomerantz and Schwaitzberg's stimuli, this requires conjunctions across the spatial location of parentheses and their orientation. Thus, if the right parenthesis faces to the right (i.e., the top two pairs of parentheses in Figure 8.10g), then the correct response is determined by the direction of the left parenthesis. When it faces to the left, the correct response is "1"; when it faces to the right, the correct response is "2." Similarly, when the right parenthesis faces to the left (i.e., the bottom two pairs of parentheses in Figure 8.10g), a response of "1" is required when the left parenthesis faces to the right and a response of "2" is required when it faces to the left. These descriptions can be thought of as a set of logical rules that could be used to determine the appropriate response.

Pomerantz and Schwaitzberg (1975) found that the levels of performance obtained for this condition depended heavily upon the spacing between parentheses (see Figure 8.5). Performance was very good when the parentheses were close together; performance was very bad when the parentheses were far apart.

First, consider the good performance when the parentheses were closely spaced. The presence of salient emergent features that were well mapped to task constraints is responsible for the high levels of performance. As illustrated in Figure 8.10g, the emergent feature of symmetry directly specified a response of "1"; the emergent feature of parallelism directly specified a response of "2." Thus, the complicated logical rules involving the conjunction of spatial location and orientation outlined before were not required because the presence of a highly salient emergent feature directly specified the appropriate response to be made. The response could be made on the basis of the visual stimulus as a whole, without consideration of its parts.

Pomerantz and Schwaitzberg (1975) found that these performance benefits disappeared as the spatial distance between parentheses was increased (see Figure 8.5). The most likely interpretation of these results is that the increased spatial distance between the parentheses decreased the salience of the emergent features that were produced, thereby eliminating the visual evidence that directly specified the correct response. The extremely poor performance at high degrees of spatial separation is most likely due to the need to apply the complicated set of logical rules. When the spatial distance between two parentheses exceeded the limit of foveal vision (approximately 2° of visual angle), the task became a complicated cognitive task, as opposed to a perceptual one. The location and orientation of each parenthesis would need to be obtained in a serial fashion and remembered, and the logical rules about conjunctions and appropriate responses would need to be applied.

### 8.5.2 Control Condition

The control condition is technically a focused task; to complete this task, the observer must consider only a part of the visual form (i.e., the facing direction of a single parenthesis, as illustrated in Figure 8.10a–d), rather than the whole. Recall that this is the experimental condition that produced a pattern of results that was very difficult to interpret in terms of perceptual objects and perceptual glue; performance was degraded, instead of improved, as the spatial distance between parentheses was increased. It is, however, fairly straightforward to interpret from the perspective of emergent features and salience. In fact, the interpretation is quite similar to that put forth for the divided attention condition.

Consider Figure 8.10a–d. The two salient emergent features (parallelism and symmetry) described previously are produced by a pair of stimuli located close in space. Most importantly, these emergent features are well mapped to the required responses in the control condition (see the textual labels under the pairs of stimuli). For example, in control conditions A and C, the emergent feature of parallelism requires a "1" response and the emergent feature of symmetry requires a "2" response. The mappings between emergent features and response are opposite for conditions B and D, but this is a minor concern since trials are blocked within the various control conditions. Thus, a salient emergent feature specifies the appropriate response; it is likely that the presence of these emergent features was responsible for the good levels of performance obtained when the parentheses were closely spaced.

The salience of these emergent features is systematically decreased as the parentheses are separated in space. At higher degrees of spatial separation, the parentheses would no longer configure to produce emergent features and the observer would need to consider the less salient individual parentheses to perform the task. These changes in the salience of the visual information that could be used in performing the task are responsible for the degradation in performance. The performance in the control condition is better than performance in the divided condition at high degrees of spatial separation because the control condition requires the consideration of only a single visual feature, as opposed to the conjunction of visual features across perceptual dimensions that is required in the divided condition.

### 8.5.3 Selective Condition

The selective attention condition is also a focused task, technically requiring an observer to consider a part of the visual form to complete the experimental task (i.e., the facing direction of a parenthesis; see Figure 8.10e and 8.10f). In contrast to the two previous conditions, however, the highly salient emergent features were not well mapped to task constraints.

Consider selective condition E (Figure 8.10e). The emergent feature of symmetry was paired with both a "1" response [")("] and a "2" response ["()"]; the emergent feature of parallelism was paired with both a "1" response ["(("] and a "2" response ["))"]. The end result is that the salient emergent features could not be used to perform the experimental task in the selective condition.

Thus, the individual parentheses had to be used to perform the task, instead of emergent features. This visual information is much less salient than the higher order emergent features that they produced. However, the emergent features did not go away; they were still visible in the visual array. This produced very low levels of signal-to-noise ratio when the parentheses were closely spaced; the task-relevant visual information (i.e., individual parentheses) was much less salient than the task-irrelevant visual information (i.e., emergent features) and performance suffered accordingly. Systematically increasing the spatial distance between parentheses reduced the salience of the irrelevant emergent features and therefore increased the relative level of salience for the visual information that was critical in performing the task.

The fact that performance for the selective and control conditions converges at the two highest degrees of spatial separation provides fairly strong evidence for this interpretation. The closely spaced parentheses produced salient emergent features that had a beneficial (control condition) or a detrimental (selective condition) impact on performance at close spatial proximity, depending upon the quality of mapping to the experimental task. At the two highest degrees of spatial separation, the parts no longer configured to produce these emergent features and the observers were essentially working with isolated parentheses. The impact of emergent features on performance disappears; the two conditions produce exactly the same intermediate level of performance.

## 8.6 Summary

This chapter has explored basic issues in visual attention and form perception. These issues are particularly important in building representations for the virtual ecologies that will support effective interaction in law-driven domains. The focus has been on a relatively restricted portion of the overall nested hierarchical structure that characterizes these domains (see Figure 6.5)—specifically, the level of geometrical forms and geometrical fragments. In terms of operational requirements, the controlling agent will need to focus on higher order relationships or properties in the domain and on individual datum. In terms of visual attention and form perception, this translates into the capability (or costs) involved in the perception of the parts of visual forms versus perception of the whole. This fundamental concern has been the primary organizational factor in the chapter; two alternative theoretical perspectives were outlined with regard to this fundamental concern.

The first theoretical perspective is predicated on the notion of perceptual objects. Graphical elements in the visual array are perceived as perceptual objects based on the laws of perceptual grouping (e.g., proximity, similarity, closure, good continuation, and symmetry). Perception of the whole takes precedence over perception of the parts; the perceptual grouping is strong and perception of the parts requires that they be "unglued" from the whole. There are inherent benefits (perception of the whole; divided attention) and costs (perception of the parts; focused attention) associated with perceptual objects. The implication for display design is that no single graphical representation can be designed to support the need of cognitive agents to consider both higher order relationships and properties and the lower level data upon which they are based. This theoretical perspective is simple and intuitive, but fails to provide reasonable interpretations of critical findings.

The second theoretical perspective provides a much more comprehensive interpretation of the empirical findings that were presented. It is predicated on the critical role of configural stimulus dimensions, emergent features, perceptual salience, and mappings to task constraints. At its core, this perspective is a triadic one (see Chapter 2). Thus, the interpretation of the results presented in the chapter are based on very specific sets of mappings between the constraints imposed by the agent (i.e., perceptual systems), the interface (i.e., properties of visual forms), and the domain or task (e.g., the correct mapping to specific responses). It leaves open the possibility that a single graphical representation can be designed to support the consideration of both higher order relationships and lower level data.

Performance at divided attention tasks depends upon two factors: (1) The emergent features produced by the display must be salient, and (2) they must tell the observer something that they need to know to accomplish the task at hand. Pomerantz (1986) makes this explicitly clear: "The point is that an emergent feature (or any type of feature) is of no use in a discrimination task unless it differentiates among the stimuli to be discriminated and is mapped into a suitable fashion onto the required response categories" (p. 16).

These principles can be scaled up for the design of configural displays for dynamic work domains. However, this can be a surprisingly difficult challenge. Pomerantz (1986) describes emergent features as "elusive," "not easily measured," "idiosyncratic," "unpredictable," and capable of producing "unusual effects" (pp. 6–13). It is important to note that these adjectives describe characteristics of stimuli that are both static in nature and extremely simple. Unlike the visual stimuli described in this chapter, the emergent features in configural displays will represent meaningful properties of complex work domains. It follows that these displays will need to be much more complex than the stimuli described in this chapter.

Another implication is that the emergent features now represent something in the work domain (i.e., performance is not dependent upon just the visual appearance of the display). Thus, the quality of the mapping between the emergent features and the domain properties that they represent introduces an additional but equally critical consideration in design. Finally, these displays will be dynamic, with the increased difficulty in design that this entails (see Section 8.4.4).

Performance at focused attention tasks is also conceptualized in terms of the same triadic considerations that were critical for divided attention tasks. The good performance for the focused control task with closely spaced parentheses was due to the presence of emergent features that were well mapped to task constraints. These emergent features were also present in the selective attention condition with closely spaced parentheses; however, in that context, they were poorly mapped to the constraints of the task. The apparent cost of focused attention in the selective task was due to the decreased salience of the critical visual features (i.e., the orientation of a parenthesis) relative to the same emergent features. The poorly mapped emergent features retained their salience, were difficult to ignore, and severely degraded performance.

Note that this is a very different interpretation from that of perceptual glue. From this perspective, information about the parts (e.g., the parentheses) is available alongside information about the whole (e.g., the emergent features). Parts never completely lose their identity relative to the whole and they can be focused upon when so desired. However, the visual information corresponding to these parts can be less distinctive or salient than the emergent features that they produce. The apparent cost for focused attention tasks is due to imbalances in salience between emergent features and lower level graphical elements. The parts can be inherently less salient than the emergent features; in some cases (e.g., the selective attention costs discussed in the current chapter), the difference will produce a cost in performance.

Pomerantz and Pristach (1989) foreshadow these conclusions:

> [W]e need not hypothesize any perceptual glue to account for the subjective cohesiveness of forms or the apparent failures of selective attention to line segments. Subjects may prefer to attend to more salient emergent features than to less salient line segments, but this is not any sort of a failure .... One implication is that line segments do not lose their perceptibility when they are embedded within configurations of the type studied here. The process of grouping involves not losses of line segments but gains of emergent features. Observers may opt to attend to these novel features, but the line segments remain accessible; the forest does not hide the trees. (p. 642)

This theoretical perspective leaves open the possibility of designing a single display format to support both focused and divided attention tasks. It suggests that if the perceptual salience of the elemental features is increased relative to the emergent features, then this potential cost may be offset. This possibility was investigated in a series of empirical studies that investigated design strategies to increase the salience of elemental features. Some of these techniques also added structural information, above and beyond perceptual salience, that was useful in offsetting these potential costs. Chapters 9, 10, and 11 will reconsider all of these issues in the context of display design.

## References

Bennett, K. B., A. L. Nagy, and J. M. Flach. 1997. Visual display. In *Handbook of Human Factors and Ergonomics,* ed. G. Salvendy. New York: John Wiley & Sons.

Boring, E. G. 1942. *Sensation and perception in the history of experimental psychology.* New York: Appleton-Century-Crofts.

Garner, W. R. 1970. The stimulus in information processing. *American Psychologist* 25:350–358.

———. 1974. *The processing of information and structure.* Hillsdale, NJ: Lawrence Erlbaum Associates.

Garner, W. R., and G. L. Felfoldy. 1970. Integrality of stimulus dimensions in various types of information processing. *Cognitive Psychology* 1:225–241.

Pomerantz, J. R. 1986. Visual form perception: An overview. In *Pattern recognition by humans and machines,* ed. H. C. Nusbaum and E. C. Schwab. Orlando, FL: Academic Press.

Pomerantz, J. R., and W. R. Garner. 1973. Stimulus configuration in selective attention tasks. *Perception & Psychophysics* 14:565–569.

Pomerantz, J. R., and M. Kubovy. 1986. Theoretical approaches to perceptual organization. In *Handbook of perception and human performance,* ed. K. Boff, L. Kaufmann, and J. Thomas. New York: John Wiley & Sons.

Pomerantz, J. R., and E. A. Pristach. 1989. Emergent features, attention, and perceptual glue in visual form perception. *Journal of Experimental Psychology: Human Perception and Performance* 15:635–649.

Pomerantz, J. R., L. C. Sager, and R. J. Stoever. 1977. Perception of wholes and of their component parts: Some configural superiority effects. *Journal of Experimental Psychology: Human Perception and Performance* 3:422–435.

Pomerantz, J. R., and S. D. Schwaitzberg. 1975. Grouping by proximity: Selective attention measures. *Perception & Psychophysics* 18:355–361.

# 9

# *Semantic Mapping versus Proximity Compatibility*

## 9.1 Introduction

> Yet within the compatibility of proximity framework, such a bar graph display has low display proximity. The bar graphs' superiority cannot be explained by redefining display proximity, as redefinition would make the concept so general it would cease to say anything more than "a display that causes better performance." (Sanderson et al. 1989, p. 196)

In this chapter we continue to examine issues in the design of analogical representations for law-driven domains, moving beyond the study of basic visual processes and into the realm of display design. Two alternative approaches to display design will be compared and contrasted, both of which are heavily influenced by the visual attention and form perception literature described in the previous chapter. Both approaches are influential. Articles describing the "semantic mapping" approach (Bennett and Flach 1992) and the "proximity compatibility" approach (Wickens and Carswell 1995) were identified as the seventh and eighth most influential papers ever published (over 50 years and 2008 articles) in the journal *Human Factors* (Salas and Cooke 2008).

These two approaches to display design are often treated very similarly in the literature. They are sometimes cited together when a design is justified, when experimental predictions are made, or when the results of experimental evaluations are interpreted (e.g., Marino and Mahan 2005; Peebles 2008). For example, Peebles (2008) states: "The relationship between display and mental proximity has been revealed in several studies (e.g., Barnett and Wickens, 1988; Bennett and Flach, 1992; Bennett, Toms, and Woods, 1993; Carswell and Wickens, 1987; Wickens and Andre, 1990; Wickens and Carswell, 1995)" (p. 86). We believe that the degree of interchangeability that is often conferred upon these two approaches is both unfortunate and misguided (e.g., Bennett and Fritz 2005).

The fundamental purpose of this chapter is to provide concrete examples of abstract principles of design that are critical for interface representations for law-driven domains (i.e., analog geometrical forms). Alternative principles are described and made concrete through detailed analyses of selected studies; predictions for performance are compared to the findings in the display design literature. A secondary purpose is to dispel the misconceptions alluded to in the previous paragraph. Each approach will be discussed in detail, including its origins in the visual attention and form perception literature and its evolution over time. We begin with a description of proximity compatibility.

## 9.2 Proximity Compatibility Principle

The proximity compatibility principle (PCP) is an approach to display design developed by Wickens and his colleagues. The original version of PCP (Barnett and Wickens 1988; Carswell and Wickens 1987; Wickens and Andre 1990) will be described first; revisions to PCP (Wickens 1992; Wickens and Carswell 1995) will be considered at the end of the chapter. The original version of PCP drew heavily upon the visual attention and form perception literature described in the previous chapter. The theoretical concept of perceptual objects assumed a critical role in PCP. This approach also emphasized the role of integral and separable perceptual dimensions (see Chapter 8, Section 8.4.1) in display design; configural perceptual dimensions were, by and large, ignored. Principles of display design and predicted performance for divided-attention tasks will be considered first.

### 9.2.1 PCP and Divided Attention

Carswell and Wickens (1987) define an object display as "any graphical technique that uses several dimensions of a single perceptual object to present multiple sources of information" (p. 511). This design technique was predicted to improve performance at divided-attention tasks due to the automatic and parallel processing of perceptual objects:

> A clear benefit of the use of these displays in multichannel information processing tasks [i.e., divided attention] emerges from the collective research on the object concept in perception … the various attributes of a single object are bound together in perceptual processing, and therefore are processed in parallel … object configurations provide an inherent constant benefit to processing of multidimensional information. (pp. 511–512)

Thus, PCP maintained that presenting multiple variables in a single perceptual object (i.e., an object display) circumvented the need to process each of these variables in a time-consuming serial fashion. The visual perceptual system accomplished this automatically and in parallel by virtue of the fact that it is built to perceive perceptual objects (see the associated discussion of perceptual grouping and Gestalt laws in Chapter 8, Section 8.3).

Dimensional integrality is the second factor contributing to predictions of improved performance for object displays and divided-attention tasks. According to the original version of PCP, object displays were composed of integral stimulus dimensions (Carswell and Wickens 1987):

> Garner (1970) distinguishes between two types of relations that can hold for a pair of dimensions. Integral dimensions are those in which specification of the level on one dimension requires that the other dimension be represented. According to this criterion, the height and width of a rectangle [i.e., object display] are integral (i.e., in order to specify the height, the rectangle must have a width; otherwise it would not be a rectangle). (p. 512)

As described in the previous chapter, integral stimulus dimensions are supposed to interact to produce new, global perceptual properties above and beyond the contributions of each stimulus dimension on its own. Thus, PCP predicted that these global properties would facilitate performance at divided-attention tasks, when the relationship between two or more variables needed to be considered in a display.

Thus, the original conceptualization of PCP drew heavily upon two concepts originally identified in the visual attention and form perception literature: perceptual objects and integral stimulus dimensions. Wickens and his colleagues coined the term "object integrality" as a form of shorthand notation to refer back to these fundamental concepts.

The original predictions of PCP for object displays and divided-attention tasks are illustrated in Figure 9.1 (Wickens and Carswell 1995). The solid line represents predicted levels of performance for high task proximity, which roughly corresponds to what has been referred to as divided-attention tasks. Object displays are conceptualized as being close, or similar, in display proximity (i.e., located on the left side of the graph's Y-axis). Thus, Figure 9.1 illustrates PCP's prediction that object displays will facilitate performance at divided attention tasks.

In contrast, PCP predicts poor performance at divided-attention tasks (solid line) for displays that are low in proximity (located on the right side of the graph). The most common example of a display that possesses low proximity is a bar graph. This format uses a unique graphical representation for each individual variable (i.e., a bar) instead of a single perceptual object. PCP explicitly linked this type of display to the concept of separable stimulus dimensions: "Separable dimensions, like separate bar graphs, do not require this joint specification" (Carswell and Wickens 1987, p. 512). These displays are referred to as "separable" or "separate" displays because of this link.

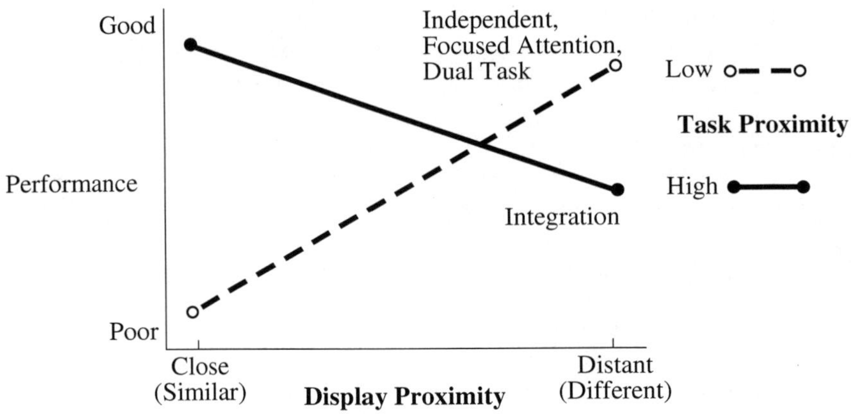


**FIGURE 9.1**
Original predictions of the proximity compatibility principle of display design. (Adapted with permission from Wickens, C. D., and C. M. Carswell. 1995. *Human Factors* 37 (3): 473–494. Copyright 1995 by the Human Factors and Ergonomics Society. All rights reserved.)

The predictions of poor performance with bar graph displays were derived directly from the concept of separable stimulus dimensions. As described in the previous chapter, no new perceptual properties are formed by the interaction of separable dimensions; no emergent properties will be present to testify with regard to relations between variables. Since these relationships are critical to the performance of divided-attention tasks, they will need to be determined by considering each variable one at a time in a serial manner. This was predicted to take time and to draw upon limited capacity resources (e.g., working memory).

### 9.2.2 PCP and Focused Attention

The opposite pattern of results was predicted for these two categories of displays when focused-attention tasks need to be performed (i.e., the dashed line representing low task proximity). PCP predicted that the processing mechanisms associated with object integrality would work against object display formats for these tasks. For example, Carswell and Wickens (1987) state that *"the various attributes of a single object are bound together in perceptual processing"* (pp. 511–512). This strongly implies a belief in the notion of perceptual glue that was discussed in the previous chapter (i.e., the parts of an object are "glued" together and extra cognitive effort is required to "unglue" them).

The negative impact of integral stimulus dimensions is made explicit (Carswell and Wickens 1987):

> The relevance of the distinction to the current issue is Garner's finding that it is difficult to focus attention directly on one of a pair of integral

> dimensions, while ignoring variation on the other (Garner and Felfoldy 1970). Given that the dimensions of an object are more likely to be integral than separable, this finding would imply at least one instance in which object displays might prove difficult to use: when focused attention to a single attribute is required (e.g., in check reading a particular value on a display). (p. 512)

Thus, object displays (close display proximity, left side of Figure 9.1) were predicted to produce poor performance at focused-attention tasks (dashed line).

In contrast, separable displays (e.g., bar graph displays, distant display proximity, right side of Figure 9.1) were predicted to improve performance at focused-attention tasks: "The second hypothesis of the principle of compatibility states that when focused attention is required separable displays will be superior to integral displays (Goettl, Wickens, and Kramer 1991, p. 1052). Separable perceptual dimensions are processed independently (unlike integral perceptual dimensions); there are no perceptual object and no perceptual glue. Therefore, information about individual variables should be readily available in bar graph displays and the observer should be able to focus on these variables effectively.

### 9.2.3 Representative PCP Study

Carswell and Wickens' (1987) study will be used to provide concrete examples illustrating the PCP approach. Observers were required to monitor two dynamic systems, each of which had three process variables: two inputs and an output. Normal system state was determined by a mathematical function (either additive or multiplicative) relating the two inputs to the output; a system failure occurred when the output deviated from the mathematical function. Thus, this qualifies as a divided-attention task: A system failure could be detected only by considering the relationship between all three variables.

Two display formats were evaluated in experiment 1. The separable display used three bar graphs with a common baseline to represent the three system variables (see Figure 9.2a). The bar graphs for the two system inputs ($I_1$ and $I_2$) were located on the left ($I_1$) and the middle ($I_2$); the bar for the system output (O) was located on the right. Therefore, this display will be referred to as the "IIO" bar graph display (using the terminology of Sanderson et al. 1989). The second display was an object display (the triangle display in Figure 9.2b). An anchor point was located at the base of the triangle (i.e., the small vertical line). The value of the first input variable determined the distance from the anchor point to the left side of the triangle ($I_1$); the value of the second input variable determined the distance to the right side ($I_2$). The height of the triangle was determined by positioning a point directly above the anchor point at a distance from the baseline that corresponded to the value of the output

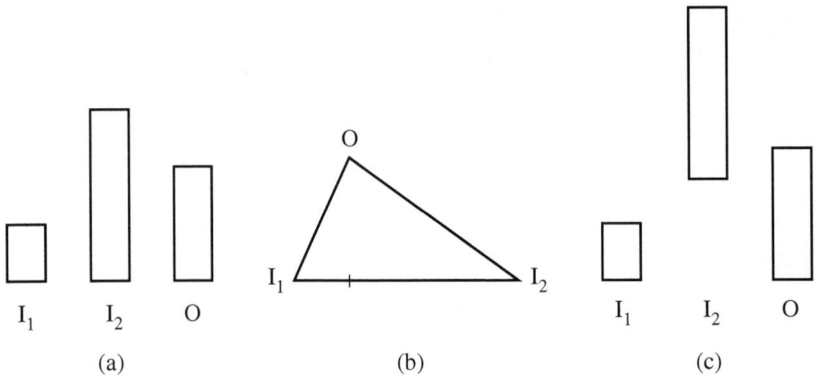


**FIGURE 9.2**
Displays evaluated in a representative PCP study. (a) Bar graph display with a common baseline. (b) Triangle object display. (c) Staggered bar graph display without a common baseline. (Adapted with permission from Carswell, C. M., and C. D. Wickens. 1987. *Ergonomics* 30:511–527. Copyright 1987 by Taylor & Francis, London. All rights reserved.)

variable (O). These three points were connected with lines to form the sides and vertices of the triangle.

Carswell and Wickens (1987) found that the object display improved the latency of performance at the divided-attention task significantly. These performance advantages were attributed to object integrality: "These data suggest that the use of the more *integral* triangle display is associated with superior performance across a number of integration [divided attention] tasks … the advantage observed with the *object* display is a fairly general one" (p. 521; emphasis added).

Observers performed a focused-attention task in experiment 2. Each of the individual variables ($I_1$, $I_2$, and O) had its own set point value; the observer's task was to monitor each variable for instances where its value crossed (i.e., became greater than or less than) the value of its set point. The same integral object display was used (Figure 9.2b); a staggered bar graph display (Figure 9.2c) replaced the original bar graph. Dependent measures of accuracy, latency, and false alarms were obtained. No significant differences were found between displays for accuracy and false alarms. However, the staggered bar graph display produced significantly faster response times than the object display. Carswell and Wickens (1987) interpret their results as follows: "This finding is consistent with other studies in our laboratory showing object-related deficits … object displays may prove to be poor choices for tasks in which independent or focused processing of some of the displayed elements must be carried out" (pp. 523–524).

### 9.2.4 Summary of PCP

The original version of PCP conceptualized issues in display design in terms of object integrality. An inherent trade-off between display and task proximity was predicted (see Figure 9.1). Each type of task (e.g., focused vs. divided attention) was viewed as requiring a different type of display (e.g., separable bar graphs vs. integral object display) for effective support; no single display format was viewed as having the capability to facilitate performance at both types of tasks.

## 9.3 Comparative Literature Review

In the early 1990s we performed a literature review of the empirical laboratory studies that had been conducted with regard to these issues in display design (Bennett and Flach 1992). A numerical index to the studies in this review is provided at the bottom of Figure 9.3. The encoding conventions used to represent the displays, outcomes, and theoretical implications of each finding are described in the top panels of Figure 9.3.

The results are organized into two figures; each figure provides the set of results for divided (Figure 9.4) and focused (Figure 9.5) attention. The three columns accumulate each finding according to the logical consistency of the outcome relative to the predictions of PCP: consistent (left), inconsistent (middle), or neutral (insignificant or mixed results, right column).

### 9.3.1 Pattern for Divided Attention

The overall pattern of results for divided attention does not provide strong support for the predictions of PCP. It was predicted that integral object displays (high display proximity) would facilitate performance at divided-attention tasks (high task proximity). A total of 54 experimental comparisons were reported for divided attention (Figure 9.4). Approximately one out of three empirical findings (19/54, 35%) was found to support this prediction. A fairly large number of findings were actually in the opposite direction: One out of five effects (11/54, 20%) indicated that separable bar graph displays (low display proximity) were statistically superior to object displays (high display proximity). The most common finding is the lack of statistically significant differences between display types (24/54, 44%).

### 9.3.2 Pattern for Focused Attention

The results for focused attention (Figure 9.5) provide even less support for the predictions of PCP. It was predicted that separable bar graph displays

**Display Icons:**

| | |
|---|---|
| Digital | Meter |
| Bar Graph | Face |
| Rectangle | Polar Coordinate |
| Staggered Bar Graph | Configural Bar Graph |
| Triangle | House |
| Coordinate | |

**Statistical & Conceptual Symbology**

Left-to-right ordering of icons represents progressively poorer performance with respective graphic formats (ordered according to reported means)

Underscoring of two icons represents statistically significant differences in performance between display formats

Highlighted icon represents a finding that is both statistically significant and meaningful in terms of the theoretical issues discussed in text

**Numerical Index to Studies**

1. Barnett and Wickens, 1988
2. Bennett, Toms, and Woods, 1993
3. Boulette, Coury, and Bezar, 1987
4. Buttigieg and Sanderson, 1991
5. Carswell and Wickens, 1987
6. Casey and Wickens, 1986
7. Coury, Boulette, and Smith, 1989
8. Goettl, Wickens, and Kramer, 1991
9. Goldsmith and Schvaneveldt, 1984
10. Jones, Wickens, and Deutsch, 1990
11. MacGregor and Slovic, 1986
12. Sanderson, Flach, Buttigieg, and Casey, 1989
13. Wickens and Andre, 1988
14. Wickens and Andre, 1990
15. Wickens, Kramer, Barnett, Carswell, Fracker, and Goettl, 1985

**FIGURE 9.3**
Key to graphical summaries of literature review presented in Figures 9.4 and 9.5. (Adapted with permission from Bennett, K. B., and J. M. Flach. 1992. *Human Factors* 34:513–533. Copyright 1992 by the Human Factors and Ergonomics Society. All rights reserved.)

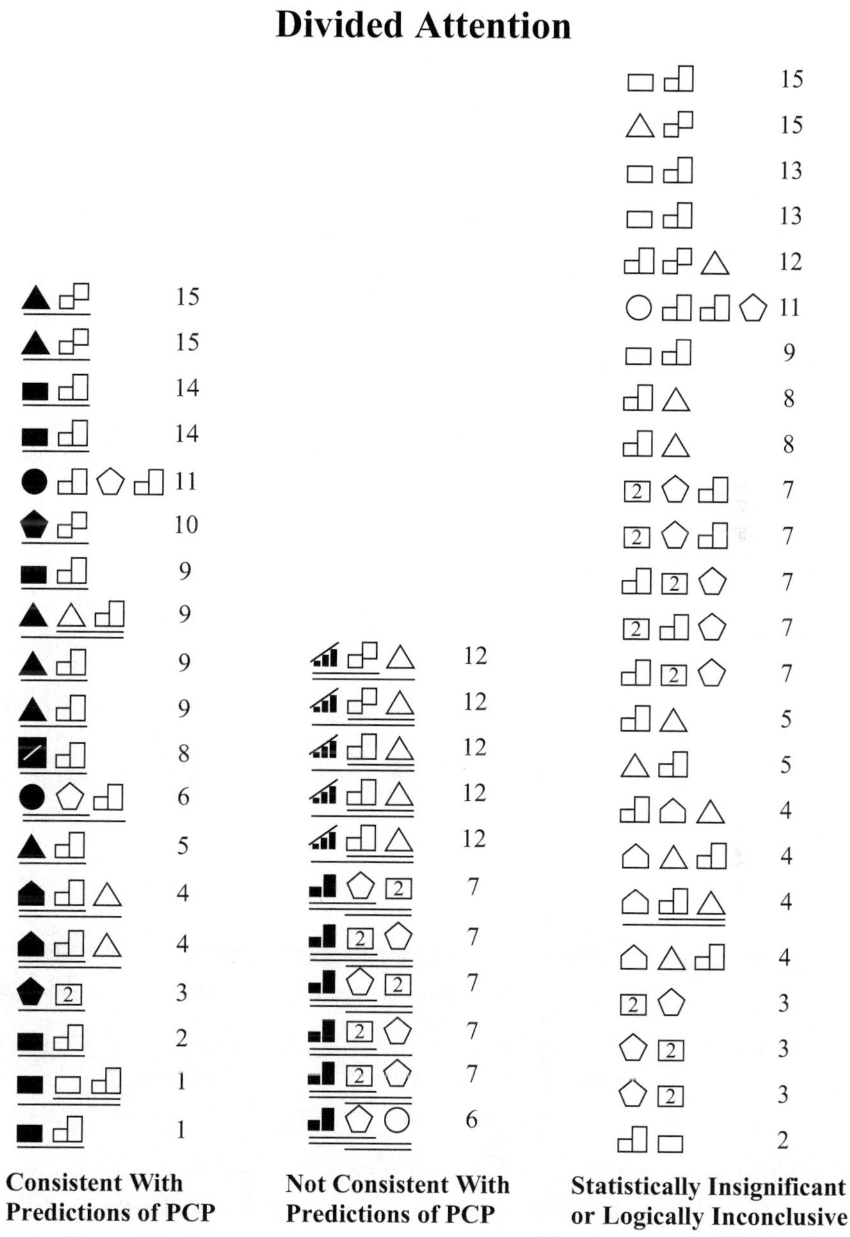


**FIGURE 9.4**
Summary of early laboratory evaluations on divided attention tasks relative to original predictions of PCP. (Adapted with permission from Bennett, K. B., and J. M. Flach. 1992. *Human Factors* 34:513–533. Copyright 1992 by the Human Factors and Ergonomics Society. All rights reserved.)

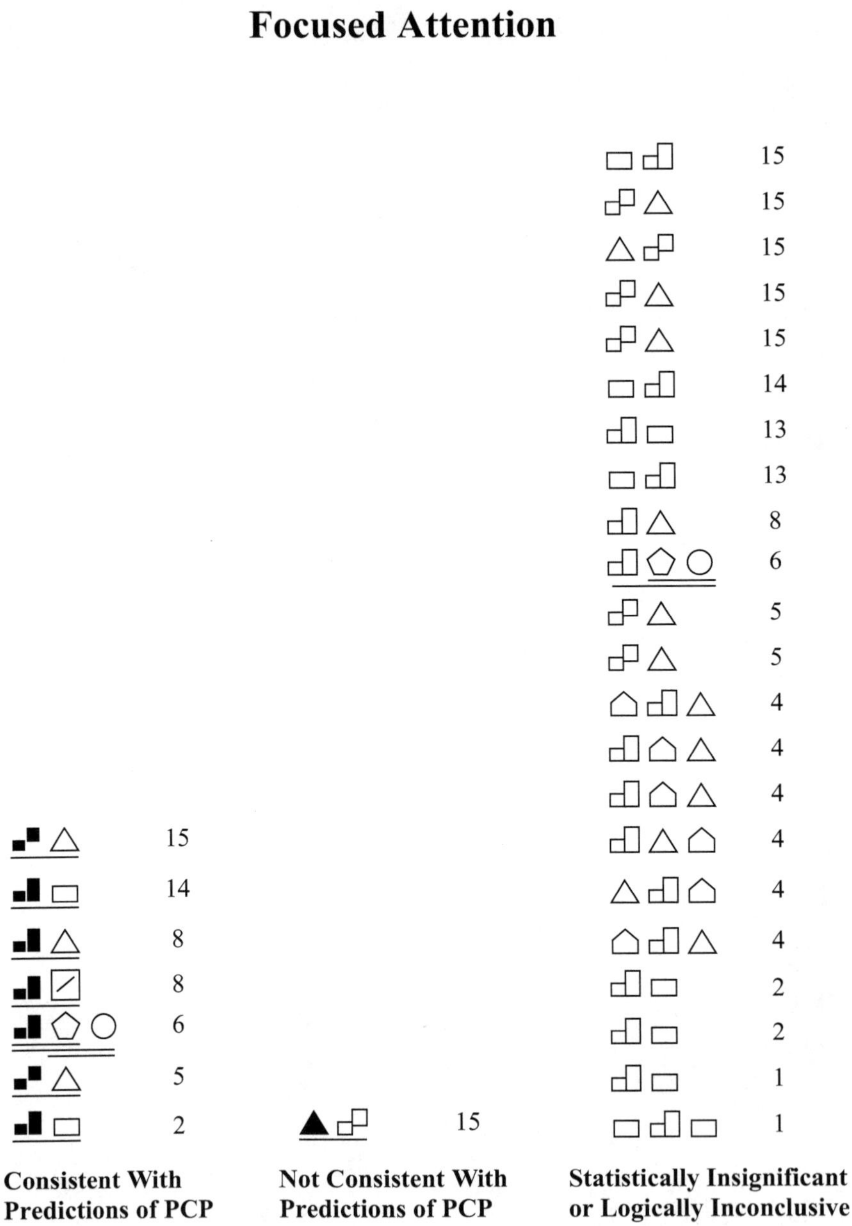


**FIGURE 9.5**
Summary of early laboratory evaluations on focused attention tasks relative to original predictions of PCP. (Adapted with permission from Bennett, K. B., and J. M. Flach. 1992. *Human Factors* 34:513–533. Copyright 1992 by the Human Factors and Ergonomics Society. All rights reserved.)

(i.e., low proximity displays) would facilitate performance at focused-attention tasks (i.e., low proximity tasks). The most common finding, by far, was a lack of statistically significant differences between display types: nearly three of four effects (22/30, 73%) fell into this category. Less than one out of four effects (7/30, 23%) was consistent with the predictions of PCP. In summary, the literature review reveals that the overall fit between the predictions of PCP and the results obtained in early laboratory investigations was not particularly good.

## 9.4 Semantic Mapping

An alternative approach to the design of geometrical form displays, which we have referred to as semantic mapping (Bennett and Flach 1992), will now be described. It is a triadic approach that draws upon the visual attention and form perception literature described in the previous chapter. In contrast to PCP, however, the emphasis is on configural stimulus dimensions (as opposed to separable and integral stimulus dimensions), emergent features, and mapping to domain constraints (as opposed to perceptual objects). The term "semantic mapping" refers to the same complex set of mappings between domain, interface, and agent that are referred to in past and future chapters. Thus, it should be considered by the reader as simply a synonym for ecological interface design.

The next two sections will describe this approach, which provides an alternative theoretical perspective for the interpretation of the pattern of results that were obtained in the literature review. Concrete and detailed examples will be provided to make the associated principles of design as clear as possible. This will include some studies that were not part of the original literature review.

The semantic mapping approach to display design is based on the triadic considerations of semiotics: The effectiveness of a display will ultimately depend upon the quality of very specific sets of mappings between three mutually interacting sets of constraints: the domain, the agent, and the display (e.g., Figure 6.6). The core problem in implementing effective displays for law-driven domains is to provide analog, visual representations that are perceived as accurate reflections of the domain constraints (its semantics or, alternatively, the affordances). Each particular representation that is chosen will produce a different set of display constraints, as defined by its spatiotemporal structure (i.e., the visual appearance of the display over time). That is, each representation will vary in terms of the degree to which it specifies the affordances of the domain (see Figure 6.6).

The concepts of configural stimulus dimensions and emergent features are critical. Thus, the term "configural" will be used to describe analog geometric displays. It is a more general adjective to use than "object" since a closed geometrical form is not required to produce emergent features. Its use also emphasizes that it is emergent features—not "objectness" per se—that control performance. The emergent features produced by configural displays will generally take the form of symmetries: equality (e.g., length, angle, area), parallel lines, colinearity, or reflection. In addition, some of the Gestalt properties (e.g., closure, good form) might also be considered as emergent features.

Ultimately, performance will depend upon the quality of mapping between the three sets of constraints. Are the critical domain constraints accurately reflected in the geometrical constraints of the display? Does the level of visual salience in the various display representations match the level of importance of the corresponding information in domain terms? Are any breaks in the domain constraints (e.g., abnormal or emergency conditions) accurately reflected by corresponding breaks in the geometrical constraints (e.g., the visual form)? That is, does the display possess a high degree of specificity?

The second consideration is the degree of attunement of the agent viewing the display. Can the agent pick up the invariants (e.g., the breaks in geometrical constraints) of the display reliably? Does the agent have sufficient knowledge about the underlying domain to properly discriminate the alternative states of the system that the display is designed to represent? How these questions are addressed will determine whether the cognitive agent will be able to obtain meaning about the underlying domain in an effective fashion. Concrete examples of these principles will be drawn from the display design literature in the following section.

### 9.4.1 Semantic Mapping and Divided Attention

The first example illustrates principles of good configural display design for the support of divided-attention tasks. Woods, Wise, and Hanes (1981) adapted the original design of Coekin (1970) in developing a safety parameter display system (SPDS) for use in power plants. This display collects over 100 individual sensor values and combines them into a single analog geometric form: an octagon. The vertices of this configural display are dynamically scaled so that the octagon is perfectly symmetrical when normal conditions exist in the plant (see Figure 9.6.a). Conversely, a developing abnormality produces systematic distortions in the symmetry of this geometric form. Particular types of abnormalities are associated with characteristic or "signature" distortions of the octagon. For example, the shape of the geometric form that characterizes a loss-of-coolant accident is illustrated in Figure 9.6.b. This display achieves a primary goal in configural display design: When the process constraints in the domain are broken (i.e., there is an abnormality), it is reflected directly in the broken geometrical constraints of the form.

This display is effective because it produces emergent features (i.e., geometric constraints) that are both salient and accurate in reflecting the underlying properties of the domain—not because it is a perceptual object. It is a great example of the fact that graphical forms will possess a "hierarchy of

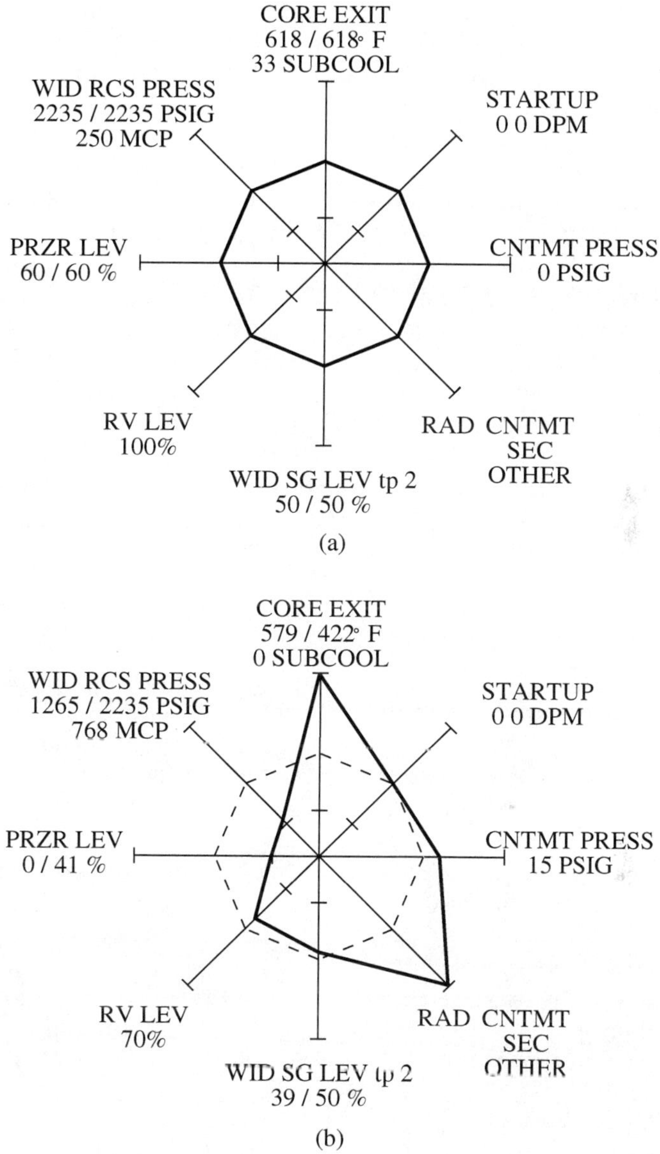


**FIGURE 9.6**
A configural display used to represent the overall status of a process control domain. (a) A perfectly symmetrical octagon represents normal operating conditions. (b) System faults are represented by characteristic distortions of the geometrical form. (Adapted from Schaefer, W. F. et al. June 23, 1987. Generating an integrated graphic display of the safety status of a complex process plant, United States Patent 4,675,147.)

nested structures, with local elements combining to produce more global patterns or symmetries" (Bennett, Nagy, and Flach 1997, p. 681). The graphical elements (e.g., a line) are at the lowest level of the hierarchy. Each line that connects the values on two adjacent axes produces a local emergent feature: orientation. Each pair of lines that connect three axes produces a contour with intermediate-level emergent features including orientation, shape (e.g., "spike" vs. "flat"), angle, and symmetry. Finally, the eight contours combine to form a closed polygon that produces higher level, global emergent features that define its overall shape.

This general display format has become a popular one (alternatively referred to as a polar coordinate, a spoke, a polygon, or a spider display). Unfortunately, it is a display format that is often used improperly or implemented ineffectively (as will be demonstrated later in the chapter). The numerous, salient, and hierarchically nested emergent features described in the previous paragraph visually specify relationships and interactions between variables. If the domain variables themselves do not have corresponding relationships (i.e., if there are poor mappings between display and geometrical constraints), then the display will be misleading and ineffective.

#### *9.4.1.1 Mappings Matter!*

The second example is one of the first empirical demonstrations of the consequences of effective and ineffective mappings. MacGregor and Slovic (1986) used a multiple cue judgment methodology: The observer's task was to consider multiple pieces of low-level data (age, 10K race times, training, etc.) and then to predict the amount of time that a runner would need to complete a marathon. Several different display formats were used to present these data, including a bar graph display, a face display, a polar coordinate display, and a deviation display. The first experiment compared performance between all four displays and found that the face display (a configural display; see Figure 9.7) produced significantly better performance.

The authors conducted a second experiment that illustrates the critical role played by mapping between geometric and domain constraints. Statistical analyses revealed that some of the information cues (e.g., 10K race results) were more useful (i.e., diagnostic or reliable) in predicting marathon times than other cues (e.g., runner's age). The authors also believed that some of the emergent features produced by the face display (e.g., the curvature of the mouth) were more salient than other emergent features (e.g., the length of the nose). This possibility was investigated by developing two versions of the face display. In the well-mapped version, the emergent features believed to be more salient (e.g., mouth curvature) were used to represent the more useful information cues (fastest 10K; see the face 1 displays in Figure 9.7). In the poorly mapped version, these same salient emergent features (mouth) were used to present the less useful information cues (age; see face 2 displays).

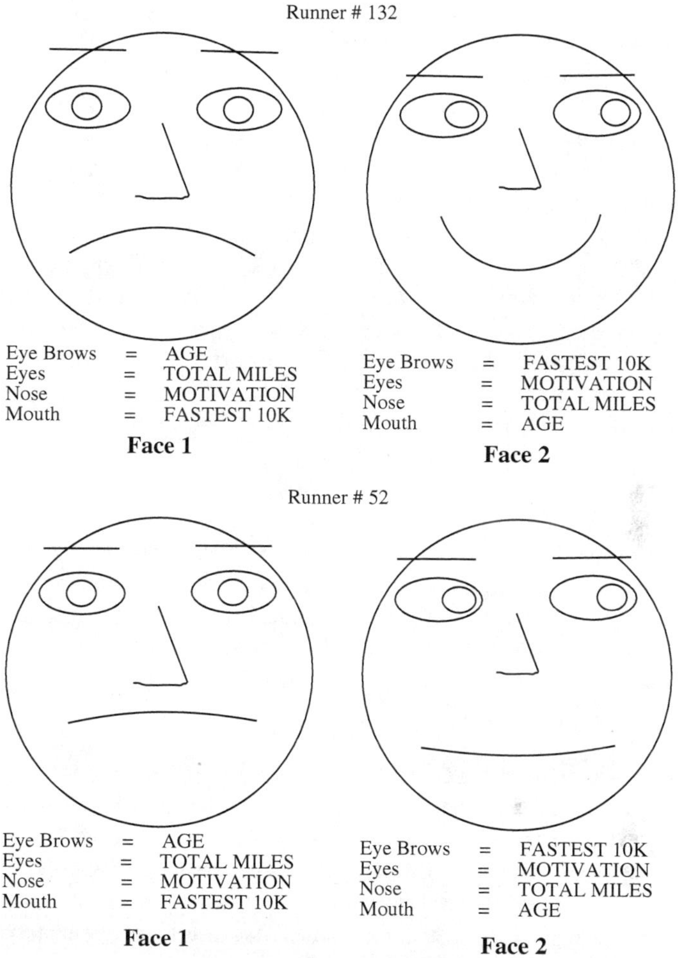


**FIGURE 9.7**
Face displays illustrating the impact of alternative mappings between informational cues and visual features (i.e., face 1 vs. face 2). (Adapted with permission from MacGregor, D., and P. Slovic. 1986. *Human–Computer Interaction* 2:179–200. Copyright 1986 by Lawrence Erlbaum Associates. All rights reserved.)

It was found that the well-mapped face display produced more accurate predictions of marathon times than any of several other formats that included bar graph and polar star forms. The poorly-mapped face display resulted in poorer performance than all the other formats. Thus, the same display format was capable of producing either the best performance or the worst performance depending upon the quality of semantic mapping. MacGregor and Slovic (1986) describe their findings in the following fashion: "Subjects exposed to the face 2 [poorly mapped] display were less able to utilize the information portrayed than were individuals receiving the face 1 [well-mapped] display" (p. 195). Overall, the authors conclude that "judgmental performance is markedly enhanced or degraded by the degree to which the display format provides the user with an organizing structure that facilitates a matching between the relative importance of information and the psychological salience of the display's graphic features" (p. 179).

These results highlight perhaps the fundamental premise of the semantic mapping approach to display design: Effective display design depends upon the quality of specific mappings between observer constraints (e.g., the salience of the emergent features; the ability of an observer to pick up information), task constraints (the nature of the work to be done), and display constraints (the spatial and temporal characteristics of the visual form). In this case, the same diagnostic information was presented using the same display format; yet, substantial variations in performance were obtained due to differences in the quality of mapping between specific emergent features and specific diagnostic cues.

As an aside, it is important that the reader not interpret this discussion as an endorsement of the face display format (it is not a recommended format). Rather, the point is that in display design, the "devil" is really and truly in the "details." Small and seemingly innocuous choices in design can have inordinately large consequences for either good or bad.

#### *9.4.1.2 Configurality, Not Object Integrality: I*

These observations constitute the fundamental principles of the semantic mapping approach and provide the framework for an alternative interpretation of the results obtained in the literature review. There were a substantial number of significant effects favoring bar graphs over geometric form displays at divided-attention tasks (11 total; see the middle column of Figure 9.4). These findings are particularly problematic for the PCP approach to display design since integral (i.e., object) displays were predicted to be superior to separable displays (i.e., bar graphs) for this type of task.

In contrast, the interpretation of these findings is relatively straightforward when one considers display design from the perspective of configurality and semantic mapping (i.e., the presence of emergent features and the degree to which they are mapped to task constraints). Ten of the eleven significant effects identified in the review will be discussed.

Sanderson et al. (1989) designed an experiment to test the possibility that configurality is a more important display design principle than object integrality for divided-attention tasks. They essentially replicated the Carswell and Wickens (1987) study investigating divided attention (i.e., experiment 1), with one important difference. The location of the $I_2$ and the O variables in the IIO bar graph display (see Figure 9.8a) were switched. The new version of the bar graph display (see Figure 9.8) is therefore referred to as the IOI display.

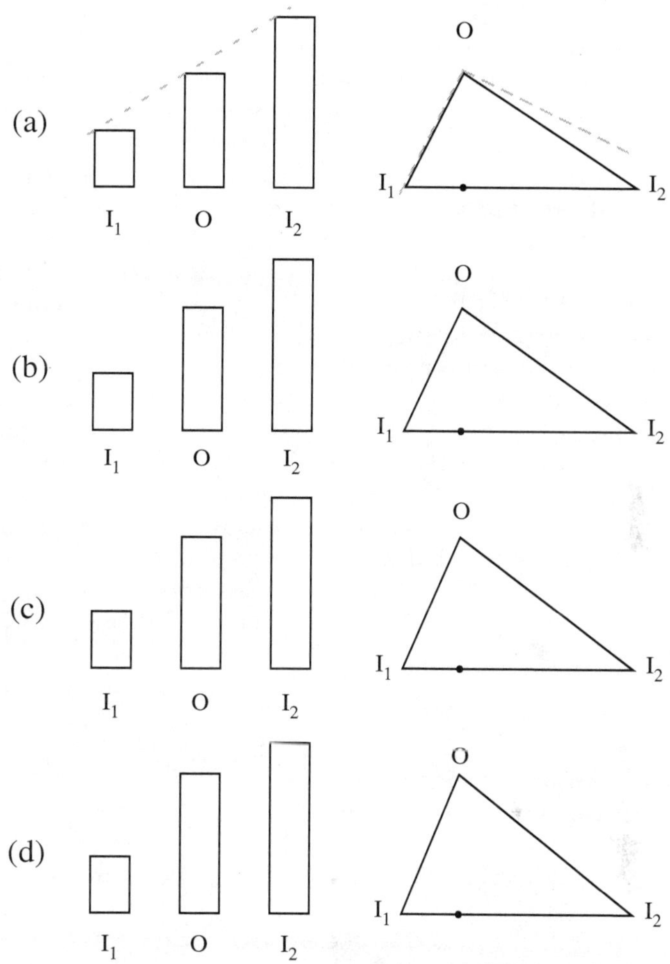


**FIGURE 9.8**
The IOI configural bar graph display (left) and a geometrical form display (right) as they would appear prior to a fault (a) and 3s (b), 6s (c), and 9s (d) after the onset of a fault. (Adapted with permission from Sanderson, P. M. et al. 1989. *Human Factors* 31:183–198. Copyright 1989 by the Human Factors and Ergonomics Society. All rights reserved.)

Sanderson et al. (1989) found that observers were far better at detecting the presence of faults with the new IOI bar graph display than with the triangle display. These performance advantages were substantial and consistent: the detection of system faults was completed in significantly less time and with significantly greater accuracy with the IOI bar graph display; this pattern of results was obtained in both experiments 1 and 2. These are the outcomes identified by the numeric label 12 in the middle column of Figure 9.4.

The interpretation of these results is based on emergent features and mapping to task constraints, rather than object integrality. The new configuration of bars in the IOI display produces a salient emergent feature that is well mapped to domain constraints. Recall that, under normal operating conditions, the value of the output variable is equal to the average of the two input values. The new arrangement of bars in the IOI display produces a linear configuration between the heights of the three bar graphs when this relationship holds true: A straight line could be drawn that would touch the top of all three of the bar graphs simultaneously. This is illustrated in Figure 9.8a by the dashed gray line at the top of the bar graphs. Sanderson et al. (1989) referred to this emergent feature as "inferred linearity" because the line itself was never actually drawn on the display.

The salience of this emergent feature and the degree to which it specifies the presence of a fault is illustrated in the left side of Figure 9.8. The four panels of the figure reproduce the dynamic changes in the IOI display 3 (9.8b), 3 (9.8c), and 9 (9.8d) seconds after the onset of a fault. The linear relationship between the heights of the bar graphs is broken as the fault progresses; the height of the middle output bar graph increases at a rate that is disproportional to the changes in the input bar graphs. This break in inferred linearity (i.e., a break in the geometrical constraints of the display) is clearly visible after only 3 seconds.

In contrast, the triangle display did not produce an emergent feature that was particularly well mapped or salient. The primary emergent feature is the size of the angle formed at the topmost vertex (i.e., the apex of the triangle). This angle only approximates a right angle under normal operating conditions (see Figure 9.8a, right side). The dashed gray lines provide a right triangle for reference purposes only (it was not present in the actual experiment). As illustrated in Figure 9.8a–d, the angle becomes progressively smaller as the fault progresses. However, these visual changes are quite difficult to detect even 9 seconds after the onset of the fault.

The results indicate that the presence of emergent features, their salience, and their mapping to task constraints are the critical factors in the design of this type of display. In this particular instance, the bar graph display produced emergent features (inferred linearity) that are more salient and well mapped to domain constraints than those produced by the triangle display (deviations in angle). This allowed observers to register, or pick up, the visual cues that signaled the onset of a fault more effectively with the bar graph display. The fact that the triangle display consists of a unitary perceptual object is far less important than the salience of the emergent features that it produces and their mapping to task demands. In Sanderson and others' words (1989): "These results demonstrate that an object display often will not support better integrated [divided] task performance (failure detection) than will a separated bar graph display. In both experiments display proximity has not served task proximity well" (p. 195).

#### *9.4.1.3 Configurality, Not Object Integrality: II*

Coury, Boulette, and Smith (1989) also found empirical results favoring a bar graph display relative to a geometric form display (a polar coordinate display) on a divided-attention task. The benefits in performance are reasonably consistent across experimental conditions: Five significant effects were found favoring a bar graph display (these are the outcomes identified by the label 7 in the middle column of Figure 9.4). These findings are also inconsistent with the predictions of PCP, but can be reasonably interpreted using the principles of the semantic mapping approach.

Participants performed a divided-attention task (multiple-cue judgment) in this study. They were presented with a display representing the value of four system variables. The experimental task was to categorize these values into one of four system states. Each state was defined by a specific range of values across the four system variables, as illustrated in Table 9.1. The experimental task clearly qualifies as a divided-attention task since the value of all four variables must be considered to determine the correct system state. Coury et al. (1989) evaluated two graphical displays and an alpha-numeric display using this classification task.

The propositional representation of the task constraints presented in Table 9.1 is precise but cumbersome. It is a great example of a fundamental point raised in Chapter 7—that the representation of a problem has a profound influence on the ease or difficulty of its solution. This representation obscures critical aspects of the task: that the relationships between some variables are critical to performance of the experimental task, while the relationships between other variables are irrelevant. The task constraints are more easily understood when they are illustrated graphically. They will be considered in the context of one of the graphical representations evaluated by Coury et al. (1989).

The bar graph display, illustrated in Figure 9.9a, had unique representations for each of the four system variables. The relationships between individual

**TABLE 9.1**

Ranges of Values for System Variables That Define Four Alternative System States

| System State | System Variable | | | |
|---|---|---|---|---|
| | Q | M | B | H |
| 1 | 25–51 | 49–75 | 0–26 | 74–100 |
| 2 | 25–51 | 49–75 | 74–100 | 0–26 |
| 3 | 49–75 | 25–51 | 0–26 | 74–100 |
| 4 | 49–75 | 25–51 | 74–100 | 0–26 |

*Source:* Adapted with permission from Coury, B. G., M. D. Boulette, and R. A. Smith. 1989. *Human Factors* 31:551–570. 

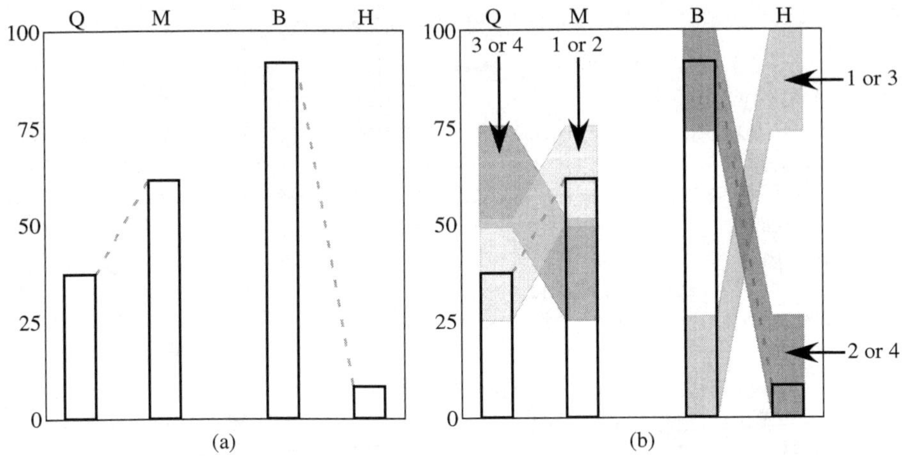


**FIGURE 9.9**
(a) A configural bar graph display that produces salient and well-mapped emergent features (inferred linearity—dashed lines) that can be used to determine system state. (b) The mapping between task constraints and geometrical constraints of the display. (Adapted with permission from Coury, B. G., M. D. Boulette, and R. A. Smith. 1989. *Human Factors* 31:551–570. Copyright 1989 by the Human Factors and Ergonomics Society. All rights reserved.)

variables and system states that must be considered for successful completion of the classification task are illustrated in Figure 9.9b. Each of the four graphical regions that have been superimposed on the bar graphs represents the constraints of the task as they are mapped into this particular representation. It is important to note that each graphical region does not correspond to a specific system state. Rather, each represents critical relationships between two pairs of variables (Q vs. M and B vs. H) that must be considered jointly to perform the classification task successfully.

First, consider the relationship between Q and M (i.e., the left pair of bar graphs in Figure 9.9b). The specific relationship between these two variables will satisfy the requirements for two system states and eliminate the remaining two system states from consideration. Thus, each superimposed graphical region represents a pair of system states and the range of values for individual variables that satisfy the associated classification rules. For example, the values of Q and M in Figure 9.9 fall within the range that satisfies either state 1 or 2 (light gray region) and eliminates states 3 and 4 (dark gray region).

Similarly, the relationship between variables B and H (the right pair of bar graphs) differentiates between another jointly specified pair of system states (i.e., state 1 or 3 versus state 2 or 4) that are also represented graphically. For example, the specific relationship between B and H falls in the region that specifies either state 2 or 4 (dark gray region). Since state 4 is eliminated by the Q and M relationship, state 2 is the correct classification.

Coury et al. (1989) obtained the best levels of performance with this display. It does not incorporate variables into a single geometric form (an object display, in PCP terminology). Once again, the interpretation from the semantic mapping perspective relies upon the presence of salient emergent features that are well mapped to task constraints. The emergent features produced by this display are the relative heights of bar graphs (see Figure 9.9a). These emergent features are emphasized by the imaginary dashed gray lines between the critical pairs of bar graphs (these lines were not actually in the displays). Note that these are the same emergent features (i.e., inferred linearity) that Sanderson et al. (1989) found to be effective.

Thus, the display produces salient emergent features that have a direct and unequivocal mapping to system states and the constraints of the categorization task. These emergent features provide a graphical "shorthand" solution to the problem of categorizing system state. When the orientation of the two inferred lines between pairs of critical variables both point upward (i.e., "/ /" or up–up), state 1 is specified. Similarly, up–down ("/ \") specifies state 2, down–up ("\ /") specifies state 3, and down–down ("\ \") specifies state 4.

The second graphical display evaluated by Coury et al. (Figure 9.10a) represents the four system variables as a geometrical form (an object display in PCP terms). This is a variation of the polar coordinate format previously described (see Figure 9.6). The primary difference is that only four poles are used, producing a four-sided polygon (see the black lines in Figure 9.10a) instead of an octagon. Note that the same values for individual variables

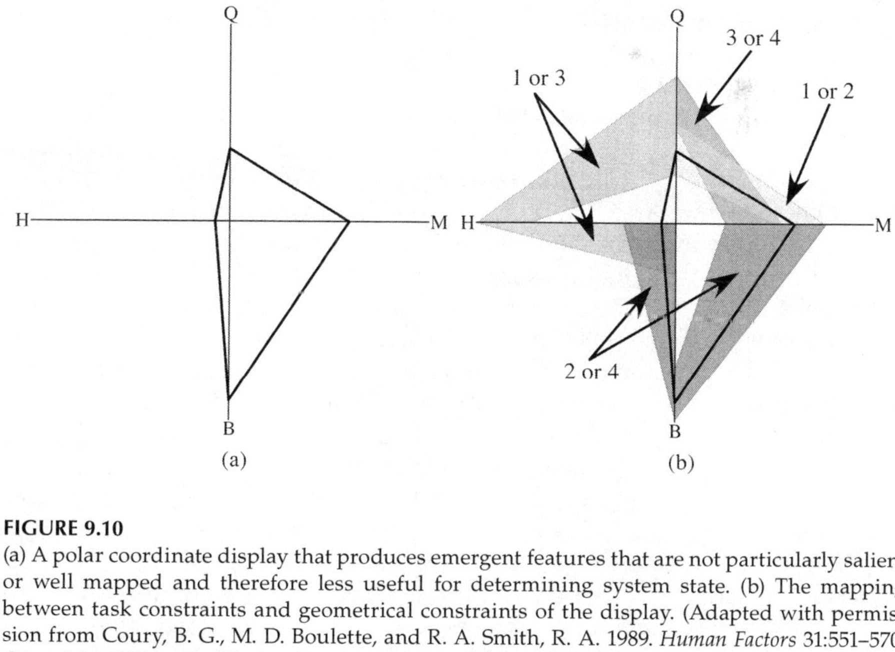


**FIGURE 9.10**
(a) A polar coordinate display that produces emergent features that are not particularly salient or well mapped and therefore less useful for determining system state. (b) The mapping between task constraints and geometrical constraints of the display. (Adapted with permission from Coury, B. G., M. D. Boulette, and R. A. Smith, R. A. 1989. *Human Factors* 31:551–570. Copyright 1989 by the Human Factors and Ergonomics Society. All rights reserved.)

are used in both Figures 9.9 and 9.10. Note also that the same general strategy and encoding conventions for illustrating the mapping between task constraints and geometrical constraints are used in Figure 9.10b. Thus, the superimposed graphical regions represent a pair of system states and the range of values for individual variables that satisfy the associated classification rules.

The numerous, salient, and hierarchically nested emergent features produced by this display format were described in Section 9.4.1. Some of these emergent features are both salient and relevant for performance of the state categorization task. For example, the required visual comparisons between the B and H variables are mapped into global emergent features of the overall geometrical form that uniquely specify system states (see Figure 9.10b): States 1 and 3 (light gray fill) are characterized by an elongated polygon that points to the left; states 2 and 4 (dark gray fill) are characterized by an elongated polygon that points down.

On the other hand, the emergent features that support the equally critical comparisons between the Q and M variables are far less discriminable. The superimposed regions in Figure 9.10b that correspond to state 1 or 2 (light gray fill) and state 3 or 4 (dark gray fill) have a high degree of overlap. Essentially, the critical visual information for this discrimination has been relegated to a relatively inconspicuous local emergent feature of the overall form: the orientation of the line connecting the Q and M poles. In other circumstances, line orientation can be a salient emergent feature (see the discussion of the mass balance indicator in the next chapter). However, the range of orientations that need to be discriminated in this particular mapping are quite small; a change of only approximately 3° can specify an alternative system state.

The problem is further exacerbated by the fact that the overlap in defined system states (i.e., a small range of variables that are acceptable for all four states, the uncertainty referred to by Coury et al. 1989) produces changes in line orientation that are approximately the same order of magnitude. The end result is that the critical emergent features produced by the display for this discrimination are neither salient nor well mapped.

A second major problem with this display is that it simultaneously produces a number of salient emergent features that are completely irrelevant for successful performance of the task. The task constraints are such that the relationships between only two pairs of variables (i.e., Q vs. M and B vs. H) are critical; all other relationships (e.g., Q vs. H) are totally irrelevant. Unfortunately, by its very nature, the polar coordinate display visually specifies many of these meaningless relationships. For example, the overall form of the intermediate and salient emergent feature produced by the two line segments that connect the H, Q, and M variables (e.g., a peak) does not map to the task constraints. The meaningless emergent features will be quite difficult to ignore (see Chapter 8, Section 8.5, for the potential negative impact of irrelevant emergent features).

In summary, Coury and others' (1989) findings of improved performance for a bar graph display on a divided-attention task, relative to a geometrical form display, are entirely consistent with the principles of semantic mapping. The bar graph display contained a set of emergent features that were salient, discriminable, and well mapped to task constraints. Although the polar coordinate display produced a wide variety of emergent features, only a few of them were both salient and directly mapped into the constraints of the task.

It is possible to design a visual display that provides an even more direct mapping between geometrical and task constraints than that provided by the bar graph display. Such a display is illustrated in Figure 9.11. The horizontal and vertical axes of this display are used to graph the difference between each of the two pairs of critically related variables. The data representation is condensed into the spatial location of a single point. The $x$ coordinate of the point is obtained by subtracting the value of H from the value of B; the $y$ coordinate is obtained by subtracting Q from M. The

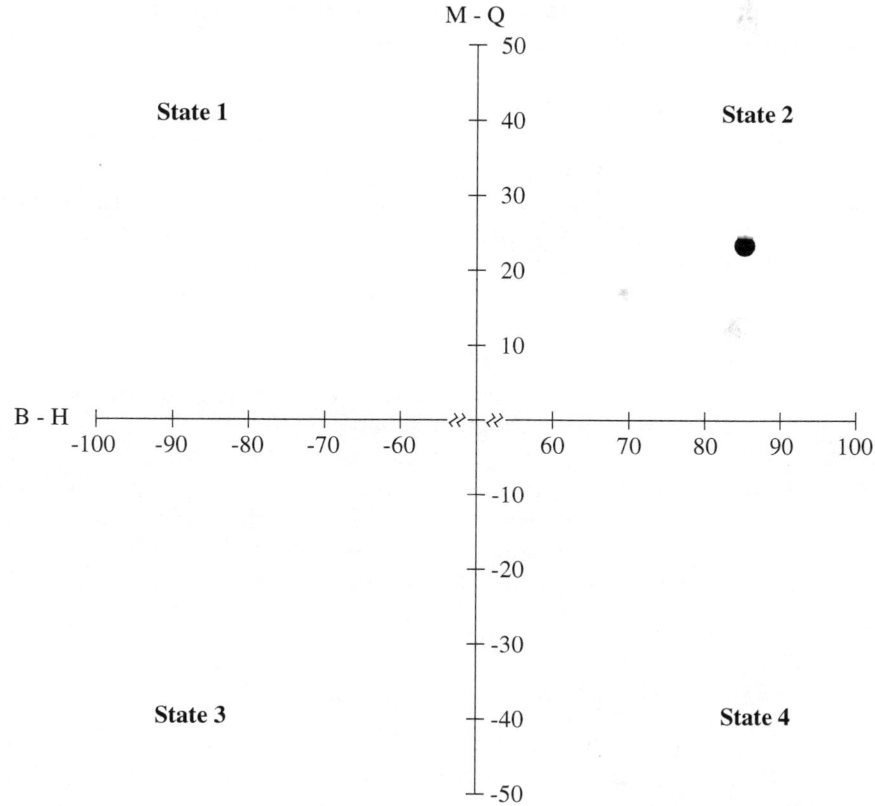


**FIGURE 9.11**
A configural coordinate display that maps the values of four system variables directly into one of four system states by virtue of the spatial location of a single point.

resulting point is then plotted in the graph. Each of the four quadrants in the graph corresponds to one of the four states of the system (see the alphanumeric labels in Figure 9.11); the spatial location of a single point directly specifies the appropriate system state. We would predict that this extremely direct mapping between geometrical and task constraints would produce significantly better performance than either the bar graph or the polar coordinate display.

#### *9.4.1.4 Summary*

The semantic mapping approach provides a very robust interpretation of the early laboratory research conducted on divided attention and display design. The large percentage of findings where separable displays produced consistently better performance than object displays (i.e., the middle column of Figure 9.4) were impossible to interpret using the PCP principles of design (i.e., perceptual objects and integral stimulus dimensions). On the other hand, the interpretation of these results is relatively straightforward when the organizing principles of configural stimulus dimensions and emergent features are applied, as outlined in the previous sections.

To reiterate the principles of the semantic mapping approach, most representational choices will produce a specific set of display constraints in the form of hierarchically nested emergent features. If these emergent features are salient (i.e., they can be picked up easily by the human observer) and if they reflect critical aspects of the task (i.e., the constraints of the work domain), then performance will be enhanced. On the other hand, if the emergent features are not salient or if they are not well mapped to domain constraints, then performance will be degraded. This is true whether the representational format is a geometrical form, a collection of bar graphs, a point in space, or any other representational form that could be devised.

Although specific analyses of the results where geometric forms produced superior performance to bar graph displays (i.e., the left-hand column of Figure 9.4) were not provided, they are all interpretable using these principles. Instances where there are no performance differences between displays (i.e., the right-hand column of Figure 9.4) indicate that the emergent features are equally salient and well mapped (or equally not salient or poorly mapped, as the case may be).

### 9.4.2 Semantic Mapping and Focused Attention

At the other end of the continuum of tasks, observers will need to obtain information from visual displays regarding the value of individual variables (i.e., focused-attention tasks). As discussed earlier, PCP predicted inevitable costs when geometric form displays (multiple variables in a single representation) are used, relative to separable displays (unique representations for each variable). These predictions were based on principles of object integrality:

dimensional integrality (coprocessed, inseparable stimulus dimensions) and perceptual objects (the parts bound together by perceptual glue).

Rather than an inevitable cost, the literature review of early laboratory studies revealed only occasional costs (see Figure 9.5). This pattern is very consistent with predictions based on configural stimulus dimensions, as detailed in the previous chapter. The lower level graphical elements (i.e., the parts of a geometrical form) interact to produce higher order emergent features that can be quite salient. However, information about these graphical elements does not disappear (as with integral stimulus dimensions); rather, it coexists alongside the emergent features. The forest does not always hide the trees; there is no perceptual glue binding the parts into a whole. Information regarding individual variables is available alongside emergent features and can be accessed when needed. In fact, observers may focus their attention at any of the various levels in the nested hierarchy at their discretion.

This pattern of results clearly supports our conceptualization of the problem in terms of configural stimulus dimensions, as opposed to object integrality. However, from the practical perspective of display design, the fundamental problem is far from being resolved. Although statistically significant performance decrements were relatively rare, the average performance for focused-attention tasks with geometric form displays was generally lower, as noted by Wickens and Carswell (1995): "But 30 of the studies showed trends in this direction, revealing significance when a meta-analysis perspective is taken" (p. 483).

We conducted a series of studies to investigate strategies that might be used to design geometrical form displays to offset these potential costs. Our original conceptualization of the problem was derived directly from the visual attention and form perception literature, which suggests that the potential for focused attention costs results from imbalances in perceptual salience. The graphical elements of a geometrical form are readily available, but can be less salient than the emergent features that they produce. Our initial studies were therefore aimed at increasing the salience of these graphical objects.

#### 9.4.2.1 Design Techniques to Offset Potential Costs

Bennett et al. (2000) investigated four design techniques to achieve this goal. Two versions of the same basic configural display were evaluated. The "baseline" display (see Figure 9.12a) mapped four system variables into a rectangular configural display (Bennett et al. 1993). Four display design techniques were applied to the baseline display to produce a "composite" version (Figure 9.12b). Three of these techniques increased the salience of individual variables by providing augmented representations.

The extender technique connected the sides (i.e., the individual variables) of the rectangular geometric form to the appropriate scale on the axes.

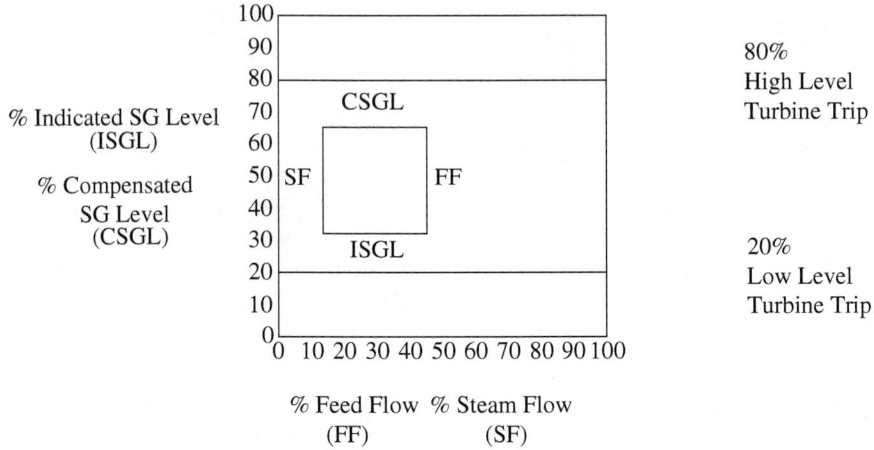


(a)

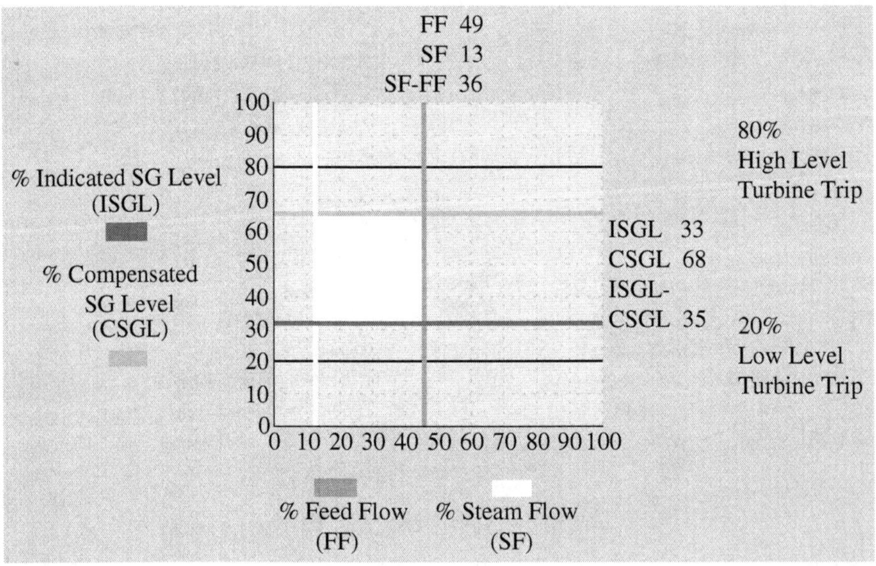


(b)

**FIGURE 9.12**
Four display design techniques used to offset potential costs with geometric form displays and focused attention tasks. (a) A configural display with no design techniques applied. (b) A configural display with all four design techniques applied. (Adapted with permission from Bennett, K. B. et al. 2000. *Human Factors* 42:287–298. Copyright 2000 by the Human Factors and Ergonomics Society. All rights reserved.)

The color/layering/separation technique color coded the four sides of the rectangle and the extenders (and applied layering techniques to stratify information visually, such as the background mat missing in Figure 9.12a).

The digital display technique provided exact values of individual variables through labels and digital values that were used to annotate the analog configural display.

The fourth design technique, the scales technique, incorporated a display grid to provide a visual context for individual values.

Bennett et al. (2000) found that the composite display significantly improved performance at the focused-attention task under the majority of experimental conditions. It produced significantly better performance for both accuracy and latency when the displays were available for inspection during the focused-attention task. It also produced significantly more accurate responses when the displays were removed from sight prior to the focused-attention task. Thus, these results strongly suggest that the four design techniques could be used to improve focused attention with geometrical form displays. It is important to note that only one of the studies (Bennett et al. 1993) in the original literature review applied any of these design techniques to the configural displays that were evaluated.

#### 9.4.2.2 Visual Structure in Focused Attention

The results of Bennett et al. (2000) demonstrate the combined utility of the four design techniques, but the relative contribution of each technique could not be uniquely identified. Bennett and Walters (2001) continued this line of investigation by teasing apart the individual contributions. The design techniques were applied individually and in combination to form a total of 10 displays. Eight of these displays were formed through a factorial combination of three techniques (scales, color/layering/separation, and bar graph/extender) applied at two levels (present or absent). The baseline display (no techniques applied) is illustrated in Figure 9.13a. These three techniques applied in isolation are illustrated in Figure 9.13b, 9.13c, and 9.13d (scales; color, layering, and separation; and bar graph and extender, respectively). The final two displays incorporated the fourth design technique of digital values. The composite display (Figure 9.13e) had all four design techniques applied. The final display (the digital display) consisted of digital values alone (Figure 9.13f). All 10 displays were visually present during performance of the focused-attention task.

The experimental manipulations and the associated results provide the basis for a much more detailed understanding of the factors that contributed to successful performance of the focused-attention task. Bennett and Walters (2001) found that three of the four design techniques produced significant improvements in performance for focused-attention tasks relative to the baseline display.

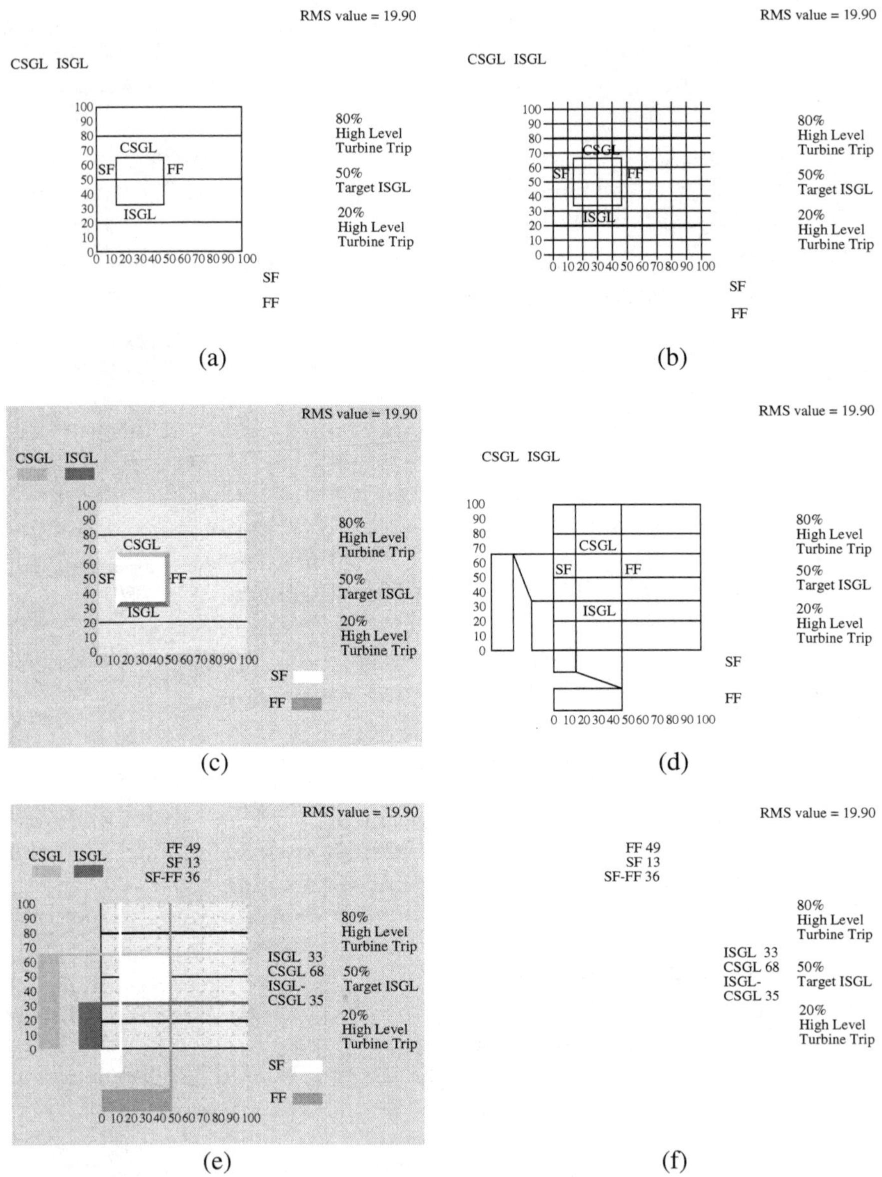


**FIGURE 9.13**
Various displays used in assessing impact of four display design techniques in offsetting potential costs. (Adapted with permission from Bennett, K. B., and B. Walters. 2001. *Human Factors* 43 (3): 415–434. Copyright 2001 by the Human Factors and Ergonomics Society. All rights reserved.)

There are several reasons why completing the focused-attention task with the baseline display was particularly poor. The primary one is that the task required several different types of mental estimation. The appropriate data marker for an individual variable (i.e., the corresponding side of the rectangle) needed to be determined. The physical location of this marker needed to be compared to the two closest numeric scale labels on the appropriate axis. A numeric estimate of the spatial distance between the data marker and these two scale labels needed to be mentally calculated. This value needed to be added to (or subtracted from) the 10's value on one of the two scale labels to obtain the final estimate.

The imprecise representation of scale (tic marks and labels only on the axes), the physical distance between the data marker and the scale (often exceeding the limits of foveal vision), and the requirement to use limited capacity working memory (numerical estimates of spatial distance, maintaining these estimates in memory, performing mental math) made performing the focused-attention task difficult with the baseline display.

Two of the design techniques (bar graph and extender and scales) significantly improved both the accuracy and latency of responses relative to the baseline display. These techniques provided additional analog visual structure (i.e., representations of scale that extended across the entire display grid; representations of individual variables spatially located next to scale markings) that was directly relevant to the focused-attention task. These techniques facilitated visual comparisons, thereby allowing powerful perceptual processes to replace one or more of the mental estimations that were required with the baseline display.

The color/layering/separation technique failed to improve performance significantly. This technique provides an important form of visual structure (chromatic contrast) that has been demonstrated to improve performance at a variety of other types of tasks (e.g., visual search). Applying this technique most certainly raised the salience of the representations of individual variables. However, the constraints of the focused-attention task require visual structure that testifies with regard to the quantitative value of an individual variable. Although color can be used in this way (see Tufte, 1990, for some excellent examples), in the present display it was used only to provide categorical information (primarily class membership or which one of four variables). Therefore, performance was not improved.

The digital values simply eliminated all of the mental estimates or extrapolations that were required to complete the focused-attention task with the baseline display. This technique was clearly the most effective: All contrasts comparing the two displays with digital values to all other displays without digital values were significant. In contrast to color coding, it offers visual structure that provides an exact match to the constraints of the task.

#### *9.4.2.3 Revised Perspective on Focused Attention*

Our initial conceptualization of the problem of supporting focused attention when designing geometrical form displays was based on insights from the visual attention and form perception literature. The original design strategy was to raise the salience of the graphical elements relative to the emergent features that they produce. However, a design strategy that clearly raised the salience of the graphical elements (i.e., color coding) was ineffective. It became clear that the issues in designing displays to support focused-attention tasks are isomorphic to the issues in designing to support divided-attention tasks.

Specifically, performance depends upon the quality of very specific mappings between the constraints of the task, the constraints of the display (including perceptual salience), and the constraints of the agent. The results of our studies on focused-attention tasks indicate that the three successful design techniques provided either additional analog visual structure or precise digital information (i.e., display constraints) that matched the constraints of the task (i.e., provided a quantitative estimate of an individual variable). Visual salience, in and of itself, was not sufficient.

## 9.5 Design Strategies in Supporting Divided and Focused Attention

One overarching issue in display design that has been addressed in this chapter is whether a single display can support a range of tasks (from focused to divided attention) or if multiple, specialized representations are required instead. The results of our studies have made it very clear that the design solution involves the combination of two of the three fundamental representational formats that were described in Chapter 6: analogical and propositional formats.

Configural displays provide analogical visual structure (i.e., geometrical constraints). As outlined in previous sections, the challenge lies in designing nested hierarchies of visual structure (i.e., global, intermediate, and local emergent features) that reflect the semantics of a domain (i.e., domain constraints). The spatiotemporal behavior of these displays will specify the affordances of a domain when they are designed properly. In turn, configural displays will provide signals that can be used to support skill-based behaviors (i.e., agent constraints). These analogical models can transform difficult cognitive tasks (i.e., reasoning about complicated goals, properties, and constraints) into relatively easy perceptual ones.

In contrast, propositional representations (i.e., digital values and alphanumeric labels) are fundamentally different representational forms that are very effective at representing detailed, precise information. Thus, if the exact value of an individual variable is needed for any purpose (e.g., monitoring a critical value, communicating its value to others, completing a checklist, or providing input to software modules), it should be represented using digital values.

The appropriate design strategy is to combine these two representational forms by annotating the geometrical form with digital values. This is made explicit in the findings of the Bennett and Walters (2001) study described earlier. The presence of an analog configural display (see Figure 9.13a–e) was necessary for successful control of a process control system (a complicated divided-attention task). On the other hand, providing digital values was clearly the most effective design strategy for improving performance at the focused-attention task. The composite display, with both types of information, was clearly the most effective display when overall performance at both divided- and focused-attention tasks is considered. Bennett and Walters conclude that "participants could select and use the specific design features in the composite display [configural display, digital values] that were appropriate for tasks at each boundary [divided- and focused-attention tasks]" (p. 431).

Hansen (1995) echoed these sentiments and takes the logic one step further: "Human factors researchers should not treat the discussion of graphical versus analytical (e.g., numerical) interfaces as an either/or issue. Instead, they should be studying ways to improve the integration of these interfaces" (p. 542). See Calcaterra and Bennett (2003) for a study investigating just how the process of annotating geometrical forms with digital values should proceed.

## 9.6 PCP Revisited

Our analysis of PCP until this point has been limited to its initial conceptualization. Wickens and Carswell (1995) proposed a revised version of PCP, noting that the original "strong form of the PCP interaction, shown in Figure la [Figure 9.1], does not emerge from many experimental results" (p. 490). A major change was the incorporation of principles of design based on configurality and emergent features (Wickens and Carswell 1995). These conceptual changes moved PCP to a closer approximation of our semantic mapping approach; the changes are the most likely source of the confusion and misinterpretation referred to in the beginning of this chapter. Despite these changes, we believe that these two approaches are fundamentally different and we will conclude this chapter by describing why.

One fundamental difference is that the revised version of PCP has retained the organizing principles of "object integration," which are identified as one of the four fundamental processing mechanisms or "forces underlying the effects observed in the PCP" (Wickens and Carswell 1995, p. 485). The concept of a perceptual object still plays a fundamental role: Wickens and

Carswell (1995) state that object integration "involves arranging information sources so that they appear to the user to be part of a single object" (p. 478). They make frequent references to object displays and to the processing benefits and costs incurred as a result. The role of integrality was reduced, particularly relative to the early conceptualization of object displays as being composed of integral stimulus dimensions (compare to Carswell and Wickens 1990).

However, numerous references are still made to integral stimulus dimensions and dimensional integrality. As the current and previous chapters indicate very clearly, we do not believe that the concepts of perceptual objects or integral stimulus dimensions play a decisive role in display design. Retaining these principles while adding principles of configurality appears to produce conceptual difficulties that unnecessarily complicate the process of building and evaluating effective displays (see the quote at the beginning of the chapter).

At a more fundamental level, the PCP and the semantic mapping approaches vary with regard to the distinction made in Chapter 2 between the dyadic and triadic approaches to semiotics. At its heart, the PCP approach is dyadic in nature. Consider the two fundamental sources of proximity. Wickens and Carswell (1995) state that "the PCP depends critically on two dimensions of proximity or similarity: perceptual proximity and processing proximity" (p. 473). Perceptual proximity is defined as "how close together two display channels conveying task-related information lie in the user's multidimensional perceptual space" (p. 473). The second source of proximity, processing proximity, is defined as "the extent to which the two or more sources are used as part of the same task. If these sources must be integrated, they have close processing proximity. If they should be processed independently, their processing proximity is low" (p. 474).

These two dimensions of proximity clearly identify PCP as a dyadic approach to display design. Consistent with Saussure's version of semiotics, the problem of display design is framed in terms of the relation between the sensory surfaces of an agent (i.e., perceptual proximity) and the internal concepts in the agent's mind (i.e., processing proximity). It is deeply rooted in the traditional information processing approach where cognition and meaning are viewed as artifacts of the mind, almost completely divorced from the situations in which they occur. In terms of Figure 2.1 in Chapter 2, PCP focuses on interpretation by studying the relationships between signifier (representations, displays) and signified (concepts, processing inside the head).

The third dimension of Peirce's triadic model, whereby meaning is established with regard to the ecology of a work domain, is virtually ignored. It is true that the dimension of processing proximity is also sometimes referred to as task proximity (e.g., Figure 9.1). This gives the impression that meaningful aspects of the work domain are considered. However, more detailed analysis of task proximity reveals that this impression is by and large misleading.

The dimensions of task proximity fall into two categories. One category couches task proximity in terms of the implications for the information processing mechanisms of the agent (e.g., integrative processing, Boolean integration, nonintegrative processing, processing similarity, independent processing). The second category couches task proximity in extremely general descriptions of the work domain (e.g., metric similarity, statistical similarity, functional similarity, temporal proximity). These dimensions simply do not do justice to work ecologies. They are general and somewhat vague; they are syntactic descriptors, not semantic ones.

In contrast, the semantic mapping approach is clearly an example of Peirce's triadic approach to semiotics described in Chapter 2. It makes a clean conceptual distinction between the constraints produced by the domain, the agent, and the display (interface), as illustrated in Figure 6.6 in Chapter 6. As described throughout this chapter, the effectiveness of a display will be determined by the quality of very specific mappings between these three sets of constraints. It has its roots in the ecological approach to cognition where a detailed understanding of the ecology (i.e., the work domain) plays a fundamental role in the design solutions that are developed. The remaining chapters in the book will provide specific examples of ecological interface design.

## References

Barnett, B. J., and C. D. Wickens. 1988. Display proximity in multicue information integration: The benefits of boxes. *Human Factors* 30:15–24.

Bennett, K. B., and J. M. Flach. 1992. Graphical displays: Implications for divided attention, focused attention, and problem solving. *Human Factors* 34:513–533.

Bennett, K. B., and H. I. Fritz. 2005. Objects and mappings: incompatible principles of display design. A critique of Marino and Mahan. *Human Factors* 47 (1): 131–137.

Bennett, K. B., A. L. Nagy, and J. M. Flach. 1997. Visual displays. In *Handbook of human factors and ergonomics,* ed. G. Salvendy. New York: John Wiley & Sons.

Bennett, K. B., M. Payne, J. Calcaterra, and B. Nittoli. 2000. An empirical comparison of alternative methodologies for the evaluation of configural displays. *Human Factors* 42 (2): 287–298.

Bennett, K. B., M. L. Toms, and D. D. Woods. 1993. Emergent features and configural elements: Designing more effective configural displays. *Human Factors* 35:71–97.

Bennett, K. B., and B. Walters. 2001. Configural display design techniques considered at multiple levels of evaluation. *Human Factors* 43 (3): 415–434.

Calcaterra, J. A., and K. B. Bennett. 2003. The placement of digital values in configural displays. *Displays* 24 (2): 85–96.

Carswell, C. M., and C. D. Wickens. 1987. Information integration and the object display. *Ergonomics* 30:511–527.

———. 1990. The perceptual interaction of graphical attributes: Configurality, stimulus homogeneity, and object integration. *Perception & Psychophysics* 47:157–168.

Coekin, J. A. 1970. An oscilloscope polar coordinate display for multi-dimensional data. *Radio and Electronic Engineer* 40:97–101.

Coury, B. G., M. D. Boulette, and R. A. Smith. 1989. Effect of uncertainty and diagnosticity on classification of multidimensional data with integral and separable displays of system status. *Human Factors* 31:551–570.

Goettl, B. P., C. D. Wickens, and A. F. Kramer. 1991. Integrated displays and the perception of graphical data. *Ergonomics* 34:1047–1063.

Hansen, J. P. 1995. An experimental investigation of configural, digital, and temporal information on process displays. *Human Factors* 37:539–552.

MacGregor, D., and P. Slovic. 1986. Graphic representation of judgmental information. *Human–Computer Interaction* 2:179–200.

Marino, C. J., and R. R. Mahan. 2005. Configural displays can improve nutrition-related decisions: An application of the proximity compatibility principle. *Human Factors* 47 (1): 121–130.

Peebles, D. 2008. The effect of emergent features on judgments of quantity in configural and separable displays. *Journal of Experimental Psychology: Applied* 14 (2): 85–100.

Salas, E., and N. J. Cooke, eds. 2008. *Best of human factors: Thirty classic contributions to human factors/ergonomics science and engineering*. Santa Monica, CA: Human Factors and Ergonomics Society.

Sanderson, P. M., J. M. Flach, M. A. Buttigieg, and E. J. Casey. 1989. Object displays do not always support better integrated task performance. *Human Factors* 31:183–198.

Schaefer, W. F. et al. June 23, 1987. Generating an integrated graphic display of the safety status of a complex process plant, United States Patent 4,675,147.

Tufte, E. R. 1990. *Envisioning information*. Cheshire, CT: Graphics Press.

Wickens, C. D. 1992. *Engineering psychology and human performance*, 2nd ed. New York: Harper Collins.

Wickens, C. D., and A. D. Andre. 1990. Proximity compatibility and information display: Effects of color, space, and objectness on information integration. *Human Factors* 32:61–78.

Wickens, C. D., and C. M. Carswell. 1995. The proximity compatibility principle: Its psychological foundation and relevance to display design. *Human Factors* 37 (3): 473–494.

Woods, D. D., J. A. Wise, and L. F. Hanes. 1981. An evaluation of nuclear power plant safety parameter display systems. Paper read at the Human Factors Society 25th Annual Meeting, at Santa Monica, CA.

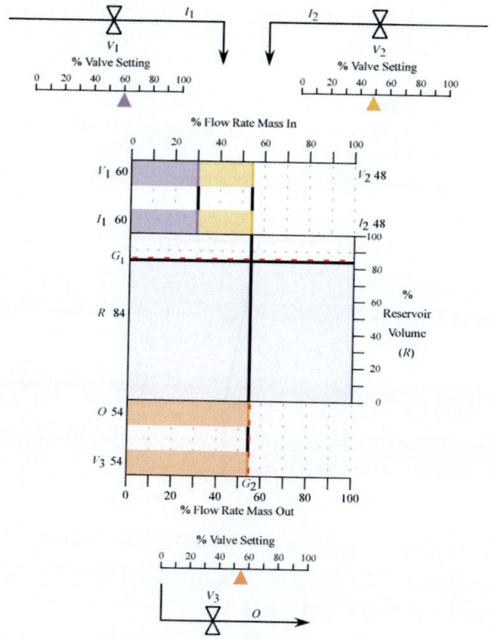


**FIGURE 10.3**
A configural display for the simple process. (Adapted with permission from Bennett, K. B., A. L. Nagy, and J. M. Flach. 1997. In *Handbook of Human Factors and Ergonomics*, ed. G. Salvendy, Copyright 1997, New York: John Wiley & Sons. All rights reserved.)

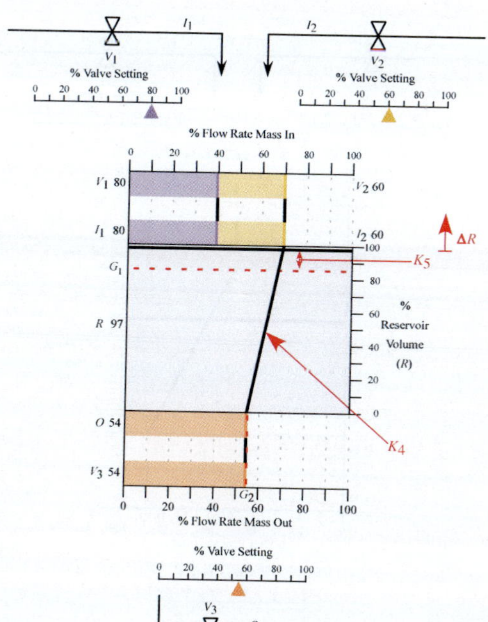


**FIGURE 10.4**
A break in the geometrical constraints of the configural display. (Adapted with permission from Bennett, K. B., A. L. Nagy, and J. M. Flach. 1997. In *Handbook of Human Factors and Ergonomics*, ed. G. Salvendy, Copyright 1997, New York: John Wiley & Sons. All rights reserved.)

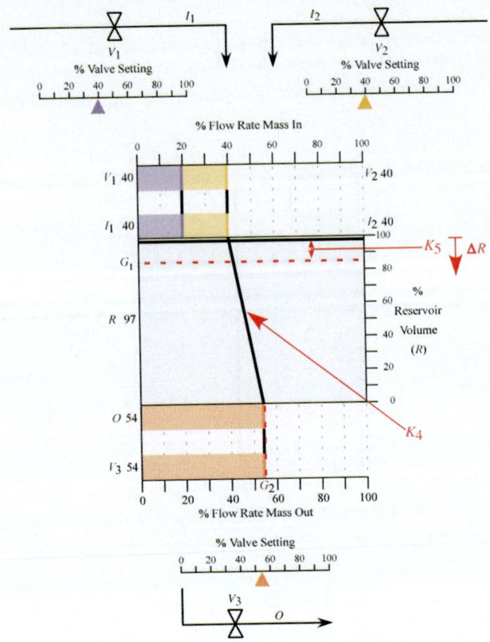

**FIGURE 10.5**
Another break in the geometrical constraints of the configural display, this time with a very different meaning. (Adapted with permission from Bennett, K. B., A. L. Nagy, and J. M. Flach. 1997. In *Handbook of Human Factors and Ergonomics*, ed. G. Salvendy, Copyright 1997, New York: John Wiley & Sons. All rights reserved.)

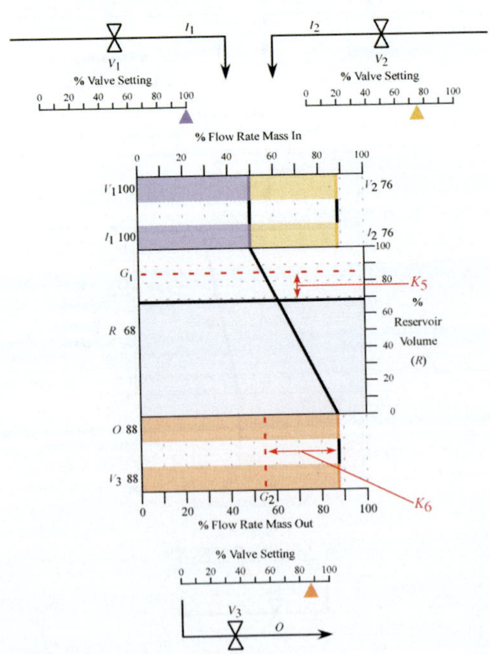

**FIGURE 10.6**
The simple system is in steady state, but a control input is required. (Adapted with permission from Bennett, K. B., A. L. Nagy, and J. M. Flach. 1997. In *Handbook of Human Factors and Ergonomics*, ed. G. Salvendy, Copyright 1997, New York: John Wiley & Sons. All rights reserved.)

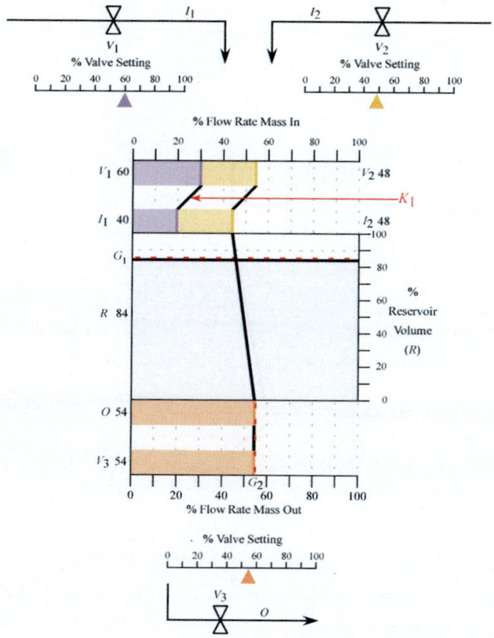


**FIGURE 10.7**
Presence and nature of a fault in the simple system is specified; it lies in the first input stream. (Adapted with permission from Bennett, K. B., A. L. Nagy, and J. M. Flach. 1997. In *Handbook of Human Factors and Ergonomics,* ed. G. Salvendy, Copyright 1997, New York: John Wiley & Sons. All rights reserved.)

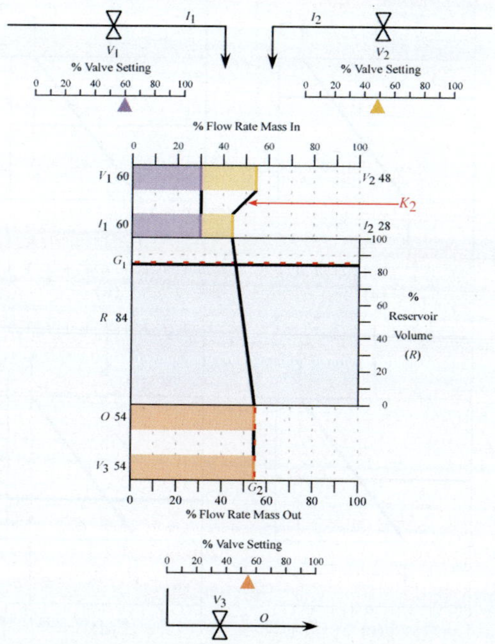


**FIGURE 10.8**
An alternative fault in the second input stream is specified. (Adapted with permission from Bennett, K. B., A. L. Nagy, and J. M. Flach. 1997. In *Handbook of Human Factors and Ergonomics,* ed. G. Salvendy, Copyright 1997, New York: John Wiley & Sons. All rights reserved.)

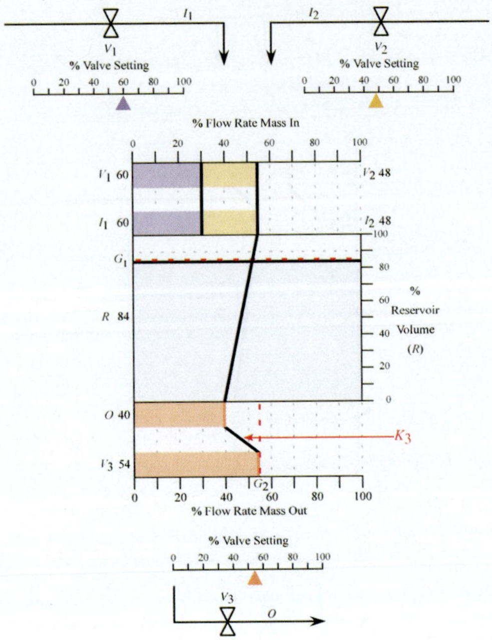

**FIGURE 10.9**
A fault in the output stream is specified. (Adapted with permission from Bennett, K. B., A. L. Nagy, and J. M. Flach. 1997. In *Handbook of Human Factors and Ergonomics*, ed. G. Salvendy, Copyright 1997, New York: John Wiley & Sons. All rights reserved.)

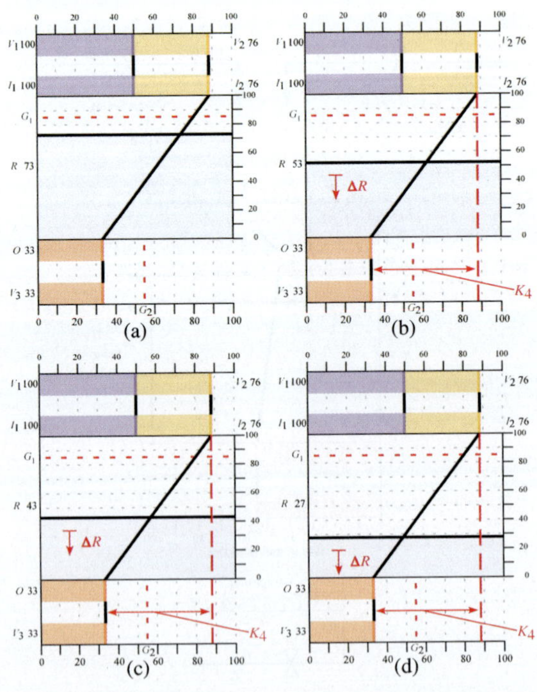

**FIGURE 10.10**
A fault in the conservation of mass is specified. (Adapted with permission from Bennett, K. B., A. L. Nagy, and J. M. Flach. 1997. In *Handbook of Human Factors and Ergonomics*, ed. G. Salvendy, Copyright 1997, New York: John Wiley & Sons. All rights reserved.)

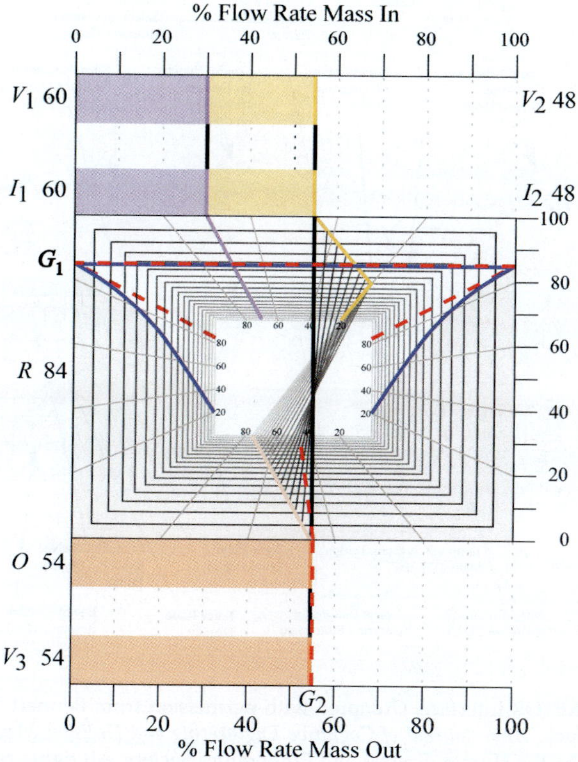


**FIGURE 10.12**
The configural display with the time tunnel design technique and the associated temporal information.

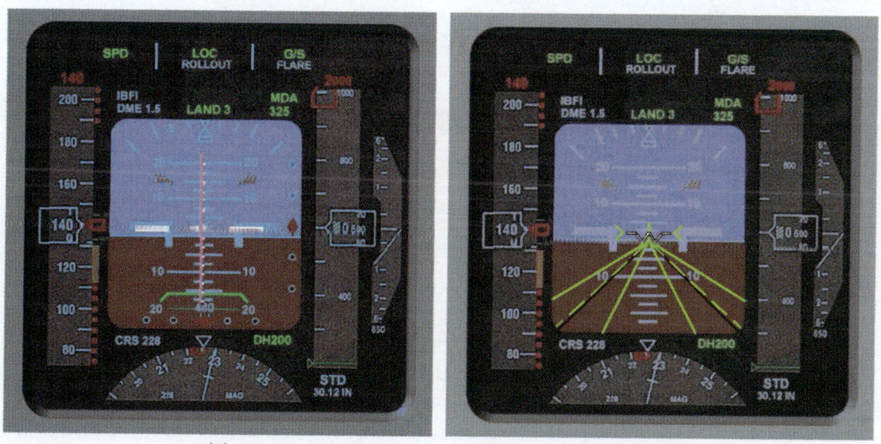


(a) (b)

**FIGURE 11.4**
These two formats, an integrated ILS display and the WrightCAD format, were evaluated in a landing task. With the ILS format (a) lateral and vertical deviations were indicated by the intersection of two bars or needles. In the WrightCAD format (b) lateral deviation was indicated by relative position of the chevrons and the aircraft symbol; vertical deviations were indicated by relative position of the splay lines against the target of 45°.

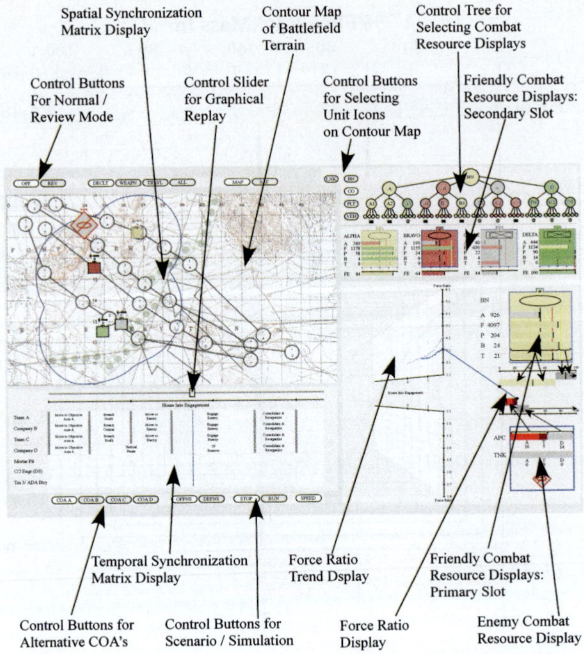


**FIGURE 14.3**
Overview of RAPTOR interface. (Adapted with permission from Bennett, K. B., S. M. Posey, and L. G. Shattuck. 2008. *Journal of Cognitive Engineering and Decision Making* 2 (4): 349–385. 

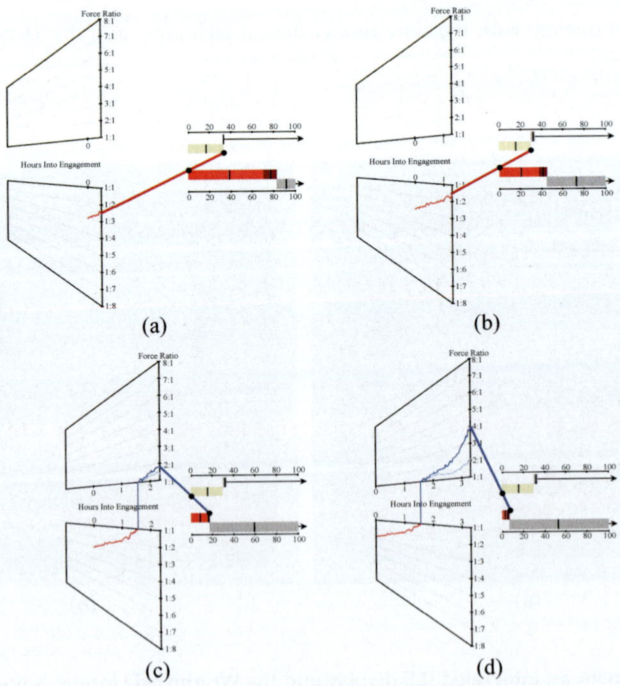


**FIGURE 14.7**
Force ratio and force ratio trend displays over time. (Adapted with permission from Bennett, K. B., S. M. Posey, and L. G. Shattuck. 2008. *Journal of Cognitive Engineering and Decision Making* 2 (4): 349–385. 

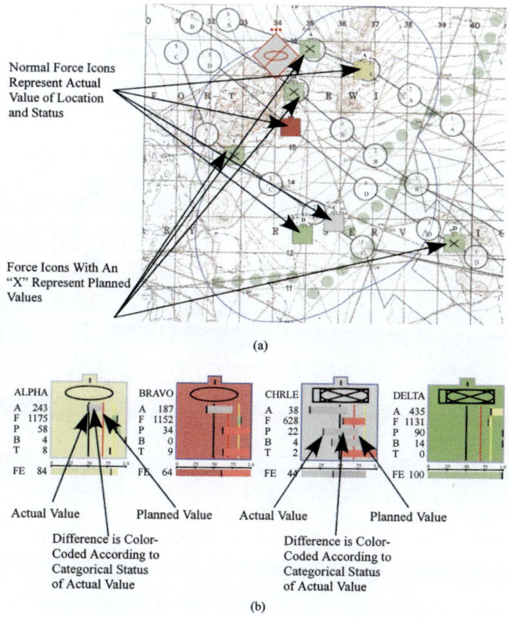


**FIGURE 14.11**
Plan review mode. (Adapted with permission from Bennett, K. B., S. M. Posey, and L. G. Shattuck. 2008. *Journal of Cognitive Engineering and Decision Making* 2 (4): 349–385. Copyright 2008 by the Human Factors and Ergonomics Society. All rights reserved.)

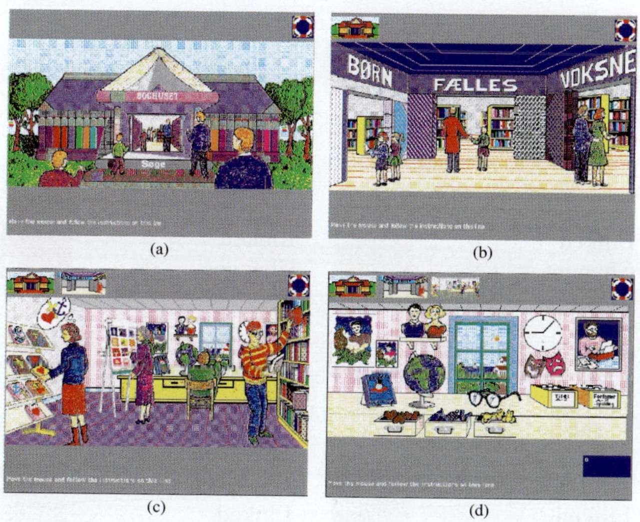


**FIGURE 15.3**
An example of the spatial structure design technique: hierarchically nested metaphors to support navigation in the BookHouse fiction retrieval system. (a) The global spatial metaphor (a virtual library; work space level). (b) A nested, intermediate spatial metaphor (i.e., the entrance hallway to select a database). (c) The search room (search strategy). (d) The analytical search room with local spatial metaphors (i.e., graphical components of interface icons) that represent search parameters. (Screenshots from the BookHouse system. Used with permission from system designer, Pejtersen, A. M. 1992. In *The Marketing of Library and Information Services 2*, ed. B. Cronin. London: ASLIB. Copyright 1992. All rights reserved.)

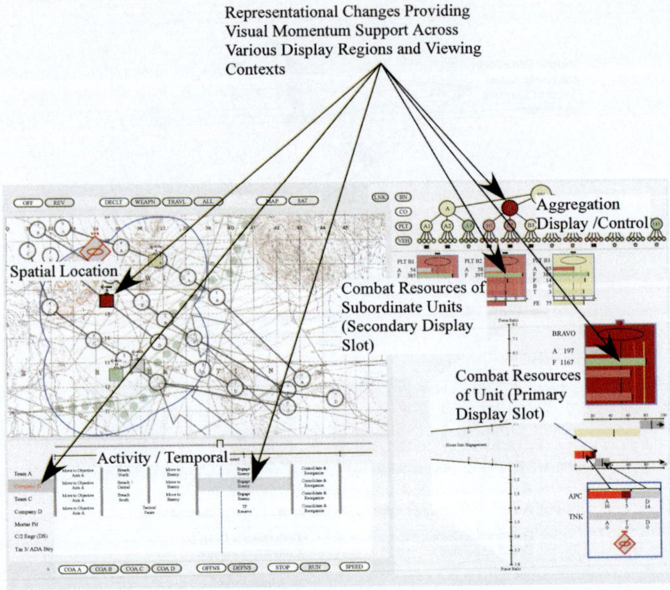


**FIGURE 15.7**
An example of the perceptual landmarks design technique: visual highlighting of unit-related information scattered across various regions of a view in the RAPTOR interface. (Adapted with permission from Bennett, K. B., Posey, S. M., and Shattuck, L. G. 2008. *Journal of Cognitive Engineering and Decision Making* 2 (4): 349–385. Copyright 2008 by the Human Factors and Ergonomics Society. All rights reserved.)

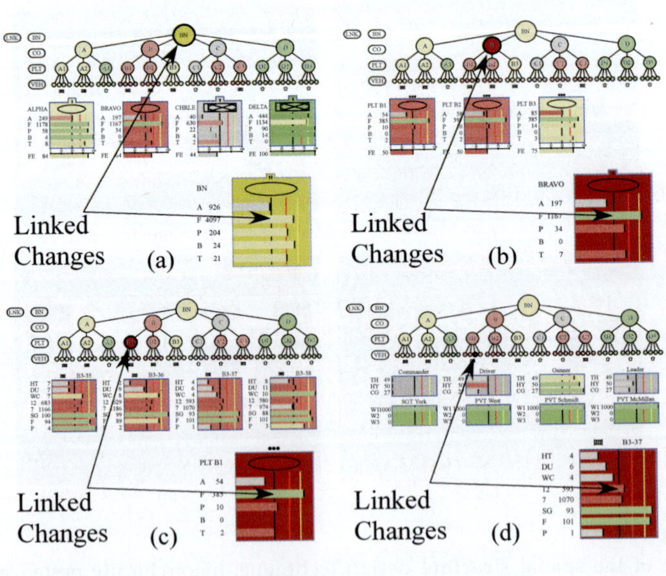


**FIGURE 15.8**
An example of the fixed format data replacement technique: the primary and secondary data slots are successively occupied by information from different units (i.e., different selective glances into the larger database). (a) Battalion level. (b) Company level. (c) Platoon level. (d) Vehicle level. (Adapted with permission from Bennett, K. B., Posey, S. M., and Shattuck, L. G. 2008. *Journal of Cognitive Engineering and Decision Making* 2 (4): 349–385. Copyright 2008 by the Human Factors and Ergonomics Society. All rights reserved.)

# 10

# *Design Tutorial: Configural Graphics for Process Control*

## 10.1 Introduction

> [P]atterns in a model configuration, as well as perceptual patterns of the physical environment, can act as signs. This is most clearly seen if externalized representations of the mental model are actually available in the form of physical models, e.g., an abacus for calculation, or in the form of graphs or other symbolic representations on paper or on visual information displays, forming artificial objects for manipulation. For display formats designed for process control, this means that rule- or skill-based control—"direct manipulation"— at a higher abstract level can be obtained if a symbolic display can be designed where there is a one-to-one mapping between the immediate appearance of the display and the properties of the process to be controlled. (Rasmussen 1986, p. 138)

This chapter provides a tutorial of ecological interface design for law-driven or correspondence-driven domains. The various principles of display and interface design described in previous chapters are woven into a coherent, concrete demonstration. A simple work domain from process control is modeled using the analytical tools (i.e., abstraction and aggregation hierarchies) of cognitive systems engineering (CSE). The process of translating these results into an effective representation is described. The analog, geometric forms in this interface are discussed in terms of emergent features and direct perception. The controls in the interface are discussed in terms of direct manipulation. The ways in which the interface can support decision making and problem solving are discussed in terms of skill-, rule-, and knowledge-based behaviors. Finally, the need for a triadic perspective in the design of interfaces for complex work domains is reemphasized through the discussion of alternative displays and the quality of the mappings between domain, interface, and agent that they produce.

## 10.2 A Simple Domain from Process Control

The process is a simple and generic one that exemplifies only some of the critical aspects of process control; it is represented graphically in the lower portion of Figure 10.1. There is a reservoir (or tank) that is filled with a fluid (e.g., coolant). The volume, or level, of the reservoir (R) is represented by the filled portion of the rectangle. Fluid enters the reservoir through the two pipes and valves located above the reservoir; fluid leaves the reservoir through the pipe and valve located below. The information in this simple process has been sorted into two categories (see the top of Figure 10.1). The term "low-level data" will be used to refer to local constraints or elemental state variables that might be measured by a sensor. The term "higher level properties" will be used to refer to more global constraints that reflect relations or interactions among variables.

### 10.2.1 Low-Level Data (Process Variables)

There are two goals associated with this simple process. First, there is a goal ($G_1$) associated with R, the level of the reservoir. The reservoir should be maintained at a relatively high level to ensure that sufficient resources are available to meet long-term demands for output flow rate (O). The second goal ($G_2$) refers to the specific rate of output flow that must be maintained. These goals are achieved and maintained by adjusting three valves ($V_1$, $V_2$, and $V_3$) that regulate flow through the system ($I_1$, $I_2$, and O). Thus, this simple process is associated with a number of elemental process variables that can be measured directly: $V_1$, $V_2$, $V_3$, $I_1$, $I_2$, O, $G_1$, $G_2$, and R (see the upper, left-hand portion of Figure 10.1).

### 10.2.2 High-Level Properties (Process Constraints)

In addition, there are relationships between these process variables that must be considered when controlling the process (see the upper, right-hand portion of Figure 10.1). The most important high-level properties are goal related: Does the actual reservoir volume level (R) match the goal of the system ($G_1$) – $K_5$? Does the actual system output flow rate (O) match the flow rate that is required ($G_2$) – $K_6$? Even for this simple process, some of the constraints or high-level properties are fairly complex. For example, an important property of the system is mass balance. The mass balance is determined by comparing the mass leaving the reservoir (O, the output flow rate) to mass entering the reservoir (the combined input flow rates of $I_1$ and $I_2$). This relationship determines the direction and the rate of change for the volume inside the reservoir ($\Delta R$). For example, if mass in and mass out are equal, then mass is balanced, $\Delta R$ will equal 0.00, and R will remain constant.

| **Low-Level Data** (process variables) | | **High-Level Properties** (process constraints) | |
|---|---|---|---|
| $T$ = time | | | |
| $V_1$ = setting for valve 1 | | $K_1 = I_1 - V_1$ | Relation between commanded flow ($V$) and actual flow ($I$ or $O$) |
| $V_2$ = setting for valve 2 | | $K_2 = I_2 - V_2$ | |
| $V_3$ = setting for valve 3 | | $K_3 = O - V_3$ | |
| $I_1$ = flow rate through valve 1 | | | |
| $I_2$ = flow rate through valve 2 | | | |
| $O$ = flow rate through valve 3 | | | |
| $R$ = volume of reservoir | | $K_4 = \Delta R = (I_1 + I_2) - O$ | Relation between reservoir volume ($R$), mass in ($I_1 + I_2$), and mass out ($O$) |
| $G_1$ = volume goal | | $K_5 = R - G_1$ | Relation between actual states ($R$, $O$) and goal states ($G_1$, $G_2$) |
| $G_2$ = output goal (demand) | | $K_6 = O - G_2$ | |

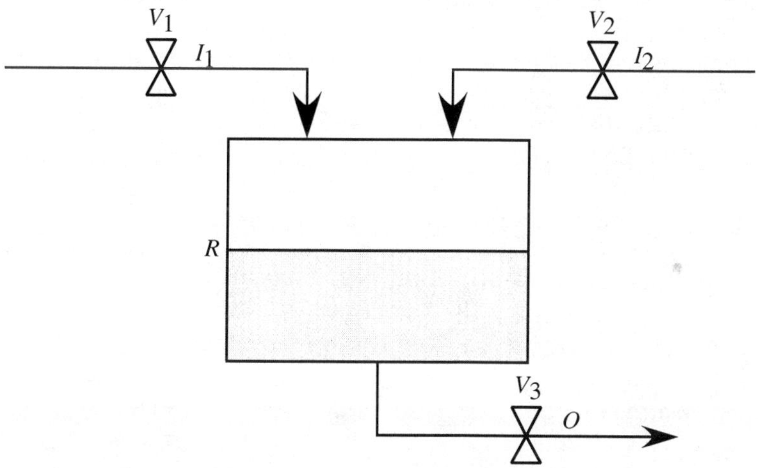


**FIGURE 10.1**
A simple work domain from process control. (Adapted with permission from Bennett, K. B., A. L. Nagy, and J. M. Flach. 1997. In *Handbook of Human Factors and Ergonomics*, ed. G. Salvendy, 

Controlling even this simple process will depend on a consideration of both high-level properties and low-level data. As the previous example indicates, decisions about process goals (e.g., maintaining a sufficient level of reservoir volume) generally require consideration of relationships between variables (whether there is a net inflow, a net outflow, or mass is balanced), as well as the values of the individual variables themselves (what the current reservoir volume is).

## 10.3 An Abstraction Hierarchy Analysis

The constraints of the simple process will be modeled using the analytical tools of CSE (primarily, the abstraction hierarchy; see Figure 10.2). As described in Chapter 3, this hierarchy has five separate levels of description, ranging from physical form to higher level purposes. The highest level of constraints refers to the functional purpose or design goals for the system. The overall purpose of this simple process is to provide coolant to a connected process. Thus, the targeted reservoir volume ($G_1$) and output flow rate ($G_2$) are located at this level. Both of these goals can be expressed in mathematical terms. For example, when the output flow rate (O) equals the output goal ($G_2$), the difference between these two values will assume a constant value (0.00). These process constraints are represented by the equations associated with the higher level properties of $K_5$ and $K_6$ in Figures 10.1 and 10.2.

The abstract functions or physical laws that govern system behavior are another important source of constraints. This level reflects the intended, proper functioning of the system. In this simple process control example, the proper function is described in terms of the flow of mass through the system. This flow is governed by the laws of nature. For example, the $K_4$ constraint reflects the law of conservation of mass. In this closed system, mass can neither be created nor destroyed; if mass enters the system, then it must be stored or it must leave. Correspondingly, any changes of mass in the reservoir ($\Delta R$) should be determined by the difference between the residual mass in ($I_1 + I_2$) and the mass out (O). $K_1$, $K_2$, and $K_3$ represent similar constraints associated with the mass flow. Flow is proportional to valve setting (this assumes a constant pressure head).

Further constraints arise as a result of the "generalized function." This level comprises the general capabilities of the system. In the present work domain, there must be a means for fluid to enter the system (i.e., a source), a means to retain the fluid within the system (i.e., a store), and a means to rid the system of fluid (i.e., a sink). One might imagine a block diagram representing these basic functions independently of the physical implementation.

The physical processes behind each general function represent another source of constraint: "physical function." This is the first level at which there is a description in terms of physical characteristics. In this case, there are two feedwater input streams, a single output stream, and a reservoir for storage. These constitute the causal connections inherent to the system. These components will possess certain functional characteristics. For example, the pipes in the system will be rated in terms of their limits for pressure, flow rates, and temperature. This is the level at which the measurement of system variables occurs: the moment-to-moment values of each variable ($V_1$, $V_2$, $V_3$, $I_1$, $I_2$, O, and R). Similarly, this is the level at which control of the system is achieved through the manipulation of these variables (e.g., changing a valve setting).

| **Abstraction Hierarchy** (Means-Ends Relations) | **Aggregation Hierarchy** (Whole-Part Relations) — Coarse Resolution → Fine Resolution |
|---|---|
| Goals, Purposes, and Constraints | $K_5 = R - G_1$<br>$K_6 = O - G_2$<br>Relation between actual states ($R$, $O$) and goal states ($G_1$, $G_2$) |
| Priority Measures and Abstract Functions | $K_4 = \Delta R = (I_1 + I_2) - O$<br>Relation between reservoir volume ($R$), mass in ($I_1$ + $I_2$), and mass out ($O$)<br>$K_1 = I_1 - V_1$<br>$K_2 = I_2 - V_2$<br>$K_3 = O - V_3$<br>Relation between commanded flow ($V$) and actual flow ($I$ or $O$) |
| General Work Activities and Functions | Two "sources"<br>A single "store"<br>A single "sink"<br>General functions that the system needs to perform, independent of physical implementation |
| Physical Activities in work, Physical Processes of Equipment | $V_1$ $V_2$ $V_3$<br>$I_1$ $I_2$ $O$<br>$R$<br>Measured variables: Mass input streams, the reservoir, the output stream |
| Appearance, Location, and Configuration of Material Objects | length, location, position size, color, etc.<br>Causal connections, length of pipes, position of valves on pipes, size of the reservoir, etc. |

**FIGURE 10.2**
An abstraction hierarchy analysis of the simple process.

Finally, the level of physical form provides information concerning the physical configuration of the system. This includes information related to length and diameter of pipes, physical location of valves on pipes, and the physical dimensions of the reservoir.

## 10.4 Direct Perception

In the following two sections, we describe the design of an interface for this simple process control work domain. Both direct perception and direct manipulation are required if the interface is to provide effective support for decision making and problem solving. The first section focuses on direct perception. The process of translating the results of a domain analysis (i.e., domain constraints) into interface representations (primarily geometrical constraints in an analog, geometrical form display) is described. The ability of this interface to support an agent working in the domain is then described in terms of skill-, rule-, and knowledge-based behavioral modes. Finally, we consider the quality of constraint matching achieved by this interface relative to other forms of representation that could be devised.

### 10.4.1 Mapping Domain Constraints into Geometrical Constraints

The abstraction hierarchy analysis described in the previous section provides information about the hierarchically nested constraints that constitute the semantics of a domain. This is essentially a model of the domain that defines the information that must be present for an individual to perform successfully. Thus, it provides a structured framework (i.e., categories of information and relationships between categories) that is essential for display design. The interface for this simple process is illustrated in Figure 10.3. A general description will be provided first, followed by a more detailed analysis in terms of the abstraction hierarchy.

The primary representations in this interface are the dynamic, analogical, and geometrical forms described in Chapters 6 through 9. The bar graphs at the top of the display represent the rates of mass flow into the reservoir. Each bar graph consists of two segments, corresponding to the two input streams. The top bar graph represents the "commanded" rates of flow (i.e., the current valve settings labeled $V_1$ and $V_2$). Thus, the combined horizontal extent of the two bar graph segments represents the "commanded" mass input: the total percentage of mass that should be flowing into the reservoir, given the valve settings made by the operator. The actual flow rates for mass in (i.e., sensor measurements) are represented by a similar set of bar graph segments ($I_1$ and $I_2$) directly below. The relationship between commanded and actual flow rates is emphasized by the bold lines that connect the segments of the

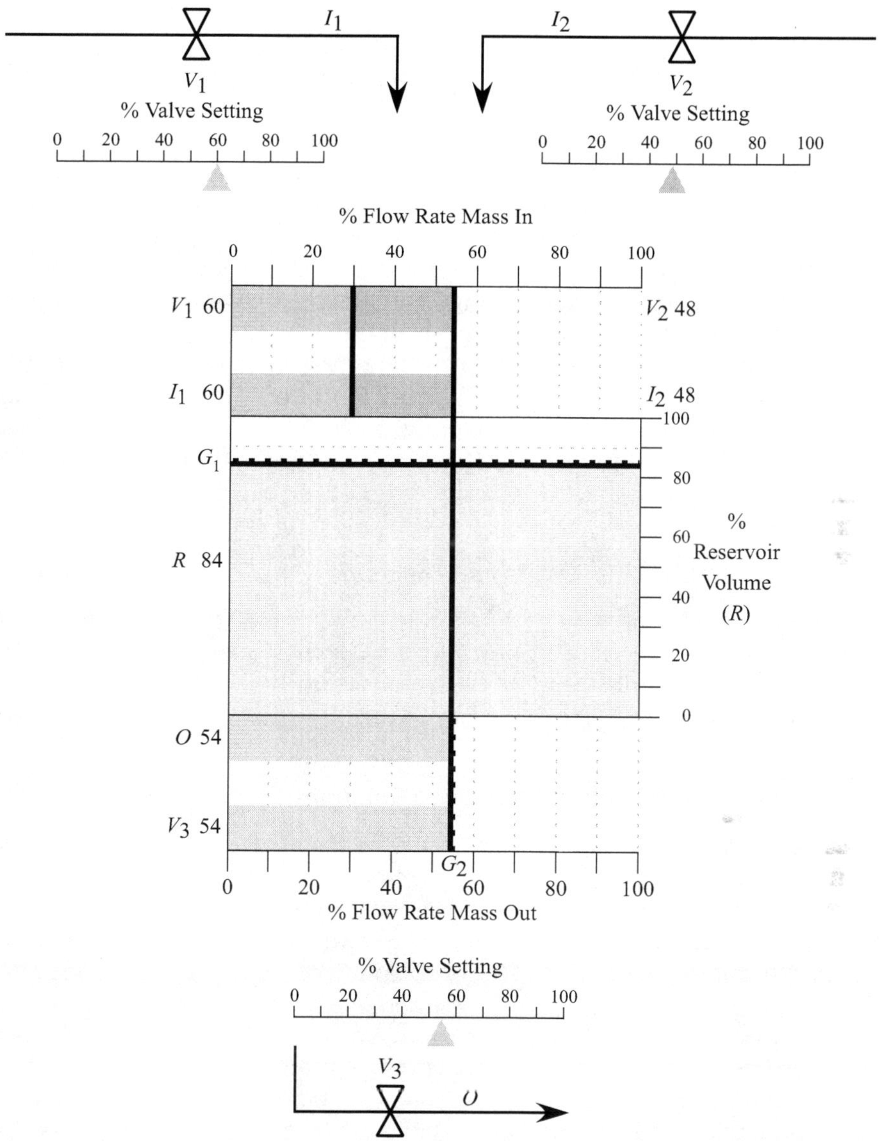


**FIGURE 10.3**
**(See color insert following page 230.)** A configural display for the simple process. (Adapted with permission from Bennett, K. B., A. L. Nagy, and J. M. Flach. 1997. In *Handbook of Human Factors and Ergonomics*, ed. G. Salvendy, Copyright 1997, New York: John Wiley & Sons. All rights reserved.)

two contribution bar graphs (i.e., the bold, vertical, and black line segments that connect $V_1$ to $I_1$ and $V_2$ to $I_2$).

The bar graphs at the bottom of the display represent the rate of mass flow out of the reservoir. The top bar graph (O) represents the actual (measured) flow rate of mass leaving the reservoir; the bottom bar graph ($V_3$) represents the commanded flow rate. The relationship between commanded and actual flow rates is emphasized by the bold line connecting these two bar graphs. The goal associated with mass output ($G_2$) is represented by the bold vertical dashed line (approximately 55%). The relationship between mass in ($I_1 + I_2$) and mass out (O) is highlighted by the bold line (i.e., the mass balance indicator) that connects the corresponding bar graphs.

Finally, the volume of the reservoir (R) is represented by the filled portion of the rectangle inside the reservoir. The associated reservoir volume goal ($G_1$) is represented by a bold horizontal dashed line (approximately 85%). A more detailed analysis of this interface and how its visual features are mapped into the abstraction hierarchy will now be provided.

#### *10.4.1.1 General Work Activities and Functions*

The general work activities and functions are related through a block diagram with the source (arrows pointing downward) located at the top, storage located at the center, and sink (arrow pointing to the right) located at the bottom.

#### *10.4.1.2 Priority Measures and Abstract Functions*

The priority measures and abstract functions are related using emergent features including equality, the resulting colinearity across the bar graphs, and the orientation of line segments. The constraints associated with mass flow ($K_1$, $K_2$, $K_3$, the relationship between commanded and actual flow rates) are represented in terms of the equality of the horizontal extent (i.e., length) of the paired bar graph segments labeled $V_1/I_1$, $V_2/I_2$, and $V_3$/O. Rather than inferred linearity (Coury, Boulette, and Smith 1989; Sanderson et al. 1989; Cleveland 1985; see Chapters 8 and 9), these visual relationships are made explicit through the bold contour lines that connect relevant segments of bar graphs. The orientation of these lines provides additional emergent features that testify with regard to the relationship between actual and commanded flow rates. In Figure 10.3 all three of these contour lines are perfectly perpendicular because the valve settings and flow rates are equal in both input streams and the output stream.

The same emergent features are used to represent the $K_4$ constraint, which describes the relationship between mass in, mass out, and the reservoir volume. The bold line that connects the actual input bar graph segments ($I_1 + I_2$) and the actual output bar graph (O) will be referred to as the "mass balance indicator." The orientation of this line is an emergent feature that specifies both the direction and rate of change of mass inside the reservoir.

As illustrated in Figure 10.3, when mass is balanced, this line will be perpendicular to the bar graphs and the reservoir volume should remain constant. When mass is not balanced, the angular deviation of the mass balance indicator line from a perpendicular orientation should be proportional to rate of change of mass in the reservoir.

#### 10.4.1.3 Goals and Purposes

Constraints at the highest level of the abstraction hierarchy (goals and purposes) are also specified directly in the configural display. Both the goal for reservoir volume ($G_1$) and the goal for output flow rate ($G_2$) are represented by the two bold dashed lines. The deviation from goal is directly specified by the spatial offset between a goal and the relevant variable. In Figure 10.3, the goals are being met.

#### 10.4.1.4 Physical Processes

There are several sources of information regarding the physical processes of equipment. This configural display, while not a direct physical analog, preserves important physical relations from the process. The schematics of the input and output streams (top and bottom) provide representations of the general locations of pipes, valves, and sensors; the presence of the reservoir and its location relative to these streams is also represented visually. There are analog and digital representations of both commanded and actual system variables. In addition, it provides a direct visual representation of the process constraints and connects these constraints in a way to make the "functional" logic of the process (i.e., the causal connections) visible within the geometric form (e.g., volume and filling).

#### 10.4.1.5 Physical Appearance, Location, and Configuration

Very little information in these displays is represented from the lowest level in the abstraction hierarchy—that of physical appearance, location, and configuration. The assumption is that the display is designed for an operator in a control room who is controlling the system remotely. Under these circumstances, this category of information is not particularly important for effective control. This information would be very important to an operator whose job involves the manual adjustment of valve settings in the field (as is sometimes the case in the petroleum industry) as opposed to a centralized control room. Similarly, physical appearance, location, and configuration would be critical to a technician repairing a broken valve.

Finally, basic physical attributes, like the location of the electronic sensor, might become critical in the diagnosis of some faults—for example, whether a leaky valve might create an electrical short in a sensor (e.g., sensor failure). This kind of interaction would be very difficult to diagnose with representations organized around purely functional relations. Sometimes, the spatial details matter. Absence of this detail in the proposed representation could be a limitation. Thus, here is a case where it might be wise to consider including a second display, such as a spatial wiring and piping diagram of the physical plant.

Note that it will generally be impossible to make all the potentially important relations salient in any single representation. Thus, most complex systems will require multiple configural graphics, each reflecting a different perspective on the complex space of possibilities.

### 10.4.2 Support for Skill-, Rule-, and Knowledge-Based Behaviors

The concepts of skill-, rule-, and knowledge-based behaviors were introduced in Chapter 5. The implications of these modes of behavior for interface design were expanded in Chapter 6. Although the need to support all three modes of behavior was mentioned, the emphasis was on skill-based behaviors. In this section we describe how the virtual ecology that was designed for this simple process control system provides support for all three modes.

#### *10.4.2.1 Skill-Based Behavior/Signals*

Recall from Chapter 5 that skill-based behaviors are defined as those activities that engage the high-capacity, sensory-motor systems associated with perception and action. As the previous section has described, the domain constraints of the simple process control system were translated into the visual appearance of analog, geometrical forms. These geometrical forms dynamically change shape as a function of the measured variables and higher order properties in the system. Thus, the pattern of visual changes in these forms over time will be specific to the events unfolding in the domain (i.e., they specify system state). These are the space–time signals required to support skill-based behavior. The agent will be able to obtain information about the state of the system directly without the need to infer, deduce, or calculate. For this reason, the perception is referred to as direct. These space–time signals provide the optical invariants necessary for effective control.

One way to think about this is that the abstract goals can be defined in display-specific terms (i.e., align the level with the goal line). As long as the process is working properly, many of the functions can be defined as operations on the analog geometry (i.e., there is a one-to-one or specific mapping between the geometric form and the state of the process). In other words, the consequences of an action are directly specified by a change in the geometric form.

#### *10.4.2.2 Rule-Based Behavior/Signs*

Rule-based behavior involves the recognition of stereotypical situations and the execution of effective procedures that have been developed through prior experience. Effective displays will provide a rich set of visual cues that serve as signs for action. Common situations will be represented by characteristic patterns of optical invariants. An expert agent will recognize the underlying system state that is associated with a particular transformational pattern. Furthermore, this recognition will activate the common sequences of activity (i.e., the procedure) appropriate for that particular set of circumstances. Examples of rule-based behavior will be discussed in the context of the simple process control system.

The configural display that was introduced in Figure 10.3 provides a mapping between domain constraints and geometrical constraints that will be a powerful representation for control under normal operating conditions, when rule-based behavior will predominate. In Figure 10.3 the display is shown with values for system variables indicating that all constraints are satisfied. The figure indicates that the flow rate is larger for the first mass input valve ($I_1$, $V_1$) than for the second ($I_2$, $V_2$) but that the two flow rates added together match the flow rate of the mass output valve (O, $V_3$). In addition, the two system goals ($G_1$ and $G_2$) are being fulfilled. In contrast, the next three figures will illustrate failures to achieve system goals. In these displays, not only is the violation of the goal easily seen, but each system variable also is seen in the context of the control requirements.

The emergent features in the configural display will provide a rich set of visual cues that facilitate the recognition of stereotypical situations. It is apparent that the $K_5$ constraint is not being met in Figure 10.4. The degree of spatial separation between the fill representing the actual level of the reservoir (97%) and the dashed line representing the goal (85%) is quite large. This emergent feature specifies a deviation from the $G_1$ goal. It is also apparent that the $K_4$ constraint is broken. The orientation of the line connecting mass in ($I_1 + I_2$) and mass out (O) is tilted to the right, thereby indicating that a positive net inflow for mass exists: Mass flowing into the reservoir (70%) is greater than mass flowing out (54%). The deviation in the orientation of this line from perpendicular is an emergent feature corresponding to the size of the difference.

These emergent features provide a rich set of visual cues that serve as signs that will trigger stereotypical action sequences to a trained agent. The display configuration illustrated in Figure 10.4 specifies the need for immediate control input. The high reservoir level (approaching its upper limit) in combination with the positive net inflow (indicating that the reservoir level will continue to rise) clearly indicates that an immediate response is required to avoid overfilling the reservoir.

This display configuration also specifies which of several potential control inputs are most appropriate. Adjustments to the settings of valves 1, 2, and 3 are all potential control inputs. The observer can see these valves in the context of the two system goals; the representation makes it clear that decreases in the settings for valves 1 and/or 2 are the appropriate control inputs to make. Although adjusting valve 3 from a value of 54 to one greater than 70 would also cause the reservoir volume to drop, it is an inappropriate control input because goal 2 would then be violated.

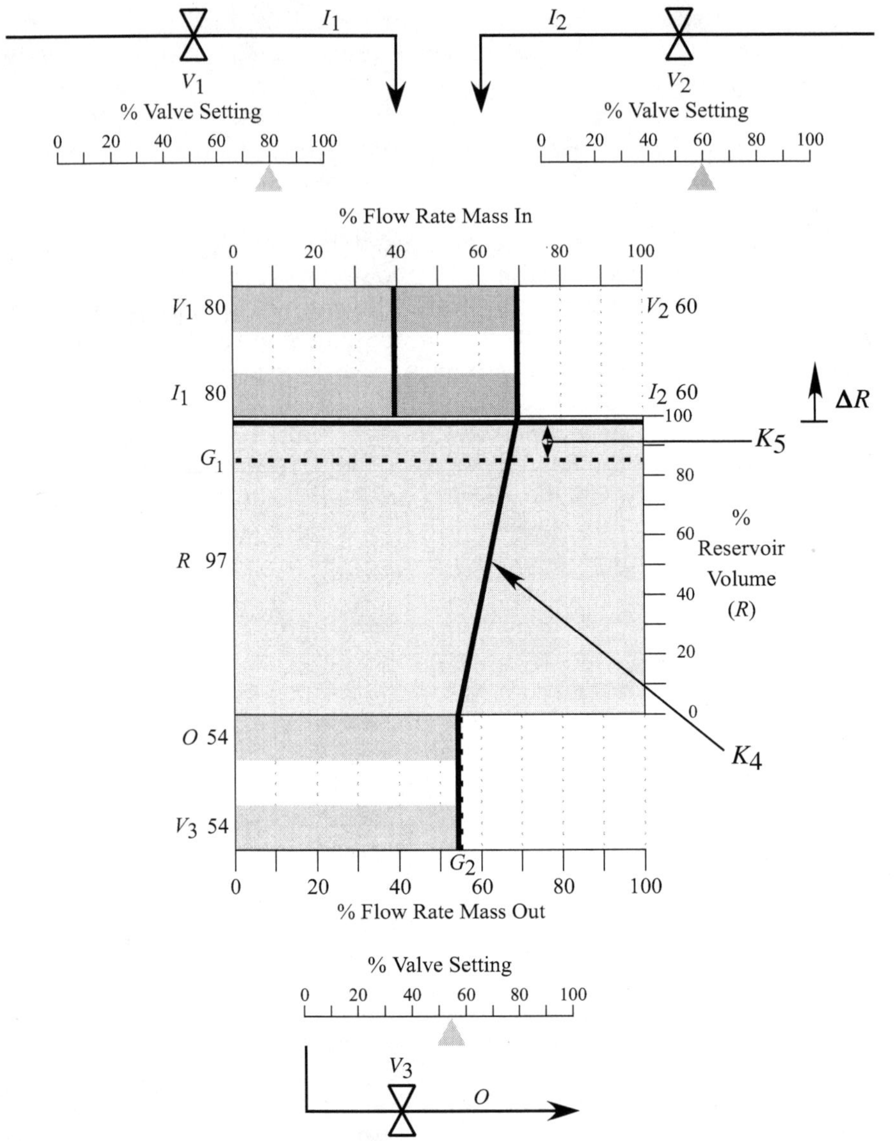


**FIGURE 10.4**

**(See color insert following page 230.)** A break in the geometrical constraints of the configural display. (Adapted with permission from Bennett, K. B., A. L. Nagy, and J. M. Flach. 1997. In *Handbook of Human Factors and Ergonomics,* ed. G. Salvendy, Copyright 1997, New York: John Wiley & Sons. All rights reserved.)

In Figure 10.5 the situation is exactly the same, with one exception: There is a negative net inflow for mass. This fact is specified by a salient emergent feature: the negative slope of the mass balance indicator line (i.e., it is tilted to the left). The reversed orientation of this line constitutes only a small change in visual appearance, but it makes a very large practical difference. An experienced operator will recognize this configuration as a clear sign that no immediate control input is required. The reservoir volume is falling because mass in is less than mass out.

Because the goal for reservoir volume ($G_1$—85%) is less than the actual value (97%), this is exactly the system configuration that is required. Of course, a control input will be needed at some point in the future (mass will need to be balanced when the reservoir level approaches the goal). The timing of this future control input will be specified by the rate of decrease in reservoir volume (as represented by the decreasing size of the rectangular fill inside the reservoir) in combination with the distance between it and the goal value (i.e., the dashed line corresponding to $G_1$).

The observer can see directly in Figure 10.6 that neither of the two system goals is being achieved. The current reservoir volume level (68%) is lower than the goal value (85%, a break in the $K_5$ constraint). The experienced operator will recognize this as a sign indicating that a positive net inflow of mass needs to be established. Furthermore, the measured mass outflow rate (88%) is greater than the goal value (55%, a break in the $K_6$ constraint). The experienced operator will perceive this as a sign that the positive net inflow needs to be established through an adjustment to valve 3 (a decrease in output). This single control input will eventually satisfy both the output requirements ($G_2$) and the volume goal ($G_1$).

Thus, in complex dynamic domains, the pattern of relationships between variables, as reflected in the geometric constraints (including emergent features), determines the significance of the data presented. This pattern ultimately provides the basis for action, even when the action hinges upon the value of an individual variable. When properly designed, configural displays will directly reflect these critical data relationships. The operators will recognize stereotypical situations that they have encountered in the past. Furthermore, specific display configurations will suggest stereotypical patterns of response to the operator—once again based on experience gleaned from the past. In this way, the display supports rule-based behavior through the provision of visual cues that serve as signs for action. There is a very clear relationship between rule-based behavior and naturalistic decision making such as recognition-primed decisions (see the associated discussion of recognition-primed decisions in Chapter 7).

#### 10.4.2.3 Knowledge-Based Behavior/Symbols

An agent will be engaged in knowledge-based behaviors when faced with situations that have not been encountered previously. Effective procedures

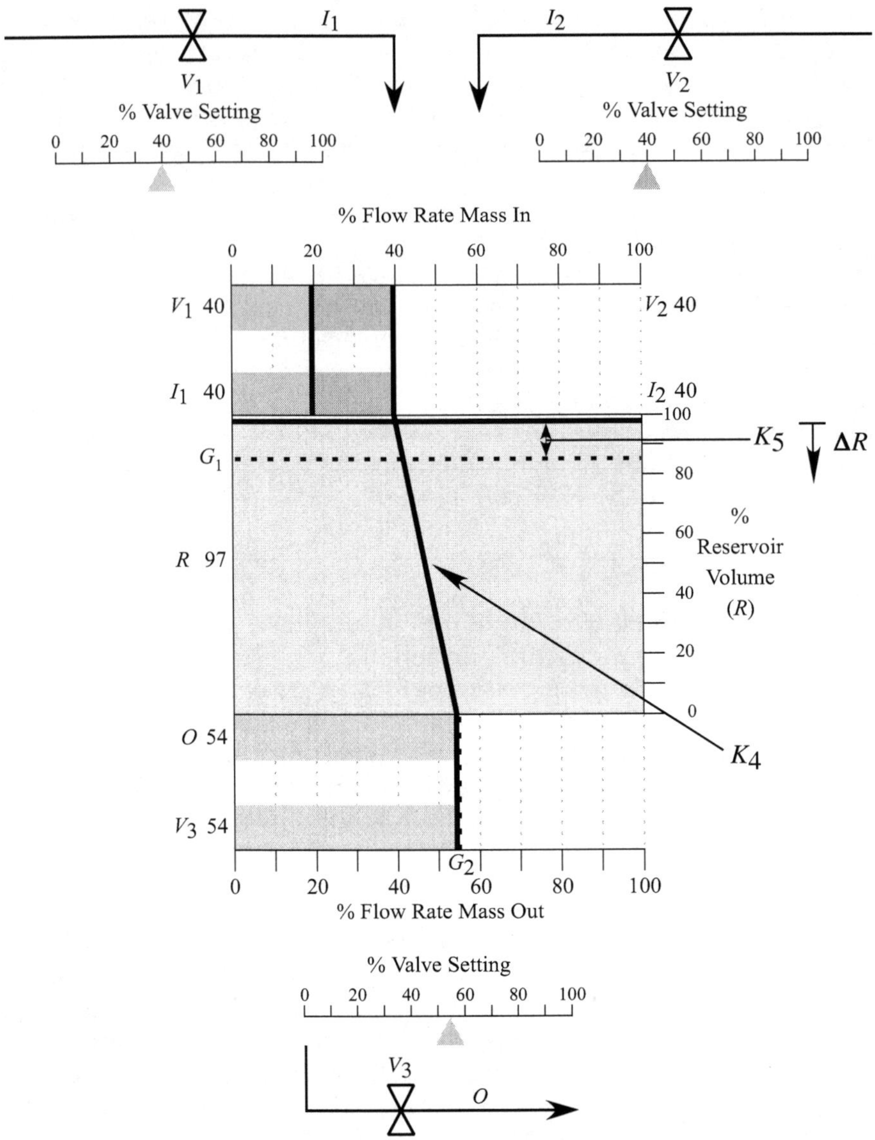


**FIGURE 10.5**
**(See color insert following page 230.)** Another break in the geometrical constraints of the configural display, this time with a very different meaning. (Adapted with permission from Bennett, K. B., A. L. Nagy, and J. M. Flach. 1997. In *Handbook of Human Factors and Ergonomics,* ed. G. Salvendy, Copyright 1997, New York: John Wiley & Sons. All rights reserved.)

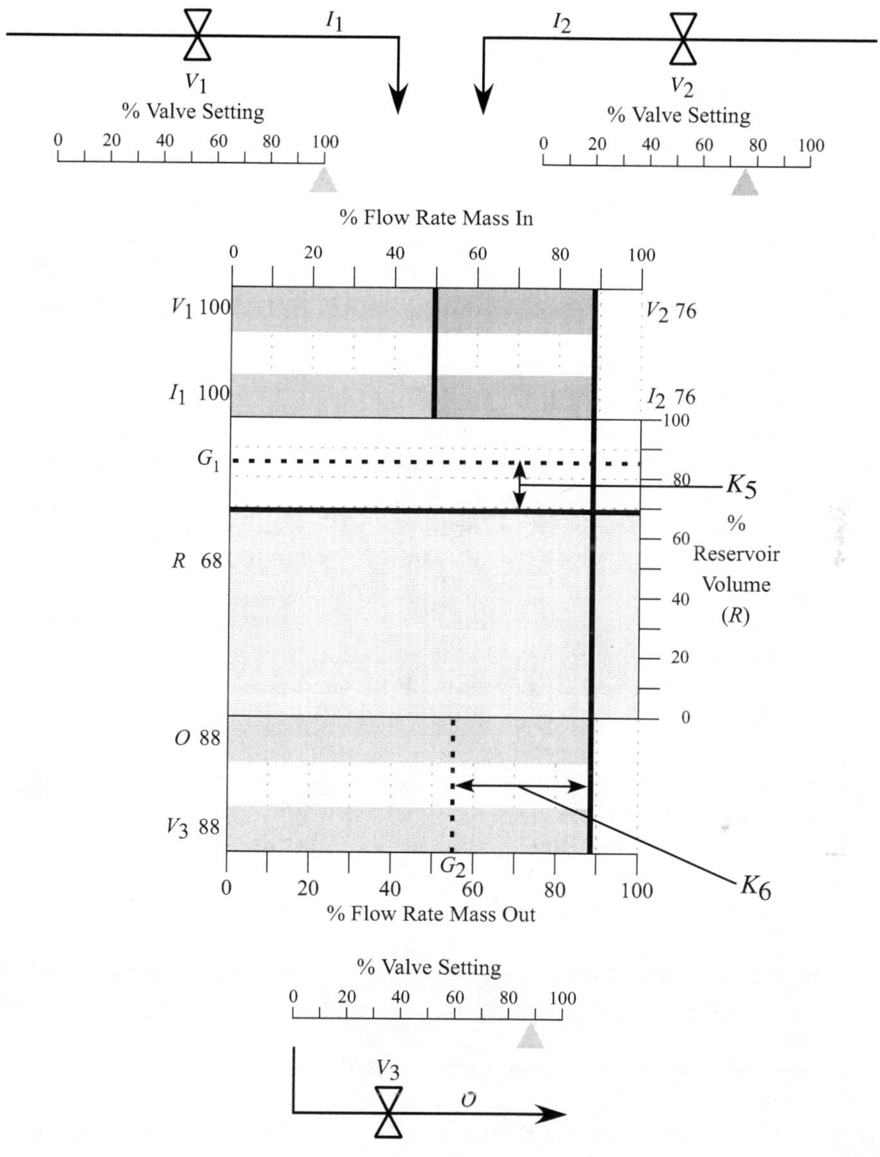


**FIGURE 10.6**
**(See color insert following page 230.)** The simple system is in steady state, but a control input is required. (Adapted with permission from Bennett, K. B., A. L. Nagy, and J. M. Flach. 1997. In *Handbook of Human Factors and Ergonomics*, ed. G. Salvendy, Copyright 1997, New York: John Wiley & Sons. All rights reserved.)

need to be invented to deal with the unanticipated events; the agent is essentially engaged in problem solving. To devise solutions, the agent needs to consider meaningful dimensions of the work domain (i.e., goals, constraints, resources). If the interface designer has done the job correctly (i.e., uncovering work domain constraints and translating them into geometric constraints), then the displays will serve as externalized models of the domain that can be used to support problem solving. When used in this fashion, the graphical displays provide symbolic information about the domain: "While signs refer to percepts and rules for action, symbols refer to concepts tied to functional properties and … are … the basis for reasoning and planning" (Rasmussen 1983, p. 260).

Several examples will be provided to illustrate knowledge-based behavior with the simple process control system. Figure 10.7 illustrates a system state where the first constraint ($K_1$) is broken; the actual flow rate ($I_1$) does not match the commanded flow rate (valve setting $V_1$). Several aspects of the display geometry specify this break in the underlying domain constraints. At a global level, there is a bow in the series of connected line segments between commanded mass in (i.e., $V_1 + V_2$) and commanded mass out (i.e., $V_3$). Normally, these line segments would be perfectly aligned (see Figure 10.3); this break in geometrical constraints provides a very salient emergent feature (nonlinearity) that specifies a system fault.

When the operator begins to reason about the nature of the fault, the displayed information is being utilized as symbols, conveying information about the structure and functionality of the underlying domain. The visual appearance of the display then provides more details about the specific nature of the fault. The operator knows that the problem is located in the input streams by the fact that the bowed segment appears at the top of the visual display. Furthermore, the operator knows that the problem is isolated in the first mass input stream by the configuration of visual elements. The width of the two leftmost bar graph segments representing the commanded ($V_1$) and actual ($I_1$) flow rates of the first input stream are not visually congruent. That is, the lower bar graph segment is smaller than the upper bar graph segment.

In contrast, the commanded and actual flow rates of the second input stream are congruent; the widths of the two rightmost bar graph segments (second input stream) are the same. The location of the fault is further specified by the orientation of the "visual contours" that connect the appropriate segments of the two contribution bar graphs. In Figure 10.7 these two visual contours are parallel to each other. This parallelism can occur only when the commanded and actual flow rates for the second input stream (i.e., $V_2$ and $I_2$) are equal.

Figure 10.8 illustrates a second system fault. The visual contour between the bar graph segments for $V_2$ and $I_2$ is in exactly the same physical location and has exactly the same orientation as it does in Figure 10.7. However, the meaning is quite different. There is still a fault, but the source of that fault

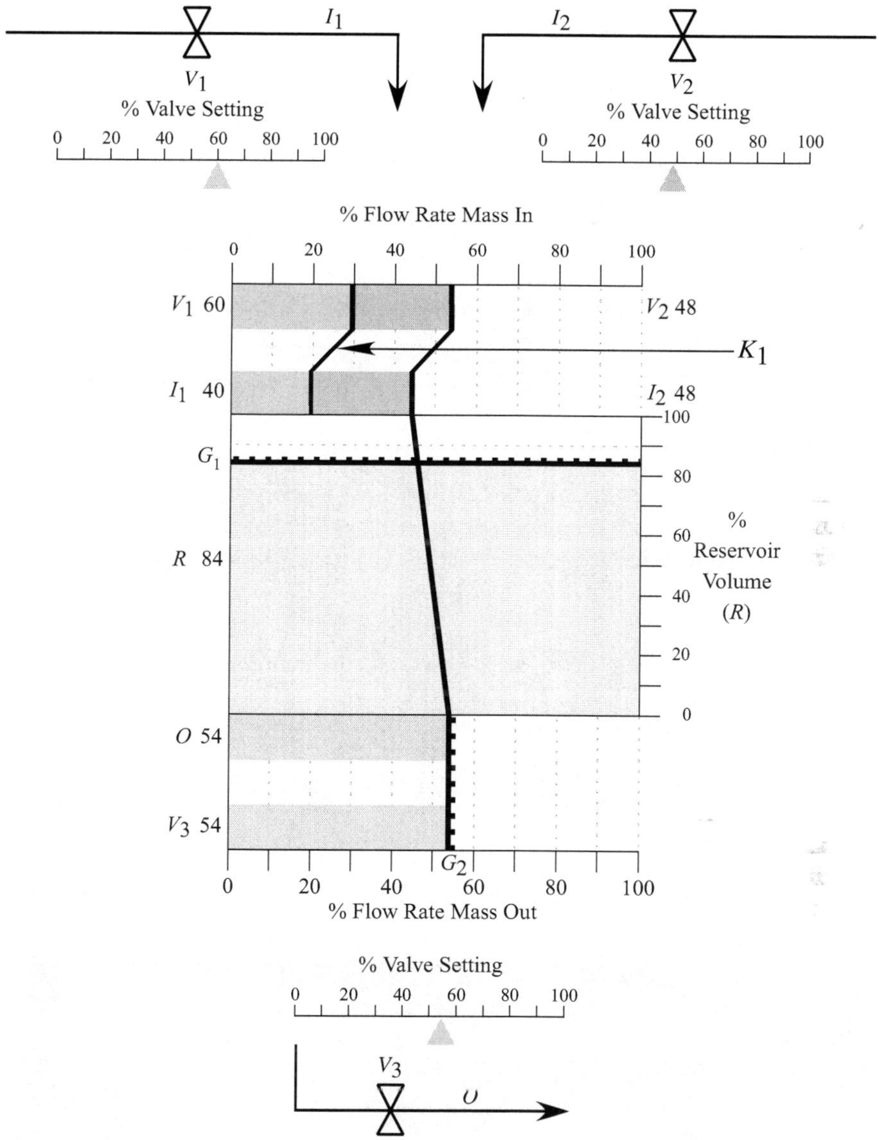


**FIGURE 10.7**
**(See color insert following page 230.)** Presence and nature of a fault in the simple system is specified; it lies in the first input stream. (Adapted with permission from Bennett, K. B., A. L. Nagy, and J. M. Flach. 1997. In *Handbook of Human Factors and Ergonomics*, ed. G. Salvendy, Copyright 1997, New York: John Wiley & Sons. All rights reserved.)

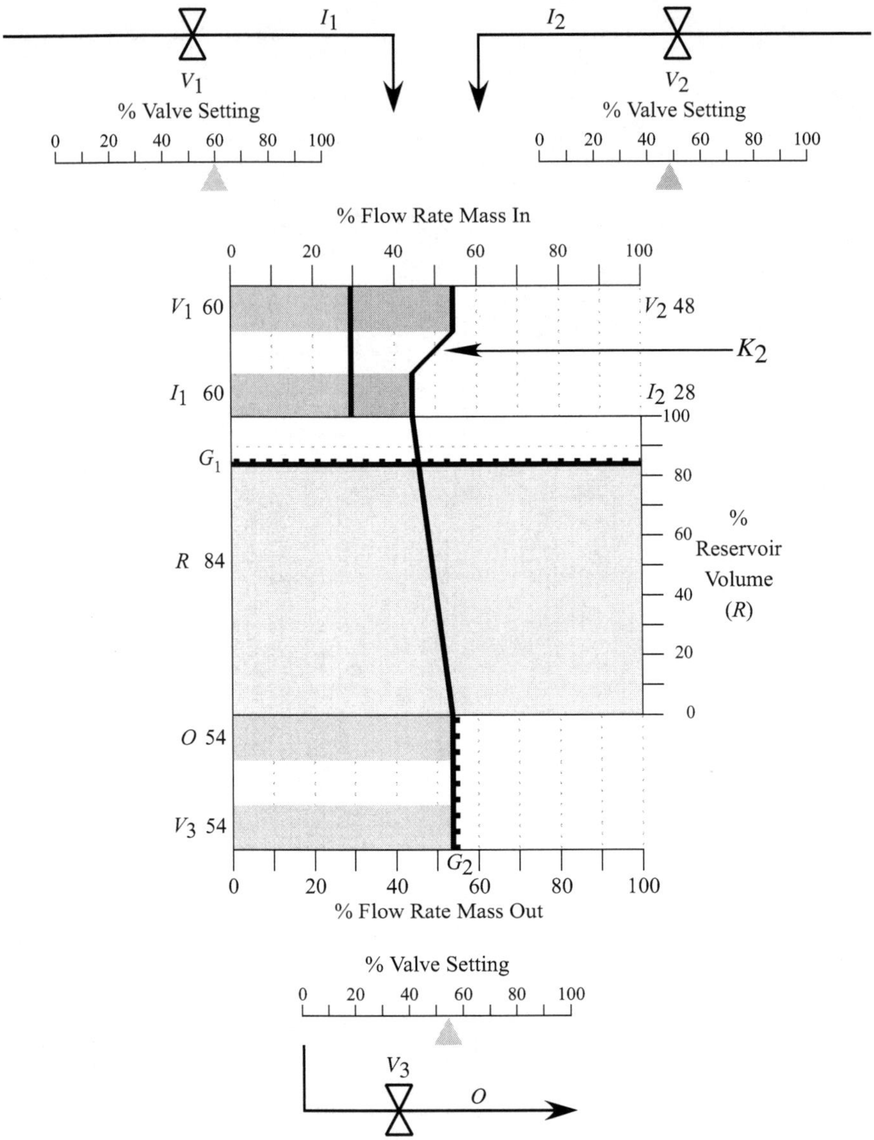


**FIGURE 10.8**
**(See color insert following page 230.)** An alternative fault in the second input stream is specified. (Adapted with permission from Bennett, K. B., A. L. Nagy, and J. M. Flach. 1997. In *Handbook of Human Factors and Ergonomics*, ed. G. Salvendy, Copyright 1997, New York: John Wiley & Sons. All rights reserved.)

lies in the second input stream, as opposed to the first. This is specified by the differences in the width of the $V_2$ and $I_2$ segments in the two contribution bar graphs and by the nonparallel orientation of the two visual contours connecting the bar graph segments. A similar mapping between geometrical constraints and domain constraints represents a fault in the $K_3$ constraint, as illustrated in Figure 10.9.

Note that the presence of a fault is revealed in Figures 10.8–10.10 but that its cause is not. A number of potential faults could be consistent with the visual discrepancies that are apparent in the display, including (1) a leak in the valve, (2) a leak in the pipe prior to the point at which the flow rate is measured, or (3) an obstruction in the pipe.

Figure 10.10 illustrates changes in the visual display (breaks in the geometrical constraints) that are associated with a different type of fault in the system (a break in the mass balance constraint, $K_4$). In Figure 10.10a, the reservoir volume (73%) is less than the goal (85%) and the operator has established a positive net inflow of mass. This is represented by the positive orientation of the mass balance indicator line (i.e., a rise from left to right). Because there is more mass entering the reservoir than leaving it (as specified by the orientation of the mass balance indicator), this system state will normally produce an increase in the volume of the reservoir over time.

If these expectations are violated, the operator will enter into knowledge-based behavioral mode and will begin interpreting the visual information presented in the display as symbols. A violation of these expectations is illustrated in Figure 10.11b–d, where the mass inventory actually decreases over time (also represented by the downward-pointing arrow located near the $\Delta R$ symbol). A trained operator would immediately begin to think in terms of a system fault upon observing this behavior. It is a clear violation of the $K_4$ system constraint, the intended proper functioning of the system. According to the law of conservation of mass, if the mass is not being stored in the reservoir under conditions of a positive net inflow, then it must be leaving the system.

Again, there are several potential explanations for this fault. The most likely explanation is that there is a leak in the reservoir itself; however, there could be a leak in the pipe between the reservoir and the point at which the flow measurement is taken. It should be noted that, while the nature of the fault can be seen (e.g., leak or blockage in feedwater line), this representation would not be very helpful in physically locating the leak within the plant (e.g., locating valve 1).

### 10.4.3 Alternative Mappings

In this section we wrap up our consideration of direct perception. We do so by considering the quality of mapping between the constraints of this simple work domain, the configural display introduced earlier in the chapter, and alternative graphical representations that could have

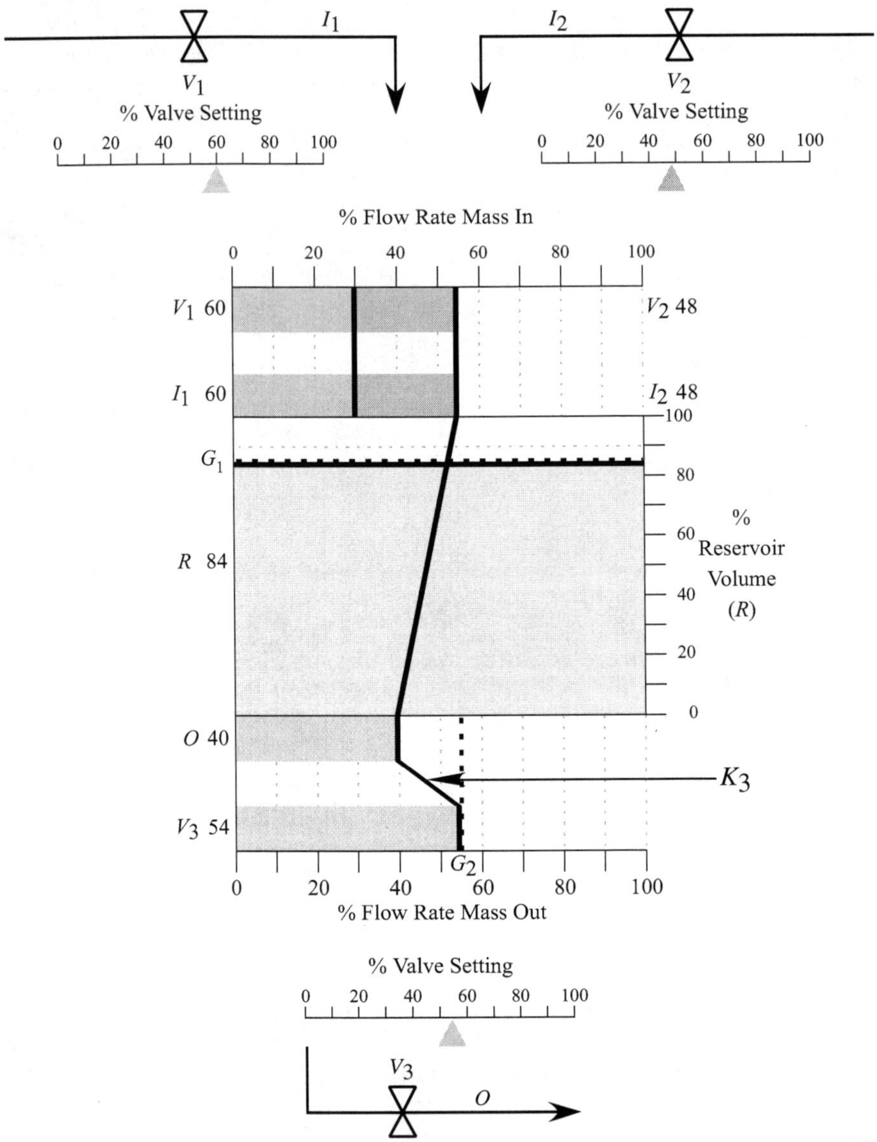


**FIGURE 10.9**
**(See color insert following page 230.)** A fault in the output stream is specified. (Adapted with permission from Bennett, K. B., A. L. Nagy, and J. M. Flach. 1997. In *Handbook of Human Factors and Ergonomics,* ed. G. Salvendy, Copyright 1997, New York: John Wiley & Sons. All rights reserved.)

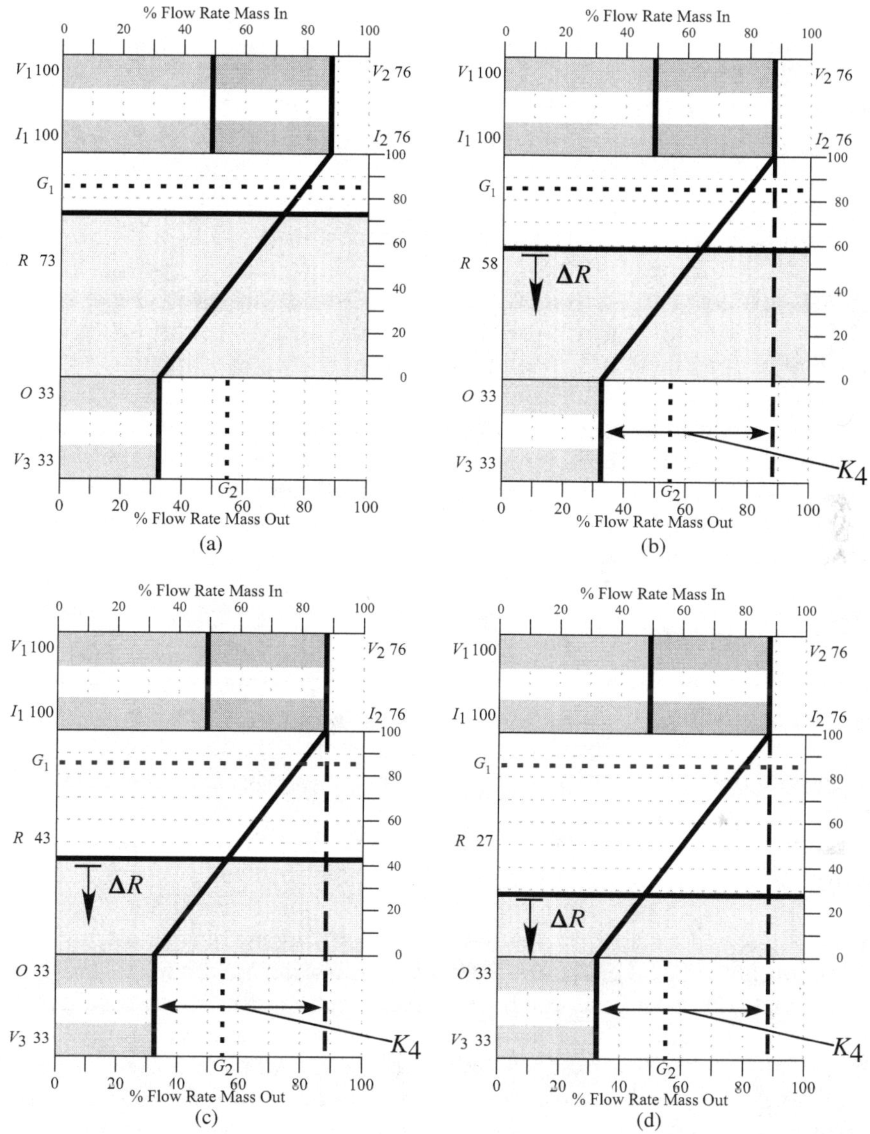


**FIGURE 10.10**
**(See color insert following page 230.)** A fault in the conservation of mass is specified. (Adapted with permission from Bennett, K. B., A. L. Nagy, and J. M. Flach. 1997. In *Handbook of Human Factors and Ergonomics*, ed. G. Salvendy, Copyright 1997, New York: John Wiley & Sons. All rights reserved.)

been devised. The focus is on the capability of these alternative displays to convey information at the various levels of abstraction (i.e., to support direct perception). In the process we will revisit the concepts of integral, configural, and separable dimensions introduced in Chapter 8. It will be demonstrated that the meaning of these concepts changes in the context of display design for complex systems when a work domain is lying behind the visual representations.

Figure 10.11 provides the six different representational types that will be analyzed. A summary of the work domain constraints is provided to the right of each display. These constraints are sorted into two categories ("P" and "D"). The P (perceived) category lists the process constraints that are represented directly in a display (that is, those constraints that can be "seen" directly). Process constraints that must be computed or inferred are placed in the D (derived) category. The theta symbol (∅) and the symbol ∫ are used to provide a shorthand notation for physical structure and functional structure, respectively, in this list. The symbol "T" stands for temporal information: the capability of a display to represent changes in a variable or property explicitly over time.

### *10.4.3.1 Separable Displays*

Figure 10.11a represents a separable display that contains a single display for each individual process variable that is measured. Each display is represented in the figure by a circle, but no special significance should be attached to the symbology: The circles could represent digital displays, bar graphs, etc.

In terms of the abstraction hierarchy, the class of displays represented by Figure 10.11a provides information only at the level of physical function; individual variables are represented directly. Thus, there is not likely to be a selective attention cost for low-level data. However, there is likely to be a divided attention cost because the observer must derive the high-level properties (note the exceptions for limited relationships that were discussed in Chapter 9). To do so, the observer must have an internalized model of the functional purpose, the abstract functions, the general functional organization, and the physical process. For example, to determine the direction (and cause) of $\Delta R$ would require detailed internal knowledge about the process since no information about physical relationships (∅) or functional properties (∫) is present in the display.

Simply adding information about high-level properties does not change the separable nature of the display. In Figure 10.11b a second separable display has been illustrated. In this display, the high-level properties (constraints) have been calculated and are displayed directly, including information related to functional purpose ($K_5$ and $K_6$) and abstract function ($K_1$, $K_2$, $K_3$ and $K_4$). This does offload some of the mental computational requirements (e.g., $\Delta R$).

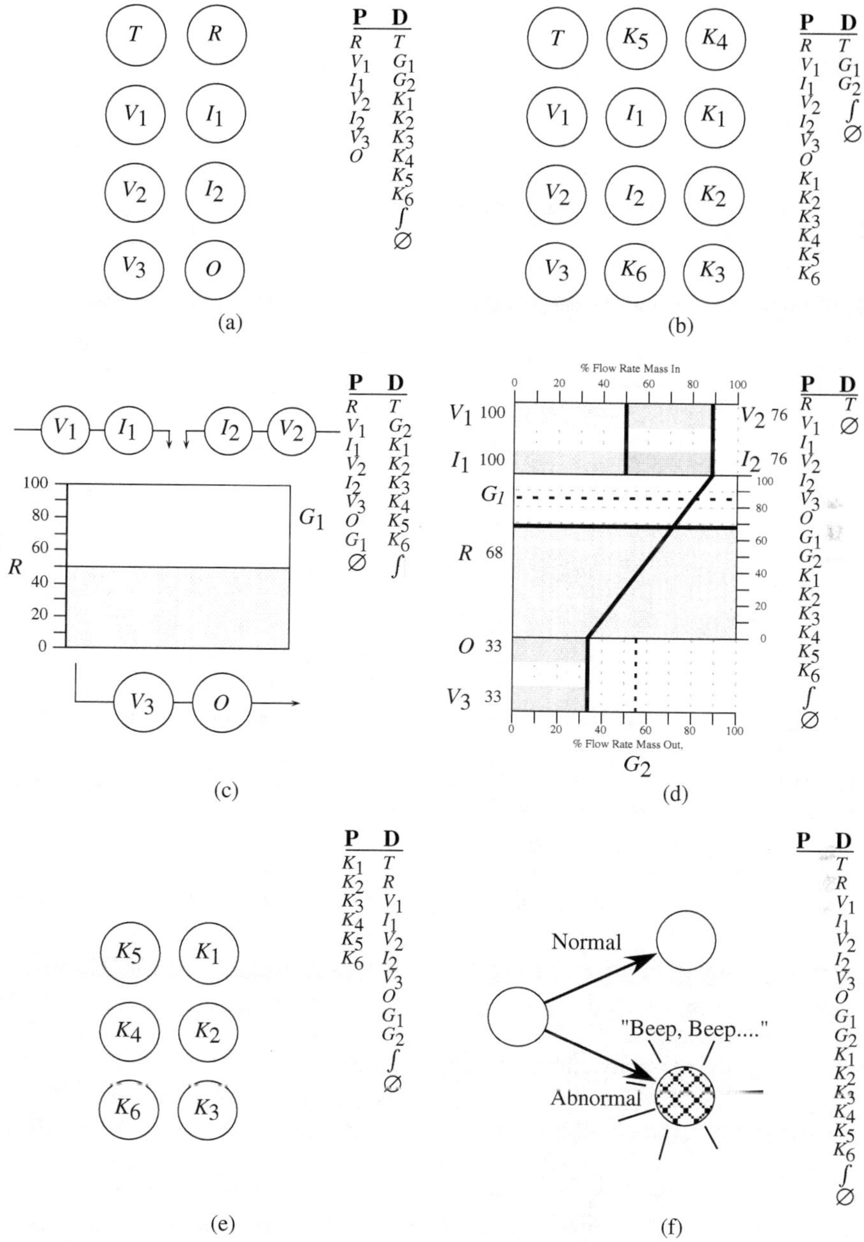


**FIGURE 10.11**
Six alternative representational formats and the associated mappings of the work domain constraints. (Adapted with permission from Bennett, K. B., A. L. Nagy, and J. M. Flach. 1997. In *Handbook of Human Factors and Ergonomics*, ed. G. Salvendy, Copyright 1997, New York: John Wiley & Sons. All rights reserved.)

However, there is still a divided attention cost. Even though the high-level properties have been calculated and incorporated into the display, the relationships among and between levels of information in the abstraction hierarchy are still not apparent. The underlying cause of a particular system state still must be derived from the separate information that is displayed. Thus, while some low-level integration is accomplished in the display, the burden for understanding the causal structure still rests in the observer's stored knowledge. Additionally, this format creates a significant danger of data overload because the potential number of display elements will increase exponentially with each new variable due to the potential relations with the other variables.

### 10.4.3.2 Configural Displays

The first configural display, illustrated in Figure 10.11c, provides a direct representation of much of the low-level data present in the display in Figure 10.11a. However, it also provides additional information that is critical to completing domain tasks: information about the physical structure of the system (∅). This type of display was introduced in the first attempts to develop electronic control and display systems and is commonplace today. The animated "mimic" display format was first introduced in STEAMER (Hollan, Hutchins, and Weitzman 1984), and issues in its design have been investigated more recently (Bennett 1993; Bennett and Madigan 1994; Bennett and Nagy 1996; Bennett and Malek 2000).

The mimic display is an excellent format for representing the generalized functions in the process. It has many of the properties of a functional flow diagram or flowchart. The elements can represent physical processes (e.g., feedwater streams); by appropriately scaling the diagram, relations at the level of physical form can be represented (e.g., relative positions of valves). Also, the moment-to-moment values of the process variables can easily be integrated within this representation. This display not only includes information with respect to generalized function, physical function, and physical form, but the organization also provides a visible model illustrating the relations across these levels of abstraction. This visual model allows the observer to "see" some of the logical constraints that link the low-level data. Thus, the current value of $I_2$ can be seen in the context of its physical function (feedwater stream 2) and its generalized function (source of mass); in fact, its relation to the functional purpose in terms of $G_1$ is also readily apparent from the representation.

Just as in the displays listed in Figure 10.11a–b, there is not likely to be a cost in selective attention with respect to the low-level data. However, although information about physical structure illustrates the causal factors that determine higher level system constraints, the burden of computing these constraints (e.g., determining mass balance) rests with the observer. Thus, what is missing in the mimic display is information about abstract function (information about the physical laws that govern normal operation).

The second configural display in Figure 10.11d, as discussed earlier in the chapter, provides information from all levels of the abstraction hierarchy in a single representation, making extensive use of the geometrical constraints of equality, parallel lines, and colinearity. While not a direct physical analog, it preserves important physical relations from the process (e.g., volume and filling). In addition, it provides a direct visual representation of the process constraints and connects these constraints in a way to make the functional logic of the process visible within the geometric form. As a result, performance for both selective (focused) and divided (integration) tasks is likely to be facilitated substantially.

#### *10.4.3.3 Integral Displays*

Figure 10.11e shows an integral mapping in which each of the process constraints are shown directly, providing information at the higher levels of abstraction. However, the low-level data must be derived. In addition, there is absolutely no information about the functional processes behind the display and therefore the display does not aid the observer in relating the higher level constraints to the physical variables. Because there would normally be a many-to-one mapping from physical variables to the higher order constraints, it would be impossible for the observer to recover information at lower levels of abstraction from this display.

Figure 10.11f shows the logical extreme of this continuum. In this display, the process variables and constraints are integrated into a single "bit" of information that indicates whether or not the process is working properly (all constraints are at their designed value). While these displays may have no divided attention costs, they do have selective attention costs and they also provide little support for problem solving when the system fails. The concept of integral stimulus dimensions has been questioned in the perceptual literature. For example, Cheng and Pachella (1984) state that "integrality may be a myth" (p. 302). However, when applied to human–machine systems design, the truth is obvious: The meaning behind an "idiot light" in one's automobile is an uncomfortable mystery that can only be resolved by the technician.

#### *10.4.3.4 Summary*

This section has focused on issues related to the quality of mapping between process constraints and display constraints. Even the simple domain that we chose for illustrative purposes has a nested structure of domain constraints: Multiple constraints are organized hierarchically both within and between levels of abstraction. The six alternative displays achieved various degrees of success in mapping these constraints. This is illustrated by the fact that these formats differ in terms of the amount of information about the underlying domain that is present (see the perceived vs. derived summaries in Figure 10.11).

The display in Figure 10.11f has the lowest quality of mapping, while the display in Figure 10.11d has the highest. The configural display in Figure 10.11d allows an individual to perceive information concerning the physical structure, functional structure, and hierarchically nested constraints in the domain directly—a capability that is not supported by the other formats in Figure 10.11b. This section has also illustrated the duality of meaning for the terms "integral," "configural," and "separable." In attention, these terms refer to the relationship between perceptual dimensions, as described in Chapter 8; in display design, they more appropriately refer to the nature of the mapping between the domain and the representation.

### 10.4.4 Temporal Information

A limitation of all of the displays discussed so far will be addressed in this last section on direct perception. The term "temporal information" refers to changes in resources, properties, and variables over time. It is represented by the symbol "T" in Figures 10.1 and 10.11. Temporal information is critical in law-driven domains (as well as others). Essentially, all physical systems have at least inertial dynamics. This implies that a requirement of control is that there be feedback of both position and rates of change (see Jagacinski and Flach, 2003, for information about order of control and implications for the required state variables).

Past system states will determine current and future system states under normal circumstances. Therefore, visualizing change over time will be essential to a complete understanding of current system states, to predicting future system states, and to choosing the appropriate control inputs. As a case in point, consider the following example from the industrial accident at Bhopal, India. On the eve of the accident, a shift change occurred and the arriving operator

> scanned the assortment of displays on the panels … . The scale on the displays ranged from 0 to 55 psig … . Tank 16 showed a pressure of 10 psig, just about in the middle of the acceptable range of 2 to 25 psig.
>
> Unknown to Dey [the operator], the pressure inside the tank was 2 psig only 40 minutes before at 10:20. But the buildup was not apparent *because no historical trace of the pressure was shown within the control room,* and the operator on the previous shift had not entered this into the log. (Casey 1993, pp. 75–76; emphasis added).

This change in pressure was critical because water was being introduced to a volatile chemical inside the tank; the chemical reaction that followed killed about 3,000 and injured tens of thousands more. Of course, a myriad of other factors led up to this disaster, but the lack of temporal information in the control room was clearly a contributing factor.

For the most part, as noted in Figure 10.11, temporal information (T) is not represented directly in any of the display formats that have been discussed so far. However, there is one exception: The slope of the line connecting input to output in the configural graphic specifies the instantaneous rate of change for the water level, if the system is operating normally (e.g., no leaks). For all the other displays presented in the previous sections, temporal information must be derived (i.e., picked up from the motion of the display elements). This is a clear limitation, since changes and trends can be critical for fault diagnosis when operators are trying to figure out how the system got into a dangerous state or for anticipating that a system is moving toward critical boundaries.

Recognizing the significance of information about the history of change for understanding complex processes, many process plants include "strip chart" displays that plot changes in the value of a variable or a resource as a function of time (e.g., Schutz 1961; Spenkelink 1990). However, these displays are sometimes physically separated from the primary displays and also suffer from the same general limitations outlined for separable displays (Figure 10.11a–b) in Section 10.4.3.1 (i.e., limited configural properties).

#### *10.4.4.1 The Time Tunnels Technique*

We conducted a series of studies to investigate how the complementary strengths and weaknesses of configural displays and trend displays could be combined (Bennett and Zimmerman 2001; Bennett, Payne, and Walters 2005). Our starting point was Hansen's (1995) work on the time tunnels display design technique. This technique involves scaling geometrical forms according to the laws of perspective geometry and then presenting them in the depth plane. The resulting "2½"-dimensional representation provides a trace of low-level data and high-level properties over time.

Figure 10.12 illustrates a variation of this technique applied to the configural display described previously. A static framework, or "perspective grid," is plotted in the depth plane of the reservoir. The outermost rectangle represents display axes that correspond to the current time frame. Each successive rectangle is scaled according to the laws of perspective geometry and plotted deeper in the depth plane to represent the display axes at a point more distant in time. Temporal information (individual variables, relationships, and goals over time) is presented within this perspective grid. Perspective trends are formed by plotting the value of individual variables at the various points in time and connecting the points in contiguous time frames. Similarly, mass balance relationships over time are represented by a series of mass balance indicator lines formed by connecting the values of steam and feed flow within a time frame.

The unfolding events depicted in Figure 10.12 will be described to make the nuances of the time tunnel display design technique clear. First, consider the relationships and properties. The initial values (located in the back of the tunnel) of both R (20%) and O (30%) were far from the associated goal values ($G_1$ and $G_2$—85 and 55%, respectively). The operator initially established a large positive net inflow; the combined input flow rate ($I_1 + I_2$) was 80% and

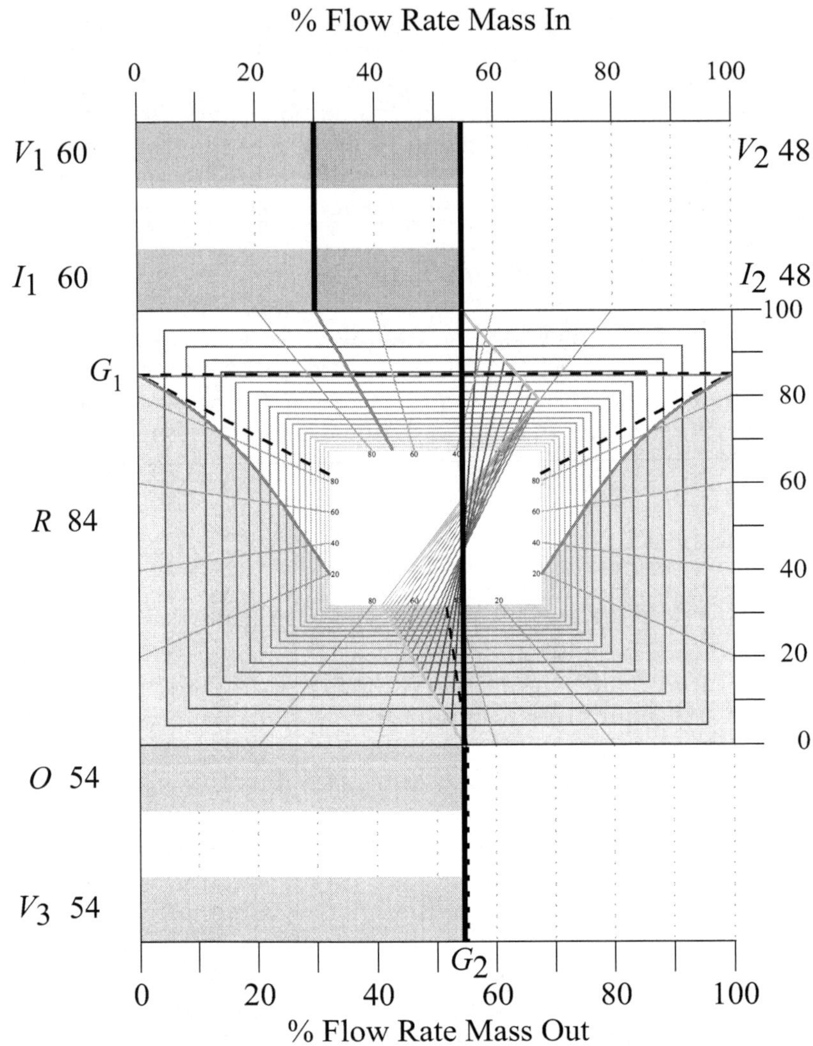


**FIGURE 10.12**
**(See color insert following page 230.)** The configural display with the time tunnel design technique and the associated temporal information.

O was 20%. This is reflected in the positive orientation of the mass balance indicator farthest back in the tunnel. The thick, black mass balance indicator associated with current time (the outermost frame of the tunnel) is perpendicular, indicating that mass is now balanced.

The transition from net inflow to mass balance is specified by the counterclockwise rotation of the mass balance indicator lines from later to earlier time frames in the tunnel. Furthermore, both goals are now being met: The volume (R) has converged on $G_1$ (85%) and the value of O has converged on $G_2$ (55%). Note that the volume level is further specified by the fill on the lower part of the tunnel walls.

The perspective trends directly specify the changes in individual variables that occurred over time. The flow rate from the first input stream ($I_1$) remained constant throughout the sequence. This is specified by the fact that the corresponding perspective trend is a straight line with an orientation that adheres to the canonical perspective geometry of the display grid. The perspective trend for the mass output variable (O) is also a straight line. However, the orientation of this line changes with respect to the perspective geometry of the display grid. Specifically, it begins far to the left of the center point of the frame farthest back in time and winds up slightly to the right of the center point in the current frame. These geometrical constraints specify a constant increase in mass out over time.

Similarly, the flow rate from the second input stream ($I_2$) remained constant in the initial frames (initial line segment) and was then decreased at a constant rate as the sequence progressed (most recent line segment). The perspective trend for volume (R) is not a straight line, nor is it a line composed of segments. This specifies varying rates of change throughout the entire sequence (higher rates initially and lower rates more recently).

The empirical laboratory evaluations of the time tunnels technique have been generally positive. Hansen (1995) investigated configural displays, trend displays, and digital values in a data monitoring task. His results were mixed, suggesting "that the spatial integration of temporal information in the time tunnel format shows promise" (p. 551). Our initial evaluations of the technique using a more realistic process control simulation were also mixed (Bennett and Zimmerman 2001). In a subsequent study (Bennett et al. 2005), we found significant performance advantages for the time tunnel technique in predicting future system states relative to a configural display with no temporal information and a traditional trend display (see Chapter 17 for a more detailed description of these results).

We have invested considerable time and effort in designing and evaluating variations of the time tunnel technique. From a theoretical standpoint, it provides the dual benefits of temporal information and configural display geometries in a single integrated representation. From a practical standpoint, it saves valuable display "real estate." This is the primary reason that we adopted a variation of the technique to use a "wall" of the tunnel in providing a perspective trend (see the force ratio trend display described in Chapter 14). On the other hand, our experience has shown that it is exceedingly difficult to combine temporal and configural geometries in an effective manner.

Finally, it should be noted that, in principle, geometries can configure in time as well as in space. This reflects Gibson's (1958/1982) construct of optical invariants. An optical invariant is a property of perspective that is invariantly related to some functionally relevant action property. For example, Langewiesche (1944) describes many invariant relations that are important for the landing approach (see Chapter 11). Thus, when display geometries are designed, it is important to consider the space–time properties of the geometry relative to the dynamic demands of the work. In some cases, it may be desirable that constraints of the dynamic process be represented as invariant properties of the display flow, rather than as properties of the static layout.

## 10.5 Direct Manipulation

As introduced in Chapter 6, the goal of developing interface support for direct manipulation is to allow the user to act directly (and perhaps "naturally") upon the objects of interest in the interface (and thereby the objects of interest in the domain). In this particular process control domain, control inputs are relatively simple, consisting of changes to the three valves. Figure 10.13

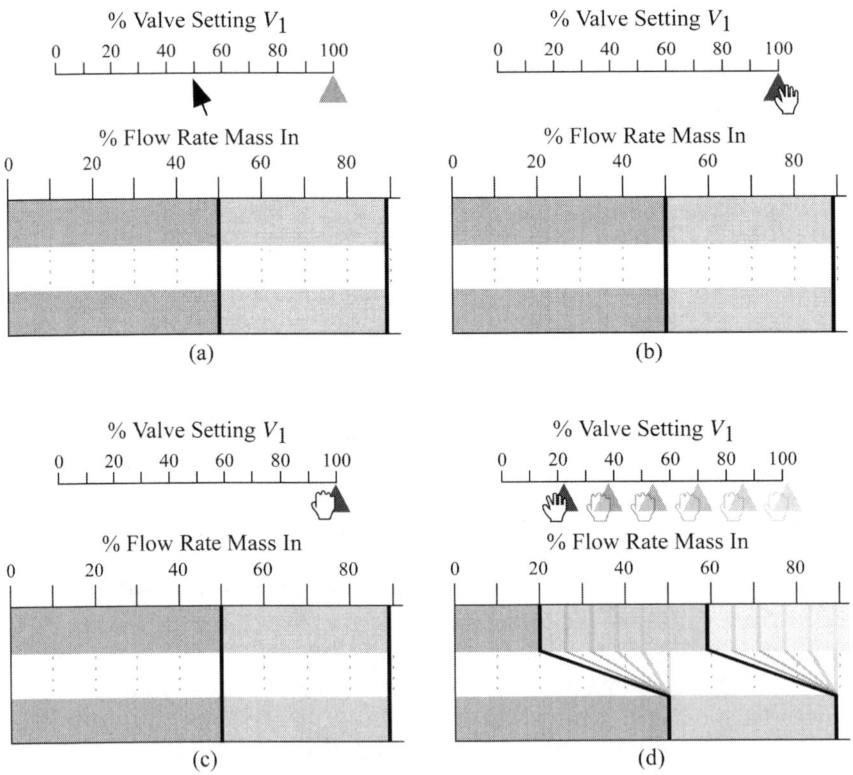


**FIGURE 10.13**
Direct manipulation of the valve setting $V_1$ in the configural display.

illustrates the interface resources that support the direct manipulation of these valve settings for the first input stream ($V_1$). In this sequence, the operator will reduce the setting for $V_1$ from its initial value of 100% to a final value of 20%.

Figure 10.13a represents the initial state of the system. To effect the change in settings, the operator selects, drags, and drops the triangularly shaped interface object at the desired setting, as illustrated in Figure 10.13a–d. The interface resources (i.e., scale, triangular object, cursor) provide continuous and analog space–time signals that support this skill-based behavior. The agent monitors the current location of the triangle relative to the desired setting; the corresponding signals guide the motor activity required to decrease the discrepancy. As described in Chapter 6, the perception–action loop is intact. The triangle serves as both a display (current setting) and a control (a means to change current setting). Contrast this to typing in a value or selecting from a pull-down menu. Note that the changes in actual input flow rates will occur after a time delay; the displays will eventually assume the configuration depicted in Figure 10.13a at the new setting.

This example illustrates several other useful techniques for implementing direct manipulation interfaces. It is critical to support the user in differentiating those interface objects that can be manipulated and those that cannot (see the expanded discussion in Chapter 13). One useful technique is to change the visual appearance of the object and the cursor when the potential for direct manipulation exists. This technique is illustrated in Figure 10.13b. The cursor changes from an arrow to an open hand when it is positioned over the triangle; the visual appearance of the triangle is highlighted (i.e., it becomes brighter and a deeper shade of color). The cursor changes again (from an open hand to a closed hand) when the agent selects the object (see Figure 10.13c), thereby symbolizing the act of manipulation. The alternative cursor shapes (Apple 2008) are metaphors that symbolically represent the potential for manipulation (open hand) and manipulation itself (closed hand).

All of the interface conventions described in the previous paragraphs are also used for valve settings $V_2$ and $V_3$. Note that the user can also drag the rightmost edge of a bar graph or a segment to change the valve setting.

It is important to note that the ability to manipulate the process by interacting directly with the graphical display using either click and grab with a mouse or a touch display is a relatively new technology. In many current systems, manipulation will be via keyboards, buttons, dials, or joysticks. In these cases, the directness of the manipulation will often depend on relations between the topology of the control action and the movement on the screen. These issues have been addressed quite thoroughly in many classical texts on human factors under the general heading of stimulus–response compatibility. In general, interface designers should do their best to maximize stimulus–response compatibility—that is, to link motion in the graphical display specifically to the topology of control actions.

## 10.6 Summary

The present chapter has provided an interface design tutorial for a simple process, bringing together many of the issues in design that were described in earlier chapters. This interface contains all three types of representations that were described in Chapter 6: analogical (dynamic, geometrical form displays), propositional (labels and digital values), and metaphorical (abstract images—the cursors). Although all three are present in the interface, the primary representational format is analogical.

These examples illustrate that properly designed analog geometrical form displays can change the fundamental type of behavior that is required of an operator under both normal and abnormal operating conditions. With traditional interfaces dominated by static graphics and propositional representations (e.g., the separable configurations illustrated in Figure 10.11), the operators are required to engage in knowledge-based behaviors. They must rely upon internal models of system structure and function (and therefore use limited capacity resources—working memory) to detect, diagnose, and correct faults (see the detailed example of the consequences in Chapter 17). As a result, the potential for errors is increased dramatically.

In contrast, properly designed configural displays present externalized models (i.e., analogies) of system structure and function through geometric constraints. This allows operators to utilize skill-based behaviors (e.g., visual perception and pattern recognition) that do not require limited-capacity resources. As a result, the potential for errors will be dramatically decreased. As Rasmussen and Vicente (1989) have noted, changing the required behavior from knowledge-based behavior to rule-based or skill-based behavior is a fundamental goal for ecological interface design.

Properly designed configural displays will also reduce the possibility of "underspecified action errors" (Rasmussen and Vicente 1989). In complex, dynamic domains, individuals can form incorrect hypotheses about the nature of the existing problem if they do not consider the relevant subsets of data (Woods 1988). Observers may focus on these incorrect hypotheses and ignore disconfirming evidence, showing a kind of "tunnel vision" (Moray 1981). Observers may also exhibit "cognitive hysteresis" and fail to revise hypotheses as the nature of the problem changes over time (Lewis and Norman 1986). Configural displays that directly reflect the semantics of a domain can reduce the probability of these types of errors by forcing an observer to consider relevant subsets of data.

We close with a very important limitation of our tutorial. On one hand, the simple process control domain that we used as the basis of our discussion is representative of this class of work domains. On the other hand, it grossly underestimates their complexity and the associated challenges that will be faced in designing interfaces for typical industrial process (e.g., power generation, chemical process control). Our goal was to demonstrate how the abstract principles and tools of cognitive systems engineering and ecological interface design translate into understandable and tangible products. In the process, we may have given the false impression that the design of these interfaces is a cut-and-dried, linear process with solutions that are either right or wrong. This is far from the truth. CSE and EID provide organizing principles that will help the designer to be "in the ballpark"; there may be more than one design solution that is equally effective. Note also that there are many more ways to do it wrong than right. See the associated discussion at the end of the next chapter.

The point of this chapter was to illustrate the logic or reasoning process that might guide display decisions, rather than to prescribe specific displays or graphical forms. Finally, we want to reemphasize the point that the goal is to effectively represent the deep structure of the process that is being managed in a way that takes maximum advantage of human capabilities (e.g., spatial reasoning). Thus, good display design always starts with an understanding of the problem being represented. There is no display technique that can compensate for a failure to understand this deep structure; in fact, a display designed around an incorrect or trivial model of a process is very likely to reinforce the misconceptions and lead to very brittle, error-prone human–machine systems.

## References

Apple. 2008. Apple human interface guidelines. Cupertino, CA: Apple, Inc.

Bennett, K. B. 1993. Encoding apparent motion in animated mimic displays. *Human Factors* 35 (4): 673–691.

Bennett, K. B., and E. Madigan. 1994. Contours and borders in animated mimic displays. *International Journal of Human–Computer Interaction* 6:47–64.

Bennett, K. B., and D. A. Malek. 2000. Evaluation of alternative waveforms for animated mimic displays. *Human Factors* 42 (3): 432–450.

Bennett, K. B., and A. L. Nagy. 1996. Spatial and temporal frequency in animated mimic displays. *Displays* 17 (1): 1–14.

Bennett, K. B., A. L. Nagy, and J. M. Flach. 1997. Visual displays. In *Handbook of human factors and ergonomics,* ed. G. Salvendy. New York: John Wiley & Sons.

Bennett, K. B., M. Payne, and B. Walters. 2005. An evaluation of a "time tunnel" display format for the presentation of temporal information. *Human Factors* 47 (2): 342–359.

Bennett, K. B., and J. Zimmerman. 2001. A preliminary investigation of the time tunnels display design technique. *Displays* 22 (5): 183–199.

Casey, S. 1993. *Set phasers on stun.* Santa Barbara, CA: Aegean Publishing Company.

Cheng, P. W., and R. G. Pachella. 1984. A psychophysical approach to dimensional separability. *Cognitive Psychology* 16:279–304.

Cleveland, W. S. 1985. *The elements of graphing data.* Belmont, CA: Wadsworth.

Coury, B. G., M. D. Boulette, and R. A. Smith. 1989. Effect of uncertainty and diagnosticity on classification of multidimensional data with integral and separable displays of system status. *Human Factors* 31:551–570.

Gibson, J. J. 1958/1982. Visually controlled locomotion and visual orientation in animals. In *British Journal of Psychology,* ed. E. Reed and R. Jones. Hillsdale, NJ: Lawrence Erlbaum Associates. Original edition, *British Journal of Psychology,* 49:182–194.

Hansen, J. P. 1995. An experimental investigation of configural, digital, and temporal information on process displays. *Human Factors* 37:539–552.

Hollan, J. D., E. L. Hutchins, and L. Weitzman. 1984. STEAMER: An interactive inspectable simulation-based training system. *AI Magazine* 5 (2): 15–27.

Jagacinski, R. J., and J. M. Flach. 2003. *Control theory for humans.* Mahwah, NJ: Lawrence Erlbaum Associates.

Langewiesche, W. 1944. *Stick and rudder.* New York: McGraw–Hill.

Lewis, C., and D. A. Norman. 1986. Designing for error. In *User centered system design,* ed. D. A. Norman and S. W. Draper. Hillsdale, NJ: Lawrence Erlbaum Associates.

Moray, N. 1981. The role of attention in the detection of errors and the diagnosis of failures in man–machine systems. In *Human detection and diagnosis of system failures,* ed. J. Rasmussen and W. B. Rouse. New York: Plenum Press.

Rasmussen, J. 1983. Skills, rules, and knowledge; signals, signs, and symbols, and other distinctions in human performance models. *IEEE Transactions on Systems, Man, and Cybernetics* SMC 13:257–266.

———. 1986. *Information processing and human–machine interaction: An approach to cognitive engineering.* New York: Elsevier.

Rasmussen, J., and K. Vicente. 1989. Coping with human errors through system design: Implications for ecological interface design. *International Journal of Man–Machine Studies* 31:517–534.

Sanderson, P. M., J. M. Flach, M. A. Buttigieg, and E. J. Casey. 1989. Object displays do not always support better integrated task performance. *Human Factors* 31:183–198.

Schutz, H. G. 1961. An evaluation of formats for graphic trend displays: Experiment II. *Human Factors* 3:99–107.

Spenkelink, G. P. J. 1990. Aiding the operator's anticipatory behavior: The design of process state information. *Applied Ergonomics* 21:199–206.

Woods, D. D. 1988. Coping with complexity: The psychology of human behavior in complex systems. In *Mental models, tasks and errors: A collection of essays to celebrate Jens Rasmussen's 60th birthday,* ed. L. P. Goodstein, H. B. Andersen, and S. E. Olsen. New York: Taylor & Francis.

# 11

# *Design Tutorial: Flying within the Field of Safe Travel*

## 11.1 Introduction

> For purposes of visual judgment in flying, then, the horizon is always as high as your eye. This comes in handy when flying near radio towers, mountains, or other airplanes; that which appears to you above your horizon is higher than you are. That which appears to you below the horizon is lower than you are. That which appears "on" the horizon (the horizon cutting through behind it) is at your altitude.
>
> In the case of airplanes, this knowledge helps avoid collisions. In mountain flying, it cuts out unnecessary worry. The more distant mountains always seem to loom higher than the near ones, and one easily gets scared into desperate climbing, using up one's reserve of fuel, when as a matter of fact one could keep cruising right across. Remember, if you can see the horizon above a mountaintop, then that mountaintop is lower than you are, however threatening it may look. (Langewiesche 1944, pp. 268–269)

In his famous book, *Stick and Rudder,* Langewiesche described how functional boundaries or constraints that are important to safe flight are directly specified as a result of the laws of dynamic optical perspective (optical flow). One important invariant relation, described in the opening quote, is the fact that the optical position of the horizon is essentially constant over change of observation point. The horizon will remain essentially at the same optical position (e.g., optical infinity) over a wide range of altitudes, as illustrated in Figure 11.1. Thus, this becomes an important referent for specifying positions of objects relative to the observer's eye height. It also helps to specify other important functional boundaries, such as the glide line, which is the furthest point that an aircraft can reach in a powerless glide.

Langewiesche (1944) observed that the glide line will be specified as

> [a] line parallel to the horizon, but, say, 10 degrees (of angular distance) below the horizon .... Any point that appears above the glide line is out of gliding range. Any point that appears below the glide line can be

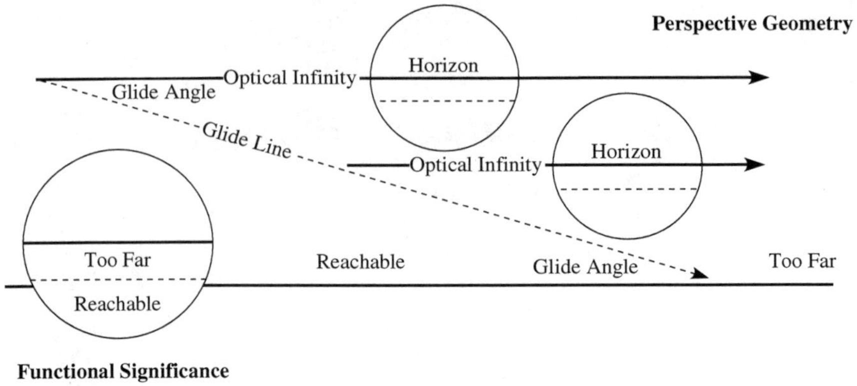


**FIGURE 11.1**
The perspective geometry of the glide angle and the functional significance of these relations.

> easily reached in a glide, will in fact be overshot unless the pilot "esses" or mushes or slideslips or otherwise steepens the glide. (p. 273)

Thus, if a pilot loses power, she should pick a spot to land that is below the glide line because it will be impossible to reach any point on the ground beyond this functional boundary. Langewiesche (1944) continues:

> How far the glide line lies below your horizon is entirely independent of your height; at any height, the glide line is the same distance (*angular* distance, in terms of degrees) below your horizon. As your height changes in the glide, both the horizon and the glide line will be at different points in the terrain below you; but *the horizon will always be at the same height as your eye; and the glide line will be the same number of degrees below the horizon; and the relation of horizon and glide line will not change.* (p. 273; emphasis added)

The emphasis in the quote from Langewiesche highlights an important insight that provided the foundation for a radical new view of perception developed by James Gibson (1966, 1979). The kernel of this new view was the possibility that important functional properties like the glide line (i.e., affordances) for controlling locomotion might be directly specified by optical invariants like the fixed angle below the horizon.

To understand why this is a radical idea, it is important to understand the prevailing view of perception (e.g., Gregory 1974). This view is based on the assumption that the important properties of objects are their Euclidean dimensions. Thus, size and distance are considered to be important invariants. Since these invariants are not preserved in the optical projection onto a two-dimensional retina, the classical view is that these properties must be recovered through inferential, cognitive processes. Gregory (1974) refers to theories that depend on an inferential process to reconstruct the Euclidean dimensions lost in the projection from a three-dimensional world as "active" theories:

> Active theories, taking a very different view, suppose that perceptions are constructed, by complex brain processes, from fleeting fragmentary scraps of data signaled by the senses and drawn from the brain's memory banks—themselves constructions from snippets from the past. On this view, normal everyday perceptions are not part of—or so directly related to—the world of external objects as we believe by common sense. On this view all perceptions are essentially fictions: fictions based on past experience selected by present sensory data. (p. xviii)

The conventional approach to perception is active in the sense that the perception of three-dimensional space is assumed to be constructed through inferential actions. The ambiguous (e.g., two dimensional) sensory information is supplemented with knowledge (e.g., about the normal size of an object) to produce a logical judgment about the "true" space. This approach is supported by ample empirical data demonstrating systematic errors in people's judgments about size and distance. However, the puzzle for conventional theories is to explain how people who are not able to judge size and distance correctly in the laboratory are able to land airplanes and skillfully move through cluttered environments.

Gibson's solution to this conundrum was to posit that the function of perception was not to answer the questions that researchers studying space perception were asking in the laboratory; rather, he posited that the function of perception was to guide action. Thus, the goal was not to judge absolute distances, but functional distances! Langeweische's observations illustrate how these are different. The pilot may not be able to judge the height or the distance of the mountains, but she can perceive the relations with the horizon that specify directly whether she can clear the mountains at the current cruising altitude. The pilot may not be able to judge the absolute distance to points on the ground, but she can perceive the angle relative to the horizon that specifies precisely those ground points that can be reached in a powerless glide from those that are too far.

The problem with the conventional approach is that space is considered to be an "empty box" that is extrinsic to or functionally independent of the organism. As Gibson (1986) noted, "I am also asking the reader to suppose that the concept of space has nothing to do with perception. Geometrical space is a pure abstraction" (p. 3). In place of this geometrical abstraction, Gibson focused on the intrinsic action or control space as articulated in the concept of the "field of safe travel" described in Gibson and Crooks's (1982) analysis of driving:

> The field of safe travel ... is a spatial field but it is not fixed in physical space. The car is moving and the field moves with the car *through* space. Its point of reference is not the stationary objects of the environment, but

> the driver himself. It is not, however, merely a subjective experience of the driver. It exists objectively as the field within which the car can safely operate, whether or not the driver is aware of it. (p. 121)

Gibson and Crooks continue that "the perceptions of the driver are organized into a number of spatial fields, each being a projection of the behavior opportunities, each being a space within which certain behavior is possible and outside of which it is impossible" (p. 134). Thus, links to the concepts of information and affordances as discussed in previous chapters begin to emerge. The link to information is reflected in the notion that this is a concrete space of possibilities, rather than an abstract geometric space. The link to affordance is reflected in the term "opportunities"—reflecting concrete action capabilities and the consequences of those actions.

Gibson, Olum, and Rosenblatt (1955) developed the formal geometry of optical flow fields that provided an analytical basis for Langeweiche's intuitions about invariant properties associated with the approach to landing. The divergence from conventional approaches to perception lies in the shift of attention from a focus on the perspective of a fixed observation point (i.e., snapshot vision) to the perspective of a moving observation point: the dynamics and invariants of flow fields. Gibson et al. (1955) provided the mathematical framework for identifying structure within the dynamic flow. Then, Gibson (1958/1982) clearly laid down the gauntlet for conventional theories of space perception:

> If the theories of space perception do not provide an adequate explanation for the visual control of locomotion in animals, let us forget about the former and pay heed to the latter. Locomotion is a biologically basic function, and if that can be accounted for then the problems of human space perception may appear in a new light. (p. 149)

Gibson (1958/1982) follows with explicit control algorithms based on the dynamics of optical flow fields:

> To begin locomotion … is to contract the muscles as to make the forward optic array flow outward. To stop locomotion is to make the flow cease. To reverse locomotion is to make it flow inward. To speed up locomotion is to make the rate of flow increase and to slow down is to make it decrease. An animal who is behaving in the corresponding ways, or, equally, an animal who so acts to obtain these kinds of optical stimulation is behaving in the corresponding ways. (p. 155)

Gibson (1958/1982) provided many other algorithms (e.g., for steering, aiming, approach without collision, etc.). However, it would be another 20 some years (until the development of interactive visual graphic displays) before these hypotheses could be systematically tested. There is now a host of empirical work that has been able to link properties of flow fields to judgments related to the control of locomotion, confirming many of Gibson's intuitions (e.g., Owen and Warren 1987; Flach and Warren 1995).

How does this story relate to the problems of interface design? There is a common misconception that the term "ecological interface design" refers to the goal of designing displays that are natural looking or naturally intuitive—like optical flow fields. Thus, some people are surprised when they see that an ecological interface for a complex process is not natural or intuitive at all. In fact, many of these displays can be quite mystifying without extensive training and practice. The motivation behind the term "ecological interface" was based on Gibson's approach to perception with an emphasis on the functional coupling between perception and action. It is based on the idea that the significance of visual information lies in its ability to specify functionally relevant relations (e.g., the field of safe travel or affordances).

Thus, the goal of ecological interface design is to create visual geometries that are specific to the functional properties of the processes being represented. Important functional properties of the processes should be mapped to salient properties of the visual representations. Functional dependencies and interactions within the processes should be specified through concrete configural properties of the display geometry. Whether the geometries for an ecological interface should look like a natural optical flow field depends on the degree to which the constraints of the process being controlled are similar to the constraints on locomotion. If they are similar—for example, in the case of aviation flight displays—then interfaces analogous to natural flow fields may be useful.

However, the challenge of ecological interface design is more often to design geometrical representations for processes where there is no natural optical flow field (e.g., process control systems, as discussed in the previous chapter). In order to satisfy the requirement of requisite variety, the complexity of the geometry will reflect the complexity of the work dynamics being represented. For complex work dynamics, the display will be no more intuitive to a novice than a chessboard. However, if the geometry is constructed well, the hope is that the process of learning will be facilitated so that the time to become relatively expert—to see the information in terms of meaningful chunks and to see good moves as the first to be considered—will be shortened.

In the remainder of this chapter we will consider some examples of flight displays chosen to illustrate the general challenge and philosophy of ecological interface design. We will begin by considering the challenge of blind flight and the evolution of the conventional attitude display. Next, we will consider a variation on the conventional attitude display designed to configure additional information regarding speed, altitude, and heading with information about attitude. Finally, we will consider a recent display concept designed to enhance the pilot's awareness of important functional relations that are often not explicitly represented in standard cockpits (i.e., total energy).

## 11.2 The Challenge of Blind Flight

Following the Wright brothers' development of the first flying machine in 1903, a major challenge to the viability of commercial aviation (e.g., airmail) was the challenge of bad weather. Repeatedly, experienced aviators would become disoriented when they lost visual contact with the ground due to weather conditions (e.g., clouds or fog). In some cases, the consequences were fatal death spirals. Thus, the challenge of "blind flight" (or fog flying) was to provide instruments that could substitute for the loss of vision when the natural visual flow fields were hidden (Ocker and Crane 1932; Previc and Ercoline 2004).

In 1927 James "Jimmie" Doolittle won the Collier Trophy in acknowledgment of an important step toward solving the problem of blind flight. On September 24, 1927, Doolittle took off from Mitchell Field, Long Island, and flew a predetermined course, returning to land where he had taken off. The remarkable thing about this was that he flew in the rear seat of a dual cockpit aircraft under a specially constructed canvas hood that prevented visual contact with the ground. It has been reported that Ben Kelsey, the safety pilot in the front seat, was tempted to grab control during the final approach, but was able to resist, allowing Doolittle to complete the blind flight. The first solo blind flight, or instrument flight, was later flown by A. F. Hegenberger at Patterson Field in Dayton, Ohio.

A critical instrument enabling Doolittle's and Hegenberger's blind flights was the Sperry artificial horizon. Using a gyroscope that maintained a constant vertical orientation, the Sperry artificial horizon display showed a fixed aircraft symbol with a representation of the horizon (blue sky and dark ground) that moved behind it to depict the relative orientation (e.g., pitch and bank angle) of the aircraft with respect to the ground. The Sperry artificial horizon is an analog to the natural horizon, showing how the horizon might appear to an observer looking at the world through a tube. When the aircraft is banked to the right, the horizon would appear to move in the opposite direction, dipping below the left wing.

However, as illustrated in Figure 11.2, another format that also preserves the invariant angular relation between the aircraft and the horizon might have been chosen. The Sperry format is called an inside–out display since it shows the view from the perspective of inside the aircraft and assumes that the fixed spatial referent for the pilot is his aircraft. The second format, called an outside–in display, maintains the artificial horizon in a fixed position and allows the aircraft symbol to move. Thus, in a bank to the right, the aircraft symbol will rotate to the right against the fixed background of the horizon. This is called outside–in because it mimics the perspective of an observer behind and outside the aircraft who maintains a fixed orientation to the horizon.

Which of these formats is better? Note that locally, in terms of the angle between the horizon and the aircraft symbol, these displays are identical.

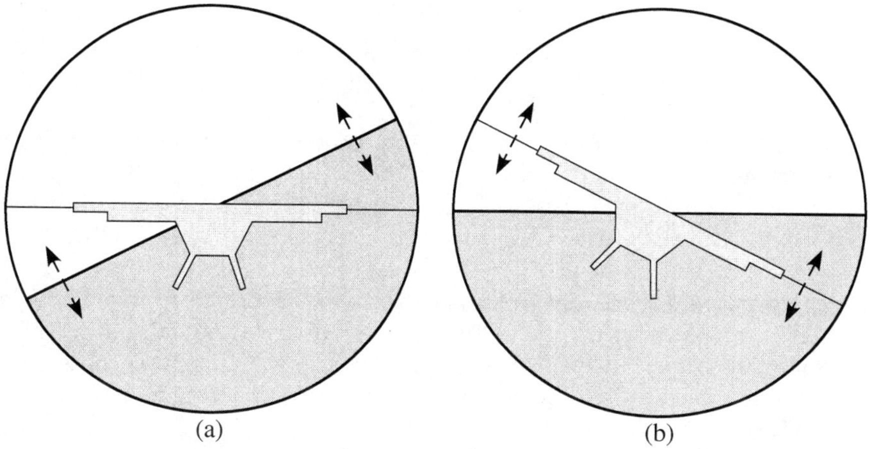


**FIGURE 11.2**
Two formats for an attitude display showing a bank to the right are illustrated. The format on the left (a) shows the format used in the original Sperry artificial horizon display. This inside–out format is the standard for most aircraft designed in the United States. The format on the right (b) shows the outside–in format. This format is used in Russian aircraft (e.g., Migs).

However, the dynamic pattern of flow and hence the potential meaningful associations are different. In essence, the question involves the relation between figure and ground in the display and figure and ground in the pilots' conceptual model. The Sperry format (inside–out) assumes that the aircraft is the pilot's conceptual ground against which the horizon "moves." This format satisfies the classical principle of pictorial realism in that the picture in the display conforms to the view through the pilot's windscreen (assuming the pilot maintains a fixed orientation to his windscreen).

On the other hand, the outside–in format assumes that the horizon is the ground against which the aircraft moves. This display satisfies the classical principle of the moving part: The part being manipulated (the aircraft) moves in a manner that is compatible with the intention (i.e., a command to bank to the right causes the right wing of the aircraft symbol to dip below the horizon).

The question of which is better remains controversial. Most pilots in the United States have been trained with the inside–out format used in the original Sperry display. However, the assumption that pilots orient to the frame of the aircraft has been called into question (Patterson et al. 1997). This research indicates that pilots tend to orient their head posture reflexively to the visual horizon. That is, when the aircraft banks to the right, pilots tend to counter-rotate their heads to maintain a fixed relation with the visual horizon. In fact, there is evidence (e.g., Roscoe and Williges 1975) that even experienced pilots will sometimes make control reversals (move the control in a direction that is opposite of their intention) with the standard inside–out format.

An elegant solution to the attitude display problem is the frequency separation concept of Fogel (1959; Johnson and Roscoe 1972). With this concept, the initial (i.e., high-frequency) response of the display to a control action or disturbance would be motion of the aircraft symbol against the fixed horizon as in the outside–in format (consistent with the principle of the moving part). However, the horizon and aircraft would slowly rotate (i.e., low-frequency response); therefore, for a steady bank, the display would eventually match the head-up view of a tilted horizon in reference to a fixed window frame consistent with the inside–out format—thus satisfying the principle of pictorial realism. Evaluations have shown that the frequency separated display resulted in more accurate control with fewer inadvertent control reversals (Roscoe and Williges 1975). Yet, as far as we know, this format remains an experimental concept that has not been implemented in any generally available aircraft cockpits.

Note that the standard attitude display also uses the horizon analogy to display information about the aircraft's pitch. When the aircraft pitches down, the horizon will rise relative to the aircraft symbol; when the aircraft pitches up, the horizon will move down, and when the aircraft is level, the horizon will be aligned with the aircraft symbol. The history of the attitude display is a great example that reflects both the subtle science and exact art of display design. Seemingly straightforward, direct analog geometries, such as the inside–out format chosen for the Sperry artificial horizon, may not provide the most intuitive representations or support the most skillful control interactions. Thus, what can we do to improve the attitude display? One possible direction that might be considered is suggested by Tufte's (1990) unconventional design strategy: "to clarify, add detail" (p. 37). This brings us to the WrightCAD concept.

## 11.3 The Wright Configural Attitude Display (WrightCAD)

The glass cockpit offers opportunities for innovative dynamical graphical representations that would not have been possible at the time of Doolittle's first instrument flight. Yet, in many cases the glass cockpit has simply recreated the traditional mechanical instruments in computerized form. There is probably good reason to do this since most pilots have learned to fly using the conventional forms and thus feel most comfortable with those formats. However, based on the discussion of the artificial horizon, it is also clear that there may be room for improvement and perhaps some of these improvements can build on the conventional instruments in a way that will not violate expectations developed with them. This was our goal for the WrightCAD display (e.g., Jacques et al. 2003; Jacques 2003).

We wondered whether it would be possible to build on the initial optical flow analogy of the Sperry display to incorporate other information about aircraft motion into a conventional attitude display. For example, could we configure other elements of natural optical flow fields with the motion of the horizon in a way that would specify additional motion states of the aircraft (e.g., altitude, speed, and heading) within a central display? There was no intention to replace the other instruments, but rather simply to enhance the attitude display in ways that would complement the other displays and enhance overall situation awareness. Figure 11.3 illustrates the WrightCAD concept. Three new elements are configured with the traditional Sperry format. These elements are depression flow lines to convey a sense of airspeed change, splay flow lines to convey a sense of altitude change, and lateral guides to convey a sense of heading direction.

The *depression lines* are simply horizontal lines moving or flowing down from the horizon line on the attitude indicator ball. These lines can be conceptualized as the way in which a series of roads or section lines perpendicular to the line of flight would appear to pilots looking out of the cockpit. Like the depression lines, the apparent motion (optical flow) of the section lines or roads increases as the airspeed increases. However, the depression lines on the WrightCAD are indexed to the nominal stall speed for the aircraft. *That is, the rate of flow of the depression lines is equal to the difference between the current airspeed and the nominal stall speed.* Thus, the depression lines slow or stop moving as the aircraft approaches a stall. In this way, the WrightCAD provides fundamental information about the safe field of travel and the task constraints (i.e., information about functional speed limits).

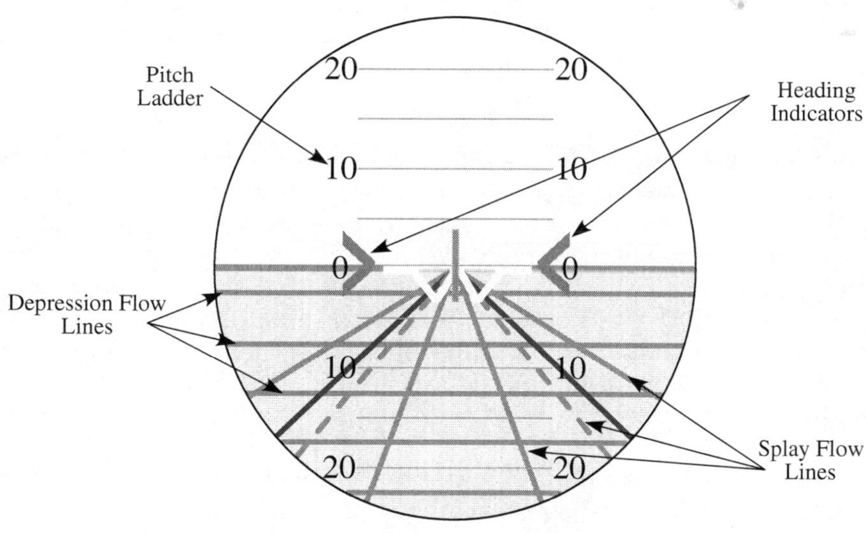


**FIGURE 11.3**
This diagram illustrates the WrightCAD concept for configuring pitch, roll, speed change, altitude change, and heading change using an optical flow analogy.

The *splay,* or slanting, lines are analogous to perspective lines converging on the horizon. On the WrightCAD, the splay lines converge on the horizon line of the attitude indicator ball. Thus, they represent the visual appearance of imaginary roads or section lines running parallel to the aircraft's heading. In flight, splay angle is the angle between these parallel roads and section lines (texture lines) and the motion path at the convergence point on the horizon. As an observer moves from higher altitudes to lower altitudes, the splay angle increases as the texture lines fan out toward the horizon. In a related way, the splay lines on the WrightCAD form a larger angle on the attitude indicator at low altitude and smaller angles at higher altitude.

The WrightCAD has several splay lines to aid in instrument approaches. The target splay line (black in Figure 11.3) is fixed at a 45° splay angle. This line represents the center of the instrument landing system (ILS) approach path. Two additional splay lines (solid gray in Figure 11.3) show the top (inside set) and bottom (outside set) of the ILS beam. The dotted gray line corresponds to the vertical position of the aircraft relative to the ILS approach path. If the aircraft is on the approach path, the dotted lines will be coincident with the target line at 45°. If the aircraft is below the target approach path, the dotted lines will be outside the target lines (splay > 45°). If the aircraft is above the target approach path, the dotted lines will be inside the target lines (splay < 45°). Thus, change of splay gives a direct indication of change in altitude that is scaled relative to the target approach path. The splay lines are similar in function to the horizontal needle in a standard ILS display. In other phases of flight, the target 45° position could be linked to other references (e.g., a specified cruising altitude or a minimum safe altitude).

The third component of the WrightCAD is the lateral or course deviation indicator. This consists of a chevron (> | <) placed on each side of a vertical mark. When the attitude indicator (fixed [own] aircraft symbol) is centered on the lateral deviation display, the aircraft is centered on glide path for the ILS. The course deviation indicator shows the aircraft's lateral position with respect to the ILS approach path. Thus, it is similar in function to the vertical needle in a standard ILS display. For other phases of flight, the chevrons could reflect a planned course. Alternatively, a sliding compass could be integrated at the top of the WrightCAD display to provide a general indication of heading.

The WrightCAD was intended to combine airspeed, pitch, roll, glide slope, and glide path/heading into configural geometry. The assumption motivating this geometry was that the configural relation might help pilots better to "see" the interplay between the states of the aircraft and how these states interact to limit the safe field of travel. The hope was that this representation might reduce the conceptual load required to understand the complex interactions that constrain flight (e.g., angle of attack) by making the interactions explicit in the representation. For example, the configural representation might help pilots to see how thrust and attitude interact to determine airspeed and flight path. In Langewiesche's (1944) terms, the goal of the configural representation is to facilitate development of the "flying instinct" (p. 56).

An alternative solution to the configural design might be a direct representation of higher order constraints, such as angle of attack. Such displays are available in some cockpits. However, these displays are generally implemented as an additional separate indicator, providing little structure to help pilots to "see" how angle of attack relates to the problem of coordinated control. With the configural representation, the elements provide a more direct mapping to both the controls (stick, rudder, and throttle) and the goal constraints (e.g., precision approach). In addition, the configural relations among the elements might make the higher order interactions among state variables more explicit. For example, the interactions between thrust and attitude in determining the speed and descent path of the aircraft (as mediated by angle of attack) might be more visible if the associated state variables were integrated into a coherent configuration.

It is important to note that although an optical flow analogy inspired the WrightCAD design, it is not a mimic or contact display. That is, the elements of the WrightCAD display are linked to aircraft states and goals in ways that are inconsistent with the behavior of the natural optical flow fields. Thus, this display was intended to function as an "ecological" interface—not because it simulates natural flow, but rather because it was an attempt to make the constraints (both higher and lower order) on the control directly visible within the representation. The motivation was to shift the burden of control from resource-limited cognitive processes to higher bandwidth perceptual processes.

Preliminary evaluations of the WrightCAD in the context of performance in landing suggest that the current design may not fulfill our hopes for this display. On the positive side, when the WrightCAD was integrated within a standard cockpit, which included a standard ILS (instrument landing display), it did no harm. That is, the additional three elements in the attitude display did not detract from its ability to provide information about pitch and roll. However, the subjective opinions suggested that the ILS instrument was the primary source for information about heading and altitude with respect to the approach targets. The chevrons and splay lines seemed to be of limited value. This is perhaps not too surprising since the pilots in the preliminary evaluations were familiar with the ILS display.

An evaluation was conducted that directly compared the WrightCAD format against a display that included a standard ILS format (crosshairs to indicate horizontal and vertical deviations from the approach targets), as illustrated in Figure 11.4. This evaluation showed nominally better performance with the ILS format. Also, subjective evaluations suggested that the integrated ILS format was more configural than the combination of splay (for vertical deviations) and the chevrons (for horizontal deviations) used in the WrightCAD format. When they used the WrightCAD format, some of the pilots indicated that it was necessary for them to switch attention

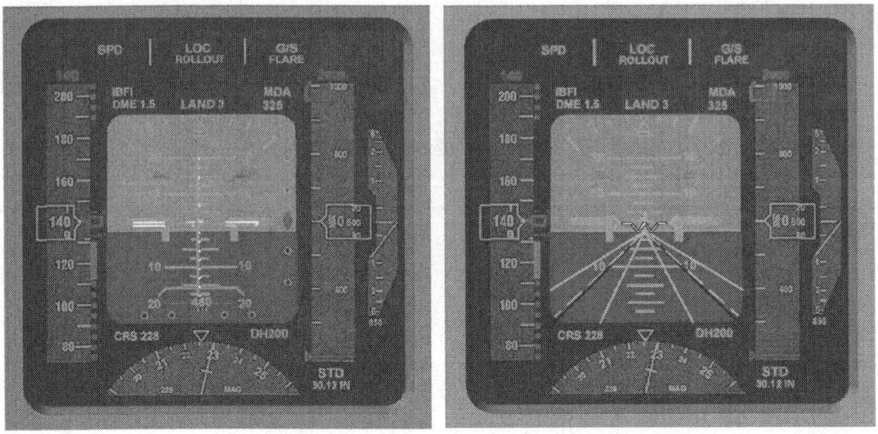


(a) (b)

**FIGURE 11.4**
**(See color insert following page 230.)** These two formats, an integrated ILS display and the WrightCAD format, were evaluated in a landing task. With the ILS format (a) lateral and vertical deviations were indicated by the intersection of two bars or needles. In the WrightCAD format (b) lateral deviation was indicated by relative position of the chevrons and the aircraft symbol; vertical deviations were indicated by relative position of the splay lines against the target of 45°.

back and forth between these elements. In other words, these elements did not configure into a single integrated chunk, as was intended. On the other hand, pilots reported that they could easily "fly" to the intersection of the ILS needles.

Pilots commented positively about the splay lines. This analogy made intuitive sense to them. They commented that these lines reminded them of a runway. One idea for better integrating the glide path and course information would be to create a virtual runway, where the magnitude of splay provided information about glide path and the symmetry of the splay lines conveyed information about course similarly to the Klopfstein synthetic runway described by Newman (1995). Another possibility is to consider ways in which speed information could be integrated into the standard ILS configuration (the cross). This might be done using the relative size of the needles; one needle might represent the target speed and the relative length of the second needle could indicate whether the current speed was over or under the target value.

At this point, we see many weaknesses in the current WrightCAD configuration and we have a new appreciation for some of the configural properties of conventional formats. A particular weakness is the use of the chevrons for direction information. These elements do not fit with the overall flow analogy. Although we are a bit humbled by the preliminary results, we do remain somewhat hopeful that the integration of the depression and splay flow lines may configure with the horizon analogy of the Sperry horizon in a way that might help overall spatial awareness and that this might reduce the current tendency toward control reversals when recovering from unusual attitudes. However, this hypothesis has not been tested empirically.

## 11.4 The Total Energy Path Display

We learned quite a bit about the process of landing while developing and evaluating the WrightCAD display (Flach et al. 2003). The evaluation of the WrightCAD was done in a CAVE VR facility and, in preparing for the evaluation, we wanted to make sure that our simulation was valid. Thus, we spent many hours testing and tweaking the simulation dynamics and the conventional instruments. A criterion that we set for the simulation was that experienced pilots should be able to land successfully on their very first attempt, using their normal procedures. In the process, we learned about the pilots' normal procedures for landing.

The procedure that most pilots follow for a straight-in approach to a typical airfield involves first setting the flaps and throttle to a fixed position. In our simulation, they only had control of the throttle. Typically, this would be set to some fixed percentage of normal cruising power (e.g., 75%). At this power, the aircraft will begin to lose altitude. With the throttle fixed, the stick becomes the primary control for following the glide path. In an instrumented approach, the stick would be used to keep the needles on the ILS aligned, as illustrated in the convention format display in Figure 11.4. Right–left deflections of the stick would cause the aircraft to bank, resulting in lateral motion in the corresponding direction. Fore–aft deflections of the stick would cause the aircraft to pitch forward or backward, respectively.

In addition to the ILS needles, a primary display for the landing task was the airspeed indicator. Pilots attempted to track a constant target airspeed, using forward deflections of the stick (elevator) to increase speed and backward deflections to reduce speed. This process is sometimes referred to as a "pitch-to-speed" strategy for flying the glide path since pitch is the primary control action and speed is the primary feedback with reference to a known target speed.

However, in talking with pilots, we learned that this was not the only strategy for landing. Several pilots indicated that a throttle-to-speed strategy was used to fly approaches to aircraft carriers or short fields. With this strategy, the fore–aft stick position was set at the initiation of the approach and the speed was tracked using the throttle as the primary control. This raised our curiosity and we began asking the aeronautical engineers at TU Delft why the different strategies existed and under what conditions one would be better than the other. In effect, we were asking which is or should be the speed control, the stick, or the throttle.

This began a path of exploration that is illustrated in Figure 11.5. The first step on this exploration was to design the CAVE VR system to handle as a conventional aircraft so that a pilot could manually achieve a soft landing using conventional instruments. We first discovered, as discussed before, that this results in a pitch-to-speed control strategy, where the throttle is set to achieve a percentage of cruising RPMs and then the stick is controlled to maintain a target airspeed (and path). When we queried the aeronautical engineers, we began to learn about energy balance.

In energy terms, landing can be thought of as a gradual reduction in total energy so that the aircraft touches down to the ground with minimal energy. Of course, the total energy is the sum of the potential energy (i.e., altitude) and the kinetic energy (i.e., speed). In a level cruise at constant altitude, the energy-in from the engines exactly matches the energy loss due to drag, so the total energy is constant. With the pitch-to-speed strategy for flying the approach, the setting of the throttle below that required for a level cruise results in a constant loss of energy. The energy-in from the engines is less than the loss of energy due to drag and gravity. If speed (kinetic energy) is kept constant, then this loss of energy will result in a constant rate of decrease in potential energy (altitude). Thus, if the throttle setting is right and the correct target speed is maintained, altitude will decrease at a constant rate (e.g., 3° glide path).

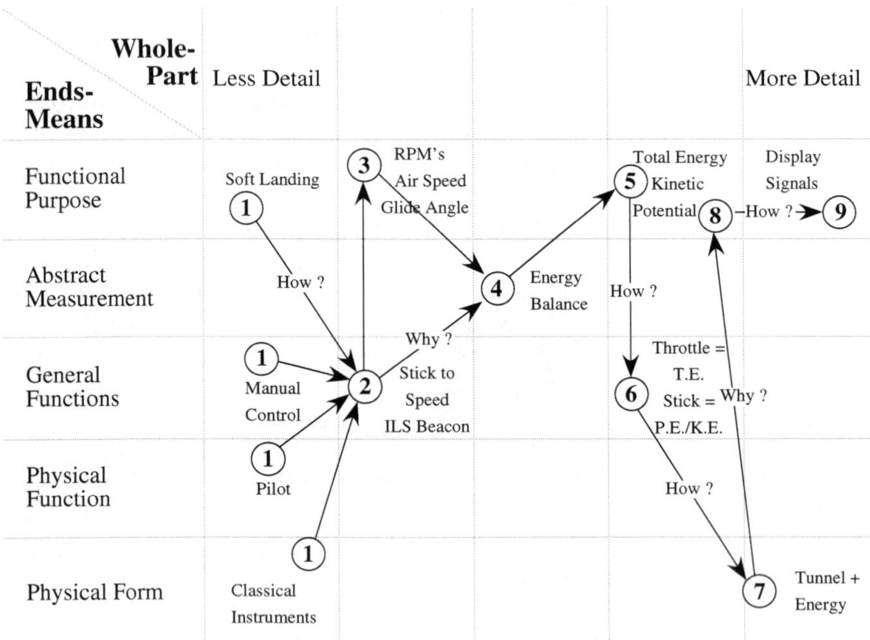


**FIGURE 11.5**
This abstraction/decomposition map shows the trajectory associated with the development of the total energy display concept.

To help us to understand the energy relations, the aeronautical engineers created the diagram in Figure 11.6 to explain the controls (throttle and stick) in energy terms. The throttle controls the total energy level. If the throttle is set higher than that needed for a level cruise (to balance drag and gravity), then energy will increase (i.e., the aircraft will gain altitude, increase speed, or both). If the throttle is set lower than that required for a level cruise, then the aircraft will lose energy (i.e., lose altitude, speed, or both). The stick (elevator) controls the distribution of energy between potential and kinetic energy. Thus, whether an increase in energy results in an increase in speed, altitude, or both depends on the position of the elevator. This reflects what Langewiesche (1944) referred to as the "law of the roller coaster." This is the ability of the aircraft to exchange potential energy (altitude) for kinetic energy (speed) by pitching forward and speeding downhill or to exchange kinetic energy for potential energy by pitching back and climbing the hill as speed is reduced.

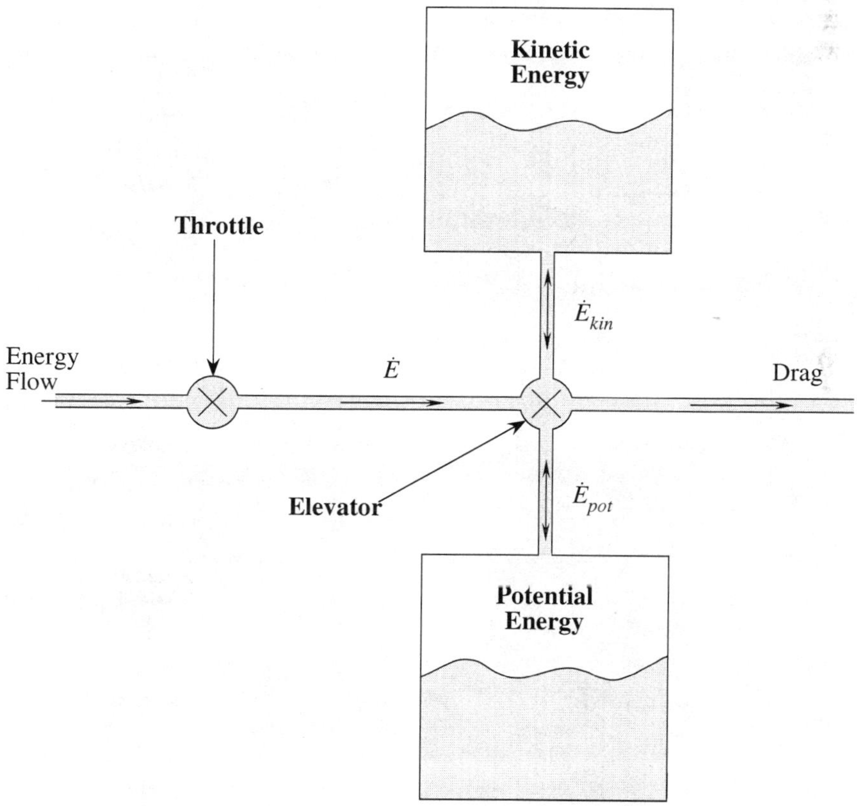


**FIGURE 11.6**
This diagram illustrates the function of the throttle and elevator in terms of energy. The throttle control adjusts the level of total energy, while the stick (elevator) determines the distribution of that energy between potential and kinetic.

In addition to framing the controls in terms of energy, it is useful to envision the landing task relative to both the glide path to the ground and the total energy path to the ground, as illustrated in Figure 11.7. The dashed line in this figure shows the total energy path to a soft touchdown and the solid line shows the glide path to a soft touchdown. The various cells in Figure 11.7 illustrate various deviations from the goal state (correct speed and correct altitude).

First, note that the diagonal from the upper left corner through the lower right corner represents situations where the aircraft is on the correct energy path (the white cells). In the upper left corner, the aircraft is high and slow. This can be corrected using the elevators (stick forward, nose down) to trade off height for speed. In the lower left corner, the aircraft is low and fast. This can be corrected using the elevators (stick back, nose up) to trade off speed for altitude. Above the diagonal toward the upper right corner, the aircraft is above the total energy target requiring an adjustment of the throttle (back). In the upper right corner, the aircraft is too high (above the target glide path) and too fast. This is illustrated as the aircraft position above the solid glide path, which equals the error in potential energy, and the distance from the glide path to the energy path,

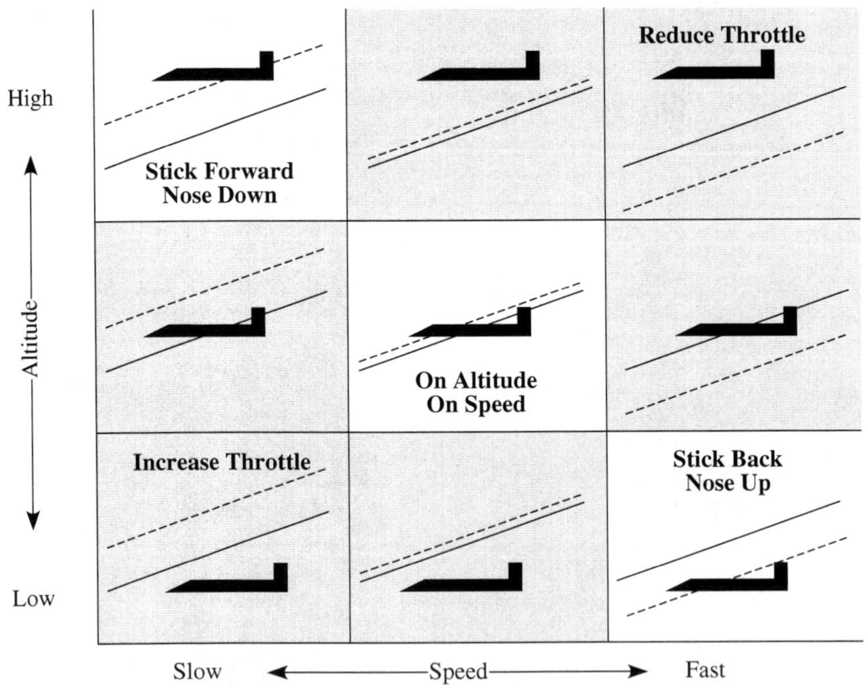


**FIGURE 11.7**
This figure illustrates the relations between the current aircraft state and both the glide path (solid line) and the total energy path (dashed line) as a function of altitude and speed, relative to the target altitudes and speeds.

which equals the error in kinetic energy, so that the distance from the aircraft to the dashed total energy path reflects the sum of these two energy components.

In the center top cell, the aircraft is on the target speed, so the total energy and path targets are aligned and the error is too much potential energy (altitude). In the center right cell, the aircraft is on the path (right altitude) so that the deviation from the energy path reflects an excess of kinetic energy (speed). The cells below the diagonal reflect situations where there is too little energy, requiring an adjustment of the throttle (forward). See if you can interpret these cells based on the description of the other cells.

Figure 11.8 illustrates the format that Amelink (2002; Amelink et al. 2005) adopted to configure a total energy path with a more standard tunnel or highway in the sky display. Position with respect to the target landing path is displayed

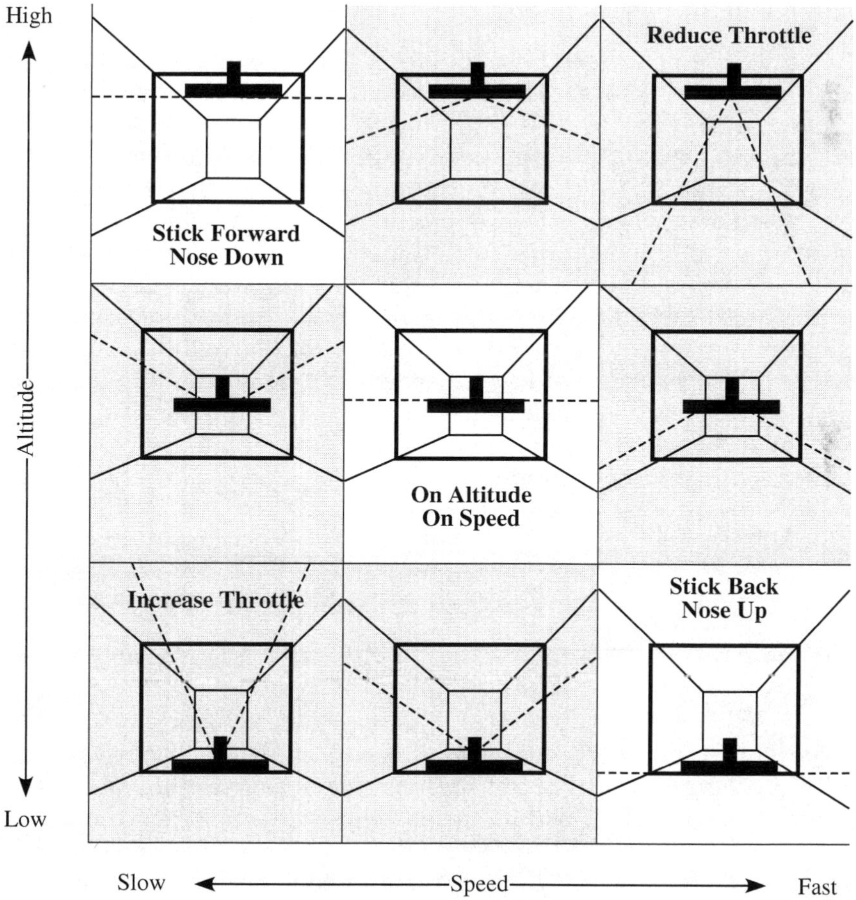


**FIGURE 11.8**
This figure illustrates a display format for the total energy display. Path is displayed as a "tunnel" or "highway" in the sky. Energy level is displayed using a "splay" analogy, where no splay (horizontal line) indicates that the aircraft is at the target total energy level.

as relative position between the aircraft symbol and the tunnel. The aircraft is on the glide path when the symbol is centered within the tunnel. Position with respect to the target total energy level is displayed using a splay analogy:

- When on the energy path, the splay lines will be horizontal.
- When above the energy path, the splay lines will appear as a roadway on the ground surface; the angles with horizontal will increase (and the inside angle will become more acute) as distance above the energy target increases.
- When below the energy path, the splay lines will appear as a row of ceiling tiles; the angles with horizontal will increase (and the inside angle will become more acute) proportional to the difference below the target energy level.

The total energy path display has not been formally evaluated, but subjective evaluations from a number of experienced pilots indicate that some confusion is created initially by the dual paths. However, with just a little experience, pilots seem to appreciate the additional information made available by this display. In particular, they begin clearly to "see" whether their throttle setting is correct—that is, whether they are meeting the energy demands.

In fact, there is a tendency for them to adapt their control strategy and to be much more active in coordinating the throttle with the stick. Instead of setting the throttle, they fly with hands on both stick and throttle, continuously tweaking the throttle to track the total energy reference while using the stick to keep the right balance of kinetic (speed) and potential (altitude) energy. This is the ideal for achieving the smoothest coordination. For example, pilot manuals typically recommend that the pitch and throttle be adjusted together. Thus, the pitch-to-speed and throttle-to-speed strategies reflect simplified suboptimal compromises to minimize the information processing load on the pilot.

## 11.5 Summary

The WrightCAD and the total energy reference path are still very much works in progress. They are not presented here as examples of successful displays, but rather to illustrate the search for effective representations—for meaning. This is truly an opportunistic and iterative process. In this case, the path began with scientific research to test hypotheses about how structures in optical flow fields might provide useful information for control of locomotion. Then, in response to opportunities associated with the glass cockpit, the question arose about whether integrating these structures into a conventional attitude display might help to reduce scanning demands and reduce tendencies toward control reversals with the standard formats.

In the process of preparing for evaluations of the WrightCAD, we learned about control strategies and this led to a deeper understanding of the control functions in relation to energy concepts. This in turn led to the concept for a total energy reference path. At this point, we still are learning about the trade-offs associated with the different control strategies. At each stage of the search, our perspective broadens, introducing greater awareness and also additional questions and uncertainties.

It would be nice if we could prescribe a linear process of stages that would be guaranteed to converge onto satisfactory design solutions. Texts such as Vicente's (1999) excellent introduction to cognitive work analysis can sometimes create the impression of such a prescription. However, this reflects the linear constraints of writing a book more than the reality of design practice. We suspect that Vicente would be one of the first to admit this. We concur with Vicente that distinctions between different perspectives such as work domain analysis, control task analysis, strategies analysis, etc. are important to understanding the layers of constraints that shape the fields of possibilities. However, our experience is that the process of exploring these possibilities is anything but a logical, linear process.

In our experience, the dynamic of design is more similar to what Schrage (2000) called "serious play." The process begins with a "seed"—maybe a specific problem or accident (e.g., difficulty flying in clouds, Three Mile Island) or perhaps an opportunity or artifact (e.g., Sperry horizon, glass cockpit). Then, from there the process tends to self-organize as insights about the work domain, the specific control demands, or specific strategies begin to emerge through an iterative process of trial and error. In the process, one begins to run into organizational constraints and human limitations. Thus, design is typically a highly dynamic, nonlinear, chaotic process.

The discipline associated with the different stages suggested by Vicente (1999) can be important for organizing and parsing the data gathered and lessons learned in this process of serious play. Vicente's distinctions become important to the retrospective process of sensemaking—in the process of parsing and organizing the experience, in the process of telling the design story, and possibly in the process of convincing people of the value of an innovation. This sensemaking process may be particularly important if we are to have any hope of generalizing from domain to domain. But this generalization is probably realized more as an openness to opportunities and ideas rather than as a driver of innovation. It enhances our ability to appreciate the next opportunity, but not necessarily our ability to generate that opportunity.

Although sometimes the seeds to this process arise from basic research, our experience is that, despite the common belief that basic research leads application, the reality is usually just the reverse. Innovation rarely originates with basic research. Basic research is most important at the evaluative stages, to test the hypotheses that emerge from the iterative processes of trial and error. The experimental discipline of basic science becomes an important protection against our natural tendencies to fall in love with our design artifacts and the natural biases associated with this infatuation. Thus, we have a "subtle science and exact art." The serious play of the designer or artist fuels the generative engine of science and innovation, while the discipline and conventions of experimental science provide the selective forces that increase the probability of a positive direction to evolution.

A final point for this chapter has to do with misconceptions about the concept of ecological interface. This label was inspired by Gibson's views about the intimate functional coupling between perception and action. The idea is that the interface should be designed with the explicit goal of linking perception with action to support successful adaptation to complex ecologies. This implies understanding the demands of those complex ecologies and linking those demands directly to dynamic flow geometries. This provides the best chance for skilled behavior that is not hampered by the limits of higher cognitive processes. Ecological does not necessarily mean natural or easy. For complex work ecologies, development of skill, even with an ecological interface, will require extensive deliberate practice. There is no shortcut!

Whether or not the WrightCAD or energy path displays are ecological depends on their ability to specify significant domain constraints directly—*not* on the fact that both use aspects of optical flow analogies. To this extent, we believe that the energy path display better meets the aspirations for an ecological interface than the WrightCAD because it specifies energy constraints in ways that help to differentiate the functions of the stick and throttle, leading to a more direct coordination of perception and action than with other, more conventional formats.

## References

Amelink, M. H. J. 2002. Visual control augmentation by presenting energy management information in the primary flight display—An ecological approach. Faculty of Aerospace Engineering, Delft University of Technology, Delft, the Netherlands.

Amelink, M. H. J., M. Mulder, M. M. van Paassen, and J. M. Flach. 2005. Theoretical foundations for a total energy-based perspective flight-path display. *International Journal of Aviation Psychology* 15 (3): 205–231.

Flach, J. M., P. Jacques, D. Patrick, M. Amelink, M. M. van Paassen, and M. Mulder. 2003. A search for meaning: A case study of the approach to landing. In *Handbook of cognitive task design,* ed. E. Hollnagel. Mahwah, NJ: Lawrence Erlbaum Associates.

Flach, J. M., and R. Warren. 1995. Low-altitude flight. In *Local application of the ecological approach to human–machine systems,* ed. P. Hancock, J. M. Flach, J. Caird, and K. J. Vicente. Hillsdale, NJ: Lawrence Erlbaum Associates.

Fogel, L. J. 1959. A new concept: The kinalog display system. *Human Factors* 1:30–37.

Gibson, J. J. 1958/1982. Visually controlled locomotion and visual orientation in animals. In *British Journal of Psychology,* ed. E. J. Reed. Hillsdale, NJ: Lawrence Erlbaum Associates. Original edition, *British Journal of Psychology* 49:182–194.

———. 1966. *The senses considered as perceptual systems.* Boston, MA: Houghton Mifflin.

———. 1979. *The ecological approach to visual perception.* Boston, MA: Houghton Mifflin.

———. 1986. *The ecological approach to visual perception.* Hillsdale, NJ: Lawrence Erlbaum Associates. Original edition, 1979.

Gibson, J. J., and L. Crooks. 1982. A theoretical field-analysis of automobile driving. In *Reasons for realism,* ed. E. J. Reed. Hillsdale, NJ: Lawrence Erlbaum Associates. Original edition, *American Journal of Psychology,* 1938, 51:453–471.

Gibson, J. J., P. Olum, and F. Rosenblatt. 1955. Parallax and perspective during aircraft landings. *American Journal of Psychology* 68:372–385.

Gregory, R. L. 1974. *Concepts and mechanisms of perception.* New York: Charles Scribner & Sons.

Jacques, P. F. 2003. The Wright configural display (WrightCAD) and the ecology of virtual instrument approaches. Department of Psychology, Wright State University, Dayton, OH.

Jacques, P. F., J. M. Flach, D. L. Patrick, and R. Green. 2003. Evaluating a configural altitude display: WrightCAD. Proceedings of the 12th International Symposium on Aviation Psychology, Dayton, OH.

Johnson, S. L., and S. N. Roscoe. 1972. What moves the airplane or the world? *Human Factors* 14:107–129.

Langewiesche, W. 1944. *Stick and rudder.* New York: McGraw–Hill.

Newman, R. L. 1995. *Head-up displays: Designing the way ahead.* Brookfield, VA: Ashgate.

Ocker, W. C., and C. J. Crane. 1932. *Blind flight in theory and practice.* San Antonio, TX: Naylor.

Owen, D. H., and R. Warren. 1987. Perception and control of self motion: Implications for vehicle simulation of self-motion. In *Ergonomics and human factors: Recent research,* ed. L. S. Mark, J. S. Warm, and R. L. Huston. New York: Springer–Verlag.

Patterson, F. R., A. J. Cacioppo, J. J. Gallimore, G. E. Hinman, and J. P. Nalepka. 1997. Aviation spatial orientation in relation to head position and attitude interpretation. *Aviation, Space, and Environmental Medicine* 68:463–471.

Previc, F. H., and W. R. Ercoline. 2004. *Spatial disorientation in flight.* Reston, VA: American Institute of Aeronautics and Astronautics.

Roscoe, S. N., and R. C. Williges. 1975. Motion relationships in aircraft attitude and guidance displays. A flight experiment. *Human Factors* 17:374– 387.

Schrage, M. 2000. *Serious playe.* Cambridge, MA: Harvard Business Press.

Tufte, E. R. 1990. *Envisioning information.* Cheshire, CT: Graphics Press.

Vicente, K. J. 1999. *Cognitive work analysis: Toward safe, productive, and healthy computer-based work.* Mahwah, NJ: Lawrence Erlbaum Associates.

# 12

# *Metaphor: Leveraging Experience*

## 12.1 Introduction

> Few would argue that images are not among the most important elements in a visual computing environment. Their impact on the presentation of a conceptual model, the tightness of the feedback loop between person and machine, and the apparent tangibility of a synthetic virtual space is greater than any other aspect of the application. (Mullet and Sano 1995, p. 170)

Many of the chapters in this book (Chapters 8, 9, 10, and 11) focus on design for narrowly defined technical domains such as process control and aviation (i.e., the right side of the continuum illustrated in Figure 6.3 in Chapter 6). In these chapters the emphasis is on making the deep structure of the control problems explicit to the human operators through analog, geometrical form displays. Thus, much of the attention is on work analysis and the use of the abstraction hierarchy as a framework for thinking about domain semantics. In this chapter we want to focus more on the other end of the semiotic dynamic: the awareness side of the equation.

From this perspective, it is possible to consider general strategies for design using metaphors that will apply broadly across a wide spectrum of work domains (as introduced in Chapter 6). These strategies may be particularly valuable for general purpose interfaces (e.g., computer and mobile device operating systems and general applications) that are designed to be used by a diverse population of users in a diverse range of work situations (the left side of the continuum in Figure 6.3). In fact, the challenge of designing effective metaphors is not necessarily restricted to computer and electronic interfaces. For example, the graphical representations of traffic signs must be designed to convey general purpose messages to extremely diverse users.

This side of the semiotic dynamic has been the primary focus for much of the classical work in human–computer interaction (HCI). One reason for this is the potential general applicability, as just noted. A second reason is that this perspective fits most naturally with the dyadic information processing approaches that emphasize internal mental processes. Our goal for this chapter is to recast some of this classical work in the context of the triadic semiotic framework.

The key to accomplishing this goal is embedded in the opening quote. When the value of images (e.g., pictorial or iconic displays, such as the desktop spatial metaphor that revolutionized personal computing) is considered, a critical element of success is the "tightness of the feedback... loop" and the "tangibility of a synthetic virtual space." In other words, these interfaces allow for an increasingly direct coupling of perception and action. Shifting the emphasis from passive information processing to active control makes the connections between the internal processes associated with awareness and the demands of problems in the ecology more explicit. It helps us to consider the constraints on awareness in the context of the triadic dynamic of abduction, where the ultimate test of awareness is successful action in an ecological niche. Thus, the focus shifts to the development of skills with respect to closing the loop successfully in specific domains.

In the same way that Rasmussen's abstraction hierarchy provided a problem-centered perspective on the triadic dynamic, Rasmussen's decision ladder provides an awareness-centered perspective on the triadic dynamic. However, in taking either perspective, it is essential to keep in mind that the dynamic itself is triadic. Thus, it is important for a problem-centered perspective to respect the constraints on awareness and for an awareness-centered perspective to respect the constraints in the work ecology.

Thus, the focus of this chapter will be to consider the general use of metaphors, such as the desktop metaphor, as viewed through the lens of the decision ladder. Why have these metaphors had such a huge impact on the quality of computers as tools to help humans solve complex problems? We think the answer is reflected in principles of skill development.

## 12.2 Spatial Metaphors and Iconic Objects

Metaphors were introduced in Chapter 6 as one of three primary representational forms and as a general interface design strategy. To reiterate briefly, the most commonly encountered example of metaphor in the interface is the icon: an object with an image (i.e., metaphor) that also incorporates the potential for action. However, note that metaphors do not necessarily require an action component and can be more global in nature than an icon (ranging from forms to settings to the workplace; see Figure 6.9, in Chapter 6, and Chapter 15 for more detailed discussion). Metaphors are ubiquitous in today's interface.

### 12.2.1 The Power of Metaphors

The success of metaphors is in part due to their compatibility with human memory. The superiority of recognition memory (i.e., recognizing a previously presented item) over recall memory (i.e., remembering all items that were previously presented without cues) is a well-established finding in the experimental psychology literature (e.g., Postman and Rau 1957). The rather impressive findings for picture recognition (Shepard 1967; Standing, Conezio, and Haber 1970) indicate that humans have a tremendous capability to recognize images. In fact, pictorial information has played a central role in prominent theories of memory (e.g., Paivio 1969, 1971, 1978). Thus, pictorial metaphors in the interface are effective partly because they leverage memory.

However, it could be argued that the use of metaphors is not simply a convenient mnemonic device; instead, the use of metaphors may be the very foundation of human rationality (e.g., Lakoff and Johnson 1980; Lakoff 1987). The general idea is that our ability to deal with abstract information is derived from our experiences in a physical world. Physical experiences (e.g., up and down; inside and outside) permeate our thinking. Thus, metaphors that tap into spatial intuitions can be particularly powerful. For example, we can talk about human feelings of happiness as being "up" and feelings of depression as being "down." We can talk about human relations in terms of "insiders" and "outsiders" and the "higher-ups" ruling "over" the "downtrodden." People in positions of power "step down" when they willingly leave those positions. Thus, consistent with the triadic approach to semiotics, Lakoff and others (e.g., Clark 1997) see cognition as "embodied." In other words, awareness is grounded through interactions with the physical ecology.

The implications of embodied cognition for design are that humans' experience with space and physical objects is an important lever for making sense of more abstract phenomena, including human and social relations, subatomic and cosmological aspects of nature, as well as information technologies. Thus, it is not surprising that spatial metaphors, in particular, will provide the potential for effective interface support in a wide range of domains. In terms of Rasmussen's decision ladder, the idea is that associations or heuristics derived from our experiences in the physical world can be the scaffolding for building up efficient skill- and rule-based associations for dealing with more abstract problems. Further, the heuristics that we have developed for testing hypotheses through physical experiences in the world (exploring space and manipulating objects) can facilitate the knowledge-based processing needed to explore more abstract domains.

### 12.2.2 The Trouble with Metaphors

> A picture is worth a thousand words. (Anonymous)
>
> Just which particular word, out of the thousand, is that interface icon supposed to refer to? (the authors)

Despite the potential benefits of incorporating metaphors into the interface, the design of effective metaphors presents a considerable challenge. As implied by the previous discussion, using a metaphor imposes a requirement that goes beyond the simple capability to recognize a picture (or a local visual metaphor). The user needs to be able to discover, unambiguously, the meaning that lies behind the image. Rogers (1989b) discusses this point in some detail:

> One of the main problems with iconic interfacing is that while on some occasions it is relatively easy to interpret the intended meaning of an icon, for others a whole range of different meanings can be attributed to a single icon each being as valid as the other. For example, an icon depicting a horizontal arrow by itself could be interpreted as "turn the page," "move the cursor to the end of the line," "scroll the page to the right" and so on. The point is that the user has no way of knowing whether an initial interpretation is correct. Only by looking in a manual or activating the icon, which may have adverse effects, can the user find out whether the understanding matches the intended meaning. Unlike verbal language, in which there are a set of syntactic and semantic rules which provide us with a means of disambiguating the meaning of verbal language, pictorial language has, as yet, no equivalent set of rules underlying its comprehension. The paradox, therefore, is that while pictorial communication has the potential of being universally understood, it does not have the rules to guide this process. (p. 106)

Mullett and Sano's (1995) discussion of visual metaphors takes this line of reasoning one step further: "The dominant position of imagery in human communication confers both the greatest opportunity for success and with [it] the greatest risk of failure" (p. 171).

This observation is consistent with our experience. There appears to be wide variation in the success of individual metaphors in today's interfaces. Some icons incorporate metaphors that are intuitive, easy to learn, and strongly suggest the meaningful aspects of the domain that they represent. Others do not. The question is how interface designers can achieve the former on a more regular basis than the later. The remainder of the chapter considers this design problem from the triadic perspective.

## 12.3 Skill Development

Perhaps it is best to frame the question of what makes a good metaphor in the context of transfer of training or skill development. Thus, the key to a good metaphor is to transfer associations developed in one domain (e.g., in the process of moving around the physical world) to a new domain (e.g., manipulating programs and information stored in a computer), with the goal of facilitating the development of skill in the new domain.

Piaget (1976) suggested that skill development involves the processes of *assimilation* and *accommodation*. He found that, in approaching novel situations, people initially depend on associations from their past experience. Thus, when someone is playing a new sport like tennis, the initial approach will be based on skills developed in other contexts (e.g., playing baseball or badminton). Piaget referred to this process as assimilation: A new situation is approached using knowledge structures developed in other domains.

In terms of the decision ladder, the rule-based and skill-based heuristics or shortcuts that worked in previous situations (see Figure 4.3 in Chapter 4 and the associated discussion) are applied in the new situation. Those heuristics and shortcuts that lead to success in the new domain will be reinforced and those heuristics that do not satisfy the demands of the new domain will be gradually eliminated and replaced by new associations discovered in interacting with the new domain. Piaget referred to this process of pruning unproductive associations and adding new associations as accommodation.

### 12.3.1 Assimilation

Thus, when seen in the context of the triadic semiotic dynamic of abduction, assimilation dominates early when one is learning a new domain. That is, the intentions, hypotheses, and expectations that shape the interactions will reflect associations from past experience in other domains. In the process of acting upon these intentions and expectations, errors and surprises will drive accommodative processes; that is, the intentions and expectations will be gradually modified to reflect the constraints of the new domain. Thus, the human develops new rule- and skill-based associations (illustrated as shortcuts across the decision ladder) in the process of interacting with the system.

The concepts of assimilation and accommodation suggest that two important facets will determine the ease of learning a new domain. The concept of assimilation suggests that the ease of learning will depend on the functional similarity between the base domain (i.e., the source of the metaphor) and the new target domain of interest. The greater the functional similarity is, the more powerful or productive will be the assimilated associations and, conversely, the less need there will be for accommodation.

Thus, the challenge for interface design is to pick a metaphor that has a high degree of functional similarity to the target domain. The problem is to determine what the significant functional similarities are because any two domains may be similar and dissimilar in many different ways. The success of the desktop metaphor for office computing is evidence that there are powerful functional similarities between the base domain of physically managing papers and files in an office and the target domain of managing data and files in a computer. Abstract functions such as organizing, retrieving, and deleting data files are easier to learn and remember when they are framed in terms of moving folders, clicking on folders, and dragging folders to the trash icon.

### 12.3.2 Accommodation

Dewey (1991) has argued quite compellingly that a primary basis for "order in thought" is generalization from experiences with action:

> Intellectual organization originates and for a time grows as an accompaniment of the organization of the acts required to realize an end, not as the result of a direct appeal to thinking power. The need of thinking to accomplish something beyond thinking is more potent than thinking for its own sake. All people at the outset, and the majority of people probably all their lives, attain ordering of thought through ordering of action. Adults normally carry on some occupation, profession, pursuit; and this furnishes the continuous axis about which their knowledge, their beliefs, and their habits of reaching and testing conclusions are organized. Observations that have to do with the efficient performance of their calling are extended and rendered precise. Information related to it is not merely amassed and then left in a heap; it is classified and subdivided so as to be available when needed. Inferences are made by most men not from purely speculative motives, but because they are involved in the efficient performance of "the duties involved in their several callings." Thus, their inferences are constantly tested by results achieved; futile and scattering methods tend to be discounted; orderly arrangements have a premium put upon them. The event, the issue, stands as a constant check on thinking that has led up to it; and this discipline by efficiency in action is the chief sanction, in practically all who are not scientific specialists, of orderliness in thought. (p. 41)

This passage is very consistent with Piaget's second dimension of skill development: accommodation. The ease of accommodation will depend on how easy it is to discover the mismatches between the base and target domains of the metaphor. There are many ways that these mismatches can be communicated. One way is to include detailed manuals and or training lectures that teach about the new domain (e.g., the new rules and procedures)—that is, to learn from reading or listening to an authoritative source. An alternative is to allow the mismatches to be discovered through trial and error—that is, to promote learning by doing (e.g., discovery learning; Bruner 1961, 1967).

It should be apparent that in safety-critical domains, such as nuclear power and aviation, it can be very risky to rely solely on learning by doing. It would be irresponsible to have a novice learn to control a nuclear power plant by trial and error. On the other hand, it should be clear that even in safety-critical domains, it is impossible to cover all contingencies in a manual or in a training course; some degree of learning by doing will always be required.

Also, based on experience with the desktop/office metaphor, with video games, and with consumer products like the iPhone® mobile digital device, it should be apparent that many people greatly prefer learning by doing. We speculate that confidence in learning by doing and impatience with learning from authority is one of the attributes that differentiates the Nintendo generation from the generation before it, who expected that reading bulky instruction manuals was a prerequisite for learning complex information technologies.

### 12.3.3 Interface Support for Learning by Doing

The power and attractiveness of learning by doing is reflected in Shneiderman's (1992) concept of direct manipulation (as noted in Chapter 6). Shneiderman suggests three principles to characterize a direct manipulation interface:

1. Continuous representation of the objects and actions of interest
2. Physical actions or presses of labeled buttons instead of complex syntax
3. Rapid incremental reversible operations whose effect on the object of interest is immediately visible (p. 205)

Shneiderman (1992) goes on to describe the beneficial attributes of direct manipulation designs:

- Novices can learn basic functionality quickly, usually through a demonstration by a more experienced user.
- Experts can work rapidly to carry out a wide range of tasks, even defining new functions and features.
- Knowledgeable intermittent users can retain operational concepts.
- Error messages are rarely needed.
- Users can immediately see if their actions are furthering their goals, and, if the actions are counterproductive, they can simply change the direction of their activity.
- Users experience less anxiety because the system is comprehensible and because actions can be reversed so easily.
- Users gain confidence and mastery because they are the initiators of action, they feel in control, and the system responses are predictable. (p. 205)

These principles and attributes have been reviewed in depth (Hutchins, Hollan, and Norman 1986). In terms of the present discussion, a key insight (often overlooked) is that many of these attributes can be summarized by simply saying that a direct manipulation (and direct perception) interface is one that facilitates accommodation in the form of trial-and-error learning (i.e., direct coupling of perception and action). However, we would like to raise some important issues with Shneiderman's prescriptions.

The first issue is the concept of "objects and actions of interest." One obvious point is that the term "direct manipulation interfaces" (Hutchins et al. 1986; Shneiderman 1992) leads to a focus on the technology of interaction (e.g., pointing and clicking). As emphasized throughout the book, direct perception in the interface is equally as important as direct manipulation. Furthermore, it is important to keep in mind that the things of interest are ultimately in the work domain, rather than in the interface. For example, we are typically worried about how the printed document will look, rather than how the document displayed on the computer screen looks. But, of course, if the document on the screen is a direct analogue of the printed document (as with WYSIWYG—what you see is what you get), then one is essentially directly manipulating the document when interacting with the display.

Further, it is important to note that in many domains the objects of interest may not be objects at all. That is, in many cases the significant aspects of the domain may involve rather abstract concepts (program, data file, application, energy) that do not have a tangible physical form at all. Thus, one of the challenges to the interface designer is to "objectify" these significant functions and relations. Thus, data, programs, and operations are objectified as spatial metaphors. Thus, this is another important aspect of choosing an appropriate metaphor. Good metaphors will typically objectify concepts and operations that are essential to problem solving (i.e., make them more concrete). For example, in Chapter 3 the concept of three numbers that sum to nine is objectified in the context of Tic-Tac-Toe as "three in a row." In the desktop metaphor, operations such as moving and deleting files are objectified in terms of clicking and dragging file icons into windows or onto other icons (e.g., the trash can icon).

A second issue with Shneiderman's prescriptions is the concept of reversible operations. This is obviously very important for learning by trial and error. For example, the home button on the iPhone mobile digital device is a great example of a reversible operation (see discussion in Chapter 13). That is, it is nearly impossible to get lost in this device's interface because, with one button press, one can instantaneously reverse the search path and return to a familiar base. This ability to reverse a search instantaneously encourages people to wander and explore the applications and learn by doing because they do not have to worry about becoming lost—no manual is required. On the other hand, this can be a great problem in many Windows-based applications, where one can easily get lost in a cascade of windows and then not be able to find the way back to retrieve critical information or to activate a particular function.

Thus, we agree wholeheartedly that reversibility is a desirable attribute; however, in many domains the reversibility of operations will be determined by physical laws, not by the decisions of the designers. For example, if an aircraft gets too far below its minimum total energy envelope, there will be no action that the pilot can take to avoid collision with the ground. This is a major challenge for safety-critical domains: to help operators to learn about safety margins without crossing them and to ensure that the operators are alerted well before they initiate a process that will lead irreversibly to catastrophic failure. In these domains, it is best that learning by trial and error be accomplished in simulators, rather than in the actual domain. In safety-critical domains, fast time simulations and predictive displays may be an important feature of an interface because they can allow trial-and-error experiments to support learning and decision making (e.g., when the predictions cross the critical boundaries in time for compensatory actions to prevent irreversible negative consequences in the target domain).

A final issue with Shneiderman's prescriptions is that the view of action is too narrowly constrained around click-and-point technologies. With the evolution of touch screens and motion sensors (e.g., the Wii), there are many options for interacting with computers that were not available when Shneiderman first introduced the concept of direct manipulation. We would suggest that the more general principle is to utilize physical gestures to manipulate the virtual objects in ways that take advantage of expectations developed through physical interactions in the world.

The flick motion to scan series of photos used in the iPhone interface (see example in Chapter 13) is a good example. Shneiderman's suggestion to avoid complex syntax is obviously right, but it is somewhat vacuous unless one can define what makes syntax complex. Are the motions used to throw a bowling ball with the Wii less complex than the button and trigger clicks and joystick motions that were used in earlier video games? We posit that complexity can only be judged relative to our expectations. A simple syntax is one that conforms to our expectations. A complex syntax is one that seems arbitrary relative to our expectations. Thus, the syntax with the older game controllers was arbitrary when seen in relation to expectations from bowling with physical balls and pins. The syntax of the Wii seems far less arbitrary and thus is much easier to learn and remember.

## 12.4 Shaping Expectations

> Imagery is at once the most obvious and least understood aspect of GUI [graphical user interface] design. (Mullet and Sano 1995, p. 175)

It is tempting to conclude that spatial forms of interaction (e.g., clicking and pointing to move through virtual spaces as in the BookHouse) are easier to learn because they are more "natural." But like Shneiderman's use of the term "complex," the term "natural" is a bit vacuous and the conclusion is tautological because we typically judge the naturalness of something by the ease of learning it. It is natural because it conforms to our expectations and, of course, if it conforms to our expectations, it is easier to learn.

**FIGURE 12.1**
Spatial metaphors used in the window control icons of two popular operating systems.

For example, the user of a Macintosh® computer finds it quite natural to close, minimize, or maximize a window using the icons in the top left corner of the window display (Figure 12.1). However, this convention seems quite arbitrary to a PC user whose experiences direct her to look in the upper right corner for icons with very different forms that represent the same functions. Thus, people whose expectations are shaped through interactions with one platform or convention find the syntax of an alternative platform to be arbitrary at best—if not obtuse!

Thus, we suggest that the key principle underlying good metaphors and, ultimately, the ease of learning and use is consistency. It is the basis for the rule- and skill-based heuristics illustrated in the decision ladder and, in fact, for all our expectations about the world. Consistency is the basis for all abductive processes: the belief or at least the hope that the past is our best guide to the future. Metaphors based on physical intuitions about space and motion are powerful because these physical intuitions are common across large populations of people. However, they are not universal. For example, the intuitions of people who have never had sight may be very different from those of the general population. But it seems clear that when information technologies began to tap into the physical intuitions shared by the vast majority of people, these technologies became much easier to use.

Computers are ubiquitous today. They are as much a part of the natural world for the Nintendo generation as are trees. Thus, conventions that seemed quite arbitrary (e.g., drop-down menus or, more specifically, the fact that the print function will be found under "file") to our generation are second nature to them. Our expectations about the world are being shaped by information technologies as much as they are by our interactions with physical objects. This was illustrated quite clearly when one of the authors, after having an iPhone mobile digital device for a month, reached up to his computer monitor to try to resize an image using the pinch gesture on the screen.

### 12.4.1 Know Thy User

This gives new significance to the old saw "know thy user." The point is that the fluidity of the triadic semiotic dynamic will depend on the experiences of the users. In designing new technologies, it is important to be aware of the expectations created by experiences with the old technologies. Long after the old technologies have faded, the expectations may still be shaping the semiotic dynamic. For example, consider the phone icon (see Figure 12.2) used to indicate that telephone services are available on the highway (U.S. Department of Transportation 2009). Note that the phone icon developed for the iPhone interface (Apple 2008) is essentially the same spatial metaphor.

This metaphor looks nothing like the majority of phones currently in use; it is a visual metaphor that reflects a much older technology. Regardless, it maps very well to the expectations of a whole generation who grew up with that old technology and the association will continue to work for a generation who never touched a classically shaped phone due to the consistency of its use. The QWERTY keyboard is another prime example; designed to slow typing due to the constraints of mechanical typewriters, this convention continues to shape the expectations of users long after the motivating constraints have been eliminated.

Another aspect of knowing thy user is that the experiences of people are diverse. We cannot afford to design for one specific person, so which user should we pick? For example, think of all the different forms of telephones that exist in the world or think of all the different shapes of people. Similarly, which single image should be chosen to represent the men's bathroom? Research on concept formation suggests that some images may represent a concept or a class of objects better than other images (e.g., Rosch 1973, 1978). For example, some birds (e.g., sparrows) are more typical of the general category of birds than other birds (e.g., chicken). This research suggests that an image such as that illustrated in Figure 12.2—one that does not necessarily

**FIGURE 12.2**
The spatial metaphor used to indicate that telephone services are available on the highway. (U.S. Department of Transportation. 2009. *Manual on Uniform Traffic Control Devices for Streets and Highways*. Washington, D.C.: Federal Highway Administration.)

**FIGURE 12.3**
The metaphor used to indicate that restaurant services are available on the highway. (U.S. Department of Transportation. 2009. *Manual on Uniform Traffic Control Devices for Streets and Highways*. Washington, D.C.: Federal Highway Administration.)

look like any particular object that it is intended to represent—may be a better representative of the general concept or class of objects than a photorealistic image.

Rogers (1989b) uses the term "exemplar" for icons that are designed to typify a particular class of function (see Figure 12.3):

> [A]n exemplar icon serves as a typical example for a general class of objects. For example, the knife and fork used in the public information sign to represent "restaurant services" .... Within the context of public services, this depiction of a very simple image is very powerful as it shows salient attributes associated with what one does in a restaurant, i.e., eating. (p. 110)

Again, it is important to appreciate that salience will always be defined relative to the expectations of a particular user group. A knife and fork may not be the most salient images for representing the function of eating in all cultures. Similarly, different cultures of technology users (Macintosh vs. PC cultures) may have expectations that were shaped by very different conventions.

## 12.5 Abduction: The Dark Side

> There is no label on any given idea or principle which says automatically, 'Use me in this situation"—as the magic cakes of Alice in Wonderland were inscribed "Eat me." The thinker has to decide, to choose; and there is always a risk, so that the prudent thinker selects warily, subject, that is, to confirmation or frustration by later events. (Dewey 1991, p. 106)

The value of a metaphor in terms of assimilation lies in the positive transfer of skills learned in a familiar domain to a new domain. However, with any metaphor, there is also likely to be negative transfer—that is, associations in the familiar domain that are not appropriate to the new domain. This fact is often painfully obvious to people who have to move back and forth across Macintosh and PC computer platforms. Some of the conventions between these platforms are similar, so expectations developed in one platform will result in positive transfer to the other platform. However, other conventions are very different, resulting in negative transfer when things do not conform to the expectations developed using the alternative platform (i.e., error, surprise, and often frustration).

### 12.5.1 Forms of Abductive Error

As Ernst Mach (1905) observed, "Knowledge and error flow from the same mental sources; only success can tell the one from the other" (p. 84). Thus, metaphors can lead to insight or they can lead us down a garden path to error and surprise and potentially irreversible negative consequences. While many cognitive researchers (e.g., Norman 1981; Rasmussen 1986) have discussed the two sides of the heuristic cognitive dynamic, Reason (1990) makes the most explicit connection between the dynamics of abductive, heuristic-based thinking and the forms of error that it invites. He associates specific forms of errors with each of the three levels (skill, rule, and knowledge) of processing within the decision ladder.

An example of a type of rule-based error is encoding deficiencies in a rule. Reason cites work by Siegler (1983) that shows that 3-year-olds fail to solve balance-scale problems correctly because they do not appreciate the role of distance from the fulcrum for achieving balance. They understand the relative weight aspect, but not the distance aspect. Eight-year-olds have no difficulty with this. Reason (1990) writes:

> The difficulty in this instance is that 5-year-olds cannot cope with manipulating two relationships at the same time. Adults, of course, do not generally retain these developmental limitations. Nevertheless, there are phases during the acquisition of complex skills when the cognitive demands of some component of the total activity screen out rule sets associated with other, equally important aspects. (p. 81)

We refer readers to Reason (1990) for detailed descriptions and discussions of each of the other error types. However, the point that we want to make is that Reason's list of error forms should be considered when designing metaphors to help anticipate the potential for negative transfer. It is essential that designers consider both sides of the assimilation coin (i.e., potential for both positive and negative transfer). A metaphor may suggest some rules from the familiar domain that are conditioned on or bounded by dimensions in the target domain that were not at issue in the familiar domain.

### 12.5.2 Minimizing Negative Transfer

There can be some instances when the goal of design is not to facilitate assimilation, but rather to minimize negative transfer. A well-known example is the international biohazard symbol (Baldwin and Runkle 1967). The visual properties of this spatial metaphor are distinctive, but they have no relationship to its meaning (i.e., a warning that a substance poses a potential threat to health). The arbitrary mapping was actually planned for the biohazard metaphor:

> To select the final symbol … the candidate symbols were tested for uniqueness by determining which had the least prior association for the viewer …. Since one of the two also obtained the lowest score in the meaningfulness test, it emerged as the one symbol best qualified as being both unique and memorable. (Baldwin and Runkle 1967, p. 265)

Thus, in the case of the biohazard spatial metaphor, the focus was on clearly differentiating the image from prior experiences, thus minimizing both positive and negative transfer.

## 12.6 Categories of Metaphors

These considerations provide the basis for a reinterpretation of traditional distinctions that have been drawn between types of metaphors, both in the HCI and other literature. Table 12.1 summarizes three categories of metaphors, the authors that have proposed them, and the labels that have been applied. Although the labels are different, an analysis of these classification schemes reveals that there is a reasonable degree of overlap in terms of the fundamental distinctions that are drawn, as well as the underlying rationale for doing so. We will provide concrete examples of each category; these examples will be drawn from both traditional HCI subject matter (i.e., electronic interfaces) and nontraditional sources (i.e., highway signs) to broaden the discussion and illustrate the generality of the points that are being made.

**TABLE 12.1**

Proposed Categories of Metaphors and Labels That Have Been Used to Describe Them

| Author | Categories of Metaphors | | |
|---|---|---|---|
| Lodding (1983) | Representational (picture) | Abstract (symbol) | Arbitrary (sign) |
| Rogers (1989a, 1989b) | Resemblance | Exemplar, symbolic | Arbitrary |
| Nadin (1988); Mullet and Sano (1995) | Iconic | Indexical | Symbolic |

The leftmost category of metaphors contains those in which the mapping between the metaphor and the domain is based on physical similarity. One classic example from computer interfaces is illustrated in Figure 12.4a. Many application programs and operating systems in the past have used an icon that physically resembled an early storage medium (i.e., a diskette) to represent the function of saving a file to disk. Some programs still use this convention, even though diskettes are rarely used as a storage medium today. An updated version of this convention is the internal hard drive icon on the desktop of a Macintosh computer. Similarly, Figure 12.4b illustrates the road sign used to represent a bus station. The metaphor in the sign bears a physical resemblance to the buses in the station. There are also several representative examples of this category in the iPhone interface, including the metaphors used in the camera, the clock, the phone, and the iPod® icons (Apple 2008).

The rightmost category of metaphors is made up of those in which the mapping between the metaphor and the domain is an arbitrary one. A representative example is the no-parking sign illustrated in Figure 12.5. The circle and intersecting diagonal line (normally colored red) is a spatial metaphor that, through convention, has become internationally recognized as one that symbolizes "prohibited" or "forbidden." Once again, the propositional representations make the meaning more precise by indicating what is prohibited (and when). This category of spatial metaphor is relatively rare in the interface (one example is the Bluetooth® wireless technology display in the iPhone interface; see later discussion).

The middle category of metaphors comprises those in which the mapping between the metaphor and the domain is defined by conceptual relations. Representative examples are found in the highway signs used to indicate the availability of fuel, as illustrated in Figure 12.6. The majority of vehicles on the road today use gasoline as fuel. The design of the metaphor (Figure 12.6a)

(a) (b)

**FIGURE 12.4**
Spatial metaphors that are designed to convey meaning through physical similarity. (U.S. Department of Transportation. 2009. *Manual on Uniform Traffic Control Devices for Streets and Highways*. Washington, D.C.: Federal Highway Administration.)

**FIGURE 12.5**
A spatial metaphor that conveys meaning through an arbitrary graphical representation. (U.S. Department of Transportation. 2009. *Manual on Uniform Traffic Control Devices for Streets and Highways*. Washington, D.C.: Federal Highway Administration.)

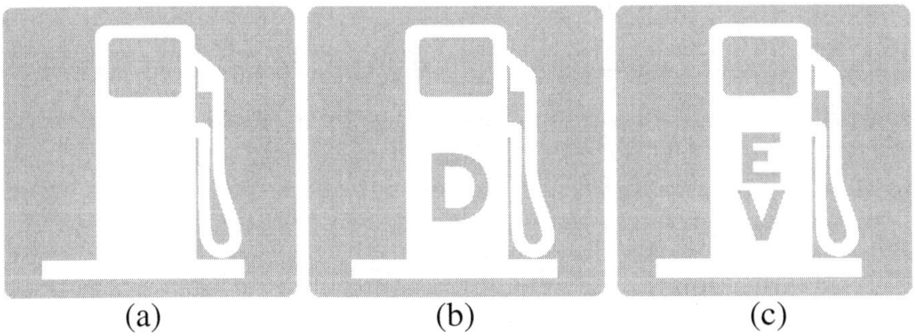


**FIGURE 12.6**
A spatial metaphor that conveys meaning through various degrees of conceptual association. (U.S. Department of Transportation. 2009. *Manual on Uniform Traffic Control Devices for Streets and Highways*. Washington, D.C.: Federal Highway Administration.)

in the associated sign bears no physical resemblance to this liquid. Rather, it looks like the device that is used to dispense it. Thus, the meaning of the sign is ultimately defined by conceptual associations. This metaphor is annotated with propositional representations (i.e., the letter "D," standing for diesel; the letters "EV" standing for electric vehicles) to represent the availability of alternative types of fuel (Figure 12.6b and 12.6c).

The meanings of these signs are established by more extensive associations, particularly in the third example: The device used to dispense fuel to electric vehicles bears no physical resemblance to the gasoline pump depicted in the sign. The meaning of this sign is established entirely by conceptual associations, including the context provided by the other two signs. Many icons in computer interfaces are representative examples of this category. For example, the preference settings icon in the iPhone interface is represented by a set of gears (see following discussion).

Although we understand the rationale that has been used to define these categories of spatial metaphors, we find several conceptual and practical difficulties. To some degree, these conceptual difficulties are reflected in the inconsistencies found in the labels of Table 12.1. A variety of different labels is applied within the same category; in one instance, the same label is used across different categories. Also, the label "indexical" is drawn from the work of Peirce (1931–1935a, 1931–1935b), but its use in describing metaphors is inconsistent with the original definition of the term. Peirce used a weather vane as an example of an indexical relation, one that incorporates a direct mapping between position of the weather vane and wind direction. Such mappings are consistent with Gibson's concept of optical invariants in which structure in the medium directly specifies a property of the ecology. Such mappings do not fit well with dyadic images of semiotics since that approach does not consider mappings to the ecology. Thus, indexical relations, as defined by Peirce, are more consistent with analog representations (see the associated discussions in Chapters 5 and 6).

More importantly, there are considerable problems with the fundamental distinctions that form the conceptual underpinnings of the different categories. For example, where does physical similarity (i.e., leftmost category) end and conceptual similarity (i.e., middle category) begin? The complex interplay between physical appearance and conceptual association was described for the fuel signs. Using other examples, the phone metaphor (Figure 12.2) and the diskette metaphor (Figure 12.4a) are clearly based on physical resemblance (i.e., left category), but they do not physically resemble the vast majority of phones and storage devices in use today (suggesting conceptual relations). Similarly, the calculator metaphor of the iPhone interface (Apple 2008) is highly abstract (i.e., its overall appearance does not resemble that of any handheld calculator; it belongs in the middle category), but it does contain graphical elements (i.e., buttons labeled with mathematical operations) representing physical components that can be found on virtually all calculators.

Furthermore, where do nonarbitrary mappings (left and middle categories) end and arbitrary mappings (right category) begin? Consider that two of the arbitrary metaphors discussed so far turn out to be not so arbitrary after all. The arbitrary biohazard symbol appears to have a long history as a Japanese family crest. A compilation (Matsuya 1972, originally published in 1913) reveals a heraldic crest that looks remarkably similar to the biohazard symbol (see Figure 12.7). The arbitrary Bluetooth symbol also has a history. The associated wireless technology was named after a Danish king (Harald "Bluetooth") who united parts of present-day Norway, Sweden, and Denmark—much like the technology unites industry groups and devices (Kardach 2005–2006). The spatial metaphor for this technology (see Figure 12.8) is a combination of the two runes that correspond to his initials. Even if one accepts the mapping in these metaphors as arbitrary, it has no impact on the ability of the metaphor to convey meaning, which is the

**FIGURE 12.7**
A Japanese heraldic crest that looks remarkably like the international biohazard symbol. (Adapted with permission from Matsuya, G. 1972. *Japanese design motifs; 4260 illustrations of heraldic crests.* Compiled by the Matsuya Piece-Goods Store. Translated with a new introduction by Fumie Adachi., Dover pictorial archive series. New York: Dover Publications, Inc. Copyright 1972 by Dover Publications, Inc. All rights reserved.)

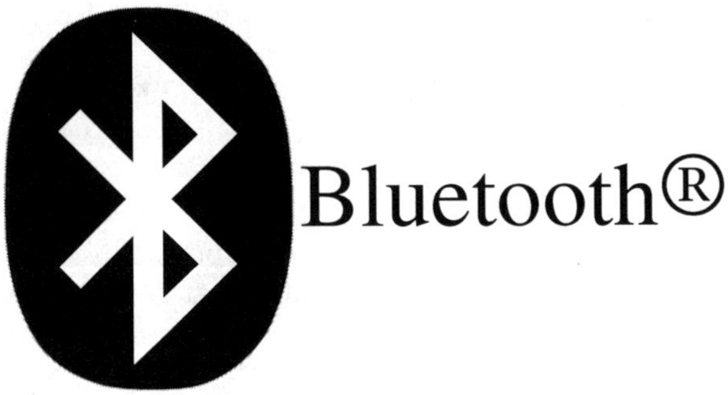


**FIGURE 12.8**
The Bluetooth® spatial metaphor. An apparently arbitrary symbol that turns out to be not so arbitrary after all. (Used with permission from the Bluetooth Special Interest Group, www.bluetooth.org. All rights reserved.)

ultimate goal. This is clearly evident in the fact that the mapping between form and meaning in all language is predominantly arbitrary.

Ultimately, the meanings of all metaphors are built on associations. Rather than discrete categories, we suggest that a continuum of mappings between metaphors reflects the degree to which the processes of assimilation and accommodation are likely to be involved in learning their meaning. Well-learned skills, such as those associated with making a phone call or listening to music, may transfer directly with a well-designed metaphor (e.g., the phone and iPod metaphors in the iPhone interface). The meaning is directly perceived through assimilation; control input to the metaphor merely confirms the intuition. Conversely, the owner of a Bluetooth headset may never know (or care about) the history of the visual appearance of the Bluetooth display.

Through accommodation, though, he or she will quickly learn to appreciate the fact that the headset will not work with the iPhone mobile digital device if the Bluetooth display is not visible.

## 12.7 Summary

> We do not approach any problem with a wholly naive or virgin mind; we approach it with certain acquired habitual modes of understanding, with a certain store of previously evolved meanings, or at least of experiences from which meanings may be educed. (Dewey 1991, p. 106)

The process of building effective metaphors still ultimately involves an inescapable element of the graphic arts. Clearly, the design of metaphors involves equal elements of art and science. We strongly recommend the work of Tufte (e.g., 1990) for examples of visual metaphors that are very effective. Perhaps the best single source available for interface metaphors is the book *Designing Visual Interfaces* by Mullet and Sano (1995). We believe it is important, however, to view these works in the context of the triadic semiotic dynamic of abduction illustrated so well by Rasmussen's decision ladder. This dynamic was described very well by Dewey (1991) in his analysis of "how we think." Dewey (1991) wrote:

> There is this double movement in all reflection: a movement from the given partial and confused data to a suggested comprehensive (or inclusive) entire situation; and back from this suggested whole—which as suggested is a meaning, an idea—to the particular facts, so as to connect these with one another and with additional facts to which the suggestion has directed attention. (p. 79)

In this respect, the design of a metaphor is a designer's suggestion about one way to integrate the potentially confusing data of a technical system or a work domain. This suggestion will be effective when it helps to make the relevant past experiences of the user salient so that they can be applied productively in the new domain as generalizations of rules and skills (assimilation) and as processes to test hypotheses directly so that the rules and skills can be adapted to the particulars of the new domain through direct interaction with minimal risk (accommodation through direct manipulation).

However, do not be surprised if metaphorical mappings that seem obvious to you (as a designer familiar with a domain) are not apparent to your users. Research in problem solving provides some evidence that the process of interpreting metaphors is likely to entail some inherent difficulties. Both Duncker (1945) and Gick and Holyoak (1980) investigated the solution of isomorphic problems (i.e., problems in different settings or contexts that involved the same solutions). Both found that problem solvers have difficulty in spontaneously recognizing structurally similar solutions in a new context. This occurred even when complete solutions to the original problem were provided and hints were given regarding the structural similarity between problems. This suggests that there can be substantial difficulties in the interpretation of metaphors. In fact, it is not difficult to find examples of symbolic association metaphors that fail in this regard (see Mullet and Sano 1995; Rogers 1989b). Great care must be taken to ensure that the higher order semantic relationships between the reference and target domains are readily apparent.

Just as is the case for analog geometrical forms (see Chapter 9, Section 9.5), a simple technique to improve the likelihood of proper interpretation substantially is to use propositional representations to annotate the metaphor. Several of the previous examples have illustrated how propositional representations can be incorporated as an integral part of the icon itself to make the meaning more precise. Propositional representations can also be used to provide a comprehensive label for the icon that appears in the interface.

One design technique is to provide a permanently visible, descriptive label that appears underneath the icon. For example, the graphical component of a file on the desktop specifies the application program that was used to generate it; the propositional representation is used to specify the content. For application icons, both the metaphor and the label are unique (e.g., application icons on the iPhone interface). An alternative technique that is often used with small icons that are part of application programs is to make these labels visible only when the cursor is positioned over the icon (a "rollover" or "tool tip").

The general approach of using propositional representations to annotate spatial metaphors has evolved into a more or less de facto standard in interface design. It takes advantage of the preciseness of language and goes a long way toward eliminating the ambiguity (see the previous quote by Rogers) associated with spatial metaphors.

It is important to realize that the processes of assimilation and accommodation are fundamental to the semiotic dynamic. As the opening quote to this section suggests, people will approach any new situation from the grounds of experience. However, the experiences within and across potential users can be extremely diverse. Thus, the choice of metaphor can be thought of as designing a filter that passes (i.e., makes salient) those heuristics that lead to success in the new domain and blocks (i.e., de-emphasizes or eliminates) those heuristics that would be counterproductive.

We suggest that the power of visual and spatial metaphors such as the desktop metaphor is that they encourage tapping into general intuitions (skill- and rule-based associations) common to a wide range of people. Thus, they facilitate the transfer of these associations and thereby make the operations of computers much more comprehensible to a wide range of users. In the end, the value might be to make the complexities of the technology functionally transparent so that users can focus their scarce knowledge processing capacity on the work (writing a book or balancing a checkbook) rather than on the technology.

Finally, it is important to reiterate that, although the emphasis in this chapter is on awareness constraints, the success of a metaphor still ultimately depends on its mapping to the work domain. A good metaphor must be cohesive in terms of the expectations of the people using it, but it must also correspond in a meaningful way to the target work or problem domain. When thinking about the mapping to the work domain, the abstraction hierarchy again provides an important guide. The levels in the abstraction hierarchy reflect some of the different ways that metaphors can map onto the work domain in terms of physical form, physical function, general function, abstract principle, or functional goal.

For example, the mapping can be in terms of physical form. That is, the metaphor reflects a specific physical object representative of the concept or function to be communicated. For example, consider the different trash can images that have been used to represent the deletion of a file in the desktop metaphor. This image has evolved to become more and more representative of a specific, typical office trash can. On the other hand, the mapping might reflect a more general physical function, in which case, a more abstract image that does not include the details of any specific object might be more appropriate.

For example, consider the three different icons for communications in the iPhone interface (Apple 2008). Three different iconic images (phone, letter, and thought bubble) represent three distinct modes (voice, e-mail, and text). The images are exemplars that reflect both the function and the mode (e.g., the old phone suggests voice, the envelope suggests e-mail, and the thought bubble suggests text, as in a cartoon—a metaphor that the iPhone interface carries into the actual display of the messaging sequence). However, they do not necessarily represent the details of any specific object. Mullet and Sano (1995) use the terms "generality" and "characterization" to refer to the level of detail in an icon's spatial metaphor. Details can be eliminated (generality) or selectively altered or exaggerated (characterization) to increase typicality with respect to a more general function.

The general functional level reflects a still more abstract relation with the target domain and it requires a little more imagination on the part of the designer. For example, the iPhone interface requires a user to navigate to one specific location to change all system and application settings. Thus, the associated icon has to represent a wide range of different functions, each with many possible different physical associations. What is the best image to reflect the diverse set of functions that can be accessed? The icon was designed using a set of gears (Apple 2008) that suggest an association with the inner workings of a complicated device (e.g., the works of a clock). Many people might not guess this at first. But the association is quickly learned and remembered. The overall desktop metaphor seems to map well at this general functional level. That is, the functions of managing electronic data files in a computer are related to the general functions of managing paper files in an office.

Note, also, that the meanings of specific icons (e.g., the file folder and the trash can) are in part a reflection of the overall desktop metaphor within which they are nested. In this respect, the coherence of an interface metaphor will depend not only on mappings at specific levels, but also on the functional links that cross levels within the abstraction hierarchy. Mullet and Sano (1995) use the term "cohesiveness" to refer to the degree to which the various icons fit together within a common style or theme.

Metaphorical relations can also be carried into higher levels of the abstraction hierarchy. For example, the choice of a compass as the spatial metaphor for the Safari® application program in Apple interfaces is a good example of a mapping at the highest functional level (i.e., to navigate). Note that at the more abstract levels, the mappings become increasingly broad and general since they have to cover more and more conceptual territory. One consequence is that the mapping becomes increasingly less specific. In these cases, the specific meaning can only be discovered through interaction. However, if the metaphor is good, then once it is discovered, the function will quickly be reinforced by the relevant associations and easily learned and remembered. When there is no obvious image to represent a function, then choosing an image that is simple and memorable may be the best option. Mullet and Sano (1991) use the term "immediacy" to refer to the visual impact of an icon.

The abstraction hierarchy provides a model of the domain structure. We hypothesize that the value of an interface metaphor will depend on its relationship to this structure. A metaphor that maps at many different levels and that reflects the constraints across levels will be more powerful. This metaphor will be more coherent and more representative. A metaphor that maps well to the structures represented by the abstraction hierarchy is likely to lead to productive thinking with respect to the domain being represented.

In terms of structural mapping within the abductive dynamic, two structures must be considered. On the one hand, there is the structure of awareness (represented by the decision ladder) and, on the other hand, there is the structure of the work domain (represented by the abstraction hierarchy). Finding the right balance between these two structures is part of the art of interface design. For those domains (typically referred to as intent driven) where an application is intended for use by a general population for a wide range of applications, we suggest that the emphasis should be on the awareness side of the equation. Consider metaphors that tap into global expectations of the population of users. For other domains (those that we have referred to as law driven), where the users are highly trained professionals and safety is critical, we suggest that the emphasis should be on the domain side of the equation. However, we caution that ignoring either side of the equation will ultimately lead to problems and inefficiencies.

## References

Apple. 2008. iPhone user guide. Cupertino, CA: Apple, Inc.

Baldwin, C. L., and R. S. Runkle. 1967. Biohazards symbol: Development of a biological hazards warning signal. *Science* 158:264–265.

Bruner, J. S. 1961. The act of discovery. *Harvard Educational Review* 31:21–32.

———. 1967. *On knowing: Essays for the left hand.* Cambridge, MA: Harvard University Press.

Clark, A. 1997. *Being there: Putting brain, body, and world together again.* Cambridge, MA: MIT Press.

Dewey, J. 1991. *How we think.* New York: Prometheus (originally published in 1910).

Duncker, K. 1945. On problem-solving. *Psychological Monographs* 58 (5): 1–113.

Gick, M. L., and K. J. Holyoak. 1980. Analogical problem solving. *Cognitive Psychology* 12:306–355.

Hutchins, E. L., J. D. Hollan, and D. A. Norman. 1986. Direct manipulation interfaces. In *User centered system design,* ed. D. A. Norman and S. W. Draper. Hillsdale, NJ: Lawrence Erlbaum Associates.

Kardach, J. 2005–2006. The naming of a technology. *Incisor* 89 (90, 92): 20–22 (11–13, 10–12).

Lakoff, G. 1987. *Women, fire and dangerous things: What categories reveal about the mind.* Chicago, IL: University of Chicago Press.

Lakoff, G., and M. Johnson. 1980. *Metaphors we live by.* Chicago, IL: University of Chicago Press.

Lodding, K. N. 1983. Iconic interfacing. *IEEE Computer Graphics and Applications* 24:11–20.

Mach, E. 1905. *Knowledge and error* (English edition, 1976). Dordrecht, the Netherlands: Reidel.

Matsuya, G. 1972. *Japanese design motifs; 4260 illustrations of heraldic crests.* Compiled by the Matsuya Piece-Goods Store. Translated, and with a new introduction, by Fumie Adachi., Dover pictorial archive series. New York: Dover Publications, Inc.

Mullet, K., and D. Sano. 1995. *Designing visual interfaces: Communication oriented techniques.* Englewood Cliffs, NJ: SunSoft Press.

Nadin, M. 1988. Interface design and evaluation—Semiotic implications. In *Advances in human–computer interaction,* ed. H. R. Hartson and D. Hix. Norwood, NJ: Ablex Publishing.

Norman, D. A. 1981. Catergorization of action slips. *Psychological Review* 88:1–15.

Paivio, A. 1969. Mental imagery in associative learning and memory. *Psychological Review* 76:241–263.

———. 1971. *Imagery and verbal processes.* New York: Holt, Rinehart and Winston.

———. 1978. Mental comparisons involving abstract attributes. *Memory & Cognition* 6:199–208.

Peirce, C. S. 1931–1935a. *Collected papers of Charles Sanders Peirce,* ed. C. Hartshorne and P. Weiss. Cambridge, MA: Harvard University Press.

———. 1931–1935b. *Collected papers of Charles Sanders Peirce,* ed. A. Burkes. Cambridge, MA: Harvard University Press.

Piaget, J. 1976. *The grasp of consciousness: Action and concept in the young child.* Madison, CT: International University Press.

Postman, L., and L. Rau. 1957. Retention as a function of the method of measurement. *Publications in Psychology. University of California* 8:217–270.

Rasmussen, J. 1986. *Information processing and human–machine interaction: An approach to cognitive engineering*. New York: Elsevier.

Reason, J. 1990. *Human error.* New York: Cambridge University Press.

Rogers, Y. 1989a. *Icon design for the user interface. International Reviews of Ergonomics* 2:129–154.

———. 1989b. *Icons at the interface: Their usefulness. Interacting with Computers* 1:105–118.

Rosch, E. H. 1973. Natural categories. *Cognitive Psychology* 4:328–350.

———. 1978. Principles of categorization. In *Cognition and categorization,* ed. E. H. Rosch and B. B. Lloyd. Hillsdale, NJ: Lawrence Erlbaum Associates.

Shepard, R. N. 1967. Recognition memory for words, sentences, and pictures. *Journal of Verbal Learning and Verbal Behavior* 6:156–163.

Shneiderman, B. 1992. *Designing the user interface: Strategies for effective human–computer interaction,* 2nd ed. Reading, MA: Addison–Wesley.

Siegler, R. S. 1983. How knowledge influences learning. *American Scientist* 71:631–638.

Standing, L., J. Conezio, and R. N. Haber. 1970. Perception and memory for pictures: Single-trial learning of 2560 visual stimuli. *Psychonomic Science* 19:73–74.

Tufte, E. R. 1990. *Envisioning information.* Cheshire, CT: Graphics Press.

U.S. Department of Transportation. 2009. *Manual on uniform traffic control devices for streets and highways.* Washington, D.C.: Federal Highway Administration.

# 13

# *Design Tutorial: Mobile Phones and PDAs*

## 13.1 Introduction

> Mac OS X is a mature technology; the iPhone is anything but … OS X 10.4 is so fundamentally good that future upgrades are likely to be on the scale of small refinements …. With the iPhone, on the other hand, Apple is heading into uncharted territory … the iPhone has no physical keyboard, and while using your finger is convenient, it's far less precise than using a mouse, trackpad, or stylus …. The iPhone interface isn't about refining and improving something that already exists. It's about completely new ideas in user-interface design …. The iPhone's 3.5 inch screen is now the biggest frontier in interface design. (Gruber 2007, p. 112)

Previous chapters have considered the defining characteristics of intent-driven domains, the advantages of using spatial metaphors as a general interface design strategy for these domains, and basic issues in the design of metaphorical representations. This chapter will expand upon these issues by considering one type of handheld consumer electronic device: mobile phones. As implied by the opening quote, the focus will be on the iPhone®. Although we can only guess at the specific design process that was used to develop this device, it would appear to have similarities to the cognitive systems engineering and ecological interface design approach. The iPhone's remarkable interface clearly incorporates important principles of design that have been discussed. Illustrating these principles in the context of intent-driven domains broadens the scope of the book and complements other excellent descriptions of this category of interfaces (e.g., the BookHouse interface; Pejtersen 1992; Rasmussen, Pejtersen, and Goodstein 1994).

We will begin with a short analysis of the trade-offs in design that are presented by this work domain. Handheld mobile computing devices like mobile phones provide a very different set of constraints and interface design trade-offs from those of the law-driven domains and traditional interfaces that were the focus of the first two tutorials. The term "mobile phone" is somewhat misleading since these devices are now capable of providing functionality that goes well beyond making a phone call. At the same time, there has been continual pressure to reduce the physical size of these devices to increase portability (at odds with increased functionality). The design problem is further complicated by the widely heterogeneous and diverse sets of skills and knowledge of the agents (a constraint often associated with intent-driven domains).

Mobile phone interfaces have evolved considerably over time in an effort to meet these conflicting demands. Early mobile phone designs were large, clunky affairs with protruding antennas (see Figure 13.1a). There was a small display at the top and widely spaced buttons arrayed over the majority of the phone. Mobile phones have become smaller, sleeker, and more sophisticated. Several additional evolutionary steps are illustrated in Figure 13.1b and 13.1c. The protruding antenna disappeared; the overall size of the device became smaller while the relative portion devoted to the display became larger. Figure 13.1d and 13.1e illustrate two more recent designs aimed at further reductions in physical size: The "flip" phone can be folded up (Figure 13.1d) and the "drawer" phone can be extended. In some designs (Figure 13.1f), there is a dedicated control button for each character in the alphabet (to support more efficient text input).

Interaction with mobile phones began as a menu-driven enterprise. The earliest menus consisted of sequential options presented as alpha-numeric text in the display, usually organized in a tree structure. The agent scrolled through each of the various options in a sequential order by pressing dedicated directional buttons and entered control input by pressing a dedicated enter button. Both displays and controls have become more sophisticated over time. The menu options are now often presented on high-resolution graphical screens with arrays of icons containing spatial metaphors. New controls allow more sophisticated forms of input. The multiple buttons required to navigate menus and enter commands (see Figure 13.1a and 13.1b) have been replaced with a single multidirectional button (i.e., the large circular-shaped button on the phones in Figure 13.1c–f).

These physical and technological changes give the impression that far more sophisticated forms of interaction have also evolved over time. However, looks can be deceiving. Although the superficial appearance of the interface to the phones depicted in Figure 13.1 clearly changed, the fundamental nature of interaction did not. The menu options were presented graphically rather than textually, but the fundamental form of interaction was still menu driven.

The introduction of the iPhone (and similar designs that followed) provided a radical departure in the philosophy of interface design. Prior to the iPhone, virtually all mobile phones had separate hardware for displays and controls (e.g., Figure 13.1). The dedicated hardware controls (i.e., physical buttons) in the iPhone have been pared down to the bare minimum. With rare exceptions, the control and display interfaces have been merged via a touch-sensitive screen. The displays on this screen also serve as "soft" controls and the agent's fingers are used as pointing devices to manipulate them.

A variety of "gesture-based" input can be recognized, including taps (single and double), drags, swipes, flicks, and pinches. These interaction conventions

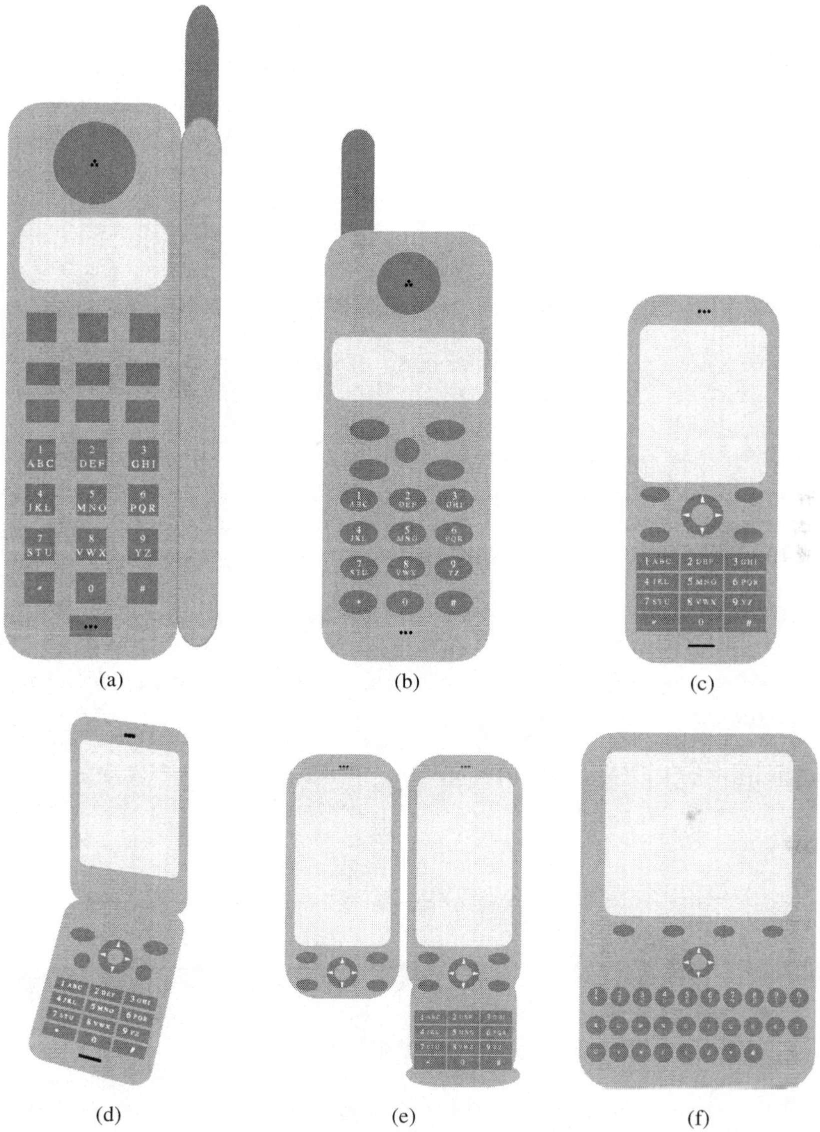

**FIGURE 13.1**
The evolution of mobile phones.

have replaced the more traditional use of menus. The iPhone will be analyzed from the perspective of cognitive systems engineering (CSE) and ecological interface design (EID) in the following sections. The discussion will be organized in terms of work domain analysis, principles of interface design (direct perception, direct manipulation, perception–action loop), and modes of behavior (skill-, rule-, and knowledge-based behaviors) that are supported.

## 13.2 Abstraction Hierarchy Analysis

An abstraction hierarchy analysis for the iPhone is presented in Figure 13.2. It represents an analysis of the original iPhone conducted at a very general level. This analysis is presented to serve several pedagogical purposes. In teaching our classes, we have found that understanding the purpose and utility of the abstraction hierarchy is facilitated by providing numerous concrete examples drawn from different types of work domains. Thus, we encourage the reader to compare the detailed contents of this analysis to those of the analysis provided for a law-driven domain in Chapter 10. Doing so will help facilitate an understanding of how the specific content in the various levels of the abstraction hierarchy varies across work domains that are located at opposite ends of the spectrum.

This example also provides the opportunity to discuss some other aspects of the abstraction hierarchy and associate work domain analyses in the context of design. As discussed in Chapter 3, an abstraction hierarchy analysis is not necessarily a "silver bullet" that is completed up front to ensure an optimal design; it is an ongoing activity conducted in parallel with design. On the other hand, if one is designing an application for the iPhone, as opposed to the device itself, then the current configuration of hardware and software (e.g., Figure 13.2 for the original iPhone) represents a harder set of constraints. For example, the second generation of the iPhone included a video recorder, true GPS, and speech-recognition technologies. Along these lines, it is illustrative to consider the suggestions that Apple provides for application developers in its iPhone human interface guidelines (HIG) document:

> Before you begin designing your application, it's essential to define precisely what your application does. A good way to do this is to craft a *product definition statement*—a concise declaration of your application's main purpose and its intended audience. Creating a product definition statement isn't merely an exercise. On the contrary, it's one of the best ways to turn a list of features into a coherent product.
>
> To begin with, spend some time defining your user audience: Are they experienced or novice, serious or casual, looking for help with a specific task or looking for entertainment? Knowing these things about your users helps you customize the user experience and user interface to their particular needs and wants. (Apple 2010, p. 35; emphasis original)

Note that these suggestions directly correspond to the dominant theme advocated in this book: To design an effective interface, one must consider both the specific constraints of the work domain (in their terms, a product definition statement) and the specific constraints of the agent (in their terms, experience, needs, wants of the user). In many ways, our goal is to provide

| | **Abstraction Hierarchy** (Means-Ends Relations) |
|---|---|
| Goals, Purposes, and Constraints | Communication, education, entertainment, safety, organization, efficiency,etc.<br>Enjoyment: Fun, delightful, compelling to use |
| Priority Measures and Abstract Functions | Flow of information, resources, messages, digital content, power, etc.<br><br>Value of completing tasks and achieving goals vs. effort and time expended<br>Cost of equipment and services vs. perceived functionality and utility, etc. |
| General Work Activities and Functions | Input (data, sound, infrared light, ambient light, gravity, electrical potential, power)<br>Output (data, sound, light, vibration, heat, power)<br>Interface (user, global and local networks, world wide web, peripheral devices)<br>Data management (conversion, processing, distribution, storage, backup)<br>Power (internal distribution, storage, conservation)<br>Operation (interface, applications, memory, files, network, security, multi-tasking)<br><br>Voice communication (mobile cell phone)<br>Multi-media entertainment (music, pictures, movies, podcasts, radio, audiobooks)<br>Internet access (information and services on the world wide web)<br>Electronic messaging (voice mail, electronic mail, text messages)<br>Digital imaging (camera); Directions and Navigation (maps, GPS)<br>Digital content management (media libraries, contacts, data, files, etc.)<br>Miscellaneous functionality (games, calendar, clock, calculator, sticky notes, etc.) |
| Physical Activities in work, Physical Processes of Equipment | Sensors: multi-touch input screen, orientation, proximity, photosensor<br>Displays: LCD output display, vibration generator, speaker, headphones<br>Interface: antenna, networks, USB, power adapter<br>Data Management: D-A/A-D conversion, iTunes<br>Power: charger, battery<br>Operation: buttons, ports, CPU, data busses, controllers, memory, power, battery, digital-analog and analog-digital converters).<br><br>Cellphone: call (respond, initiate, hold, conference, speaker, mute, save, clear); voice message management (greeting, listen, delete, save, replay)<br>Multi-media: select player and media type; play, stop, pause, fast forward, rewind<br>Internet activities: search, navigate, content, quality, input, path, revisit sites<br>Messaging activities: receive, read, compose, store, send, record, threads<br>Imaging activities: take photographs and movies<br>Digital content: search, purchase, import, construct, augment, store, rate, organize, access, export, delete, synchronize, backup, parental control<br>Applications activities: navigate, select / use alternative applications, customize |
| Appearance, Location, and Configuration of Material Objects | Overall size, weight, and color of device<br>Physical dimensions, pixel resolution, luminance, and chromaticity of the screen<br>Size of font, physical configuration of ports, jacks, etc. |

**FIGURE 13.2**
Abstraction hierarchy analysis for Apple, Inc.'s iPhone®.

more concrete principles and analytic tools that allow the designer to be more precise in defining these sources of constraint and their implications for interface design.

## 13.3 The iPhone Interface

Apple appears to have navigated the "uncharted territory" referred to in the introductory quote (Gruber 2007) quite well: The iPhone interface is elegant, effective, and fun to use. In this section we examine this interface in terms of the design principles that were introduced in Chapter 6, including direct perception, direct manipulation, and an intact perception–action loop.

### 13.3.1 Direct Perception

> You can't assume that users have the time (or can spare the attention) to figure out how your application works. Therefore, you should strive to make your application instantly understandable to users. (Apple 2010, p. 37)

This suggestion from the iPhone HIG essentially corresponds to one of two fundamental principles of interface design that we have outlined in the book: the concept of direct perception. This refers to the ability of an interface to support the human agent in obtaining meaningful information about the domain through powerful perceptual skills. The goal of direct perception is to ensure that the affordances of a domain (i.e., the possibilities for action) are represented in the interface and easily obtained. In Chapter 6 we described the structure of an interface in terms of a nested hierarchy of invariants, including information at the levels of forms, views, and work space (see Figure 6.9). It is this structure that we will use to organize the discussion of direct perception in the iPhone.

#### *13.3.1.1 Forms Level*

> When possible, model your application's objects and actions on objects and actions in the real world. This technique especially helps novice users quickly grasp how your application works ... examine the task your application performs to see if there are natural metaphors you can use. Bear in mind, though, .... Unless the metaphors you choose are likely to be recognized by most of your users, including them will increase confusion instead of decrease it. (Apple 2010, p. 31)

This suggestion from the iPhone HIG mirrors the recommendations for metaphor design that were discussed in detail in Chapter 12. To reiterate briefly, the primary challenge is to invoke the correct concept or knowledge regarding the underlying functionality (i.e., to support assimilation); success or failure revolves around the extent to which the spatial metaphor provides a direct and effective semantic link. Do the visual features of the metaphor suggest appropriate concepts and activities?

The iPhone interface accomplishes this design goal quite nicely (in fact, many of the examples discussed in Chapter 12 were specifically chosen from the iPhone). The icons in the interface contain spatial metaphors that are perceptually distinctive (e.g., differences exist in the structure, color, and other features), employ common visual themes, and possess a high degree of visual abstraction and representativeness. In short, these metaphors provide effective representations for the various application programs and modes; they provide effective semantic links to the affordances of the domain.

As discussed previously, one potential difficulty with the use of metaphors in the interface is their inherent ambiguity (i.e., the potential for multiple interpretations of the underlying meaning). The iPhone designers solved this problem by placing a constantly visible label under each icon. These labels are visible, but not overly conspicuous—an important attribute because the need to consult these labels will disappear as the agent gains experience with the interface. This represents another example of the use of propositional representations (see Chapters 6 and 9) to annotate graphical representations in the interface.

#### *13.3.1.2 Views Level*

The next level up in the nested hierarchy of graphical forms in the interface is the level of views. This level corresponds to the information that is present within a single screen. Sets of spatial metaphors (i.e., the forms level described in the previous paragraph) are nested within a view. The primary design consideration at this level involves the placement of spatial metaphors within a view—that is, their arrangement within a setting. One technique that can be used to provide a higher order degree of organization is a semantic theme: that is, a setting that provides a context for the placement of individual metaphors so that they fit together logically and consistently (e.g., see the rooms of a virtual library illustrated in Chapter 15, Figure 15.3).

A second technique to provide this organization at the views level is spatial structure; this is the technique employed in the iPhone. The various spatial structures that are described in the iPhone HIG (Apple 2010) are illustrated in Figure 13.3a. There is a mechanical "home button" located at the bottom of the interface. The status bar at the top contains basic information (e.g., connection, time, battery power). There is a general display area in the middle that is used for presentation of applications or objects within an application. The remaining two structures, the navigation bar and the tab bar/toolbar, are variable both in terms of what they contain and whether or not they are present in the interface.

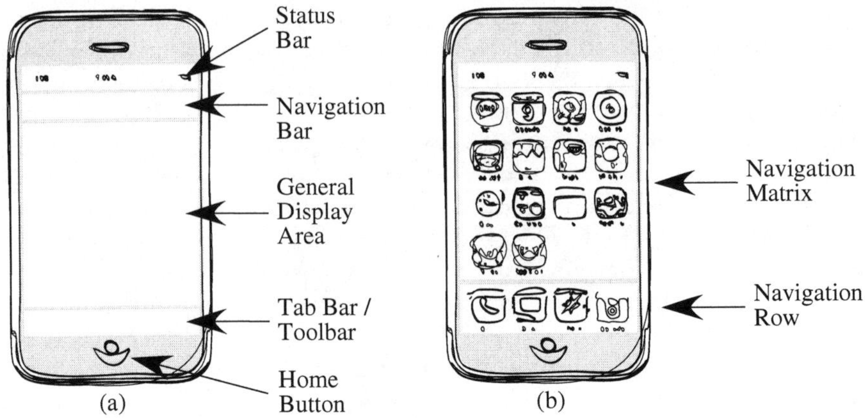


**FIGURE 13.3**
Spatial structure provided in the iPhone interface at the view level. (a) Spatial structures as defined in the iPhone HIG. (b) Additional spatial structures that we have defined to facilitate discussions of navigation.

We will define two additional spatial structures to streamline the ensuing discussions. Figure 13.3b illustrates the iPhone in the "home" configuration. The "navigation matrix" is a spatial structure formed by the rows and columns of icons with metaphors that represent applications in the home configuration. We rename the tab bar as the "navigation row" to reflect its purpose in the examples that we provide (i.e., it is a second spatial structure that contains icons with metaphors that represent either applications or modes within an application that can be navigated to).

#### *13.3.1.3 Work Space Level*

The work space level is the highest in the interface; all of the views in an interface are nested within it. The fundamental issues in design at this level revolve around the global spatial layout and navigation within it. What is the spatial layout of views in the work space? What are the navigational paths between them? We illustrate these considerations with a representative navigational sequence in Figure 13.4.

The navigation matrix and row in the home configuration (see Figure 13.4a) contain a number of icons with metaphors that represent software applications that can be navigated to. The agent taps on an icon (Figure 13.4b) to effect navigation (in this case, the Phone application). As illustrated in Figure 13.4c, the navigation matrix is then replaced with the interface objects associated with a mode of the selected application (in this case, the keypad mode of the phone application).

Furthermore, the icons in the navigation row are replaced with other icons that represent modes within the selected application. Control input to an application mode icon produces navigation to an alternative view associated with that mode, as illustrated in Figure 13.4d (contacts mode), Figure 13.4e (recents mode), and Figure 13.4f (voice mail mode). Control input to the home button at any point returns the agent to the home configuration (Figure 13.4a).

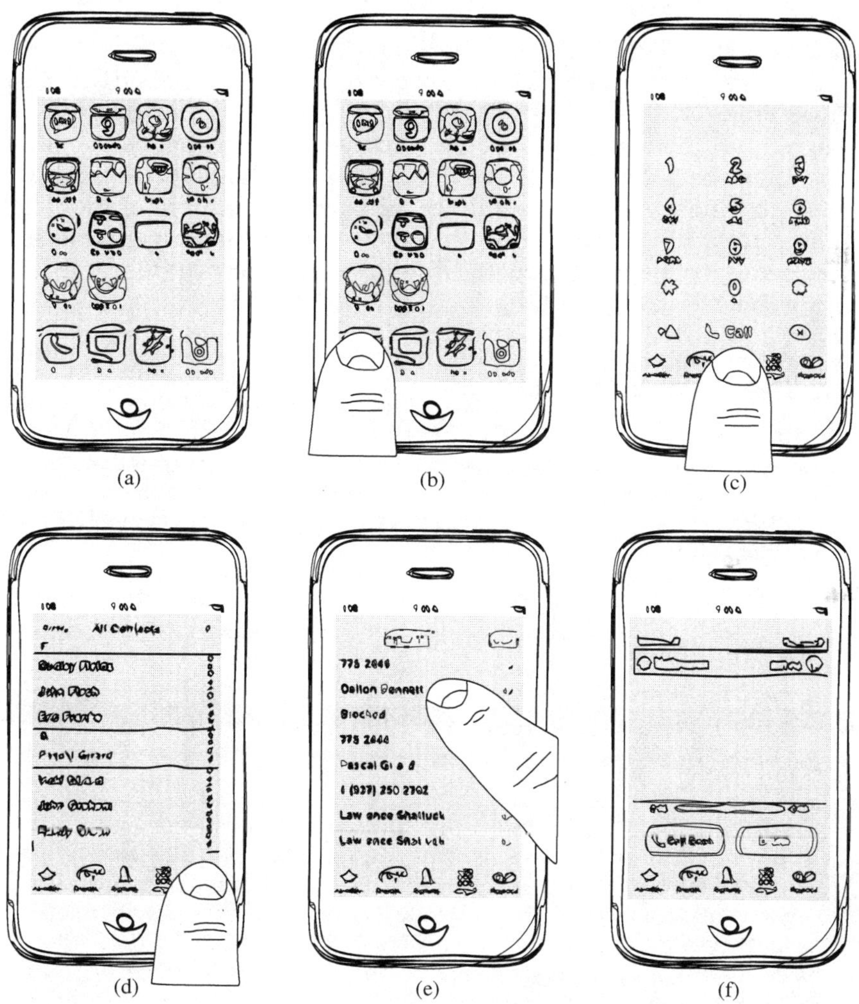


**FIGURE 13.4**
A representative example of navigation in the iPhone interface. (a) The default or home configuration, used for navigation between applications. (b) Control input (i.e., tapping) applied to an application icon in the navigation row. (c) The resulting navigation to an application view (i.e., the phone application in keypad mode). (d–f) Within-application navigation to views associated with alternative application modes (contacts, recents, voice mail).

The sequence illustrated in Figure 13.4 is representative of navigation in some of the core applications of the iPhone (i.e., phone, iPod, iTunes, clock, and app store). As this example illustrates, there are only two basic levels of interaction. One level occurs between applications using the navigation matrix, the navigation row (application icons), and the home button. The second level of interaction occurs within applications using the navigation row (application mode icons). The details of mode navigation do change somewhat for other Apple applications (predominantly, controls in the navigation bar), but the basic observation of just two levels of interaction holds true.

In summary, this section has described a hierarchically nested set of spatial metaphors that were devised for an intent-driven domain. Note that this design solution provides an exact parallel to the hierarchically nested analogical representations for law-driven domains that were described in Chapters 8 through 11. Note also that the spatial structures of the iPhone interface and their contributions to effective interaction are examined in greater detail in Chapter 15, where issues in navigation (e.g., visual momentum) are considered.

### 13.3.2 Direct Manipulation

> Direct manipulation means that people feel they are controlling something tangible, not abstract. The benefit of following the principle of direct manipulation is that users more readily understand the results of their actions when they can directly manipulate the objects involved. (Apple 2010, p. 31)

Without doubt, the most innovative aspect of the iPhone interface lies in its implementation of direct manipulation. There are no command lines, no pull-down menus, no cursors, and no mechanical pointing devices (e.g., touch pad, track ball, mouse, or stylus) in the iPhone interface. Instead, the fingers of the agent are used to manipulate objects in the interface directly.

> People use their fingers to operate the unique multi-touch interface of iPhone OS-based devices, tapping, flicking, and pinching to select, navigate, and read Web content and use applications. There are real advantages to using fingers to operate a device: They are always available, they are capable of many different movements, and they give users a sense of immediacy and connection to the device that's impossible to achieve with an external input device, such as a mouse. (Apple 2010, p. 41)

Thus, the iPhone accommodates one very important form of "gesture-based" input via its touch-sensitive screen. The various types of gestures that are supported by the iPhone interface will now be considered in greater detail.

### *13.3.2.1 Tap*

The fundamental method of providing control input to the iPhone is the single "tap." Consider the examples provided in Figure 13.4. A tap on an application icon (Figure 13.4b) allows the agent to navigate between applications. A tap on a mode icon allows an agent to navigate between application modes. A tap on various objects in the interface (e.g., phone number) produces the execution of a command (e.g., dialing the number). The tap is used universally in all applications and modes; it is both the simplest gesture that the iPhone recognizes and the one that is required most often. This gesture replaces the double click of the mouse often required in traditional interfaces and it serves to streamline interaction by removing an intermediate step.

### *13.3.2.2 Flick*

The "flick" is a gesture that allows a group of objects to be scrolled through quickly. This gesture will be discussed in the context of the software application for storing and viewing photographs. Tapping the icon representing a photo album (Figure 13.5a) produces a matrix of small thumbnails representing the photos in the album (Figure 13.5b). The agent positions a finger at the bottom (see Figure 13.5b) or top of the matrix, then quickly extends (i.e., flicks) it across the face of the iPhone (see Figure 13.5c) to execute the gesture. The rate at which the objects are scrolled is determined by the speed of the flick. This gesture is extremely useful and is available for use when there is a list of items that need to be searched (e.g., contacts, recent phone calls, favorite phone numbers, songs, etc.).

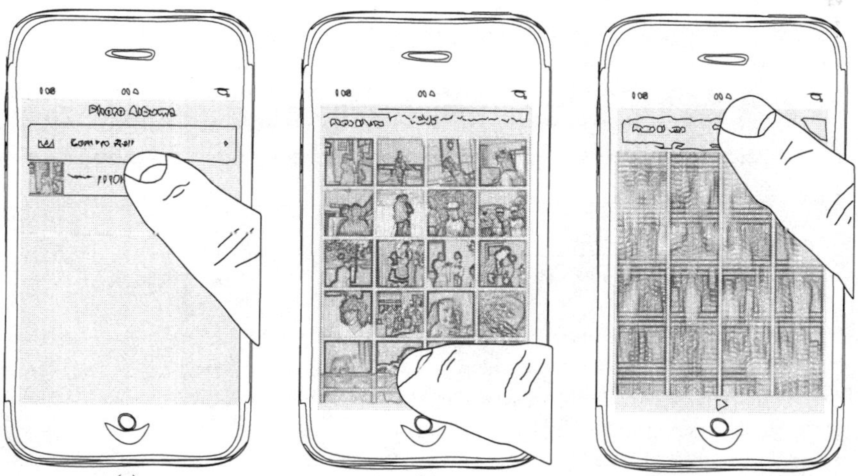

(a) (b) (c)

**FIGURE 13.5**
The "flick" gesture used for quick scrolling through collections of interface items.

### *13.3.2.3 Double Tap*

The "double-tap" gesture (two taps in quick succession) is very similar to the double-click control input that is a staple of traditional human–computer interfaces. In the iPhone, however, this gesture is used on a fairly limited basis. The primary use is to execute a zoom function that allows close-up viewing of photographs. A single tap on the thumbnail of a photograph in an album (e.g., Figure 13.5b) brings up the photo for viewing (Figure 13.6a, landscape mode). Two taps on a photo in quick succession (i.e., a double tap, see Figure 13.6a) results in the enlargement of the photo (Figure 13.6b). The location of the double tap determines the focus point of the zoom (in this case, to the left of the subject's face). The amount of zoom executed by a double tap is predetermined and constant. A second double tap, at any subsequent point, returns the photo to its original size.

### *13.3.2.4 Drag*

The "drag" is a familiar form of direct manipulation in traditional interfaces that is also present in the iPhone interface. The agent positions a finger on the iPhone's display and then moves the finger across the display to change the position of an object on the screen. Figure 13.7 illustrates this gesture being used to view successive photos in an album. The photo must be at its normal size (i.e., not zoomed), as illustrated in Figure 13.7a. The agent positions a finger on the right side of the photo and then drags it across the iPhone display to see the next photo in the album (see Figure 13.7a–e). The original photo slides off the screen to the left and is systematically replaced by the ensuing photo.

A complementary gesture (dragging from the left to the right) allows the agent to view the previous photo. The drag gesture is also used to provide a slower, more controlled method (relative to the flick gesture) to scroll through lists (e.g., contacts, recent phone calls, photograph thumbnails, songs, etc.). This gesture is also used in combination with a graphical software "slider" (an object with a predetermined and constrained motion path) to unlock or to power down the phone. The "swipe" gesture is a close variant of the drag

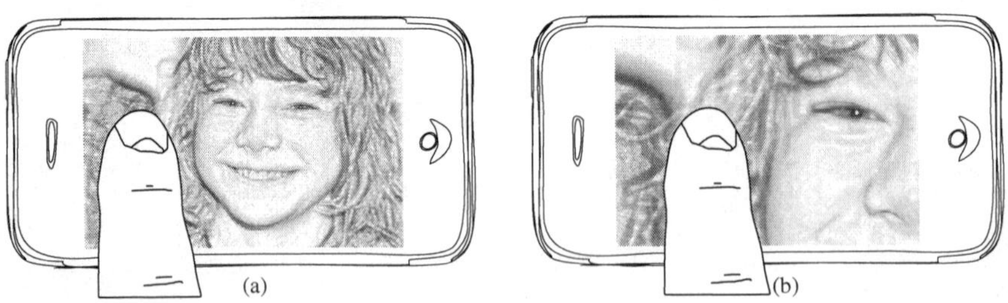


**FIGURE 13.6**
The "double tap" gesture (a and b, used to magnify photos).

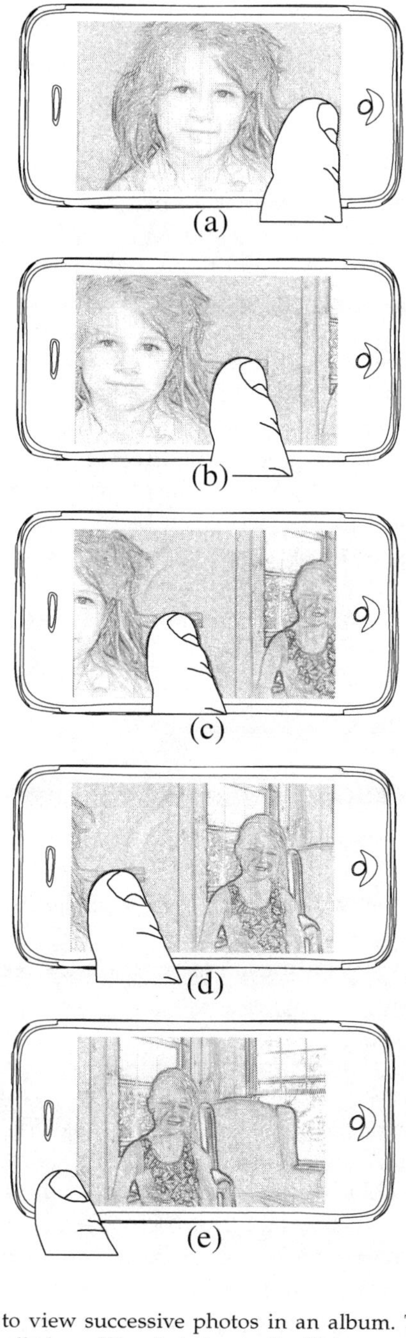


**FIGURE 13.7**
The "drag" gesture used to view successive photos in an album. This gesture is also used for slower, but more controlled scrolling (relative to the flick gesture) through collections of interface items.

gesture, which is used less frequently (e.g., swiping across an e-mail in a list produces the delete button).

#### *13.3.2.5 Pinch In/Pinch Out*

The pinch-in and pinch-out gestures are used to produce variable zooming into and out of the interface. Consider Figure 13.8a. A photo in an album has been selected and is currently being shown in the default configuration (i.e., the entire photo is shown). The agent can execute the pinch-out gesture to zoom in to the photograph by a variable amount. To do so, the agent positions the thumb and index finger together at the focus point of the zoom. A systematic increase in the spatial distance between the finger and thumb (i.e., pinching out) serves as control input to execute the zoom, as illustrated by Figure 13.8a–c. The agent duplicates this gesture in Figure 13.8d and 13.8e to achieve maximum magnification. The inverse of this gesture (i.e., a pinch in) serves as control input to zoom out of the picture. These complementary gestures are not limited to the photo application. For example, it is a useful feature for increasing the magnification of a Web page during Internet browsing: The high resolution of the screen produces a very small font size, which sometimes needs to be enlarged for easier reading.

## 13.4 Support for Various Modes of Behavior

The previous section described the physical appearance of the iPhone interface and the nature of the interaction that occurs with it. The goal of this section is to reinterpret this interaction in terms of the three primary modes of behavior that were outlined in Chapter 5. This is important because all three modes of behavior need to be supported in an interface.

### 13.4.1 Skill-Based Behavior

An effective interface will be designed to engage powerful perception–action skills to the extent that is possible (and, conversely, to avoid activities that draw upon limited-capacity resources such as working memory). The iPhone's support for skill-based behavior has been the focus of the previous sections, which have detailed the role of direct perception and direct manipulation in the interface.

To reiterate, the perception–action loop is clearly intact: The agent directly perceives and directly manipulates objects to operate the iPhone. Direct perception is supported through nested spatial structures in the interface, ranging from global (i.e., application and mode level) to local (icons with spatial metaphors) levels of structure. The affordances for action are specified,

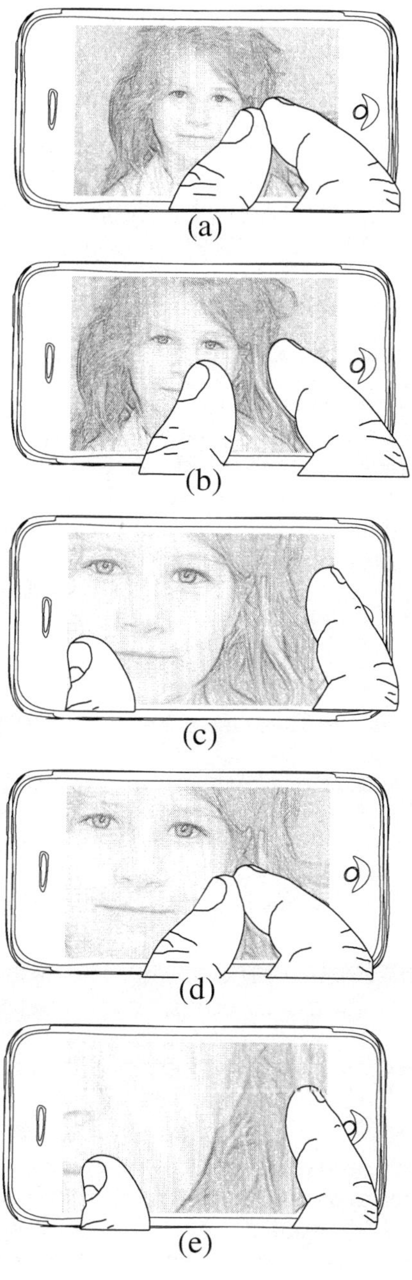

**FIGURE 13.8**
The "pinch-out" gesture is used to magnify or zoom into portions of the screen. The inverse gesture, the "pinch-in" gesture, is used to zoom out.

thereby leveraging the human's powerful capability to pick up visual information. The space–time signals produced by the interface also support direct control input. The agent is not describing the actions to be taken to an intermediary (i.e., command line interface); nor is the agent indirectly specifying actions (i.e., pull-down menu). Control inputs are implemented through the direct manipulation of objects in the interface (e.g., tapping, swiping, and flicking the objects on the screen). Furthermore, the agent is not required to execute these control inputs using intermediate tools (mouse, cursor, stylus, etc.); the fingers are the input devices.

### 13.4.2 Rule-Based Behavior

The iPhone's interface will also provide a rich set of signs for action that support rule-based behavior by the skilled agent. An agent will quickly become familiar with the iPhone interface, including global and local metaphors and other objects of interest for the assorted application modes (e.g., controls). Skilled agents will also have developed fairly specific knowledge regarding how these resources can be used (i.e., sequence of activities) to accomplish various goals. Under these circumstances, the interface will provide a set of signs that will be used to guide these action sequences.

Consider the following scenario where the agent has the goal of executing a phone call to an old friend. This general activity sequence (i.e., placing a phone call) is a common one that has been executed many times in the past and is therefore well learned. The agent recalls the general location (lower left) and the visual appearance of the phone icon, locates it, and taps on it (Figure 13.9a). The phone application was exited in keypad mode and therefore returns to that mode (Figure 13.9b). The agent needs to enter the contact mode, locates the appropriate icon and taps it. A contact list appears with entries beginning with the letter M (Figure 13.9c); the agent taps the letter "G" in a "letter index" search tool (i.e., the column of single letters located on the right-hand side of the screen). The desired contact is the first one encountered under the "G" heading and the agent taps on the contact object (Figure 13.9d) to bring up the associated information (Figure 13.9e) and then taps on the correct phone number (i.e., "home") to execute the call (Figure 13.9f).

In this case, the agent is executing a series of actions that have been completed many times in the past. The steps in this general activity sequence are well known (including viable alternatives); their successful execution relies upon signs in the interface that specify context-specific variations. The iPhone interface objects that serve as signs in the previous example include the following: phone application icon, contacts mode icon, contact list index tabs, letter index search tool, contact labels, and number labels. It is important to note that the exact sequence of activities to be executed is not totally predictable, even for well-learned sequences like this particular example. This is because there will always be some uncontrollable variation in the

**FIGURE 13.9**
The graphical representation of a well-learned sequence of activities (placing a phone call) to illustrate how interface structures can serve as signs to support rule-based behavior.

context (e.g., the application mode visited last and the entry visited last will not always match the agent's intentions).

The degree of familiarity will vary with the application, the task, and the agent's experience. In some cases, the agent may have memorized both the sequence of activities and the specific appearance and location of the relevant interface objects (e.g., home button, phone icon). In other cases, the agent will have only a general idea of the appearance and location of the appropriate interface object and will therefore need to search among alternatives in a particular region (e.g., the correct mode icon in the mode bar). The visual appearance of the icons' spatial metaphors will trigger the recognition of the correct application, mode, and sequence of activities to take. The cues or signs provided by the interface are successful because they remind the agent of what needs to be done (i.e., they support recognition memory), instead of forcing the agent to remember the relevant information for interaction (i.e., recall memory, as required by command line interfaces).

### 13.4.3 Knowledge-Based Behavior

All agents will, at some point, be engaged in problem-solving activities (i.e., knowledge-based behavior) as they try to accomplish novel tasks with the iPhone. Sometimes, the agent will have an intention and wonder whether or not it can be accomplished. Sometimes, the agent will have a well-formulated goal and the knowledge that it can be done, but only vague recollections of the specific activity sequences that are required. Sometimes, the agent will just spontaneously realize that some activity is possible with no prior intent. A good interface should support these problem-solving activities.

Interfaces for traditional mobile phones often do not provide this support. Historically, the manual that came with a traditional phone was thick, complicated, and entirely indispensable; it was often the only way to find out about the device's functionality and to determine the steps (sometimes quite convoluted) that were necessary to use it. Furthermore, having successfully solved the problem before was no guarantee that the manual would not need to be consulted again: the convoluted steps were often just too difficult to remember. The bottom line is that the functional capabilities of traditional mobile phones were grossly underutilized simply because they were buried within an ineffective interface design.

We are not aware of any formal, published studies that have investigated the usability of the iPhone. However, our experience with the iPhone indicates that it provides a quantum leap in usability relative to traditional phones. Problem solving (i.e., knowledge-based behavior) is supported exceptionally well. Although there is a manual for the iPhone, we have not found it necessary to consult (except for rare technical issues such as checking battery-charging practices). In essence, the iPhone interface replaces the manual and we will now describe some of the ways in which it supports knowledge-based behavior.

A major factor in the effectiveness of the iPhone interface is the simple navigational structure that it provides. These features are described in detail in Chapter 15 and will only be briefly summarized here. As described previously, there are only two levels of interaction: application or mode. There are no complicated, hierarchical menu structures in which to get lost. The agent can normally see the space of possible destinations at any point in time and is never more than one button press away from starting over. An agent using the iPhone often knows where he or she is, has been, and needs to go, in contrast to traditional mobile phones (e.g., Ziefle and Bay 2006).

The second major factor contributing to the effectiveness of the iPhone interface in supporting knowledge-based behaviors is the fact that the affordances of the domain are directly specified. The functionality of the iPhone is available to be perceived directly through the local spatial metaphors and other objects in the interface. These visual representations support appropriate interpretations by relating functionality to familiar objects, thereby leveraging preexisting concepts and knowledge (i.e., assimilation). The alphanumeric labels specify the underlying meaning of icons' spatial metaphors more directly. Thus, the capabilities of the iPhone are not hidden in cryptic commands or pull-down menus; much of what can be done with the iPhone can be seen directly via the metaphors and objects on the screen.

The combination of direct perception (i.e., simple spatial structures for navigation and direct specification of domain affordances) and direct manipulation (i.e., natural, effective, gesture-based control input) provides very effective problem-solving support. Rather than consulting a manual to determine how to accomplish a task, the ease of use and simplicity of the interface encourage agents to explore what is possible with the iPhone in a more direct fashion. One simply navigates to an application and explores the various modes for potential solutions. In fact, it has been our experience that these interface resources support what has been referred to as discovery learning (Bruner 1961, 1967). That is, early stages in the learning process were dominated by an exploration of the interface and the associated functionality without any particular intention in mind (i.e., to discover just what kinds of activities were possible). Using the iPhone is pleasurable; frustration is rare. The interface plays a large role in this experience.

## 13.5 Broader Implications for Interface Design

Without question, the iPhone is an extremely innovative product. It has managed to export a number of technologies out of the research laboratory, make them work together well (including synchronization with computer-based applications and information), and make them available to the general public in a commercial product. The most obvious example is the iPhone's use of gesture-based, multitouch input. Research on this topic began as early as the mid-1960s (e.g., Teitelman 1964). Buxton (2009) provides an historical overview and identifies key factors in multitouch implementations (e.g., input device, contact, position, pressure, multipoint, force vectors, etc.) that have been discovered. Apple's implementation is elegant in its simplicity, but it represents only the tip of the iceberg in terms of what is potentially available for multitouch input.

Similarly, researchers in a variety of disciplines have investigated alternative forms of multimodal input (e.g., auditory/speech, haptic, body, physiology, etc.). Durlach and Mavor (1995) provide an early summary of these "natural user interfaces" in the context of virtual reality. More recently, Norman (2010) identifies some of the design challenges that these interfaces entail (see following discussion). Although we have chosen to focus on traditional interfaces in this book, these alternative forms of interaction certainly hold promise for the future.

At this point, the iPhone is not the only multitouch phone available. As a result, some interesting differences in underlying philosophies of interface design are beginning to emerge. One of the most prominent differences lies in the role of menus in the interface. There are no menus in the iPhone interface and no developer tools to implement them (the lone exception is a basic edit menu). In contrast, the Android™, a new mobile operating system, allows limited, short menus. Some additional discussion is warranted since this difference in philosophy extends well beyond mobile phone interfaces.

### 13.5.1 To Menu or Not to Menu? (Is That the Question? Is There a Rub?)

Elsewhere in the book, we have hinted at potential difficulties that can arise from the use of menus in the interface. From one perspective, menu selection might be considered the functional equivalent of describing an action to the computer intermediary, as opposed to acting on the interface as a model world (Hutchins, Hollan, and Norman 1986). From this perspective, there is not much difference between typing a command into a command line interface and selecting it in a menu. In either case, the perception–action loop is broken: Although the means by which the user is making the request to the computer intermediary have changed, the user is not directly manipulating the objects of interest on the screen.

The alternative is to design the interface so that the affordances of the domain are directly visible in the objects on the screen (i.e., direct perception), control input can be executed via these objects (i.e., direct manipulation), and the perception–action loop is intact. Examples of this design philosophy are provided in the previous discussion of the iPhone interface, the RAPTOR interface (next chapter), and process control (Chapter 10).

One might argue that menus impose additional interactional overhead relative to this type of interface: The agent must remember where a command is located (i.e., which of several pull-down menu contains it and where it is located with regard to other commands that are present) and physically select it (this can prove annoyingly difficult when cascading submenus are involved and can produce the "one-off" error in selection from long menus). These arguments are certainly true when the menus are complex and deep, contributing also to navigational problems such as the "getting lost" phenomenon (see Chapter 15).

Manjoo (2010) emphasizes these points and suggests that menus may be outdated:

> The Android platform is much looser in this regard. Its interface guidelines don't discourage hidden menus: "All but the simplest applications have menus," the interface guide tells developers. In other words, under Android's design philosophy, *menus are a natural consequence of complexity*—and the more powerful a program, the more likely it is to be stuffed with hidden menus. That's a familiar view of computing, one deeply tied to the interface on the standard PC—after all, every program on your laptop or desktop hides much of its functionality under menus, too.
>
> But that philosophy feels outmoded. We're increasingly abandoning desktop programs for most of our computing needs, and we're replacing them with Web apps or mobile apps that are much more straightforward to use. I rarely reach for menu bars anymore; the programs I use most often these days—Chrome, Gmail, Google Maps, Microsoft Office 2007, and nearly everything on my iPhone—present most of their functions on the main screen. (Manjoo 2010; emphasis added)

On the other hand, there are potential problems with a design philosophy that does not use menus. One lies in determining exactly what actions can be performed at any point in time. Norman (personal communication, April 4, 2010) observes that the absence of menus can force the interaction to be "a puzzling exercise to figure out how to do an operation. (Do I change pages? Touch some secret location? Swipe left to right, right to left, up or down? Do a long touch, a short one, a double touch? Maybe I should shake the device, which on some apps does a new search. The graphical user interface made things visible for a reason.)" In traditional interfaces, this problem can be alleviated somewhat by providing visual or auditory feedback (i.e., mouse-overs, tool tips, or even static visual properties) that specifies which objects offer the potential for manipulation. However, implementing this strategy with multitouch technology becomes more difficult since "finger-overs" or "finger-tips" are not an option.

One might also offer some counter arguments in support of menus. Norman (personal communication, April 4, 2010) believes that "menus are an important part of the GUI's visibility argument; all possible actions are findable (visible), so you don't have to memorize. Also, the menu item is quite specific, so typing errors are not possible." In fact, menus ultimately function much like the spatial metaphors of the icons visible on the screen: They provide visible, precise affordances and support exploration of an application's capabilities. Furthermore, in contrast to the position taken by Manjoo, the need for menus may really be "a natural consequence of complexity." Specifically, an extensive suite of commands and options (e.g., Photoshop) might necessitate the kind of organization and access provided by logical and well-designed menus.

In the end, the difference between these two design philosophies might be more superficial than one might first assume. The success of either philosophy might come down to the quality of implementation, as opposed to an inherently right or wrong way to design interfaces. Consider the following quote by Manjoo (2010):

> *The essential problem is that Android's interface is much less intuitive than the iPhone's.* Much of the OS's functionality is hidden—Android can do a lot, but unlike the iPhone it keeps many of its options stuffed in menu bars. As a result, the Nexus One asks new users to climb a steeper learning curve. You've got to poke around every program to find out how to do its most basic tasks. Even once you've learned the easy stuff, the OS is still a struggle—it takes several steps to do something on Android that you can do in one step on the iPhone ... (emphasis added)

We believe that Manjoo has identified the core attribute of an effective interface: whether or not it is intuitive to use. However, intuitiveness will be determined by a much more fundamental consideration than just the presence or absence of menus in the interface. The intuitiveness of an interface will ultimately be determined by the extent to which a full range of potential actions is available—actions that are tuned to the specifics of the situational context. This is a point that is made explicit in the iPhone HIG:

> A good way to achieve focus is to determine what's most important in each context. As you decide what to display in each screen always ask yourself, Is this critical information or functionality users need right now? ... If not, decide if the information or functionality is critical in a different context or if it's not that important after all. (Apple 2010, p. 39)

Thus, the advice that Apple gives to application developers (and, presumably, the approach taken by the iPhone design team) is essentially the same message that we have expressed throughout this book: The quality of interface design begins with an effective work domain analysis. The CSE and EID approaches offer a structured framework and analytical tools for conducting work domain analyses—providing an alternative perspective that is potentially more effective than the simple (but true) advice offered by Apple. The abstraction and aggregation hierarchies provide a structured approach to modeling the categories of information and functional relationships of the work domain; the decision ladder provides a structured approach to modeling the decision-making process and shortcuts that are (or can be) used by experts. Although we certainly lean toward a design philosophy that avoids menus when possible, we also believe that effective design can be achieved with either design philosophy (or some combination thereof) when this structured knowledge serves as the basis for design.

## 13.6 Summary

Gruber expresses the opinion (see the quote at the beginning of the chapter) that the iPhone interface incorporates "completely new ideas in user-interface design." This statement could certainly be debated since the principles of design that are evident in the iPhone's interface have been around for a long time. For example, the interface design goals of direct perception and direct manipulation have been well articulated for decades (e.g., Hutchins et al. 1986). On the other hand, Gruber's observations are well founded in the sense that no other single commercial product has ever managed to realize these abstract concepts in the same stunning fashion.

In contrast, we share Gruber's opinion that interfaces like the iPhone represent the "biggest frontier in interface design." The iPhone, along with other devices that have similarly innovative natural user interface solutions (e.g., the Wii gaming platform), provides a quantum leap forward and represents the wave of the future in interface design. At some point in the not so distant future, keyboards and mice (at least the dedicated hardware versions to which we have grown accustomed) may no longer be a standard part of desktop or portable computing solutions. As a case in point, consider the recently introduced iPad®. Gruber's take? Three years later he forcefully reinforced his original point (Gruber 2010):

> That brings us to the iPad. Initial reaction to it has been polarized, as is so often the case with Apple products. Some say it's a big iPod touch. Others say it's the beginning of a revolution in personal computing. As a pundit, I'm supposed to explain how the truth lies somewhere between these two extremes. But I can't. The iPad really is The Big One: Apple's reconception of personal computing.
>
> The designers and engineers at Apple aren't magicians; they're artisans. (p. 100)

The futuristic interface portrayed in the movie *Minority Report* (Spielberg 2002), including wall-sized, virtual displays hanging in thin air that are walked around and through and manipulated by whole-body gestures (based on John Underkoffler's work at the MIT Media Lab), may not be as far off as it once seemed. The feasibly of such a reality is no longer in doubt. The question is whether our science is subtle enough and our art exact enough to implement it wisely.

## References

Apple. 2010. iPhone human interface guidelines (user experience). Cupertino, CA: Apple, Inc.

Bruner, J. S. 1961. The act of discovery. *Harvard Educational Review* 31:21–32.

———. 1967. *On knowing: Essays for the left hand.* Cambridge, MA: Harvard University Press.

Buxton, W. 2009. Multi-touch systems that I have known and loved. http://www.billbuxton.com/multitouchOverview.html.

Durlach, N. I., and A. S. Mavor, eds. 1995. *Virtual reality.* Washington, D.C.: National Academy Press.

Gruber, J. 2007. The new frontier. *Macworld* September: 112.

———. 2010. Apple's constant iterations. *Macworld* April: 100.

Hutchins, E. L., J. D. Hollan, and D. A. Norman. 1986. Direct manipulation interfaces. In *User-centered system design,* ed. D. A. Norman and S. W. Draper. Hillsdale, NJ: Lawrence Erlbaum Associates.

Manjoo, F. 2010. What's wrong with Android (if Google fixes one simple thing, its operating system will surpass the iPhone's). http://www.slate.com/id/2244165/

Norman, D. A. 2010. Natural user interfaces are not natural. *Interactions* XVII (3).

Pejtersen, A. M. 1992. The BookHouse: An icon-based database system for fiction retrieval in public libraries. In *The marketing of library and information services 2,* ed. B. Cronin. London, England: ASLIB.

Rasmussen, J., A. M. Pejtersen, and L. P. Goodstein. 1994. *Cognitive systems engineering.* New York: John Wiley & Sons.

Spielberg, S. 2002. *Minority report.* 20th Century Fox.

Teitelman, W. 1964. Real-time recognition of hand-drawn characters. In *Proceedings of the October 27–29, 1964, Fall Joint Computer Conference, part I,* pp. 559–575. San Francisco, CA: ACM.

Ziefle, M., and S. Bay. 2006. How to overcome disorientation in mobile phone menus: A comparison of two different types of navigation aids. *Human–Computer Interaction* 21:393–433.

# 14

# *Design Tutorial: Command and Control*

## 14.1 Introduction

> We must insist that the designers of these [military command *and* control] systems have appropriate respect for the expertise of proficient operators and ensure that their systems and interfaces do not compromise this expertise. We must find ways to present operators with displays that will make situation assessment easier and more accurate. We also want displays that will make it easier for operators to assess options in order to discover potential problems. In other words, we want to build decision support systems that enhance recognitional as well as analytical decision strategies. (Klein 1989b, p. 64)
>
> I could see that decision ... the display made that decision abundantly clear to me. (Postevaluation comments of experienced Army officer with regard to the RAPTOR interface)

In the previous tutorials we have considered work domains located at the endpoints of the continuum described in Chapter 6. In this tutorial we will explore issues in ecological interface design for work domains located in the middle of this continuum. These work domains are characterized by interaction that is driven, in equal amounts, by law-driven constraints on the one hand and by intent-driven constraints on the other (e.g., hospitals, offices, and manufacturing plants; see Figure 6.3, in Chapter 6, and Rasmussen, Pejtersen, and Goodstein 1994).

Military command and control is a particularly good example of these "intermediate" domains. There are law-driven constraints that arise from an extensive technological core (e.g., weaponry, sensors, communication, etc.) and also intent-driven constraints. The juxtaposed set of intentions that exist between military organizations (i.e., the two sides involved in a conflict) is by far the most obvious example. However, intent also plays a substantial role within military organizations, especially for the United States military. For example, during tactical engagements, lower level leaders base their actions upon an interpretation of the commander's intent statement in mission orders (e.g., Klein 1994). Thus, both sources of constraints contribute equally to the patterns of events and activities that unfold in these intermediate domains.

In recent years, a number of research efforts have applied the CSE (cognitive systems engineering) perspective to military settings (Rasmussen 1998; Burns, Bisantz, and Roth 2004; Burns, Bryant, and Chalmers 2005; Naikar, Moylan, and Pearce 2006; Potter et al. 2002; Potter, Gualtieri, and Elm 2003). Unlike the fairly well-defined interface design strategies for domains that fall at either of the two ends of the continuum (see previous design tutorials), cognitive systems engineering (CSE) researchers are in the process of determining which interface design strategy (or perhaps strategies) will be most appropriate for intermediate domains. This tutorial will describe the products of a research program that has had the goal of developing an ecological interface to support mobile army decision makers during tactical operations. Some additional background will be provided to set the stage for these developments.

Military command and control is characterized by all of the classical dimensions of work domains, including complexity, inherent risk, dynamics, and uncertainty. A major source of the complexity and uncertainty is an intelligent adversary. The task force commander pursues mission objectives by marshaling forces, resources, and opportunities so that combat power is maximized and available for delivery at an appropriate point in space and time. Task-force-level command and control has historically occurred at a tactical operations center located in the field—a semimobile assortment of trailers, trucks, equipment, and staff.

However, both the physical location where these activities occur and the technological systems that support them have undergone dramatic changes in recent years. Most commanders now direct tactical operations from fighting vehicles located at forward positions in the battlefield. This provides additional challenges to effective command: There is no longer direct access to the resources (staff, information systems, etc.) that are located in the tactical operations center.

The army has developed several computerized systems (e.g., Force XXI Battle Command Brigade and below [FBCB2]) designed to provide decision support for mobile commanders. These systems are in various stages of development, refinement, and implementation. There are some indications that commanders and their staffs continue to make critical battlefield errors that compromise mission objectives (e.g., Prevou 1995). For example, the Task Force Army Warfighting Experiment (Army 1997b) indicated very clearly that the FBCB2 interface and related technology contributed directly to poor decision making. Commanders and their staffs were often inundated by a large amount of data and the way in which they were presented, particularly during combat situations when high stress and heavy workloads were imposed.

In approximately 2000, a collaborative research program was initiated by Wright State University, the United States Military Academy, and the army research laboratory (ARL FEDLAB, Advanced Displays and Interactive

Displays Consortium) to develop a prototype interface capable of providing more effective decision support for mobile army commanders. This interface has been referred to as RAPTOR (representation aiding portrayal of tactical operations resources). This research program was primarily funded through ARL's Collaborative Technology Alliances research program (Advanced Decision Architectures).

The RAPTOR interface provides one solution to the challenges presented by the intermediate category of domains. Concrete examples of the application of the CSE/EID approach and the RAPTOR interface will be provided.

## 14.2 Abstraction Hierarchy Analysis of Military Command and Control

As described previously, a fundamental premise of the CSE approach is that a detailed understanding of the work domain is crucial. A number of work domain analyses were conducted that focused on army tactical operations at the battalion level and below. These analyses were completed in cooperation with a large number of subject matter experts, including project investigators (a colonel and two majors), active duty personnel (e.g., battalion commanders, intelligence officers, and Army ROTC cadre), and army research laboratory personnel (participants in the Advanced Decision Architectures Collaborative Technology Alliance Consortium). Furthermore, army commanders (brigade and battalion) were observed during large-scale field exercises and army publications were consulted. The overall results indicate that the scope, complexity, and severity of the challenges presented by this domain are staggering. The results of the abstraction hierarchy analysis are summarized in Figure 14.1.

### 14.2.1 Goals, Purposes, and Constraints

An army battalion's purpose is to conduct tactical land-based warfare operations. General goals are set by the mission objectives obtained from the unit above (i.e., division/brigade). A commander and his staff will develop mission statements that further specify these goals; the mission statement is ultimately translated into more specific goals in the form of operation orders for lower level units. An important component of these mission plans is the commander's intent (see more detailed description in the decision-making section later). Overall, the goal is to achieve mission objectives through efficient execution.

There are several outside, real-world constraints on the ways in which these goals should be accomplished. The resources of the battalion (e.g., equipment, personnel) are finite and valuable; their expenditure must be

| **Abstraction Hierarchy** (Means-Ends Relations) | **Aggregation Hierarchy** (Whole-Part Relations) Coarse Resolution | | | | Fine Resolution |
|---|---|---|---|---|---|
| Goals, Purposes, and Constraints | Mission plans and objectives, collateral damage, public perception, etc. | | | | |
| Priority Measures and Abstract Functions | | Flow of resources, relative military force (force ratio), value of mission objectives vs. resource expenditure, probability of success / failure, etc. | | | |
| General Work Activities and Functions | | | Source, store, & sink. Tactical functions (command, control, maneuver, service support, air defense, intelligence, fire support, mobility & survivability), etc. | | |
| Physical Activities in work, Physical Processes of Equipment | | | | Number of vehicles (speed, maneuverability), weapons (power, range), and sensors (sensitivity, range); terrain (avenues of approach), etc. | |
| Appearance, Location, and Configuration of Material Objects | | | | | Physical location of units, physical characteristics of terrain and weather, etc. |

**FIGURE 14.1**
Abstraction and aggregation hierarchies for army tactical operations at battalion echelon. (Adapted with permission from Bennett, K. B., S. M. Posey, and L. G. Shattuck. 2008. *Journal of Cognitive Engineering and Decision Making* 2 (4): 349–385. Copyright 2008 by the Human Factors and Ergonomics Society. All rights reserved.)

minimized. Tactical operations must comply with military laws (e.g., the Geneva Convention) that specify how these operations should be conducted (i.e., rules of engagement). Military personnel and civil populations must be protected to the extent possible; collateral damage should be minimized. Activities should comply with local laws and customs to the extent possible. Political and public opinion must be considered.

### 14.2.2 Priority Measures and Abstract Functions

The primary abstract function for tactical operations is the property of combat power (see also Potter et al. 2002). Combat power is the military "force" or potential that can be applied by a unit at a particular location at a particular point in time. This power is determined by a variety of factors, including tangible (e.g., the number and type of equipment, personnel, ammunition, etc.) and intangible (e.g., morale, leadership, initiative, etc.) resources. Combat power is a fluctuating commodity; resources are continually flowing into (e.g., logistic reinforcements) and out of (e.g., expended in tactical engagements) the system.

The priority measures at this level include several important considerations and difficult trade-offs. How valuable or important is the strategic objective in terms of higher order initiatives? How many of the finite resources should be expended to achieve that objective? What is the probability of success, given the commitment of these resources?

### 14.2.3 General Work Activities and Functions

Descriptions at this level are extensive for a battalion, including maneuver (e.g., position forces, control terrain), fire support (field artillery, close air support, electronic warfare), air defense (protect from enemy aircraft and missile attacks), intelligence (collect information about enemy, weather, geography, etc.), mobility and survivability (eliminate obstacles impeding movement, establish protected fighting positions), and combat service support (arm, fuel, and fix equipment; logistics). Communication is a general function that cuts across all others.

The work domain analyses reported here, however, focus on the general functions and activities of command and control. Command includes establishing commander intent, visualizing future battlefield states, formulating concepts of operations, assigning missions, ranking priorities, allocating resources, conducting risk assessments, monitoring current status, and anticipating change. Control includes computing requirements, defining limits, allocating means, monitoring status, monitoring performance, projecting change, and developing specific instructions from general guidance.

### 14.2.4 Physical Activities in Work; Physical Processes of Equipment

The resources and activities of the battalion at this level are extensive. One critical type of information at this level includes the functional characteristics of the battlefield equipment. The primary vehicles for friendly forces are the Abrams tank and the Bradley fighting vehicle. Each of these combat vehicles has functional specifications that include maximum speed, maximum traversable slope, cruising range, weapons (weapon type, number of weapons, destructive power, disruptive power, range), vulnerability/armor, radiation signatures, ammunition capacity, crew requirements, etc. Other types of equipment include artillery, mortar, sensors, helicopters, communications, unmanned aerial vehicles (UAVs), etc. This equipment will have important functional characteristics. The enemy's equipment also possesses a similar set of functional characteristics; the differences between friendly and enemy forces in terms of these functional characteristics play an important role in tactical operations. Functional aspects of the terrain (e.g., avenues of approach) are also at this level.

### 14.2.5 Appearance, Location, and Configuration of Material Objects

The physical characteristics of the battlefield play an especially critical role in land-based tactical operations. A primary consideration is the battlefield terrain; this places a very stringent set of constraints upon what can and cannot be done. For example, mountains and rivers are natural barriers inhibiting movement. Knowledge with regard to the physical location of friendly and enemy troops, equipment, weapons, and sensors is critical. Physical factors associated with the weather are also important (e.g., the presence of clouds can interfere with the provision of close air support).

### 14.2.6 Aggregation Hierarchy

A complementary analytical tool is the aggregation hierarchy. This tool is used to provide models of the "part–whole" structure of a domain: the different grains of resolution (from coarse to fine) that need to be considered. Unlike the abstraction hierarchy, there is not a generally appropriate number of categories in the aggregation hierarchy. One dimension of aggregation in the battalion can be seen in Figure 14.1 (coarser levels on the left; finer levels on the right). A second dimension not listed in Figure 14.1 is the hierarchical organizational structure of the battalion. A battalion is typically composed of from three to six units that constitute lower echelon levels (in the present chapter, we assume four: Companies A, B, C, and D). Each of these companies is further divided into three units at a lower echelon level (i.e., 1st, 2nd, and 3rd Platoons). Finally, each platoon consists of a specific configuration of combat vehicles, resources, and personnel.

## 14.3 Decision Making

Classic explanations of decision making have viewed this process as a highly analytical and cognitively intensive activity. Goals are reasonably well defined; alternatives are carefully weighed and deliberately prioritized in terms of the probability of success. The work domain analyses indicated that this traditional conceptualization of problem solving corresponds very closely to the initial stages of planning for army tactical operations. The army refers to this as the military decision-making process (MDMP) and it has been the traditional focus within the military.

### 14.3.1 Military Decision-Making Process (or Analytical Process)

The MDMP is a fairly lengthy process that occurs prior to the onset of a tactical engagement. Figure 14.2 provides a summary of MDMP mapped onto the decision ladder. (Note that the traditional labels for states and activities in the decision ladder have been refined to reflect the military domain; see Rasmussen 1998.) Each activity in the ladder will be described in greater detail.

#### *14.3.1.1 Situation Analysis*

Decision making formally begins with a mission statement received from a higher echelon. In practice, however, an enormous amount of information is gathered before the tactical operation even begins. The activities and products of the situation analysis phase (i.e., the left leg of the decision ladder in Figure 14.2) provide a necessary foundation for both effective initial planning and subsequent execution. The amount of raw data available is staggering. For example, extremely detailed information about friendly combat resources (e.g., the inner temperature of an individual artillery gun's bore or the gallons of gas in an individual vehicle) can be obtained in near real time via telemetry.

As has been previously noted (Woods 1991), a fundamental problem is to convert these raw data into meaningful information. A partial listing of information products that are routinely prepared by army personnel during data analysis and conditioning are listed in Figure 14.2. These products include the essential elements of friendly information (EEFI—how to prevent the enemy from seeing me), the friendly forces information requirements (FFIR—how I see myself), the priority intelligence requirements (PIR—how I see the enemy), the commander's critical information requirements (CCIR—mission-related information needed by the commander to make decisions), the mission, enemy, troops, terrain and weather, and time available (METT-T—fundamental information about the engagement), the modified combined obstacles overlay (MCOO—terrain analysis), and intelligence preparation of the battlefield (IPB—a thorough analysis of enemy and terrain).

#### *14.3.1.2 Develop Courses of Action*

The commander and his staff consider these and many other factors in developing, evaluating, and choosing between alternative courses of action (COAs). The four primary activities (mission analysis, commander's guidance, COA development, and COA analysis) are illustrated at the top of the decision ladder in Figure 14.2. As the annotations suggest, these activities are quite extensive. Although they are listed in loose chronological order

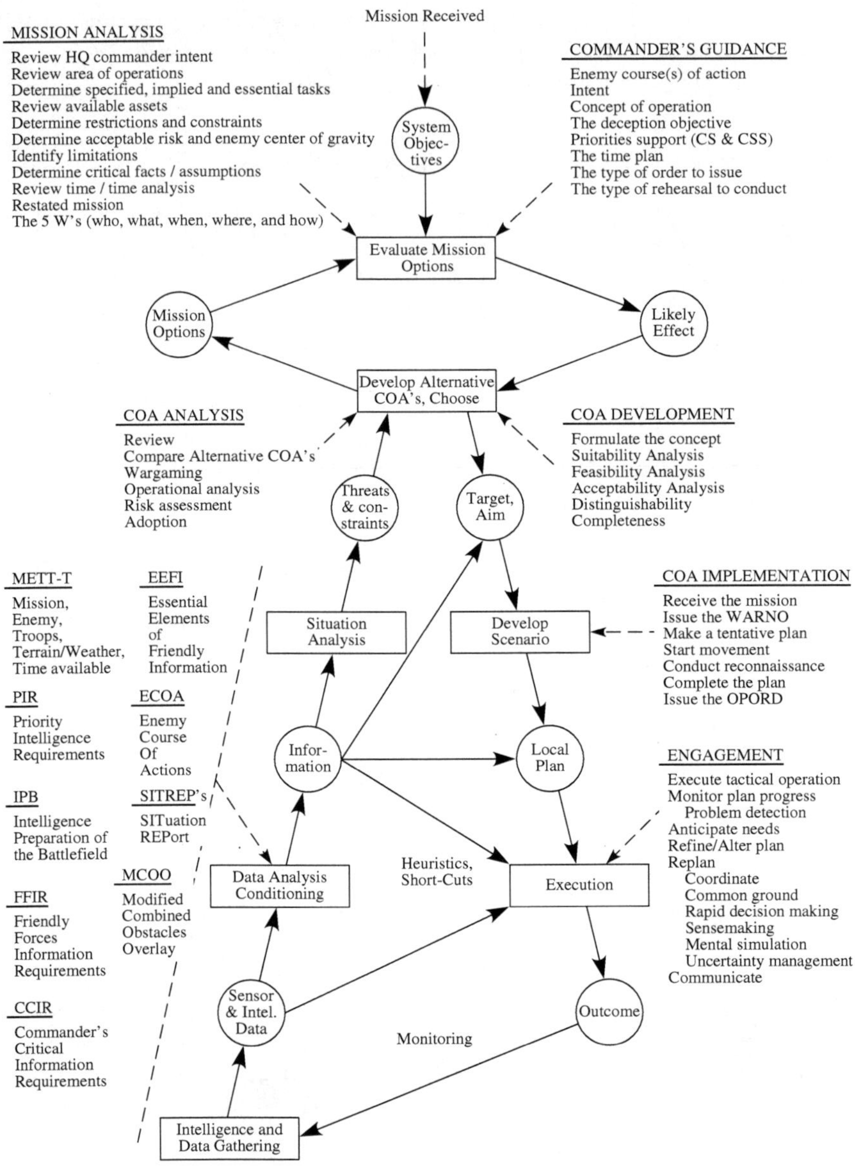


**FIGURE 14.2**
Decision ladder for army tactical operations at battalion echelon. (Adapted with permission from Bennett, K. B., S. M. Posey, and L. G. Shattuck. 2008. *Journal of Cognitive Engineering and Decision Making* 2 (4): 349–385. Copyright 2008 by the Human Factors and Ergonomics Society. All rights reserved.)

(early activities in the upper left and clockwise around to later activities), there will be multiple iterations in this loop when COAs for a battalion are developed.

It is important to emphasize that the MDMP is a deliberate and exhaustive exercise that closely mirrors classical explanations of decision making. The commander and his staff are making value judgments regarding the ultimate worth of the objective, the probability of success or failure, and the associated costs. They are working with incomplete and potentially misleading information and must consider a number of factors (e.g., descriptions of the size of the force to be encountered, the various phases of the battle, objectives to be taken, movement across physical terrain, resources to be expended, and a final state to be achieved).

Typically, not one but several alternative COAs will be devised, accepted, and prioritized. Each COA can be fairly complex and can have several preplanned variations (branches and sequels). Descriptions of the potential courses of action that could be taken by the enemy will also be developed, including the most likely and the most dangerous courses of action. In a very real sense, the goal of this overall activity is to consider all factors and available options and then to determine the COA that has the highest probability of success.

An important component of the mission statement is a section referred to as the "commander's intent" statement. The army (1997a, pp. 1–34) defines commander's intent in the following fashion:

> A clear, concise statement of what the force must do to succeed with respect to the enemy and the terrain and to the desired end state. It provides the link between the mission and the concept of operations by stating the key tasks that, along with the mission, are the basis for subordinates to exercise initiative when unanticipated opportunities arise or when the original concept of operations no longer applies.

Klein (1994) collected and analyzed a total of 35 mission statements from army training exercises at the brigade and battalion level. His analysis provides a script for effective commander's intent statements. There are seven categories of information in the script: (1) purpose of the mission—higher level goals, (2) mission objective—image of the desired outcome, (3) plan sequence, (4) rationale for the plan, (5) key decisions, (6) antigoals (outcomes to be avoided), and (7) constraints and considerations.

#### 14.3.1.3 Planning and Execution

The planning and execution phase of an engagement is initiated when the battalion commander and his staff issue a mission statement that is conveyed to lower echelon leaders. This represents movement down the right leg of the decision ladder in Figure 14.2. The mission statement is complete in the sense that critical information is specified (see previous discussion). However, this mission statement (and the associated COA) should not be confused with a plan for the engagement. The guidance it contains is fairly general in nature and it is quite short (76–200 words, Klein 1993). It is the responsibility of the lower echelon commanders (in this case, the company commanders) to determine the details of how the mission gets accomplished. The lower level commanders interpret the higher level commander's intent and generate the specific details required to fill in the mission plan. This division of responsibility and authority provides an interesting contrast to other military organizations, where plans are implemented primarily in a top-down manner (e.g., the former Soviet Union).

Thus, the primary goal of the next stage of activity (develop scenario) is to implement the COA through the development of a mission plan. The resulting mission plans can be quite detailed and complex. Each course of action might include mission goals, detailed plans for lower level units (e.g., routes of ingress and egress, activities, synchronization points), levels of enemy resistance expected, and acceptable levels of resource expenditures. The mission plan is then communicated to lower level units (i.e., companies and platoons) for execution through an operation order (OPORD; see Figure 14.2).

### 14.3.2 Intuitive Decision Making (or Naturalistic Decision Making)

The next activity is to execute the plan (see the bottom of the right leg in Figure 14.2). There is a need, obviously, for decision making and problem solving at this stage. However, the deliberate, analytic processes used prior to an engagement (i.e., MDMP) are too time consuming for use during the actual engagement. This is primarily due to the combination of extreme time pressure and uncertainty that occurs (i.e., the "fog" of war).

The army has recently recognized a second category of decision making. Although initially referred to as the combat decision-making process, it is now referred to as "intuitive" decision making (Army 2003). This distinction parallels recent developments in the decision-making literature, with the intuitive decision making corresponding to the naturalistic (e.g., recognition-primed decision [RPD], Klein 1989a) approaches discussed in Chapters 4 and 7. The commanders utilize perceptual cues and their prior experience to recognize situations, develop alternative solutions, and mentally simulate potential outcomes to determine an alternative course of action.

The goal at the onset of a tactical engagement will be to complete the mission according to the plan (or set of plans) developed. During the execution phase, the primary locus of control shifts from higher level commanders to lower level leaders and troops who are fighting the battle. Ultimately, lower level leaders will base their actions upon their understanding of the current battlefield situation and the commander's intent as expressed in the mission's operation order. Plans often need to be revised, especially when there is an intelligence adversary. In fact, changes to a mission plan are probably the norm, rather than the exception. A change may be as simple as a minor modification of the current COA. A minor modification is defined as one in which the alterations to the plan involve no changes in goal priorities and no additional changes in coordination between units. These minor modifications will occur spontaneously.

In other cases, entirely new mission plans must be developed. We will refer to this as "replanning." Under these circumstances, leaders might well be trying to determine what the appropriate goal should be, based on their assessment of the current context. In essence, commanders are forced to reenter the problem-solving activities at the top of the decision ladder where values and priorities must be considered and traded off and new courses of action must be determined. They are under extreme time pressure, stress, and risk. Consistent with the principles of RPD, it is very likely that commanders will not be searching for an optimal solution, but rather will be considering a very small number of potential solutions that could work.

It is also important to note that the reason for replanning efforts may not always be the failure of a current plan or the lack of appropriate branches or sequels. In fact, replanning might be needed because the plan is succeeding far better than expected; new opportunities are recognized, but substantial changes in the original plans are needed to seize them. A Warfighter exercise at Ft. Drum provided a particularly interesting example along these lines: A new COA was needed because the original plan was working too well (enemy resistance was far less than expected), as opposed to failing. In this case, the original mission goals and plans were changed to seize an opportunity that was presented. The new COA maximized their combat power, took advantage of their earlier successes, and allowed them to take an alternative objective quickly. In this sense, replanning is driven by both error (a deviation from objectives) and surprise (a deviation from expectations). This reflects the semiotic dynamic of abduction and the joint operation of assimilation and accommodation processes.

The final observation regarding replanning is that this is a course of events that commanders do not undertake lightly. If replanning is required, that means that the initial understanding of the ways in which the tactical engagement would unfold was incorrect. A commander and his staff now need to "go back to the drawing board" to try to do a better job than the first time around. It will be somewhat more difficult because there will be greater time pressure. A decision to issue new operations orders also entails a great deal of overhead activity (e.g., communicating new mission plans to all units).

## 14.4 Direct Perception

An ecological interface was developed to provide support for some of these complicated decisions. An annotated overview of the RAPTOR interface is provided in Figure 14.3. One goal was to allow army decision makers to

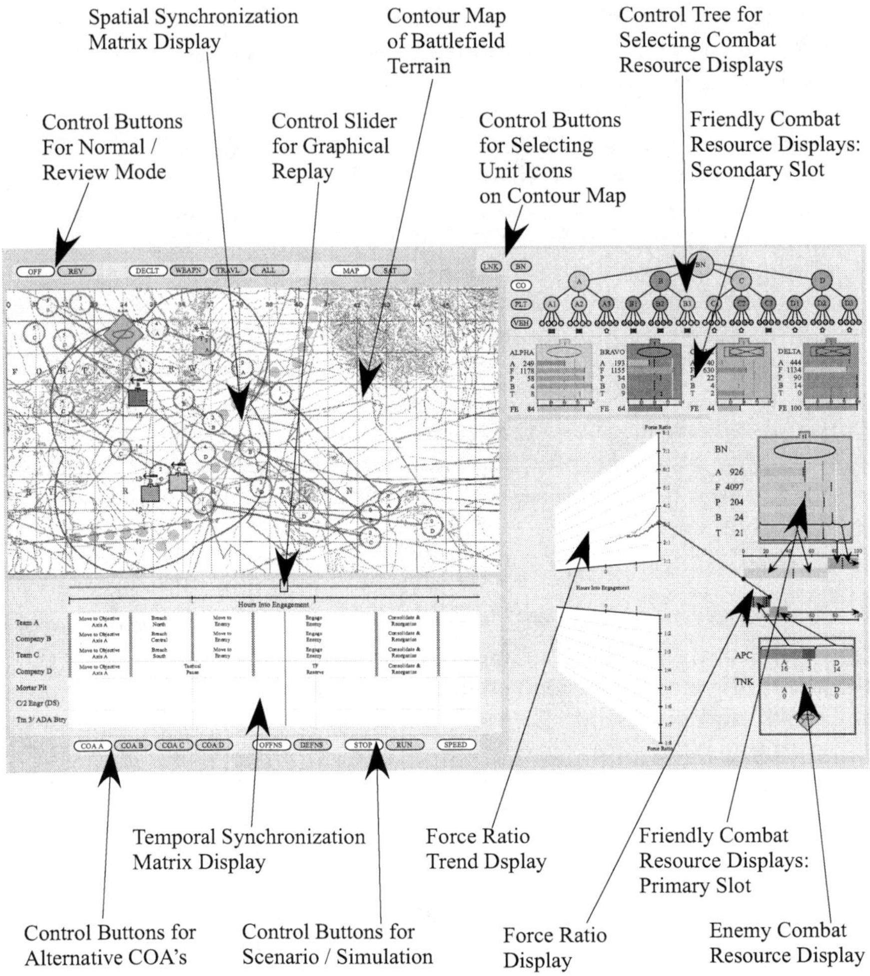


**FIGURE 14.3**
**(See color insert following page 230.)** Overview of RAPTOR interface. (Adapted with permission from Bennett, K. B., S. M. Posey, and L. G. Shattuck. 2008. *Journal of Cognitive Engineering and Decision Making* 2 (4): 349–385. Copyright 2008 by the Human Factors and Ergonomics Society. All rights reserved.)

perceive critical situational factors directly (direct perception), as opposed to reasoning about them. The interface also allows army decision makers to act directly on objects to execute control input (direct manipulation). The overall goal was to transform the interaction requirements associated with decision making and problem solving during tactical operations from cognitive activities (requiring limited-capacity resources such as working memory) to perceptual–motor activities (using powerful and virtually unlimited perceptual resources).

Creating effective graphical representations for a domain is a substantial design challenge that requires consideration of visual forms, domain constraints, processing capabilities and limitations of the human visual system, creativity, and art. The major displays in the RAPTOR interface will be described now, beginning with additional details of the work domain analysis that are relevant to its design.

### 14.4.1 Friendly Combat Resources Display

The work domain analyses indicated that one of the primary requirements for effective tactical decision making is to monitor the current level of friendly combat resources. A unit's primary resources are its tanks and Bradleys, as well as the ammunition, fuel, and personnel required to operate them. A single graphical format was developed to represent these resources at each of the echelon levels. A primary consideration in the design of this format is that the individual combat parameters are essentially independent; changes in their values can be correlated (e.g., fuel and ammunition expenditures in an offensive scenario), but do not necessarily have to be (ammunition, but not fuel, in a defensive scenario). Thus, the proper design choice is independent graphical representations of each parameter (e.g., bar graphs), as opposed to a combined representation (e.g., a single geometric form for all five variables) that produces irrelevant emergent features. See Chapter 9 for a more detailed discussion of these and related issues.

The graphical format for friendly combat resources is illustrated at the company level in Figure 14.4. The primary representational format consists of horizontal, analogical bar graphs (one for each combat resource). The base of each bar graph is located on the left edge (0%); the horizontal extent of the bar graph (emphasized by a short vertical line—the analog percentage indicator) provides an analog indication of the percentage for that resource (100% is located on the right edge). These bar graphs are also color coded (green, amber, red, and black) to represent the categorical status of the associated resource. Each color corresponds to a category of resource percentages (100–85%, 84–70%, 69–50%, and <49%, respectively) defined by army convention. Note that analog graphical representations of the boundaries between these categories are also represented in the display. For example, the boundary between red and black categorical status is represented by the thin vertical line located at 50% and extending behind the bar graphs.

Several other representational conventions were also used in the display. The categorical status of the unit as a whole is represented by the background color code of the entire display. Alphanumeric representations were used to present exact values for combat resources. These are the single-character labels (e.g., "T" for tanks) and the digital values that appear on the left side of the display. The digital values provide absolute numbers, not percentages. Additional information regarding the unit's name, size, type, and the amount of time since the last update of information (i.e., uncertainty) in the display (represented by the vividness of the outline around the display) is also incorporated.

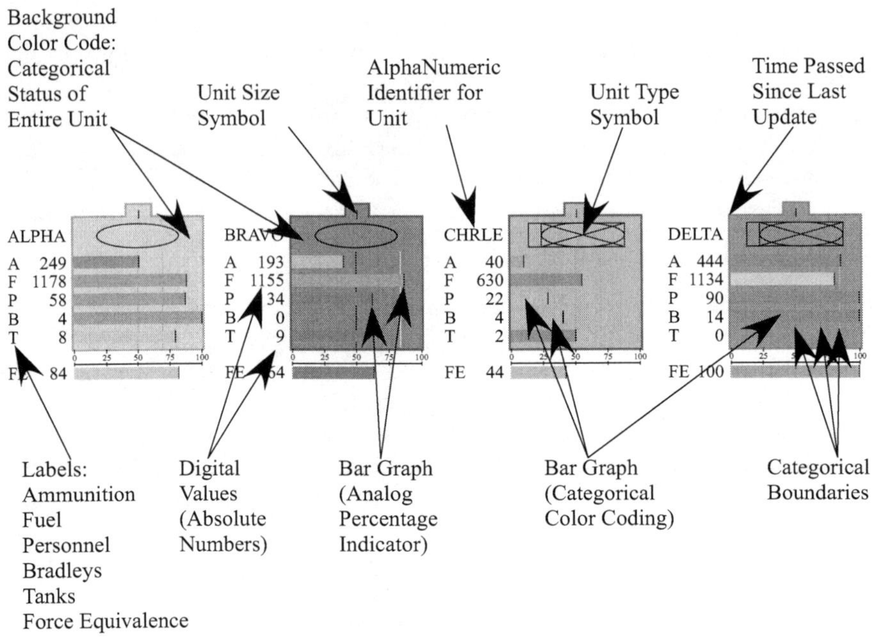


**FIGURE 14.4**
Friendly combat resources display. (Adapted with permission from Bennett, K. B., S. M. Posey, and L. G. Shattuck. 2008. *Journal of Cognitive Engineering and Decision Making* 2 (4): 349–385. Copyright 2008 by the Human Factors and Ergonomics Society. All rights reserved.)

In summary, all three representation types described in Chapter 6 are used in this display: analogical, metaphorical, and propositional. Together, these representations provide commanders and leaders with support for a variety of informational needs. The analog bar graphs provide relatively precise representations of the value of each combat resource. More importantly, they provide analog graphical representations that are particularly useful in determining patterns and relationships (e.g., the value of parameters relative to each other or to boundaries).

Propositional representations include the labels and digital values. The digital values provide support when precise values are needed (e.g., when providing other personnel with "slant" summary reports). The unit type symbols are metaphors that represent the nature of the unit. For example, the oval shape symbolizes the treads of a tank. The categorical color coding is probably the most salient information in the display; it supports commanders in "spot checking" or loosely monitoring the overall status of the unit or a combat parameter.

### 14.4.2 Enemy Combat Resources Display

The domain analyses revealed that estimates of enemy combat resources are also needed. These estimates are obtained from a variety of sources (e.g., satellite imagery, UAVs and surveillance aircraft, spotters, battlefield reports, intelligence estimates). There is, of course, a higher degree of uncertainty in these estimates relative to friendly combat resources.

Army intelligence officers were consulted to determine exactly the type and resolution of information that should be incorporated into an enemy combat resources display. They indicated that the primary concern was the status of enemy combat vehicles (i.e., tanks and personnel carriers). Furthermore, the grain of resolution was fairly course, involving three different categories of information. The first category comprises enemy vehicles that have been observed and verified as being alive (A) and dangerous. The second category is composed of enemy vehicles that have been engaged and disabled (D). The third category comprises template (T) enemy vehicles—those that are likely to be in the area of engagement (based on intelligence analyses), but have not yet been observed.

The enemy combat resources display represents this information using the same general kinds of representations as those in the friendly combat resources display: analog, metaphor, proposition, and categorical. The primary representation format is a horizontal contribution (or stacked) bar graph (see Figure 14.5). Each segment of a bar graph represents a portion of the combined resources.

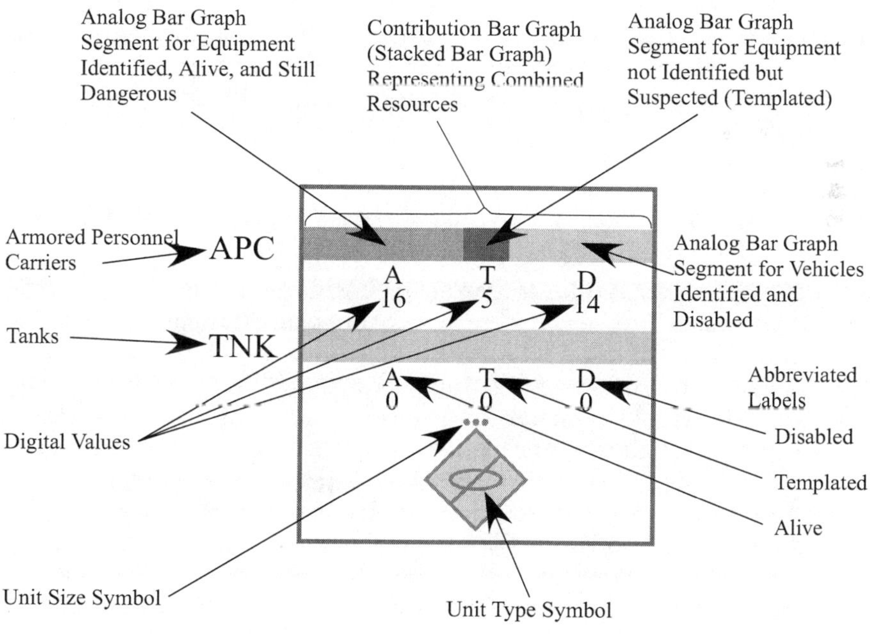


**FIGURE 14.5**
Enemy combat resources display. (Adapted with permission from Bennett, K. B., S. M. Posey, and L. G. Shattuck. 2008. *Journal of Cognitive Engineering and Decision Making* 2 (4): 349–385. Copyright 2008 by the Human Factors and Ergonomics Society. All rights reserved.)

Consider the top contribution bar graph, which represents information regarding enemy personnel carriers. The left, middle, and right bar graph segments provide an analog representation of the percentage of vehicles that are alive, template, and disabled, respectively. They are also color coded (bright red, dull red, and gray, respectively). The analog graphics are annotated with labels and digital values that provide exact values of the number of vehicles in each category (and assorted other information). The bottom contribution bar graph represents tanks. The lack of A and T segments indicates that all tanks have been disabled (or that the unit had none initially). Metaphors are also used to denote the type of unit.

### 14.4.3 Force Ratio Display

The domain analyses revealed that a fundamental consideration in tactical operations is "force ratio": the relative amount of combat power that exists between two opposing forces at any point in time. Force ratio is considered throughout a tactical operation. It is a primary consideration during the planning stages. For example, army doctrine dictates that a force ratio of six to one or better is needed for a friendly unit considering an offensive attack against a well-fortified and dug-in enemy. Force ratio is also a primary consideration during a tactical engagement. Commanders and their staffs develop detailed estimates of how force ratio should change during the course of an engagement. Commanders monitor force ratio to assess progress (or a lack of progress) toward tactical goals during an engagement. Thus, force ratio is a critical piece of information that testifies with regard to decisions to initiate, continue, alter (e.g., choose another course of action), or abort a mission.

A simplified estimate of force ratio was devised in concert with army subject matter experts. As described earlier, combat power is broadly defined and includes both tangible and intangible factors. The primary tangible contributors to combat power (tanks and armored personnel carriers) were used to compute estimates of force and force ratio. Numerical estimates of the military force of individual friendly and enemy combat vehicles were obtained (Army 1999). Military force for a unit is estimated by taking the number of operable vehicles, multiplying by the appropriate constant, and summing across the two vehicle types. This numerical estimate will be referred to as force equivalence, primarily to retain the distinction between it and the broader concept of power. An estimate of the force ratio between two opposing forces is obtained by dividing the larger of the two force equivalences by the smaller. An eventual goal is to devise more comprehensive estimations of force equivalence in the future (including, for example, artillery, aviation, morale, etc.).

The force ratio display is illustrated on the right in Figure 14.6. The primary graphical format is the contribution bar graph. There are two of these, aligned on the left-hand side with the origin of the axes; one is for friendly force equivalence (top) and one is for enemy force equivalence (bottom). The friendly

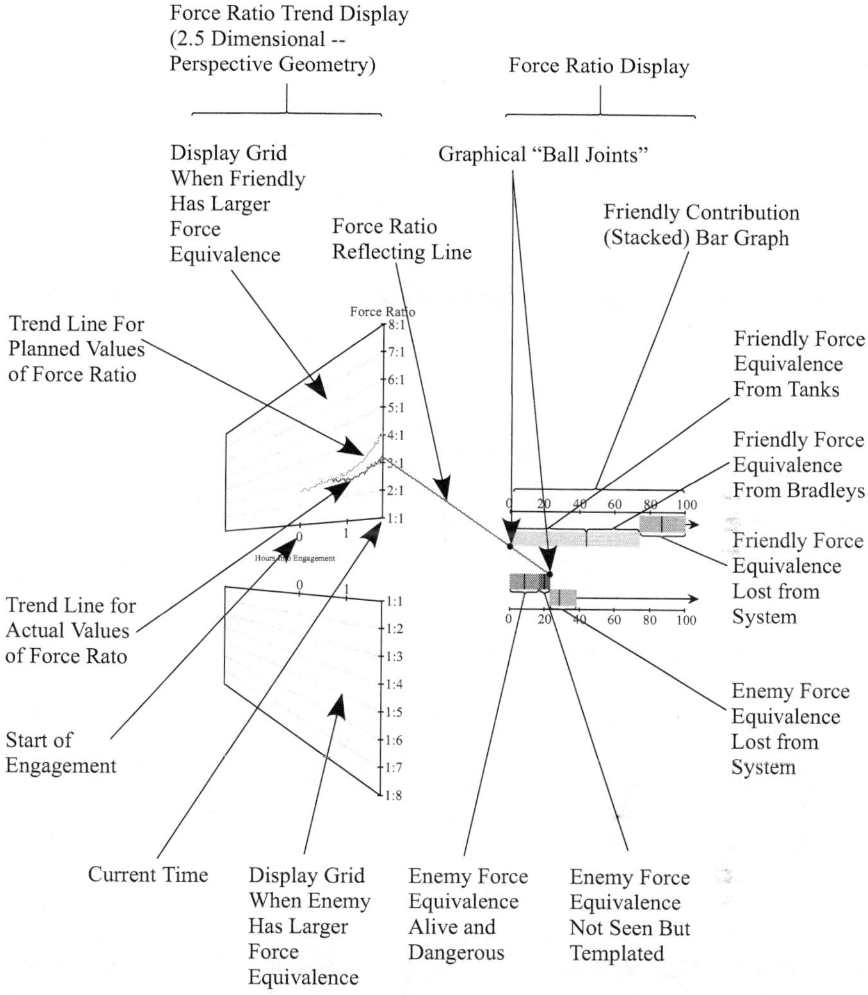


**FIGURE 14.6**
Force ratio and force ratio trend displays. (Adapted with permission from Bennett, K. B., S. M. Posey, and L. G. Shattuck. 2008. *Journal of Cognitive Engineering and Decision Making* 2 (4): 349–385. Copyright 2008 by the Human Factors and Ergonomics Society. All rights reserved.)

contribution bar graph contains two segments on the left (tanks and Bradleys, respectively). These segments represent the force equivalence of current, available resources and are color coded according to the resource's categorical status. The two segments on the right (offset vertically and upward) represent disabled tanks and Bradleys (i.e., the military force that has exited the system). The enemy contribution bar graph has four segments on the left: enemy tanks and personnel carriers that are alive (left two segments) and enemy tanks and personnel carriers that are in the template (right two segments). The two segments on the right (lower, offset) represent disabled vehicles.

Unlike the two previous displays (friendly and enemy resources), the two variables (force equivalence) represented in this display interact in a lawful fashion to define a higher order property (force ratio). This relationship is emphasized by the force ratio reflecting line that connects the force ratio and the force ratio trend displays in Figure 14.6. This line is connected to the two bar graphs, as emphasized by the graphical "ball" joints (Vicente 1991). The reflecting line intersects the scale of the force ratio trend display at the exact spot that corresponds to the current value of force ratio.

This is ensured by the following geometrical properties. Changes in the horizontal extent of the smaller bar graph (the enemy bar graph in Figure 14.6) push (or pull) the end point of the line, thereby changing its orientation. Changes in the horizontal extent of the larger bar graph (the friendly bar graph in Figure 14.6) push (or pull) the entire force ratio trend display (the left graph in Figure 14.6) away (or toward) the force ratio display. Thus, the spatial location of the force ratio trend display is variable; the distance between it and the force ratio display is specified by the horizontal extent of the larger bar graph in the force ratio display.

This is an example of a configural display that produces emergent features (see Chapters 8 through 11). The most salient emergent feature produced by the force ratio display is the orientation of the force ratio connecting line (see Figure 14.6), which dynamically changes as a function of the relationship between friendly and enemy force equivalence. An example of the dynamic behavior of this display over time is provided in Figure 14.7. A fundamentally different configuration is provided for illustrative purposes. The enemy force equivalence is initially greater than the friendly force equivalence Figure 14.7a). Therefore, the force reflecting line is now anchored at the bottom right corner of the friendly bar graph and the upper left corner of the enemy bar graph; the distance between the force ratio and the trend display is equivalent to the length of the enemy bar graph (alive and template segments).

Figure 14.7b illustrates the effect of substantial losses incurred by the enemy approximately 1 hour later. The enemy force ratio is substantially smaller, but still greater than the friendly force ratio; the force ratio trend display has been drawn successively closer as a result. The enemy losses continue to be substantial over the next hour, as illustrated in Figure 14.7c. The force ratio has tipped toward the friendly side. This is reflected in the orientation of the connection line: It is now anchored on the upper right of the enemy bar graph and passes through the lower left of the friendly bar graph. In addition, the distance between the force ratio trend display and the force ratio display is now determined by the length of the friendly bar graph. The enemy losses continue in Figure 14.7d; the diminishing length of the enemy force equivalence bar graph pulls the orientation of the connecting line upward.

### 14.4.4 Force Ratio Trend Display

The force ratio trend display (left side of Figure 14.6 and Figure 14.7) illustrates the actual and planned values of force ratio over time, as illustrated

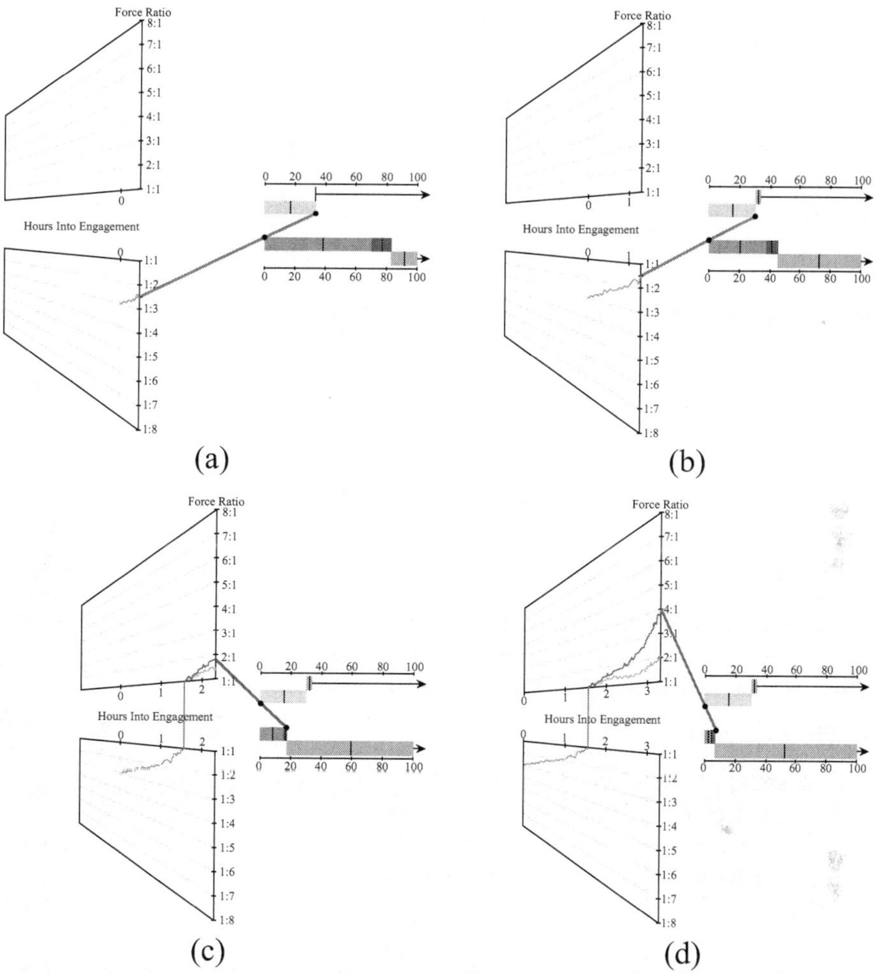


**FIGURE 14.7**
**(See color insert following page 230.)** Force ratio and force ratio trend displays over time. (Adapted with permission from Bennett, K. B., S. M. Posey, and L. G. Shattuck. 2008. *Journal of Cognitive Engineering and Decision Making* 2 (4): 349–385. Copyright 2008 by the Human Factors and Ergonomics Society. All rights reserved.)

in the previous example. A few additional points are in order. The display is scaled using the laws of perspective geometry (toward a vanishing point to the left). This is a variation of the time tunnel design technique (Bennett and Zimmerman 2001; Bennett, Payne, and Walters 2005; Hansen 1995) that produces substantial savings in display real estate. Trend lines for both actual value and planned values of force ratio can be plotted on the display grids. These lines provide several emergent features that should be useful to commanders.

The trend lines are an analog representation of the values of planned and actual force ratio over time. This historical trace specifies both the direction and the rate of change for force ratio across the engagement. The degree of spatial separation between the planned and actual trend lines provides an analog property that visually specifies discrepancy from plan (an important consideration identified in the domain analyses). This visual property could serve as an early warning that alternative courses of action need to be considered or replanning needs to be initiated.

#### *14.4.4.1 A Brief Digression*

Several years ago Rasmussen (personal communication, December 7, 1999) kindly reviewed some process control graphical displays of ours and remarked that "display designs just seem to appear from the blue sky like works of art. I think it would be very productive and influential if you could describe how you selected the visual designs." No time seems better than the present, since a design solution that requires an entire display to change positions dynamically certainly seems to be an odd choice. Before settling on this design for displaying the combination of force ratio and trend displays, we considered a number of alternatives. Due to the complexity of the information that is being displayed, each design involved substantial trade-offs.

One of our first designs is represented in Figure 14.8. This design has the advantage of display axes that remain stationary. The bar graph representing larger force equivalence is located on the bottom and the bar graph representing smaller force equivalence is located on the top. The reflecting line is connected to the upper left (larger force) and lower right (smaller force) corners of the bar graphs; the geometries involved ensure that it intersects the force ratio axis at the appropriate point.

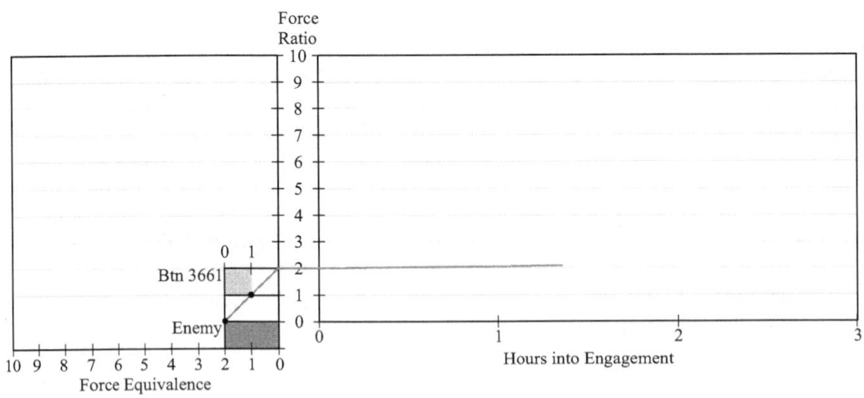


**FIGURE 14.8**
An early alternative design for the combined force ratio and force ratio trend displays.

However, the advantage of the stationary display axes is offset by some substantial disadvantages. First, it provides an inconsistent representation of similar domain properties. Increases in equivalence for the smaller force (upper bar graph) produce an expansion of the bar graph to the right. However, increases in equivalence for the larger force (bottom bar graph) produce an expansion to the left. The latter convention not only is inconsistent, but also violates a population stereotype. Second, the location of the bar graphs that represent friendly and enemy force equivalence is variable: They must switch places as a function of relative size. Third, the location of the graphical origin of the smaller force equivalence is also variable (specified by the horizontal extent of the larger bar graph). Fourth, the friendly and enemy combat resource displays cannot be integrated with these displays (which the current solution does allow; see Figure 14.3).

All other alternatives that we devised suffered from some combination of these or other disadvantages. The current design seems to be the best alternative; a moving display seems a small price to pay for an otherwise consistent and conventional format.

### 14.4.5 Spatial Synchronization Matrix Display

The domain analyses revealed that there are substantial requirements to coordinate and synchronize the activities of the various units. The land-based nature of army tactical operations places a premium on spatial considerations; the physical characteristics of the terrain (e.g., mountains) place critical constraints on what can and cannot be done. The location of friendly and enemy forces with respect to critical features of the battlefield terrain is therefore an extremely important consideration. Was the enemy initially found in the physical location that intelligence sources had indicated? Are friendly forces in appropriate physical locations relative to the enemy? Are friendly forces arrayed in appropriate physical locations relative to one another? What potential actions are supported by terrain features? What potential actions are limited by terrain features?

The spatial synchronization matrix display illustrates a number of spatial constraints (see Figure 14.9) that are critical to land-based tactical operations. The primary component is a contour map providing an analog spatial representation of the physical characteristics of the battlefield terrain (i.e., the contour lines representing changes in elevation). Although not pictured, the capability to view this terrain via satellite imagery (and to toggle between views) has been incorporated. Explicit representations of key spatial synchronization requirements have also been incorporated. A synchronization point is a location in space (i.e., a physical location on the battlefield) that a friendly unit must occupy (usually at a particular point in time; see the following complementary discussion). A synchronization point is represented in the display by a labeled circle. The letter inside the circle indicates the unit; the number refers to successive synchronization points for that unit.

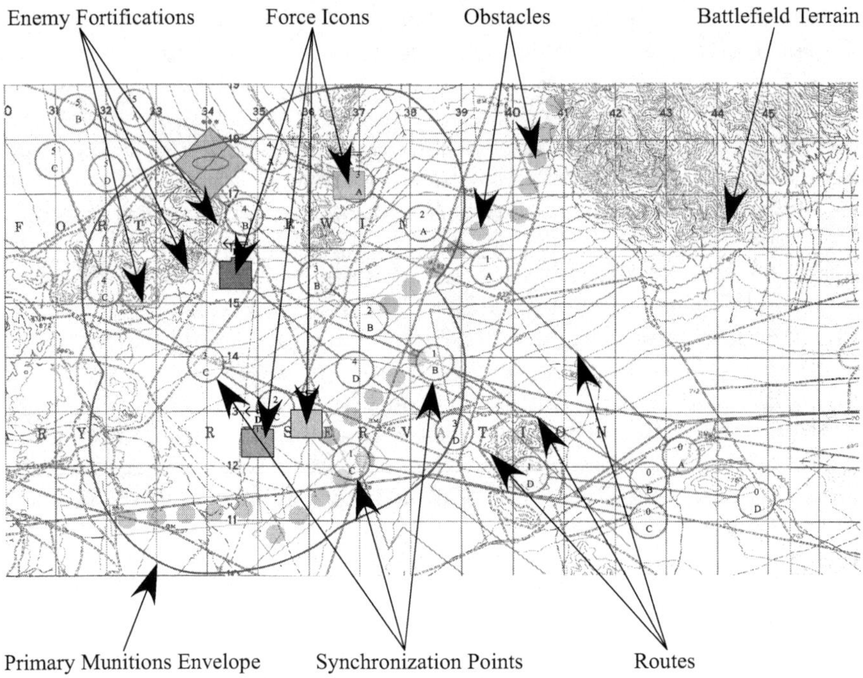


**FIGURE 14.9**
Spatial synchronization matrix display. (Adapted with permission from Bennett, K. B., S. M. Posey, and L. G. Shattuck. 2008. *Journal of Cognitive Engineering and Decision Making* 2 (4): 349–385. Copyright 2008 by the Human Factors and Ergonomics Society. All rights reserved.)

The planned spatial route for each unit in the mission is represented by the activity segments (lines) that connect the synchronization points.

Thus, the spatial synchronization requirements are situated in the context of the battlefield using analog graphical representations. Additional spatial information in the display includes transparent icons representing friendly force locations, arcs representing primary weapons envelopes, obstacles, and enemy units and fortifications.

### 14.4.6 Temporal Synchronization Matrix Display

There is also a need to coordinate the activities of the various units across time. These activities will have different initiation times and will require different amounts of time for their completion. A number of critical temporal synchronization points (e.g., movement to a geographical point by a specific time; completion of a coordinated set of activities by multiple units at a specific point in time) are typically included in a plan. These temporal

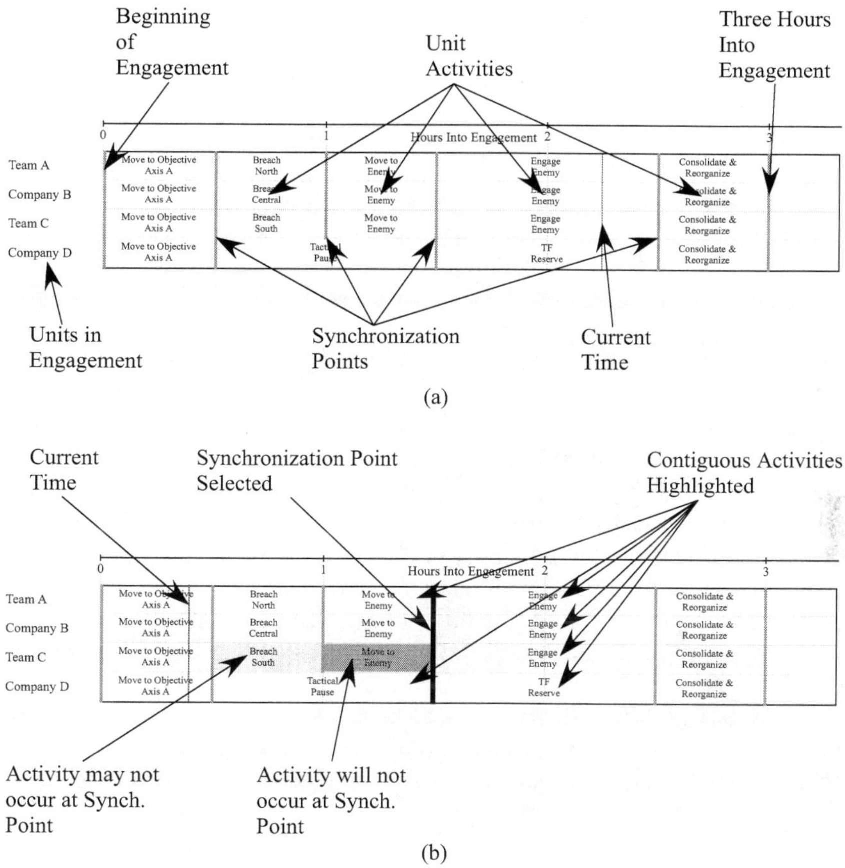


**FIGURE 14.10**
Temporal synchronization matrix display. (Adapted with permission from Bennett, K. B., S. M. Posey, and L. G. Shattuck. 2008. *Journal of Cognitive Engineering and Decision Making* 2 (4): 349–385. Copyright 2008 by the Human Factors and Ergonomics Society. All rights reserved.)

considerations become extremely important when critical events require that these activities culminate simultaneously (e.g., breaching fortifications or obstacles).

The temporal synchronization display explicitly illustrates some of the temporal synchronization requirements that were identified (see Figure 14.10a). Time is represented in the X axis of the matrix, ranging from the initiation of the engagement (0) to a point 3 hours later in time (3). The various units involved in the tactical operation are represented along the Y axis of the matrix (e.g., Company B). A row in the matrix graphically illustrates the sequence of activities that are planned for each unit (e.g., breach north) and an analog representation of the amount of time that each activity should take (the horizontal size of the cell). In addition, temporal synchronization points (i.e., the points in time where there is a requirement to coordinate the activities of these units) are illustrated graphically by the thick gray lines that run vertically through the display. The present time in the engagement is illustrated by the thin vertical line.

Visual changes in the display indicate the status of various activities with regard to their associated temporal synchronization requirements. For example, in Figure 14.10b, team C's activity of "breach south" has been highlighted by a change of color (yellow in the actual display). This indicates that this unit is in danger of not completing the activity on time. Similarly, the next activity for the same unit ("move to enemy") has also been highlighted (red in the actual display). This indicates that the unit will not complete the activity on time (e.g., the unit cannot travel fast enough to reach the destination by the designated time).

The army currently represents synchronization requirements via an alpha-numeric table in the operations order for a mission—clearly not a particularly effective representation. The spatial and temporal matrices devised for the RAPTOR interface provide analog graphical representations of the constraints that are related to these synchronization requirements. Although these two displays have been described separately, critical events often need to be synchronized in both space and time simultaneously. Therefore, these two displays have been designed to work together in a complementary fashion. For example, positioning the cursor over one of the two displays will produce visual rollover effects in the associated symbol in the other display; if the leader places the cursor over a synchronization point in the temporal synchronization matrix, the corresponding visual information is highlighted in the spatial synchronization display.

### 14.4.7 Plan Review Mode and Displays

The work domain analyses included the observation of army commanders who were participating in large-scale training exercises. Prior to these observations, it was fully expected that commanders and leaders would monitor the progress of a mission. What came as a surprise was the extent to which that was true. In practical terms, commanders were often observed to ask the same fundamental question—"Where am I relative to plan?"—in a variety of different ways and with respect to a variety of different resources.

Thus, commanders monitored not just the actual status of combat resources, but also the actual status within the context of planned mission activities and resource expenditures. To make informed command and control decisions, commanders and leaders need to monitor planned and actual progress in terms of space, time, objectives, resources, goals, intentions, and courses of action for both friendly and enemy forces. At the present time, there is very little computerized support to assist leaders in this regard.

A plan review interface mode was developed to meet this need. This mode can be toggled on and off by pointing and clicking on the "REV" button (upper left

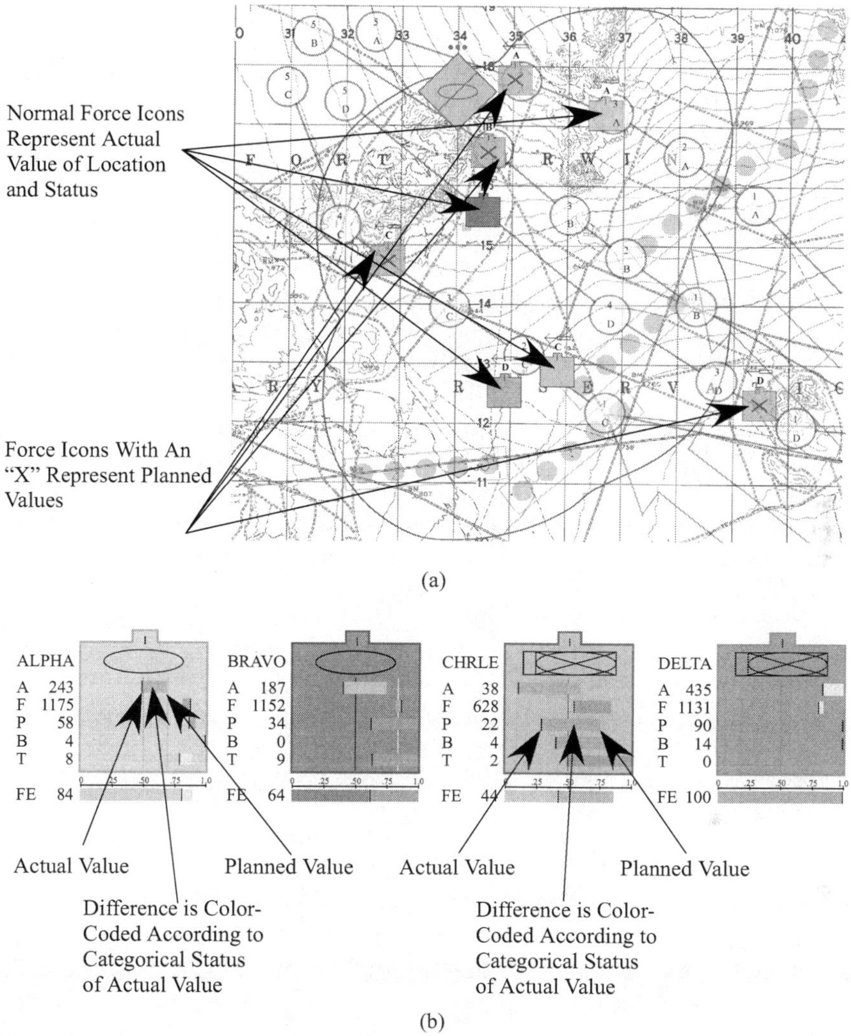


**FIGURE 14.11**
**(See color insert following page 230.)** Plan review mode. (Adapted with permission from Bennett, K. B., S. M. Posey, and L. G. Shattuck. 2008. *Journal of Cognitive Engineering and Decision Making* 2 (4): 349–385. Copyright 2008 by the Human Factors and Ergonomics Society. All rights reserved.)

corner in Figure 14.3). Two primary changes occur (see Figure 14.11). Normally, small, transparent "force" icons are used to represent the actual spatial location and overall categorical status of combat units, vehicles, and soldiers (i.e., various dimensions of military force) on the terrain map. For example, the force icons depicted in Figure 14.9 represent this information at the company level. The first change is that an additional plan force icon will appear for any actual force icon on the map (see the plan icons, which are designated with a black "X" in Figure 14.11a). Each plan icon represents the planned physical location and the planned categorical strength of combat power (indicated by color coding). Deviations from plan are specified by differences in spatial location or color between the planned and actual force icons.

The second change involves the format of the friendly combat resource displays. A floating bar graph segment appears for each combat resource (see Figure 14.11b). This floating segment specifies a range of values that corresponds to the difference between the planned and the actual value for a parameter. Thus, the width of this segment specifies the degree of discrepancy. Furthermore, the color of the segment specifies the direction of the discrepancy by assuming the categorical color code of the actual value. This results in a direct specification of the amount and the direction of deviation from plan for all combat parameters. For example, the presence of large black and red floating segments is a very clear and prominent indication that Company C has deviated substantially from plan and in a negative fashion. Conversely, large yellow or green floating segments would specify that the current operations are going better than planned.

Representing deviations between planned and actual values will facilitate a leader's capability to recognize that a discrepancy requires alternative preplanned courses of action or replanning. Leaders will be alerted to the fact that the battle is deviating from plan earlier in the engagement and can therefore be more proactive in adapting plans to meet the particular needs of the present context.

### 14.4.8 Alternative Course of Action Display

The domain analyses revealed that commanders usually develop multiple COAs during the analytic planning stage prior to a tactical engagement. One COA will be chosen as the single best alternative; other COAs reflect the ways in which the actual battle is most likely to deviate from that plan. Adopting an alternative COA can be dictated by battlefield events. For example, the commander's critical information requirements (CCIRs) and mission statements often contain a description of specific events that will trigger a preplanned response (a branch or sequel in the original plan). This corresponds to the heuristics (shortcuts) on the decision ladder; a particular knowledge state (situation analysis leg) can mandate a preplanned or well-known response (the solid arrows pointing to the planning/execution leg). Several interface resources were provided to support the commander in making these very important decisions.

The graphical replay slider (see Figure 14.3) allows the commander the opportunity to review the status of the current COA. The leader can point, click, and drag the slider along the horizontal extent of the track to provide either a historical "replay" of the tactical operation as it unfolded across time or a "preplay" of future events as they are envisioned in the current COA. The displayed information changes to review mode when the slider is selected (i.e., both planned and actual mission information is displayed). Dragging the slider to the left (from current time to initiation) provides a continuously updated "rewind" of all displayed information regarding the engagement. Dragging the slider to the right (from the initiation time) provides a continuously updated "replay" of the engagement. The displayed information changes when the slider continues past the current time in an engagement; only planned information is then displayed since there is no corresponding actual information.

There are no limits on the number of rewinds, replays, or preplays. The slider springs back to the location corresponding to current time when it is released. Thus, the graphical replay slider allows the leader to review past, present, and future battlefield activities with regard to a number of critical factors, including planned versus actual spatial synchronization, temporal synchronization, and expenditure of combat resources.

The commander may also review the alternative COAs and their fit with the current context. An alternative COA can be made temporarily visible in both the temporal and spatial matrix displays by positioning the cursor over the appropriate button at the bottom of the interface (see Figure 14.3). Graphical representations of the new COA replace the old COA; elements of the new COA that are different from the old COA will be highlighted in red. This allows the leader to review an alternative COA in light of current spatial, temporal, and resource constraints. Moving the cursor away from the button results in the reappearance of the original COA. If a leader decides that an alternative course of action is more appropriate, this decision could be implemented by clicking on the button (and then verifying that choice). This would initiate the process of communicating this change in plan to lower level units.

## 14.5 Direct Manipulation

As described previously, one goal of design from the CSE and ecological interface design (EID) perspectives is to build an interface that maintains an intact perception–action loop. Thus, dragging an icon into the trash can involves continuous space–time signals (graphical representations of both object and target that are physically located in the display space), skill-based behaviors (e.g., visual perception and action), and an intact perception–action loop (user monitors progress toward goal and adjusts movements based on discrepancies from goal). In contrast, the continuous space–time signals are missing with a command line or a pull-down menu; the perception–action loop is broken, not intact. The interface resources designed to support direct manipulation (i.e., provide an intact perception–action loop) in RAPTOR will now be described.

### 14.5.1 Synchronization Points

The spatial and temporal synchronization matrix displays provide simple examples of direct manipulation. The visual representations of the synchronization points in the temporal matrix (i.e., the bold vertical lines in Figure 14.10a) can be manipulated directly to adjust the timing of activities. For example, if one unit is lagging behind and is clearly not going to make a synchronization point on time (as illustrated in Figure 14.10b), the leader can point, click, and drag the appropriate vertical line to the right, thereby establishing a later synchronization time. Similarly, a leader can point, click, and drag the graphical representations of the spatial synchronization points (i.e., the circles in Figure 14.9) to alter the point in space to which a unit maneuvers. These actions constitute the minor modification of an existing plan. Ultimately, these modifications could be made visible in shared displays to facilitate communication and thereby the synchronization of activities across various units.

### 14.5.2 Levels of Aggregation

As described in Chapter 3, experts regularly exhibit changes in the span of attention (i.e., the grain of resolution) with regard to the domain within which they are working. The aggregation hierarchy is an analytical tool that models the landscape upon which this zooming in and out occurs. Not surprisingly, this was an important consideration in the design of the RAPTOR interface, particularly with regard to the critical requirement to track the status of friendly units during a tactical operation. These units are organized in a nested hierarchy ranging from battalion (1), companies (4), and platoons (3 in each company for a total of 12) all the way down to individual vehicles and soldiers—a classic part–whole structure. The agent must track the associated combat resources and a variety of related information (e.g., spatial location, spatial synchronization points, routes of travel, activities, temporal synchronization points, etc.) associated with these units.

Collectively, these requirements introduced a number of design challenges to the RAPTOR interface. The trade-offs and solutions are described in greater detail in Chapter 15. Here, we focus on the role that direct manipulation played.

#### *14.5.2.1 Aggregation Control/Display*

The primary interface mechanism for controlling changes in the level of aggregation for friendly forces is the aggregation control/display, illustrated in Figure 14.12. The tree display graphically illustrates the part–whole relationships between the various echelon levels of the battalion. The currently selected node, representing a unit, is highlighted (i.e., increased coloration, outline). Positioning the mouse over a different node temporarily changes a variety of unit-specific information displayed throughout the interface (e.g.,

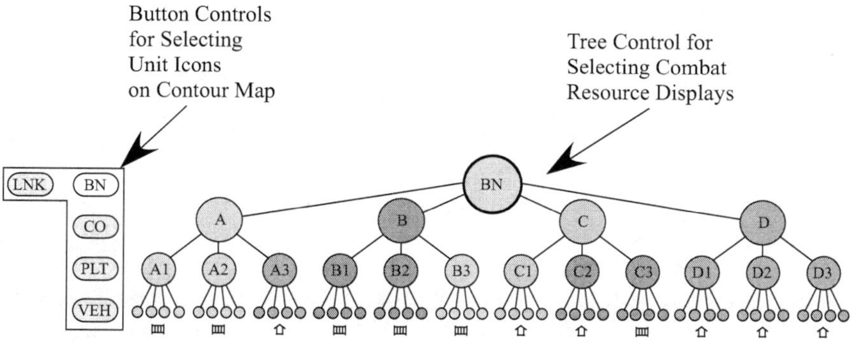


**FIGURE 14.12**
Aggregation control/display for changing the level of aggregation of displayed units.

the levels of unit information displayed in the primary and secondary display slots; see Chapter 15 for a more detailed description). Clicking on a node makes these changes permanent. Thus, the capability to control the span of attention regarding friendly units easily and efficiently (ranging from the battalion down to an individual vehicle) is supported via direct manipulation.

### *14.5.2.2 Control of Force Icons*

Direct manipulation was also used to control one particularly important facet of the unit-specific information referred to previously: the physical location of units, vehicles, and personnel on the battlefield terrain (i.e., the force icons depicted in Figures 14.9 and 14.11). Some interesting design challenges were associated with this critical information. The commander needs to change the span of attention with regard to levels of aggregation. At some point the spatial location of entire units might be of interest (e.g., Company A) and at other points the spatial location of the vehicles associated with Platoon D3 might need to be considered (or even individual soldiers).

These information needs are at odds with the primary purpose of the information displayed on the map (e.g., routes of egress); the traditional solution (i.e., to put all vehicles on the map) both hides this terrain information and produces substantial clutter (a recurring complaint about current electronic battlefield maps). Thus, control mechanisms are needed that allow the commander selectively to choose which units, vehicles, or personnel are currently on the battlefield terrain map.

These control mechanisms were implemented in a variety of ways. Direct manipulation of any visible force icon (i.e., point and click) replaces it with all icons at the next level down in the hierarchical tree structure. For example, clicking on a company force icon replaces it with the three platoon icons for that company. All other force icons currently on the screen remain visible. This provides commanders with the capability to "drill-down" selectively into branches within the overall aggregation structure.

Direct manipulation also allows switching to a completely different level of aggregation (or to a specific unit) to be accomplished easily. Sets of force icons at a particular level of aggregation can be made visible via manipulation of the control buttons located next to the aggregation control/display (see Figure 14.12). For example, pointing and clicking on the PLT button places the 12 platoon force icons on the terrain map and removes all of the other icons previously on the map. In contrast, an isolated unit can be made visible via direct manipulation of any combat resource display in the primary or secondary slots (i.e., pointing and clicking on a display places that unit's force icon on the map and removes all others). Finally, any portion of the tree structure (see Figure 14.12) can be selected via the point–click–drag–release input convention (just like selecting files on a desktop) and the corresponding force icons will then replace all others currently on the screen.

All of the manipulations used to control levels of aggregation described thus far are independent of the more general control exerted by the aggregation control/display described in the previous section. Activation of the LNK control button (see Figure 14.12) simultaneously links changes in the composition of force icons on the battlefield terrain with this more general control mechanism. For example, selecting the battalion combat resource display while the link button is enabled replaces all existing force icons on the terrain map with the battalion's. Collectively, the interface resources described in this and the previous section provide strong support for a fundamental requirement of interaction with complex systems: Direct manipulation allows the commander to change the span of attention to different levels in the aggregation hierarchy easily and directly while maintaining high levels of visual momentum.

In summary, RAPTOR embraces direct manipulation to the fullest extent. There is no command line; there are no pull-down menus. All potential actions by the commander are executed directly upon objects in the interface. Thus, the interface merges displays and controls, provides continuous space–time signals and ensures an intact perception–action loop.

## 14.6 Skill-Based Behavior

The fundamental goal of interface design is to leverage powerful and natural human skills of perception (e.g., obtaining visual information) and action (e.g., manipulating objects). Stated alternatively, the interface should be designed so that domain practitioners are required to use limited-capacity mental capabilities (e.g., mental calculations that require working memory) only when it is absolutely necessary. The previous sections describing the RAPTOR interface in terms of direct perception, direct manipulation, and the perception–action loop provide a detailed account of the resources that were designed to support these skill-based behaviors.

## 14.7 Rule-Based Behavior

Effective displays should also provide graphical representations that produce a rich set of signs to support rule-based behavior. A properly designed interface will convey the current state of a domain directly to the practitioner through its spatiotemporal behavior, thus providing support for the recognition of prototypical system states. In turn, a properly designed interface will suggest the actions that are appropriate (i.e., shortcuts on the decision ladder)—at least to expert agents who have previously learned the actions and the associated perceptual cues (Rasmussen 1983).

Consider one example involving the enemy combat resource display (see Figure 14.5). Imagine that the display indicates that a large percentage of enemy vehicles have remained in the template category late into the course of an engagement. The critical display feature (i.e., the middle segment remaining large) serves as an explicit sign specifying a dangerous possibility: The enemy may well be engaged in a feint maneuver (i.e., the larger enemy unit that is supposed to be here is actually poised to wreak havoc elsewhere). An experienced commander would see the perceptual cues provided by the display and realize that a sequence of activities to test that possibility would need to be performed.

Other examples of visual cues providing support for rule-based behavior include the following discrepancies between planned and actual values: force ratio over time (force ratio trend display; Figures 14.6 and 14.7), physical locations of units (spatial synchronization matrix in review mode; Figure 14.11), timing of activities and events (temporal synchronization matrix; Figure 14.10), resource levels (friendly combat resources displays in review mode; Figure 14.11), and general progress (alternative COA displays).

Thus, the displays in RAPTOR provide a rich set of signs that facilitate a leader's ability to recognize prototypical situations and that suggest appropriate actions (e.g., further investigations to clarify the situation, the adoption of preplanned branches or sequels, or the need for more extensive replanning). Leaders will be alerted to the fact that the battle is deviating from plan earlier in the engagement and will therefore be more proactive in their reactions to meet the particular needs of the present context.

## 14.8 Knowledge-Based Behavior

When plans fail or need to be revised substantially, as is often the case with an intelligent adversary, commanders and leaders are faced with novel situations (i.e., circumstances outside those considered prior to the engagement) and will therefore be engaged in problem solving. Lewis, Klein, Klinger, Thunholm, Bennett, Smith and Obradovich (2004) summarize observations on knowledge-based behaviors in tactical operations:

> Replanning [knowledge-based behavior or problem solving] draws on a variety of cognitive activities that emerge in naturalistic settings .... It draws on problem detection (to determine that a plan needs to be modified), coordination (to ensure the teamwork needed in modifying the plan), common ground (to ensure that the units involved do not hold conflicting views of the situation), rapid decision making (to judge whether, when and how to revise the plan), sensemaking (to appraise the affordances of friendly forces in a dynamic setting), mental simulation (to gauge if the revised plan is likely to be effective), and uncertainty management (to handle the confusion of modifying a plan in progress). (p. 7)

To support these activities, the displays should provide symbolic representations of critical information at all levels in the abstraction hierarchy. These displays will serve as external models that allow critical constraints and relationships to be perceived directly, thereby providing the graphical explanations that are necessary for effective problem solving. This symbolic content at each level of the abstraction hierarchy will be briefly described for the RAPTOR interface. These are graphical representations of the information listed in Figure 14.1.

The tangible information at the level of goals, purposes, and constraints (highest level in the hierarchy) are mission plans and objectives. The spatial matrix, temporal matrix, and alternative COA displays contain explicit representations of plans; all other displays contain implicit representations in terms of planned versus actual progress.

The primary representations at the level of abstract function and priority measures are the force ratio and the force ratio trend displays. These displays show the ebb and flow of military force (as estimated by force equivalence) for friendly and enemy forces.

The level of general functions is a closed-loop control system that involves the monitoring of troop activities (comparing current states to planned goals) in order to guide tactical decision making. Input to the monitoring function includes feedback information about troop strengths and movements that can be compared to spatial and temporal goals (and/or expectations). The tactical decision function involves choices about whether to continue with the planned attack or to deviate due to unexpected dangers or opportunities that evolve during the engagement.

There are numerous representations of information at the level of physical processes and activities. The friendly and enemy combat resource displays represent measured values of combat parameters (e.g., tanks, Bradleys, ammunition, fuel, and personnel). Other information at this level includes the weapons envelope for the primary munitions of both friendly and enemy vehicles and a variety of unit-related information (i.e., identification, type, and size symbols).

Information at the lowest level of the abstraction hierarchy (physical form and configuration) includes the physical characteristics of the battlefield terrain and the physical location of a unit on this terrain.

## 14.9 Evaluation

The RAPTOR interface was evaluated continually throughout the entire time course of its development (over 10 years). Informal evaluations were obtained on a regular basis during two annual meetings attended by army and civilian researchers working in government labs and active army personnel. In both cases, invaluable feedback on the interface was obtained from researchers and practitioners who were very familiar with the domain.

Several formal empirical evaluations were also completed. All of these studies incorporated realistic tactical scenarios that were developed in cooperation with experienced army practitioners. All of these studies also used experienced army officers as participants. The majority of the studies compared performance with the RAPTOR interface to experimental versions of existing army interfaces. The primary evaluations will be described in chronological order.

### 14.9.1 Direct Perception

The first round of evaluation utilized controlled laboratory experiments designed to assess the effectiveness of selected displays and controls in the RAPTOR interface. For example, Talcott et al. (2007) evaluated the capability of the RAPTOR interface to support a critical activity: obtaining the status of friendly combat resources. A qualitative simulation of a traditional offensive engagement (i.e., "force on force") was developed; the status of combat resources (e.g., ammunition) was summarized at three different points in time and at three different echelon levels (battalion, company, and platoon). An experimental, baseline interface was designed to replicate the visual appearance and selected functionality of an existing army interface: Force XXI Battle Command Brigade and below.

The participants were asked to report on the status of combat resources (e.g. the number of tanks) at different echelon levels (e.g., company) using the RAPTOR and FBCB2 interfaces. The accuracy and latency of their responses were recorded. Talcott et al. (2007) summarize the results:

> The superior performance of the RAPTOR interface was present in all assessment categories (quantitative, categorical, and needs), dependent

> variables (accuracy, latency), and echelon levels (battalion, company, platoon). Fifteen statistical comparisons between these two interfaces (main effects or simple main effects) were significant and all of them favored the RAPTOR interface. These results clearly indicate that the RAPTOR interface provided better support for obtaining friendly combat resources than did the FBCB2 interface. (p. 131)

### 14.9.2 Situation Assessment; Decision Making

Subsequent empirical studies expanded the scope of evaluation, moving away from the study of simple information extraction and toward the study of more complicated activities that need to be supported in tactical operations. More varied and more realistic scenarios were developed to accommodate this change in focus.

One study examined two versions of RAPTOR: an enhanced version and a baseline version (without the force ratio, force ratio trend, and review mode). Experienced army officers assumed the role of battalion commander and viewed a dynamic, authentic scenario. The scenario was paused at six different points that coincided with critical events and the participants were required to complete two questions: (1) "Please verbally describe the situation as you understand it." (2) "What actions, if any, would you take at this time?"

Participants who used the enhanced RAPTOR interface exhibited a greater tendency to produce specific references to plans and operations orders. Participants were also extremely enthusiastic about RAPTOR. One participant stated, "I could see that decision … the display made that decision abundantly clear to me." No more eloquent description of the goals of direct perception as a goal of interface design is possible. A second participant expressed the opinion that RAPTOR was "much more useful than some of the other interfaces I've … used … in Iraq."

A final capstone evaluation was recently completed (Hall 2009; Hall, Shattuck, and Bennett, 2010). The RAPTOR interface and a baseline interface (modeled on the functionality of the FBCB2 interface) were evaluated; 16 experienced officers participated. Three complex, realistic, dynamic, and interactive scenarios were developed using commercial simulation software. One scenario was used during training (defensive) and two (offensive, counterinsurgency) were used during testing. Performance data were obtained on the following dependent measures:

- Situation reports (SITREPs). Participants filled out 22-item status reports on the current state of the engagement (e.g., physical locations, combat resources, historical activities of friendly and enemy elements).
- Commander's critical information requirements (CCIRs). Participants monitored the engagement for events that were designated as being particularly important to mission success (e.g., friendly activities, enemy activities, environment).

- Decision points. Participants monitored the engagement for critical events that were designated as "triggers" to activate alternative courses of action.
- Requests for information. This was the number of times that a participant was forced to access information contained in the written operations order.
- Workload. Participants reported their subjective workload.

The results indicate that performance with the RAPTOR interface was dramatically better than performance with the baseline interface for all but one dependent measure. Overall, RAPTOR participants produced significantly better performance for SITREPs and CCIR monitoring, less need to access the written operations orders, and lower perceived workload.

To justify the claim that performance was dramatically better with the RAPTOR interface, we will consider a representative set of results in greater detail. The results for situation reports are illustrated in Figure 14.13. Participants using the RAPTOR interface completed these reports in approximately half the time (means = 3.30 vs. 6.05 min) and with approximately twice the accuracy (means = 98 vs. 54%) relative to those participants using

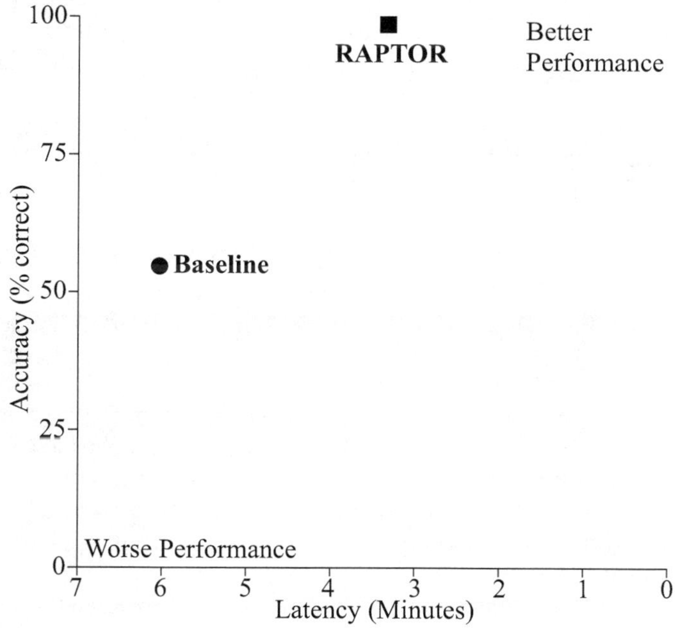


**FIGURE 14.13**
Situation report performance for the RAPTOR and baseline interfaces. (Hall, D. S. 2009. RAPTOR: An empirical evaluation of an ecological interface designed to increase warfighter cognitive performance, Department of Operations Research, Naval Postgraduate School, Monterey, CA.)

the baseline interface. These are not trivial differences in performance—not average differences expressed in milliseconds or tenths of a percent. They would appear to be meaningful since they were obtained with experienced officers engaged in realistic scenarios. Moreover, they probably underestimate real performance differences: The officers were not sleep deprived and were not under stress (i.e., being shot at)—real-world conditions that would presumably enhance performance benefits for the RAPTOR interface (because it leverages more resilient, skill-based behaviors).

## 14.10 Summary

There is a tremendous need for ecological interfaces to provide effective decision support for military command and control during tactical operations. Military personnel are subjected to work conditions that can have a serious impact on their capability to perform effectively. They experience sleep deprivation, extreme physical fatigue, and high degrees of stress (often involving life or death). It is under these conditions that the benefits of the CSE and EID approaches (the leveraging of powerful perception–action skills during decision making and problem solving) are likely to be most beneficial.

The RAPTOR interface, developed from the CSE and EID design frameworks, represents a good start toward meeting these requirements. The principles of direct perception, direct manipulation, and the perception–action loop have been applied to the development of an interface that should contribute to reducing the significant problem of managing many degrees of freedom and large amounts of data. Allowing leaders to "see" constraints, opportunities, and solutions directly in the interface and to act upon them ensures that leaders are likely to have better understandings of the dynamic and stressful conditions arising in the battlefield and therefore make faster and more effective decisions.

However, one must be cautious with respect to any speculations with respect to lifting the "fog of war." It is important to appreciate that much of the uncertainty in war results from the quality of the information available (e.g., accuracy of intelligence reports about the enemy troop strengths and positions). A concern has been raised about whether a high-quality representation of bad data will lead decision makers "down a garden path." So far, empirical investigations of this issue have not supported this concern (St-Cyr 2005; Reising and Sanderson 2002a, 2002b). In fact, the opposite is typically the case: A well organized display tends to reveal measurement and sensor failures due to changes in the display that violate constraints designed into the display. After all, what is the alternative if the data are poor—design poor displays so people will ignore them? This is completely unacceptable (somewhat akin to "throwing out the baby with the bath water").

Constraints on measurement are also part of the domain and there are many ways to represent the uncertainty graphically (e.g., error bars). The recommendation is that, if uncertainty is an issue, then it should be addressed explicitly in the design of the graphic. For example, one way that uncertainty is explicitly represented in RAPTOR is the brightness of the outline around the icons (Figure 14.4), which fades with time since the last update for that unit. This reflects the constraint that uncertainty with regard to the actual values will grow with the time since the last report.

The interface design strategies illustrated in RAPTOR are directly relevant for researchers developing computerized military decision support. The general approach and the principles of design that have emerged are useful for all intermediate domains (i.e., both intent- and law-driven sources of constraints). For example, see Talcott et al. (2007) for a brief discussion of how these principles could be applied to flexible manufacturing.

## References

Army. 1997a. Field manual (FM) no. 101-5-1. Operational terms and graphics. Washington, D.C.: Headquarters Department of the Army.

———. 1997b. NTC rotation 97-06: Initial impressions report advanced warfighting experiment. Fort Leavenworth, KS: Center for Army Lessons Learned.

———. 1999. Battle book/ST 100-3. Ft. Leavenworth, KS: U.S. Army Command and General Staff College.

———. 2003. Field manual (FM) no. 6-0. Mission command: Command and control of army forces. Washington, D.C.: Department of the Army.

Barnett, V., and T. Lewis. 1984. *Outliers in statistical data.* Chichester, England: John Wiley & Sons.

Bennett, K. B., M. Payne, and B. Walters. 2005. An evaluation of a "time tunnel" display format for the presentation of temporal information. *Human Factors* 47 (2): 342–359.

Bennett, K. B., S. M. Posey, and L. G. Shattuck. 2008. Ecological interface design for military command and control. *Journal of Cognitive Engineering and Decision Making* 2 (4): 349–385

Bennett, K. B., and J. Zimmerman. 2001. A preliminary investigation of the time tunnels display design technique. *Displays* 22 (5): 183–199.

Burns, C. M., A. M. Bisantz, and E. M. Roth. 2004. Lessons from a comparison of work models: Representational choices and their implications. *Human Factors* 46:711–727.

Burns, C. M., D. B. Bryant, and B. A. Chalmers. 2005. Boundary, purpose and values in work domain models: Models of naval command and control. *IEEE Transactions on Systems, Man and Cybernetics Part A* 35:603–616.

Hall, D. S. 2009. RAPTOR: An empirical evaluation of an ecological interface designed to increase warfighter cognitive performance, Department of Operations Research, Naval Postgraduate School, Monterey, CA.

Hall, D. S., L. G. Shattuck, and K. B. Bennett. 2010. Evaluation of an ecological interface design for military command and control. Manuscript under review.

Hansen, J. P. 1995. An experimental investigation of configural, digital, and temporal information on process displays. *Human Factors* 37:539–552.

Klein, G. A. 1989a. Recognition-primed decisions. *Advances in Man–Machine Systems Research* 5:47–92.

———. 1989b. Strategies of decision making. *Military Review* May: 56–64.

———. 1993. Characteristics of commander's intent statements. Arlington, VA: U.S. Army Research Institute for the Behavioral and Social Sciences.

———. 1994. A script for the commander's intent statement. In *Science of command and control: Part III Coping with change*, ed. A. H. Levis and I. S. Levis. Fairfax, VA: AFCEA International Press.

Lewis, R., G. Klein, D. Klinger, P. Thunholm, K.B. Bennett, P. Smith, and J. Obradovich. 2004. Preliminary account of replanning in the Army unit of action command post. Collaborative Technology Alliances, Advanced Decision Architectures, U.S. Army Research Laboratory.

Naikar, N., A. Moylan, and B. Pearce. 2006. Analyzing activity in complex systems with cognitive work analysis: Concepts, guidelines, and case study for control task analysis. *Theoretical Issues in Ergonomics Science* 7:371–394.

Potter, S. S., W. C. Elm, E. M. Roth, J. W. Gualtieri, and J. R. Easter. 2002. Using intermediate design artifacts to bridge the gap between cognitive analysis and cognitive engineering. In *Cognitive systems engineering in military aviation environments: Avoiding cogminutia fragmentosa!* ed. M. McNeese and M. A. Vidulich: A report produced under the auspices of the Technical Cooperation Program Technical Panel HUM TP-7 human factors in aircraft environments (HSIAC-SOAR-2002-01). Wright Patterson Air Force Base: Human Systems Information Analysis Center.

Potter, S. S., J. W. Gualtieri, and W. C. Elm. 2003. Case studies: Applied cognitive work analysis in the design of innovative decision support. In *Cognitive task design*, ed. E. Hollnagel. New York: Lawrence Erlbaum Associates.

Prevou, M. 1995. The battle command support system: A command and control system for Force XXI (monograph). Ft. Leavenworth, KS: U.S. Army School of Advanced Military Studies.

Rasmussen, J. 1983. Skills, rules, and knowledge; signals, signs, and symbols, and other distinctions in human performance models. *IEEE Transactions on Systems, Man, and Cybernetics* SMC-13:257–266.

———. 1998. *Ecological interface design for complex systems: An example: SEAD–UAV systems.* Smorum, Denmark: HURECON.

Rasmussen, J., A. M. Pejtersen, and L. P. Goodstein. 1994. *Cognitive systems engineering.* New York: John Wiley & Sons.

Reising, D. C., and P. M. Sanderson. 2002a. Work domain analysis and sensors I: Principles and simple example. *International Journal of Human–Computer Studies* 56:569–596.

———. 2002b. Work domain analysis and sensors II: Pasteurizer II case study. *International Journal of Human–Computer Studies* 56:597–637.

St-Cyr, O. 2005. Sensor noise and ecological interface design: Effects of noise magnitude on operators' performance and control strategies. Dissertation, University of Toronto.

Talcott, C. P., K. B. Bennett, S. G. Martinez, L. Shattuck, and C. Stansifer. 2007. Perception–action icons: An interface design strategy for intermediate domains. *Human Factors* 49 (1): 120–135.

Vicente, K. J. 1991. Supporting knowledge-based behavior through ecological interface design. Urbana-Champaign: Engineering Psychology Research Laboratory, Department of Mechanical Engineering, University of Illinois.

Woods, D. D. 1991. The cognitive engineering of problem representations. In *Human–computer interaction and complex systems,* ed. G. R. S. Weir and J. L. Alty. London: Academic Press.

# 15

# *Design Principles: Visual Momentum*

## 15.1 Introduction

> An area that has received little attention in the design of computer-based display systems is how the user integrates data across successive displays (cf. Badre 1982)....Failure to consider the requirements for effective across-display user information processing can produce devastating effects on user performance....The "getting lost" and "keyhole" phenomena are not inevitable consequences of using computer-based displays; neither do they represent human limitations (for example, short-term memory) that must be compensated for through memory aids or walls of VDUs. Across-display processing difficulties are the result of a failure to consider man and computer together as a cognitive system. (Woods 1984, pp. 229–230)

As emphasized throughout this book, the domains within which human agents are asked to work are complicated ones. Today the computer is a window into large, complex work spaces and databases. As a direct consequence, the interfaces that provide decision support for these agents will also need to be complicated (i.e., will need to address the requisite variety). All of the information that is needed for control will rarely be contained on one screen and the human agent will often be required to navigate through a set of screens.

The potential problems that can be introduced by these navigational requirements were clearly illustrated during observations we made as part of the human system integration evaluation team for air force exercises at Nellis AFB to evaluate new technology for combat air operation centers in 2006. The tanker officer, whose job was to coordinate refueling requests not previously scheduled in the air tasking order, indicated that he could not complete any pairing requests (scheduling a tanker to meet with an aircraft for refueling) made that day. It turns out that the newly designed software suite required a number of work-arounds to complete this task. Navigation across several windows was required; opening some windows completely locked out view of or access to other windows.

Thus, in order to get information to answer queries on one window, the operator had to close the window, navigate back to an earlier window (often several layers back) to get the information, and then find his way back to the window with the query. He often had to jot down the information using pen and paper so that he did not forget it during navigation back to the query (which often included interruptions from other demands). At the end of the 8-hour shift, no pairings requests had been satisfied. As the tanker officer observed, "My world operates at 15 miles per minute." He explained that for every minute delay in making a satisfactory pairing, the separation between the aircraft needing fuel and the tanker might increase by as much as 15 miles.

As this example clearly illustrates, navigational requirements introduce an additional set of constraints on human–computer interaction. The agent must know her current location within the work space, the space of possible locations to which she might navigate, and the navigational resources that will allow her to do so. This introduces the problem of navigating through what can be a vast information space. This involves navigating within the interface (from one representation or window to another) and navigating within the functional space (exploring the space of possibilities in the work domain). Because of the law of requisite variety, it will typically not be wise to minimize (i.e., trivialize) the complexity of the work domain. However, designers should make every effort not to add to this complexity by creating clumsy organizations of information in the interface.

Support for these navigational requirements is often missing or inadequate. A commonplace electronic device, the mobile phone, will be used to provide another illustration of the problem and a more precise description of some of the consequences that arise. As described in Chapter 13, the functional capabilities of mobile phones have increased over time while their physical size has decreased. The combination of increased functionality and decreased display surface produces an excellent example of the "keyhole" effect described by Woods (1984). All of the relevant information required to operate a cell phone cannot possibly be displayed in parallel; the agent must selectively view small portions of information serially, over time, through the limited keyhole produced by the small display surface.

Historically, cell phone interfaces have provided very little support for dealing with these navigational problems. Control of mobile phones has traditionally been menu driven; these menus were sequential, linear, and alphanumeric in the earliest cell phones. The overall structure of the menu system (some variation of a tree structure) and its complexity (many branches of varying depth) exacerbate the problem. Although the physical characteristics of the interface have evolved (e.g., increased graphics capabilities), the basic navigational problems remain the same. There is very little contextual information that assists the agent in identifying where he or she is within this complex structure.

The end result is what Woods (1984) referred to as the "getting lost" phenomenon: "[U]sers often do not know where they are, where they came from, or where they have to go next" (Ziefle and Bay 2006, p. 395). This is often exacerbated by the fact that the functional needs and expectations of users are often not considered in the design of the hierarchical organization of the menus. That is, common functions may be "hidden" in the menu structure where they are not expected to be and where they require users to waste extra keystrokes to get to them.

Woods (1984) was one of the first researchers to call attention explicitly to these challenges. A quarter of a century has passed since the publication of that article. Despite significant advances in interface technologies (or perhaps due to them!), problems like the keyhole effect and the getting lost phenomenon have become even more commonplace. Cell phone interfaces are but one example. Being lost in the Web pages of the Internet, the screens of a process control interface, or the menus of consumer electronics are quite common situations in today's electronic age.

## 15.2 Visual Momentum

The fundamental problem in interface design that has been described thus far can be conceptualized, in an overly simplified fashion, as a problem in parallel versus serial presentation of information. The combination of domain complexity and limited display real estate produces the requirement for the serial presentation of information and the associated burden that is placed upon the agent to integrate information across successive views.

Woods (1984) coined the term "visual momentum" (VM) to refer to the extent to which an interface supports the agent in dealing with these problems: "a measure of the user's ability to extract and integrate information across displays, in other words, as a measure of the distribution of attention" (p. 231). Woods makes a comparison between visual momentum in the interface and scene transitions in film editing. An interface with high visual momentum will support transitions between views that are similar to "a good cut from one screen or view to another in film editing" (Woods 1984, p. 231). In contrast, an interface with low visual momentum "is like a bad cut in film editing—one that confuses the viewer or delays comprehension. Each transition to a new display then becomes an act of total replacement ... both display content and structure are independent of previous 'glances' into the data base" (Woods 1984, p. 231).

In some sense, this is consistent with the idea of positive transfer discussed in Chapter 12, except here we are not talking about transfer between two different domains, but rather from one window or view within the same domain to another view. Thus, when visual momentum is high, the expectations and habits created in one window or view are easily realized in the second window, leading to smoothly coordinated transitions. When visual momentum is low, the expectations created in one view are violated in the second view, potentially leading to clumsy transitions (e.g., surprise, error, and frustration).

### 15.2.1 Design Strategies

Woods (1984; Watts-Perotti and Woods 1999) proposed a number of general design techniques that can be used to increase visual momentum; these are presented in Figure 15.1. The relative position of a technique on this continuum represents its potential to increase visual momentum ranging from low (left) to high (right). Each will now be considered in turn.

#### *15.2.1.1 Fixed Format Data Replacement*

The first technique is referred to as the "fixed format data replacement" technique. A simple and common example of this technique is spatial dedication. The various objects (e.g., menus, commands, categories of information, display frames, labels, etc.) in the interface should be assigned to specific and consistent physical locations across views. This design convention makes accessing information and effecting control input simpler: Agents will learn where specific objects and the associated information are located as they gain experience with an interface. This strategy might also be described as consistent mapping. As discussed in Chapter 4, consistent mapping can lead to the development of automatic processing (or rule-based shortcuts across the decision ladder) that can improve the efficiency of human performance.

However, the signature variation of this design technique is one that allows an agent to change the informational content of a display selectively

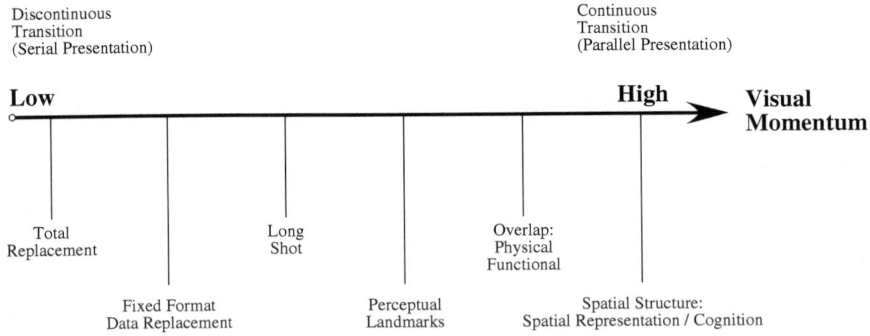


**FIGURE 15.1**
Design techniques that can be used to increase visual momentum in the interface. (Adapted with permission from Woods, D. D. 1984. *International Journal of Man–Machine Studies* 21:229–244. Copyright 1984 by Elsevier Limited. All rights reserved.)

without changing the viewing context within which it resides (e.g., general format, axes, units, labels). For example, when data plots from two different experiments or experimental conditions are compared, visual momentum can be improved if the data are plotted on common scales. This can be particularly useful when switching back and forth between two slides in a PowerPoint-style presentation. In this case, the differences in the data can be very salient as apparent motion against the fixed background scales. Thus, the agent can change the current "selective glance" into a larger set of data or information.

Visual momentum is increased in at least two ways. First, the new informational content is available in parallel within the original viewing context (i.e., the agent does not need to navigate to an alternative view, remember the new data, and then return to consider what they mean). Second, the transition between old and new information is facilitated by the common viewing context; no additional cognitive effort is required for reorientation and the focus can be on changes in the data instead (see a similar logic put forth by Tufte, 1990, with regard to the "small multiples" technique). This technique represents one very effective way to deal with the trade-offs imposed by the combination of limited display real estate and large amounts of data.

#### 15.2.1.2 Long Shot

The long-shot design technique is used to provide an overview of display relationships. Lower level elements are organized into a coherent whole that explicitly specifies structural relationships (i.e., how the lower level elements relate to the whole as well as how they relate to each other). Information about the elements may need to be abbreviated (e.g., overall status may be represented). One example of a long shot is the "site map" of an Internet Web site that specifies its overall structure. Another example is a cartographic map providing key features of the ecology and the spatial relationships between them (e.g., orientation, direction, distance).

For a long shot to work it must "explicitly incorporate a set of interdisplay relationships that are important to the user's tasks to be portrayed in the long shot. Merely summarizing data is insufficient for effective across-display information extraction" (Woods 1984, p. 236). Moreover, to be truly effective, a long shot "must also be seen in parallel with detailed views" (Watts-Perotti and Woods 1999, p. 274). This technique supports visual momentum because an abstract representation of the global "space of possibilities" of the interface is provided.

#### 15.2.1.3 Perceptual Landmarks

A third design strategy with the potential to support visual momentum is to provide perceptual landmarks in the interface. Siegel and White (1975) described real-world landmarks in the following manner:

> Landmarks are unique configurations of perceptual events (patterns). They identify a specific geographical location. The intersection of Broadway and 42nd Street is as much a landmark as the Prudential Center in Boston … . These landmarks are the strategic foci to and from which the person moves or travels …. We are going to the park. We are coming from home. (p. 23)

This concept is equally relevant to display and interface design. Perceptual landmarks in the interface are unique perceptual configurations that (1) occupy dedicated spatial locations and (2) are associated with meaningful aspects of the underlying work domain. Perceptual landmarks at the work space level will increase visual momentum by facilitating navigation between views: "Clear landmarks help the viewer to integrate successive displays by providing an easily discernible feature which anchors the transition, and which provides a relative frame of reference to establish relationships across displays" (Woods 1984, p. 236).

Perceptual landmarks at lower levels in the interface will include visual elements associated with both the viewing context (e.g., nondata elements such as display axes) and the data (e.g., bar graphs). For example, a set of common baseline or normative data might be included in plots from multiple experiments or conditions as a common reference to facilitate comparison across the different plots. This might be particularly valuable if the data are being compared across different scale transformations. The baseline data creates an anchor to help visualize the mapping from one scale to another and as a reference for comparing data in one plot against data in another. These landmarks will serve as visual features that guide exploratory visual attention within views.

It is important to note that the various techniques can be combined to enhance display effectiveness. For example, making a particular location salient as a landmark may be very useful in relating a reduced scale long-shot display with one's current position within a multiwindow environment. In fact, this is often done by making the current position a salient landmark within the long-shot display (e.g., "you are here" indication).

#### 15.2.1.4 Display Overlap

The display overlap design strategy involves the use of a supplemental viewing context to frame the presentation of information. This context facilitates understanding and interpretation by providing additional information that testifies with regard to physical contiguity, functional similarity, or other meaningful relations. One illustrative example comes from cartography: An area of interest (e.g., the state of Ohio) is presented in a detailed map situated within a larger, abbreviated context (e.g., partial maps of the surrounding states). A second example is the use of overlays, where contextual viewing information is superimposed onto an existing display framework.

For example, interactive computerized weather maps may have the capability to overlay several different types of information (e.g., precipitation, clouds, temperature, wind speed, UV index, etc.) over the same geographical area.

In summary, the display overlap technique places information of interest within a larger viewing context that explicitly illustrates meaningful relations. Visual momentum is increased because relational information is explicitly available in parallel (i.e., within the currently visible display or view) as opposed to serially (i.e., searching for the interconnections across other displays or views, remembering it, and returning to the original display or view).

#### *15.2.1.5 Spatial Structure (Spatial Representation/Cognition)*

The final two design techniques to increase visual momentum described by Woods (i.e., spatial representation and spatial cognition) appear to be very closely related and will be treated together. As described throughout this book, the process of graphical interface design generally involves incorporating (or imposing, in the case of nonspatial data) a degree of spatial structure and organization into the controls and displays of an interface. As a result, the powerful perception–action skills of the human agent are leveraged and "the process of finding data becomes a more automatic perceptual function rather than a limited-capacity thinking function" (Woods 1984, p. 239). This is, of course, a major focus of the book.

The spatial structure design technique can also refer to a more comprehensive use of spatial organization (Woods 1984):

> Spatial organization translates the normative user internal model into a perceptual map. The user sees, rather than remembers, the organization of data in the system and can move within the system just as he moves in an actual spatial layout .... One spatial access technique is to organize the data base as a topology and then provide the viewer with a mechanism to move through the space ... inter-display movements can be conceptualized as itineraries or paths through the space ... the user's perceptual and attentional skills ... can be supported ... by constructing a conceptual or virtual space. (pp. 238–240)

Using spatial structure in this fashion leverages powerful perception, action, and cognitive skills; interacting with an interface becomes much like interacting with a real ecology (e.g., an environment). Thus, the skills developed in interacting with physical ecologies can be transferred to facilitate interactions with information technologies.

### 15.2.2 Interface Levels

The previous descriptions of design strategies for increasing visual momentum were quite abstract. This was necessary because they are relevant to design across the various levels of the interface introduced in

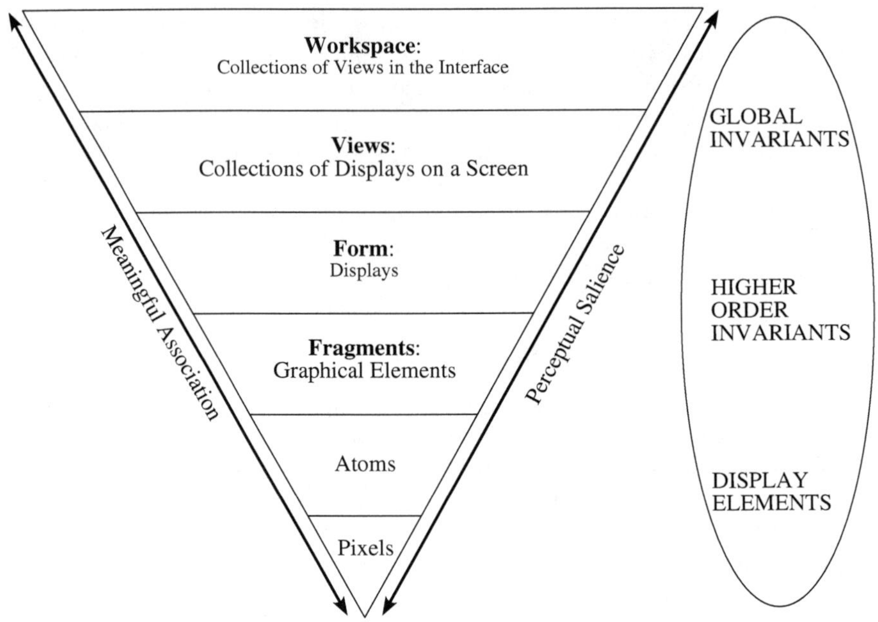


**FIGURE 15.2**
The hierarchically nested levels of structure that comprise an interface. (Woods, D. D. 1997. The Theory and Practice of Representational Design in the Computer Medium. Columbus: Ohio State University, Cognitive Systems Engineering Laboratory.)

Chapter 6 (Woods 1997; see Figure 15.2 for a summary of these levels). The remainder of this chapter will provide concrete examples that illustrate these visual momentum techniques and their application at the top three levels of the interface.

## 15.3 VM in the Work Space

The need to support visual momentum is clear when an interface has multiple screens. Navigation at the work space level involves transitions between views: One view (i.e., a collection of displays on a screen) is replaced with another view. The degree of visual momentum in an interface is determined by the "smoothness" of these segues. Does the agent know where they are, where they have been, and where they might want to go? Does the interface prepare the agent (i.e., set up an appropriate set of expectations) for this transition by perceptually and/or cognitively orienting the agent to the particulars of the new destination? We will now examine the use of the various techniques to increase visual momentum at the work space level using concrete examples from two previously discussed interfaces (BookHouse and iPhone®).

### 15.3.1 The BookHouse Interface

> The visual design of interfaces for the casual user must be based on a conception of mental models that the user has adopted during other familiar activities .... Consequently the visual design of the BookHouse interface is based on a pictorial representation of concepts and objects that are familiar to the general public .... The BookHouse metaphor is a functional analogy to a local library .... The user "walks" through rooms with different arrangements of books and people .... It gives a familiar context for the identification of tools to use for the operational actions to be taken. It exploits the flexible display capabilities of computers to relate both information in and about the database, as well as the various means for communicating with the database to a location in a virtual space. (Rasmussen, Pejtersen, and Goodstein 1994, pp. 288–289)

The first example will be the interface designed for the BookHouse system (Pejtersen 1980, 1988, 1992). This system is used in a library setting to assist patrons in finding a book of fiction. The process of searching for a book of fiction is equated to acts of navigation through a virtual library. The visual appearance of the interface during a prototypical search is illustrated in Figure 15.3. The initial view (Figure 15.3a) reinforces the global metaphor: The most visible object on the screen is a building constructed from books (i.e., the virtual library). The agents in the view are looking toward or entering the library, suggesting that the first step required to find a book is to enter the virtual library.

The agent enters the virtual library by clicking upon the entrance and subsequently arrives at a second view illustrated in Figure 15.3b. This view portrays an entrance hallway with thresholds to three adjoining rooms. Each room corresponds to a fundamentally different category of search (i.e., book holdings, terms) that is specialized for children (left room), adults (right room), or both (center room). The visual appearance of the local spatial metaphors (i.e., the patrons) graphically reinforces these choices. The act of navigating through one of these three doorways (via point and click) constitutes a decision to perform a particular kind of search.

A third view is illustrated in Figure 15.3c, the result of navigating through the rightmost room (a control input specifying a search of the adult section). This view depicts a room in the virtual library with patrons engaged in various types of activities. Each combination of patron and activity represents an alternative search strategy that can be executed. The four strategies are to search by (1) analogy (patron on the left), (2) random browsing (patron on the right), (3) analytical (patron at desk), and (4) visual thesaurus (patron at easel). Choosing one of these patrons (i.e., a point and click on a virtual patron) constitutes a decision to execute the associated search strategy.

A fourth view, the result of a control input specifying an analytical search, is illustrated in Figure 15.3d. This view depicts a specific area in the virtual library. The local spatial metaphors in this room (e.g., the globe, the clock,

(a) (b) (c) (d)

**FIGURE 15.3**
**(See color insert following page 230.)** An example of the spatial structure design technique: hierarchically nested metaphors to support navigation in the BookHouse fiction retrieval system. (a) The global spatial metaphor (a virtual library; work space level). (b) A nested, intermediate spatial metaphor (i.e., the entrance hallway to select a database). (c) The search room (search strategy). (d) The analytical search room with local spatial metaphors (i.e., graphical components of interface icons) that represent search parameters. (Screenshots from the BookHouse system. Used with permission from system designer, Pejtersen, A. M. 1992. In *The Marketing of Library and Information Services 2*, ed. B. Cronin. London: ASLIB. Copyright 1992. All rights reserved.)

the eyeglass icons) represent the various dimensions that can be specified during the process of conducting an analytical search (e.g., geographical setting, time period, font size). The extensive use of design strategies to increase visual momentum in this interface will now be described.

#### *15.3.1.1 Spatial Structure*

The BookHouse interface provides a clear example of the most powerful technique for visual momentum: spatial structure. It provides a nested hierarchy of spatial metaphors ranging from global (virtual library, work space level) and intermediate (areas or rooms in the virtual library, views level) metaphors to local (objects in the rooms, forms level) metaphors. This provides a set of overarching spatial constraints that serve to organize the interaction.

The interface supports the casual user by leveraging preexisting knowledge and skills. The actions that are required to interact with the system (i.e., book searches) are related to commonplace, natural activities carried out constantly in everyday life (e.g., navigation through buildings). Different activities that are required to conduct a book search (e.g., choose a database or search strategy) are associated with particular areas within the virtual ecology (i.e., the various rooms). Furthermore, the specific actions required to conduct a search (e.g., specify search parameters) are associated with local metaphors in the interface (e.g., globes, glasses, clocks) that are also commonplace in everyday life.

#### *15.3.1.2 Display Overlap*

A key design feature supporting visual momentum in the BookHouse interface is the "navigation bar" located in the top portion of the screen. This area of the interface is used to capture an agent's navigational path through the virtual library: Each time the agent leaves an area of the virtual library a small-scale replica of that view (i.e., a "replica" spatial metaphor) is added to the navigation bar. For example, in Figure 15.3 there are no replica metaphors in the navigation bar initially (15.3a) and there are three replica metaphors after navigation to three rooms has occurred (15.3d). Thus, the navigation bar is a display that provides a historical record of the unfolding search.

This is a particularly creative instantiation of a more general display overlap technique known as the "breadcrumbs" strategy (referring to the trail of breadcrumbs left by Hansel and Gretel). Pardue et al. (2009) describe perhaps the most familiar use of this technique, one that is encountered in Web browsers:

> With the problem of becoming disoriented in large information spaces, orientation cues have been used to guide users, and found to be important for effective navigation of spatial maps, for example (Burigat and Chittaro 2008). One way to provide orientation cues in navigational hierarchies is with breadcrumbs (Bernstein 1988). A breadcrumb is a metaphor describing the practice of marking the path the user has taken. In Web-based hypertext systems, the taken path has come to be represented by location breadcrumbs, defined as a text-based visual representation of a user's location within the navigational structure (Teng 2003). Location breadcrumbs are usually implemented as a list of links, each separated by a character such as an arrow or ">" to indicate the direction of the navigation. For example, you are here: Home > Grocery > Pasta & Grains > Pasta > Spaghetti. (p. 235)

Thus, the navigation bar (and, more generally speaking, the breadcrumbs technique) is an excellent example of the display overlap technique. The replica metaphors represent successive "waypoints" in the agent's navigational path through the virtual library. This provides one perspective of the spatial relations within the virtual ecology: a serially ordered, linearly arranged sequence of spatial metaphors representing the physical interconnections between rooms. This constitutes a context-sensitive "map" that, in principle, is very similar to the cartographic overlap technique described earlier (i.e., state map in an atlas); it provides information about the structural, physical relationships between views.

#### *15.3.1.3 Perceptual Landmarks*

Notice that the breadcrumbs in the BookHouse navigation bar are graphical, as opposed to textual. Notice also that these breadcrumbs pull "double duty" since they also serve as controls; the agent can navigate back to a previously visited room by pointing and clicking on a replica metaphor (icon) in the navigation bar. This feature allows the agent to change the nature of a search quickly by returning to any of the previously visited areas.

Thus, the navigation bar illustrates an excellent implementation of the perceptual landmarks design technique to increase visual momentum. The most prominent perceptual landmarks of the interface are those associated with the areas of the virtual library that can be navigated to (i.e., alternative views in the interface). Each of these physical locations (e.g., the entrance hallway) has a distinctive visual appearance (e.g., the three thresholds in the entrance hallway).

The replica metaphors in the navigation bar preserve these distinctive visual features, thereby providing a set of perceptual landmarks. They facilitate transitions between views (i.e., support visual momentum) by providing a visual preview of potential destinations in the interface that will appear if the icon (i.e., breadcrumb) is clicked. This preview orients the agent, both visually and cognitively, to the ensuing transition, thereby supporting transitions between views. These perceptual landmarks specify where the agent is in the virtual ecology, where he or she has been, and where he or she might go back to.

#### *15.3.1.4 Fixed Format Data Replacement*

The navigation bar also qualifies as an example of the fixed format data replacement technique. It is located in a particular area of the interface (i.e., spatial dedication) and therefore limits the effort required to locate the associated information. Furthermore, the data elements that are displayed in the navigation bar (i.e., the replica metaphors) are variable since the sequence of metaphors that appears is dependent upon the specific navigational path of the agent (i.e., data replacement). This technique supports visual momentum by providing one solution to the inherent trade-off between large amounts of data and limited display real estate (and the associated keyhole phenomenon).

An alternative solution would have been to use the long-shot design technique to show the spatial relationships between views via a "floor plan" of the virtual library. This would have provided additional spatial information, including orientation and distance of the rooms relative to each other, but would also have required larger amounts of valuable display real estate. The navigation bar constitutes an effective compromise: Key spatial relationships are retained (mitigating the keyhole phenomenon) while valuable display real estate is conserved.

### 15.3.2 The iPhone Interface

The second set of examples illustrating support for visual momentum at the work space level will be drawn from the iPhone interface. The fundamental aspects of iPhone navigation were described in Chapter 13. To reiterate briefly, between-application navigation is accomplished via the navigation matrix (e.g., the 4 × 4 matrix of application metaphors in the middle of the screen in Figure 13.3 in Chapter 13), the navigation row (e.g., the row of metaphors at the bottom of the screen in Figure 13.3), and the home button. Within-application navigation (i.e., between application modes) is accomplished via mode metaphors that replace the application metaphors in the navigation row (see Figure 13.9).

One aspect of the iPhone interface not discussed in Chapter 13 is the capability to add applications and additional home screens. An iPhone interface with one home screen and one navigation matrix (with 13 application metaphors) is illustrated in Figure 15.4a. A new home screen can be created either automatically (e.g., by exceeding 16 applications) or manually. The latter method is accomplished by pressing, holding, and then dragging a metaphor off the right edge of the screen. In Figure 15.4b the agent has applied this control input to the App Store icon and a new, second home screen has been created with a navigation matrix containing one metaphor (see Figure 15.4b).

Transitions between multiple home screens is achieved via the drag gesture described in Chapter 13: The agent positions a finger on one side of the currently visible matrix and then drags it across the iPhone display to pull a new matrix into view. For example, a left-to-right drag across the new home screen depicted in Figure 15.4b would result in navigation back to the original home screen depicted in Figure 15.4c. Note the changes in this home screen (i.e., no App Store metaphor). The iPhone interface will now be analyzed in terms of the design strategies to support visual momentum.

#### *15.3.2.1 Spatial Structure*

The single consistent real-world "cover story" that unifies the hierarchically nested metaphors in the BookHouse interface was discussed earlier. Mullet and Sano (1995) point out the possibility that the success of this approach may be the exception rather than the rule:

> The most spectacular failing of simplicity is often seen in those products trying most earnestly to simplify the GUI for non-technical users.

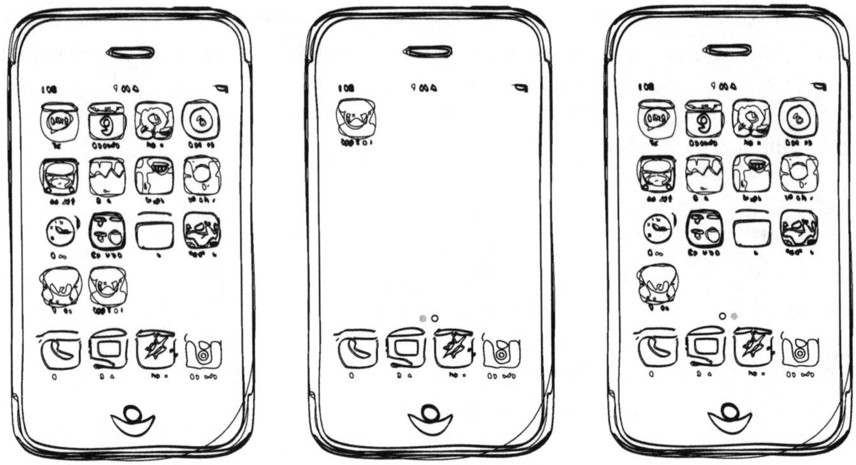

(a) (b) (c)

**FIGURE 15.4**
Navigation in iPhone interface via multiple home screens (i.e., between-application navigation). (a) The original interface configuration with one home screen. (b) Manual addition of a new home screen and the appearance of the dot indicator for multiple home screens. (c) Navigation back to original home screen (note change in dot indicator display).

> Applications … attempting to leverage users' knowledge about the physical world through a "3D Office" (or 3D world) metaphor are beginning to reach the marketplace. This approach has always been something of a rite of passage for GUI designers …. The extremely literal translation of the "real" world … virtually ensures that users will find the resulting environments cumbersome and inefficient. (p.36)

The iPhone interface avoids these difficulties by providing a nested hierarchy of spatial structure and organization that is more general in nature. The spatial structure at the global level (i.e., work space) consists of the multiple home screens. The spatial structure at an intermediate level (i.e., views) consists of portions of the screen dedicated to navigational requirements (e.g., navigational matrix and row, home button) as well as the idiosyncratic structure of the various application modes that can be navigated to. The spatial structure at a local level (i.e., forms) consists of the icons and the spatial metaphors that represent these applications and modes (along with various other types of interface objects).

Collectively, these structures in the interface provide a set of overarching spatial constraints that support navigation within the iPhone work space. To be sure, it is unlike the spatial structure provided in the BookHouse in the sense that it is not necessarily tied to a real-world setting. On the other hand, in many ways the spatial structure and navigational resources in the iPhone and BookHouse interfaces are very similar. The home screens of the iPhone provide global structure that is conceptually similar to that provided by the library; the various application modes provide intermediate spatial structure much like the alternative rooms in the library. The local metaphors serve the same functional purposes in both interfaces. Thus, interaction with both devices is equated to navigation within a spatial structure; the differences in implementation are one of degree, rather than substance.

#### *15.3.2.2 Long Shot*

One of the positive features of the iPhone is the open architecture that allows third-party applications to be added; literally thousands of them are available. The total number of applications that can reside on the iPhone at any one point in time is fairly large (between approximately 100 and 200, depending upon the model), but the number of application icons that can be viewed simultaneously in a single navigation matrix is limited to a total of 16. Thus, the iPhone has increased functionality (i.e., a larger database to glance into) and a greater potential for keyhole effects than most other mobile phones.

The long-shot design technique is used to offset keyhole effects in the iPhone. Additional navigation requirements are imposed when the number of applications on the iPhone exceeds 20 (the maximum number of applications supported by a single home page). Figure 15.4 and the associated text describe the process of creating and navigating between additional home pages. Note that when a second home screen was created (see Figure 15.4b), a new visual indicator appeared in the iPhone interface (compare to Figure 15.4a) located between the navigation matrix and the navigation row. This will be referred to as the "dot indicator" display since it comprises a series of dots.

The number of dots indicates the number of home screens that are currently available (in this case, two); the highlighted dot indicates which of those home screens is the currently active home screen (in the case of Figure 15.4b, the second home screen). Thus, the dot indicator is one instantiation of the long-shot technique used to increase visual momentum in the iPhone. It provides a summary display that informs the agent with regard to how many home screens are available and where the agent is currently located within that space of possibilities.

#### *15.3.2.3 Fixed Format Data Replacement*

The fixed format data replacement design technique is utilized extensively in the iPhone to mitigate the keyhole effect. The most obvious use of this technique lies in the navigation matrix. This interface structure is spatially dedicated (i.e., consistently located in the same area of the interface); its content is variable. Navigation to a different home screen produces a clear example of data replacement in the navigation matrix; the application metaphors of the old home screen are replaced with those of the new home screen (see Figure 15.4). Although somewhat less obvious, the same technique is also used in the navigation row. The application icons in this spatial structure are consistent across all home screens. However, these application icons are replaced with mode icons when navigation to one of the core applications of the iPhone (i.e., Phone, iPod, iTunes, Clock, and App Store) occurs. Thus, the fixed format data replacement technique is used to increase visual momentum by providing alternative, manageable glances into a larger database of applications.

#### 15.3.2.4 Perceptual Landmarks

The spatial metaphors located within the navigation matrix and the navigation row provide excellent examples of the perceptual landmarks design technique. They serve exactly the same purpose as the replica metaphors in the navigation bar of the BookHouse: They are landmarks that represent the places in the interface that can be navigated to. In this case, these places are the screens of the various applications, modes, and the associated functionality. These spatial metaphors facilitate transitions between views by providing a conceptual "preview" of potential destinations, thereby orienting the agent to the ensuing transition, supporting transitions between views, and increasing visual momentum.

Note that although the functional purpose is similar, the underlying nature of these perceptual landmarks is qualitatively different from those that appear in the BookHouse. The visual appearance of the replica metaphors is based on physical similarity to the associated destination. In contrast, the perceptual landmarks provided in the iPhone are symbolic references to the associated view. Thus, the initial success or failure of these landmarks in supporting visual momentum depends upon the issues in metaphor design discussed in Chapter 12 and elsewhere (i.e., assimilation).

### 15.3.3 Spatial Cognition/Way-Finding

Both of these examples illustrate the spatial structure design strategy to increase visual momentum at the work space level. The ultimate goal is to support the powerful, perhaps innate, processes associated with spatial cognition. This technique uses interface resources to impose a high degree of organized spatial structure that will allow agents to navigate between the views in a work space much like they navigate through man-made (e.g., a familiar building) or natural ecologies (e.g., from home to work).

The general cognitive skills into which this design technique might tap have been studied extensively. Terms used to describe knowledge about an ecology include cognitive maps (e.g., Tolman 1948; Kosslyn, Pick, and Fariello 1974; Kitchin 1994), spatial knowledge (e.g., Siegel and White 1975; Herman and Siegel 1978), spatial thinking (Gauvain 1993), and environmental cognition (e.g., Evans 1980).

More specific factors involved in the capability to navigate through the ecology have been identified in the literature on way-finding (e.g., Lynch 1960; Passini 1980; Chen and Stanney 1999; Hutchins 1995). One of the major concepts identified in this literature is the distinction between three different types of spatial knowledge: landmark, route, and survey (e.g., Siegel and White 1975).

Landmark knowledge is a term used to describe an agent's memory for specific locations or features within the ecology. As previously described, this refers to aspects of a specific location in the environment that are distinctive in their own right (e.g., the red water tower) or that have become distinctive by virtue of their importance in navigation (e.g., the street sign for Arthur Avenue).

Route knowledge refers to an agent's knowledge about a specific path through the ecology. This includes a start point, an end point, and a collection of landmarks, turns, and distances that define that particular path.

Survey knowledge refers to an agent's knowledge about the overall layout of the ecology. This includes a comprehensive representation of the landmarks, their spatial locations within the ecology, and directional relationships and relative distances between them, as well as other topographic, topological, and spatial relationships inherent to the ecology.

This literature has focused primarily on the role of internal knowledge (e.g., memory for landmarks, specific routes, and the ecology as a whole) and how it is used in the process of navigating within an ecology. Note that navigational information can also be represented externally (e.g., landmarks, driving instructions, and a map). In fact, one might consider the degree of support (i.e., external representations) for these three types of spatial knowledge as a set of criteria to judge the effectiveness of the spatial structure design technique. The ways in which these two interfaces support spatial cognition as applied to way-finding (i.e., landmark, route, and survey knowledge) will now be discussed.

#### 15.3.3.1 Landmark Knowledge

One goal in supporting navigation at the work space level is to provide external support for landmark knowledge. This goal translates into the requirement to provide visual landmarks that represent the various views that can be navigated to (and signs to the associated functionality located there). As discussed in the previous sections describing perceptual landmarks, the replica metaphors in the BookHouse navigation bar and the local metaphors (icons) in the iPhone navigation matrix and navigation row accomplish this goal. They facilitate navigation directly and close the perception–action loop: They specify destinations (serving as a display) and ways to get there (serving as controls). They also serve as visual reminders of the associated functionality (supporting more effective recognition memory, as opposed to recall memory). Finally, they support the formation of internal landmark knowledge about the layout of the spatial structure of the interface.

#### 15.3.3.2 Route Knowledge

A second goal in supporting navigation at the work space level is to provide external support for route knowledge. This goal translates into the provision of visual evidence that testifies with regard to the sequence of views (i.e., the navigational routes) that will be required to complete a particular goal or task. Both interfaces provide external resources that serve to specify these "navigational routes."

The BookHouse interface specifies route knowledge by equating general search options and specific types of searches to specific navigational paths through the virtual library. In some cases (e.g., Figure 15.3a and 15.3b), the next "turn" to take is literally represented as a navigational path in the virtual library (i.e., a door to enter). In other cases (e.g., Figure 15.3c), key portions of the current view (i.e., the desk and surrounding objects) serve as signs that mimic the structure of an adjoining virtual room (Figure 15.3d) that will be visited if control input is applied (i.e., clicking on the desk). Finally, the sequence of breadcrumb metaphors in the navigational bar directly specifies the exact route required to execute a particular search.

The need to support route knowledge in the iPhone interface is greatly diminished. All navigational paths are extremely simple, due to the "wide and shallow" spatial structure of the interface. Any route to any destination will never require that more than two levels of spatial structure be visited (application, mode). The local spatial metaphors in the navigational matrix and row serve as signs that symbolically represent these locations. They support the agent in recognizing the next "turn" or "destination" required.

Collectively, these interface resources provide external reminders of potential routes that are possible when the agent is learning the interface (i.e., they support navigation in their own right). These external representations will also accelerate the formation of route knowledge: Agents will quickly learn the interconnections between structures in the interface (i.e., interface landmarks) and the particular sequence of traversals through them (i.e., "turns") required to effect the associated functionality.

#### 15.3.3.3 Survey Knowledge

A third goal in supporting navigation at the work space level is to provide external support for survey knowledge. This translates into the need for information about the overarching spatial structure of the entire interface, including relationships between the various views. The agents must be provided with visual clues with regard to this overarching structure, including where they have been, where they are currently located (e.g., "you are here"), and where they might be in the future.

Survey knowledge is specified directly by external representations in the iPhone interface. The multiple home screens (and the unique navigation matrix associated with each screen) specify the space of possibilities for the device at the application level. These screens provide the functional equivalent of an atlas with multiple maps and destinations that can be navigated to. The dot indicator provides an overview of this space of possibilities and where agents are currently located (metaphorically speaking, at which page of the atlas the agent is currently looking). The navigation row serves a similar functional purpose. It provides a map of the space of possibilities within an application when it is populated with application mode metaphors. The broad and shallow nature of the high-level spatial structure (i.e., navigation that is constrained to one of two levels—either between or within applications) further contributes to agents' spatial cognition about the device.

One final and very important aspect of the design of the iPhone interface needs to be emphasized, both in terms of its contribution to increased levels of visual momentum and its overall effectiveness. The spatial structure of the interface is extremely broad and shallow (i.e., horizontal), with only two basic levels: application and mode. This stands in sharp contrast to the traditional control structures in other mobile phones and applications (i.e., vertical) that have complicated branches and multiple levels of depth. This design decision increases visual momentum by eliminating one of the primary contributing factors: the getting lost phenomenon.

For example, while getting lost in a traditional mobile phone interface is commonplace, if not inevitable (e.g., Ziefle and Bay 2006), getting lost in the iPhone interface is virtually impossible. The agent is never more than one button press from an alternative mode or an alternative application. The home button of the iPhone means that the person is effectively never lost. With one step, he or she can return to a familiar screen. We believe that, by reducing the risk of getting lost, this design encourages exploration and thus facilitates the accommodative processes of learning by doing.

Survey knowledge is also specified directly in the BookHouse interface. The use of the location breadcrumbs navigational support technique, in combination with an overarching, real-world metaphor, provides an interesting contrast in style to the iPhone. There is no overview display that specifies the overall layout of the virtual library. However, the graphical breadcrumbs in the navigation bar (i.e., the replica metaphors) clearly support the agent in "looking behind" and assessing the navigational structure of the interface that has been traversed.

These breadcrumbs also function as controls, thereby allowing navigational shortcuts within the spatial structure of the interface to occur (i.e., quickly redefine the search strategy at multiple levels: change database, strategy, or parameters). There is also some support provided for "looking ahead" in the navigational structure. The options for the next step in specifying a particular search (e.g., identify a search strategy) are visible in the form of metaphors in the screen (e.g., patron–activity pairs that specify a search strategy; see Figure 15.3).

## 15.4 VM in Views

The examples that have been provided so far illustrate the most obvious need to support visual momentum in interface design: when the visual transitions occur between the views of a work space and the ensuing data replacement is total. In fact, the need to support visual momentum also occurs at lower levels in the interface. Woods (1984) makes this point explicit:

> It is important to note that across-display integration can refer to two types of display transitions: successive views across different units within the data base (*different fields of view within a single representation*) and successive shifts in the kind of view or representation of a single data unit (or, for that matter, shifts in representation across the entire data base). (p. 235; emphasis added)

The reason that there is less appreciation for the need to support visual momentum at lower levels in the interface lies in our phenomenological experience of a stable and unbounded visual world (Gibson 1966):

> By stability is meant the fact that it does not seem to move when one turns his eyes or turns himself around … and that it does not seem to tilt when one inclines his head. By unboundedness is meant the fact that it does not seem to have anything like a circular or oval window frame. The phenomenal world seems to stay put, to remain upright, and to surround one completely. This experience is what a theory of perception must explain. (p. 253)

Thus, the subjective impression that one experiences when looking at a computer screen (i.e., a view) is one of parallel data presentation. However, this impression does not match the anatomical and physiological facts. There is a blind spot and the eyes are in continual motion (i.e., tremors, saccades, compensatory and pursuit movements). Furthermore, the visual system is capable of registering only a small portion of the surrounding world at a particular point in time. The forward-facing placement of the eyes limits the field of view to an oval that is approximately 180° horizontally and 150° vertically (Gibson 1966). Visual acuity within this field of view varies substantially across foveal and peripheral vision (Hochberg 1964): "Only in the foveal region of the retinal image [approximately 2°, or the size of a thumbnail held at arm's length] can details be made out. As we depart from the central fovea, visual acuity deteriorates alarmingly; by 5°, it has dropped 50 per cent" (p. 25).

From this discussion it is clear that the visual system can only obtain information about the world by scanning it in a serial fashion: coordinated movements of the body, head, and eye produce a sweeping of the external world via a series of successive fixations (i.e., exploratory visual attention). Thus, the problem of visual momentum is also applicable to lower levels of the interface (see Figure 15.2). The fundamental design challenge is how to guide successive fixations; the solution involves the provision of effective visual structures (Gibson 1966):

> How are the exploratory shifts of fixation guided or controlled? What causes the eyes to move in one direction rather than another, and to stop at one part of the array instead of another? The answer can only be that interesting structures in the array, and interesting bits of structure, particularly motions, *draw* the foveas toward them. Once it is admitted that the variables of optical structure contain information or meaning, that they specify what their sources afford, this hypothesis becomes reasonable. Certain loci in the array contain more information than others. The peripheral retina registers such a locus, the brain resonates vaguely, and the eye is turned. Subjectively we say that something "catches" our attention. Then the orienting reactions begin and continue until the retino-neuro-muscular system achieves a state of "clearness," and the brain resonates precisely. The focusing of the lens is just such a process and it, of course, accompanies fixation. (p. 260)

### 15.4.1 The RAPTOR Interface

In this section we illustrate how graphical structure can be used to increase the degree of visual momentum at the level of a view in the interface (i.e., within a computer screen). The discussion is organized in terms of the various techniques and the examples are drawn from the RAPTOR interface (see Chapter 14).

#### *15.4.1.1 Spatial Dedication; Layering and Separation*

A well-designed view will consistently place specific types of information (e.g., purpose, function, or mode) in specific locations on the screen (i.e., spatial dedication). The spatial regions of the RAPTOR interface and their informational content are illustrated in Figure 15.5. These regions are perceptually grouped via layering and separation techniques: Light-gray mats are placed on a dark-gray background fill. This increases visual momentum by providing global visual structures that guide the agent's visual attention toward broadly defined categories of information within the view.

#### *15.4.1.2 Long Shot*

The aggregation control/display (see Figure 15.6) utilizes the long-shot display technique to represent the organizational structure of friendly forces at the battalion level. The battalion (BN) is composed of four companies (A, B, C, D); each company is composed of three platoons (1, 2, 3); each platoon has four combat vehicles (represented by standard metaphors; Army 1999).

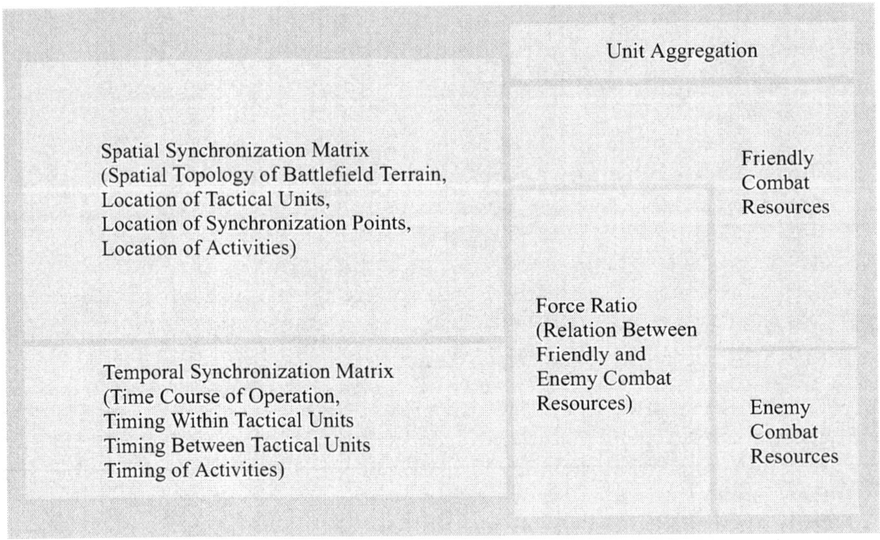


**FIGURE 15.5**
Perceptual grouping, global visual structures, and spatially dedicated information in the RAPTOR interface.

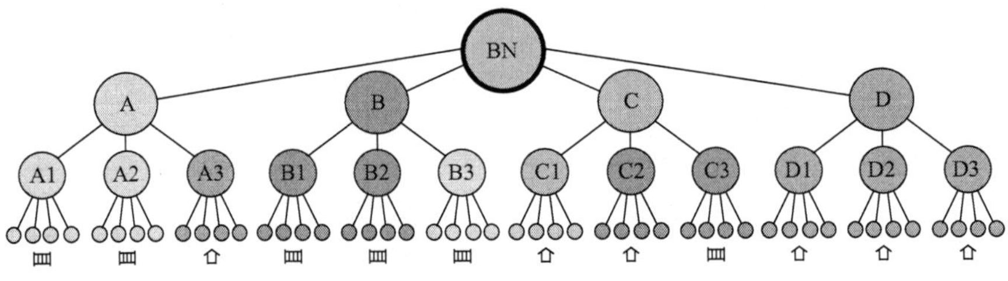


**FIGURE 15.6**
An example of the long-shot design technique: the aggregation control/display of the RAPTOR interface. (Adapted with permission from Bennett, K. B., Posey, S. M., and Shattuck, L. G. 2008. *Journal of Cognitive Engineering and Decision Making* 2 (4): 349–385. Copyright 2008 by the Human Factors and Ergonomics Society. All rights reserved.)

The tree structure captures the hierarchically nested, whole–parts organization of these combat units. The circular nodes in the tree are color coded to reflect the categorical status of the unit or vehicle (green, amber, red, and black, representing successively fewer resources). The currently activated node is perceptually differentiated from the others (higher contrast; outline).

Thus, this is a classic example of the long-shot display technique since it "provides an overview of the display structure as well as summary status data. It is a map of the relationships among data that can be seen in more detailed displays and acts to funnel the viewer's attention to the 'important' details" (Woods 1984, p. 236).

### 15.4.1.3 Perceptual Landmarks

The nodes of this long-shot display also serve as controls: Manipulating a node produces systematic changes in the information that is displayed across the view. For example, in Figure 15.7 the agent has positioned the cursor over the node associated with the unit of interest (i.e., Company B). One set of subsequent visual changes involves the highlighting of all information about that unit, which is scattered across the different regions of the view. This includes the combat resource information that appears in the primary slot of the friendly combat resource display, the spatial location that appears in the spatial synchronization matrix, and the current activity that appears in the temporal synchronization matrix (see Figure 15.7).

The visual highlighting of this related but scattered information constitutes an example of the perceptual landmark design technique to improve visual momentum within a view. Specifically, the highlighted representations

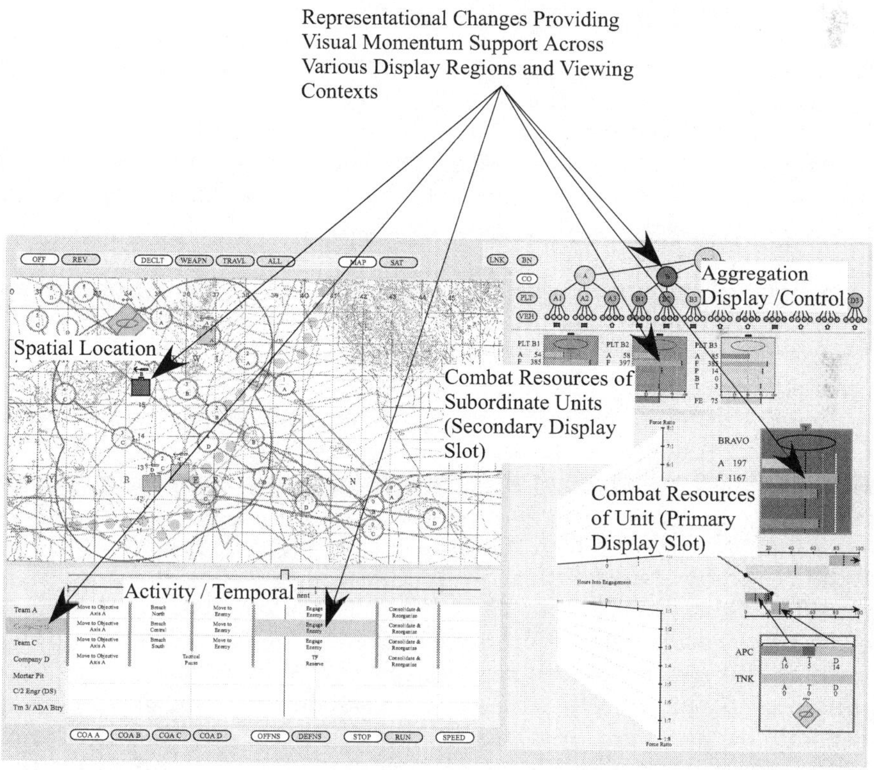


**FIGURE 15.7**
**(See color insert following page 230.)** An example of the perceptual landmarks design technique: visual highlighting of unit-related information scattered across various regions of a view in the RAPTOR interface. (Adapted with permission from Bennett, K. B., Posey, S. M., and Shattuck, L. G. 2008. *Journal of Cognitive Engineering and Decision Making* 2 (4): 349–385. Copyright 2008 by the Human Factors and Ergonomics Society. All rights reserved.)

provide visual structure that will be effective in coordinating successive visual fixations. Without these visual changes, an agent would need to engage in laborious search activities to locate this information within each of the various viewing contexts. In terms of visual search, this would often require the agent to differentiate between target representations and nontarget distracters that are very similar in visual appearance. In contrast, the visual changes in the RAPTOR interface assist the search process by providing physical differentiation between targets and distracters (i.e., by providing perceptual landmarks).

It is important to note that the selection of a node determines what will be shown in a display and, perhaps more importantly, what will not be shown. Thus, each choice can be considered a kind of filter. This is important because one of the common problems with tactical displays is data overload. If all of the basic elements were always displayed on the map, the amount of information could make if very difficult to see anything in particular.

#### 15.4.1.4 Fixed Format Data Replacement

The primary design technique used to support visual momentum for viewing friendly combat resources in the RAPTOR interface is the fixed format data replacement technique. The same fixed display format is used to represent a wide variety of domain objects (organizational units, vehicles, personnel). This supports visual momentum by providing a consistent viewing context that avoids the need for cognitive reorientation to a variety of alternative display formats.

The data replacement component of the technique plays a substantial role in increasing visual momentum. The information requirements associated with friendly combat resources are yet another excellent example of the keyhole effect. These resources are both expansive and structurally complex (i.e., they are distributed across the hierarchically nested organizational structure of the battalion). If one counts units and vehicles only (ignoring individual personnel), there are a total of 65 successive "glances" into the database that need to be supported. This corresponds to one glance for each node in the aggregation display (see Figure 15.6). The simultaneous presentation of 65 displays is simply not an option.

The data replacement technique is implemented via the primary and secondary display slots (see Figure 15.7). Positioning the cursor over a node (i.e., a rollover) in the aggregation control/display places the combat resources of the associated unit in the primary display slot and the combat resources of its subordinate units in the secondary display slot. For example, the primary display slot in each successive graph in Figure 15.8 is occupied by the combat resources for the battalion (Figure 15.8a), a company (Figure 15.8b), a platoon (Figure 15.8c), and a vehicle (Figure 15.8d). Thus, the displays support the agent in selectively narrowing his or her span of attention from the battalion level to the level of an individual vehicle and its personnel; the design strategy increases visual momentum by supporting quick and efficient transitions

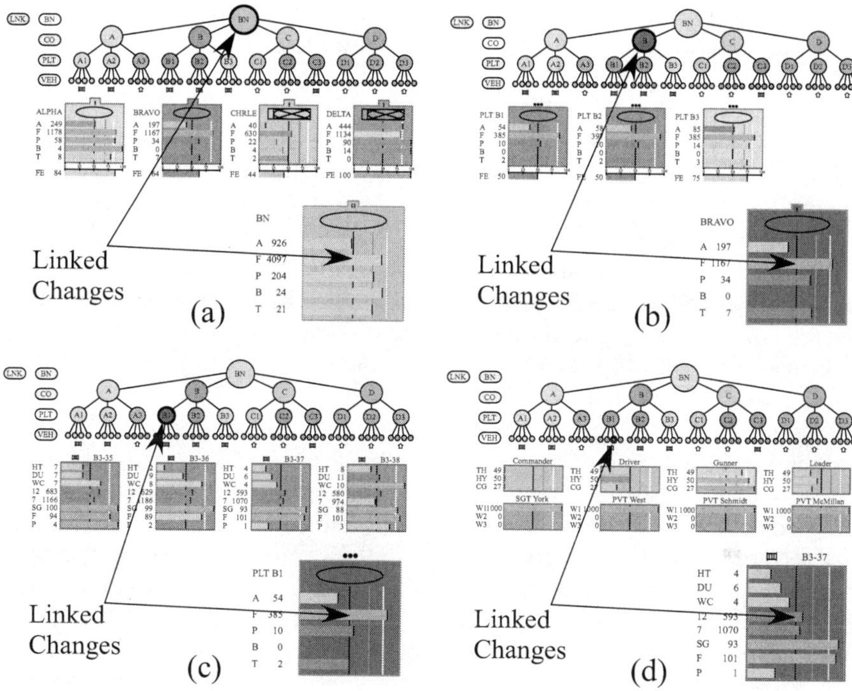


**FIGURE 15.8**
**(See color insert following page 230.)** An example of the fixed format data replacement technique: the primary and secondary data slots are successively occupied by information from different units (i.e., different selective glances into the larger database). (a) Battalion level. (b) Company level. (c) Platoon level. (d) Vehicle level. (Adapted with permission from Bennett, K. B., Posey, S. M., and Shattuck, L. G. 2008. *Journal of Cognitive Engineering and Decision Making* 2 (4): 349–385. Copyright 2008 by the Human Factors and Ergonomics Society. All rights reserved.)

between alternative glances into the overall database. The getting lost phenomenon is avoided by highlighting the particular glance currently displayed in the long shot provided by the aggregation display (i.e., the active node).

### *15.4.1.5 Overlap*

The primary and secondary displays discussed in the previous section also utilize the display overlap technique. The overlap in this case is functional in nature, involving variation along the whole–parts dimension. The simultaneous visual display of combat resources for both a unit and its subordinate units constitutes parallel data presentation across multiple levels of aggregation. This will be useful since the status of individual subordinate units will often vary about the mean of the higher unit (see Figure 15.8a) and these variations may have important implications for planning and execution.

Thus, this example illustrates how visual momentum can be increased by the display overlap technique through a "widening of the keyhole" through which the agent is able to view the underlying database.

A second example of the display overlap technique incorporates functional variation along the means–end dimension (i.e., the abstraction hierarchy). Figure 15.9 illustrates three different displays in the RAPTOR interface (see Chapter 14 for more detailed explanations): the friendly and enemy combat resources displays (top and bottom, right), the force ratio display (middle right), and the force ratio trend display (left). The information presented in these three displays spans a number of levels in the abstraction hierarchy (see Figure 14.1 in Chapter 14). The friendly and enemy combat resources provide information at the level of physical processes (e.g., number of vehicles). The graphical containers and physical connectors (i.e., the arrows and the force ratio connecting line) provide information at the level of general function (i.e., source, store, sink). The force ratio display provides information at the level of abstract function (flow of resources, relative military force). The force ratio trend display provides information regarding goals (e.g., planned vs. actual force ratio values over time).

The placement of these three displays in the same view (as opposed to different screens), in close physical proximity (as opposed to distant parts of the screen), and with explicit graphical connectors between displays (as opposed to without) serves to increase visual momentum via the display overlap technique. However, the increases in visual momentum supported by this design go beyond the obvious benefit of helping an agent locate related information that is physically scattered across multiple displays in a view. As discussed in Chapter 3, the categories of information in the abstraction hierarchy (and the relations between information in these categories) provide the basis for reasoning about a work domain in terms of ends (i.e., goals) and means (resources available to achieve those goals). In other words, the spatial organization is designed to support navigation in a semantic functional space, heightening situation awareness.

## 15.5 VM in Forms

In this final section we will consider issues in increasing visual momentum at the level of visual forms (i.e., visual analogies and metaphors). The process of supporting visual momentum at the level of visual forms is very similar to the process of designing effective displays (see Chapters 7 through 12). Techniques such as layering and separation, spatial dedication, and local/global emergent features can be used to produce a display with a hierarchically nested visual structure: "variables of optical structure [that] contain information or meaning, that ... specify what their sources afford" (Gibson 1966, p. 260).

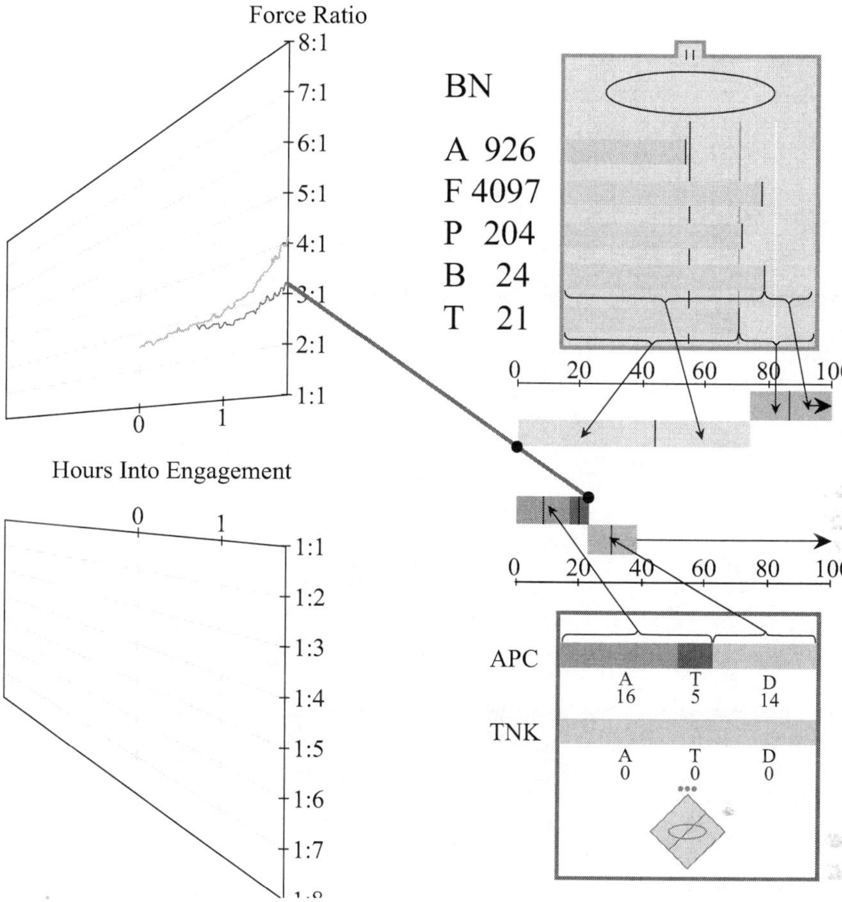


**FIGURE 15.9**
An example of the display overlap technique: simultaneous and parallel presentation of interconnections between goal-related, functional, and physical information appearing in different displays (i.e., the provision of overlapping visual contexts across multiple levels of the abstraction hierarchy). (Adapted with permission from Bennett, K. B., Posey, S. M., and Shattuck, L. G. 2008. *Journal of Cognitive Engineering and Decision Making* 2 (4): 349–385. Copyright 2008 by the Human Factors and Ergonomics Society. All rights reserved.)

### 15.5.1 A Representative Configural Display

The configural display designed for the simple process control system (see Chapter 10) will be used to make the discussion concrete. The two versions of this display illustrated in Figure 15.10 present exactly the same underlying information. However, the display on the left provides a nested hierarchy of visual structure that supports visual momentum while the display on the right provides a flat visual structure that does not. This section will be reasonably short since it provides a complementary discussion of fundamental issues that have been explored in this chapter and in earlier chapters on display design.

### 15.5.1.1 Fixed Format Data Replacement

The principles of layering, separation, and spatial dedication and their use in supporting global shifts in visual attention were described earlier (see Section 15.4.1.1). The same principles apply at the form level, only at a smaller scale. The configural display in Figure 15.10a has two global, spatially dedicated areas (i.e., data vs. context), which are perceptually segregated (e.g., background mats) in the display and therefore provide support for fundamental acts of exploratory visual attention. Spatial dedication is also used extensively at lower levels of spatial structure; domain variables, labels, axes, etc. are presented at spatially dedicated locations within the display grid.

### 15.5.1.2 Perceptual Landmarks

Perceptual landmarks at the level of form are the unique visual features produced by the low-level graphical elements of a display. In this particular type of display, the most important perceptual landmarks are the local emergent features produced by the graphical elements of a display (e.g., the horizontal linear extent of the bar graphs, the orientation of the mass indicator balance). They have many of the same characteristics as their counterparts in

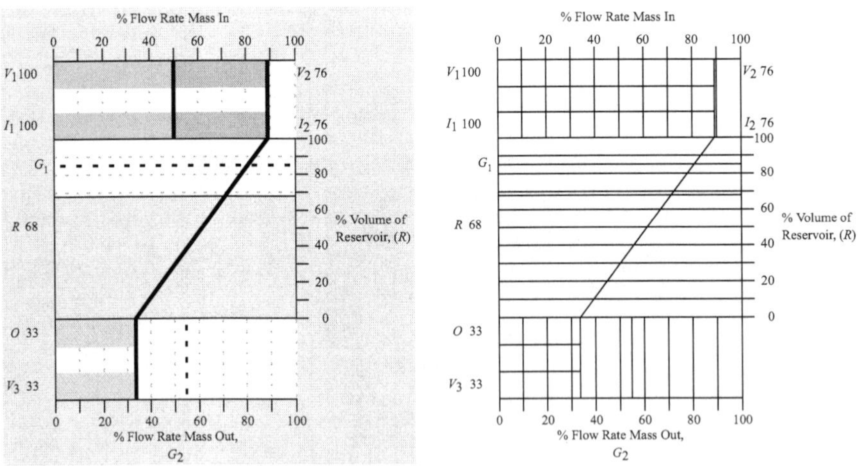


**FIGURE 15.10**
Alternative versions of a configural display from process control illustrating the use of visual momentum techniques at the level of form. (Adapted with permission from Bennett, K. B., A. L. Nagy, and J. M. Flach. 1997. In *Handbook of Human Factors and Ergonomics*, ed. G. Salvendy. Copyright 1997 New York: John Wiley & Sons. All rights reserved.)

large-scale ecologies: They appear in dedicated spatial locations (e.g., bar graphs for mass in and mass out are located at the top and bottom) and they are associated with meaning (e.g., orientation of the connecting line specifies mass balance).

The conceptual similarity is clearly evident in the original definition of landmarks in the ecology: "unique configurations of perceptual events (patterns)" (Siegel and White 1975, p. 23), which easily could have appeared in Chapter 8 as a definition of emergent features. These perceptual landmarks are an important part of the "interesting and meaningful" bits of graphical structure that guide successive visual fixations and assist the agent in focusing on informationally rich (and meaningful) areas of the display.

#### *15.5.1.3 Long Shot*

The long-shot design technique, as it applies to the level of visual forms, refers to the global perceptual patterns, or invariants (e.g., symmetry and parallelism), produced by a display. These global configural relations are meaningful (when the display has been designed effectively) and provide what amounts to a map of the ecology; particular patterns of global configural relations will map into particular states of the system. See Chapter 9 (Figure 9.6) for related discussions of the polar graphic display of Woods, Wise, and Hanes (1981) and Chapter 10 (Figures 10.3 through 10.10) for related discussions of the display depicted in Figure 15.10.

#### *15.5.1.4 Overlap*

The display depicted in Figure 15.10 illustrates the design technique of functional overlap very clearly. It combines some information from all five levels of the abstraction hierarchy into a single display. The levels of abstraction for the underlying domain (i.e., Figure 10.2) and their mapping into the properties of the display were described at length in Chapter 10. All of the benefits for visual momentum arising from the simultaneous presentation of information from multiple levels of the abstraction hierarchy that were described in Section 15.4.1.5 and elsewhere throughout the book are relevant here. Again, the key in this case is supporting navigation over a functional landscape.

#### *15.5.1.5 Section Summary*

The design principles outlined in Chapters 8 through 11 can be reinterpreted in terms of visual momentum. The most important properties in the domain (e.g., mass balance) will be represented by the most salient global emergent features. The contribution of critical individual variables (e.g., mass in and out) to the overall pattern will be represented by local emergent features (e.g., linear extent of the bar graphs). The resulting optical structures will provide perceptual landmarks or destinations that guide visual explorations (e.g., peripheral information obtained from these landmarks will guide subsequent saccades and ensuing fixations) and support exploratory visual attention. Thus, increased visual momentum, implemented through the hierarchically nested visual structure in a well-designed visual display, will facilitate an observer's ability to locate and extract the various kinds of information needed for effective task performance.

## 15.6 Summary

The combination of limited display real estate and large amounts of domain information present a nearly universal challenge in interface design. The keyhole effect and the getting lost phenomenon are but some of the consequences. Woods' landmark article (1984) provides an excellent theoretical overview of both the problem and potential solutions. Although this article is cited often, our impression is that the implications for interface design are not always fully appreciated. Our goal in writing this chapter was to make these implications more explicit.

The techniques for increasing visual momentum are applicable to a variety of levels in the interface. However, the physical manifestation of each technique can be quite different for these levels. A convenient line of demarcation occurs between the work space level and those below it. Increasing visual momentum at the work space level can be conceptualized as providing spatial structures that facilitate navigation. Increasing visual momentum at levels below can be conceptualized as providing spatial structures that support exploratory visual attention.

We have provided additional theoretical positioning of these issues with regard to larger sets of related literatures (i.e., way-finding, spatial cognition, visual attention). We have also provided concrete examples from the interfaces and displays that were previously discussed. In the end, we hope we have conveyed the ways in which these important concepts can be used to improve the quality of displays and interfaces.

Finally, we want to emphasize that, at a fundamental level, one principle—the semantic mapping principle—is the ultimate guide for how we organize information within configural graphical displays and within multiwindow interfaces. The point is that the organization of a configural display, the organization of multiwindow interfaces, or the organization of menus should reflect both the constraints of the work domain and the preferences, values, and expectations of the users. A goal for interface designers is to make sure that navigating in the interface does not get in the way of navigating through the functional space. The interface should not add a layer of complexity on top of the functional complexity, as is often the case with steep hierarchical menu interfaces to many applications.

In a very real sense, the interface is the map to the functional work space. Our goal is to design a map that makes it easier for people to get where they want to go. Thus, it has to correspond with the space of possibilities (i.e., the ecology) in an interesting way, it has to be readable or coherent (i.e., map to the beliefs and expectations of the user), and it should support learning by doing so that, over time, mismatches between situations and awareness are reduced.

## References

Army. 1999. Department of Defense interface standard (MIL-STD) no. 2525B common warfighting symbology. Washington, D.C.: Headquarters Department of the Army.

Bennett, K. B., A. L. Nagy, and J. M. Flach. 1997. In *Handbook of human factors and ergonomics,* ed. G. Salvendy. New York: John Wiley & Sons.

Bennett, K. B., S. M. Posey, and L. G. Shattuck. 2008. Ecological interface design for military command and control. *Journal of Cognitive Engineering and Decision Making* 2 (4): 349–385.

Chen, J. L., and K. M. Stanney. 1999. A theoretical model of way-finding in virtual environments: Proposed strategies for navigational aiding. *Presence* 8 (6): 671–685.

Evans, G. W. 1980. Environmental cognition. *Psychological Bulletin* 88 (2): 259–287.

Gauvain, M. 1993. The development of spatial thinking in everyday activity. *Developmental Review* 13:92–121.

Gibson, J. J. 1966. *The senses considered as perceptual systems.* Boston, MA: Houghton Mifflin.

Herman, J. F., and A. W. Siegel. 1978. Development of cognitive mapping of the large-scale environment. *Journal of Experimental Child Psychology* 26 (3): 389–406.

Hochberg, J. E. 1964. *Perception, foundations of modern psychology.* Englewood Cliffs, NJ: Prentice Hall, Inc.

Hutchins, E. L. 1995. *Cognition in the wild.* Cambridge, MA: MIT Press.

Kitchin, R. M. 1994. Cognitive maps: What are they and why study them? *Journal of Environmental Psychology* 14:1–19.

Kosslyn, S. M., H. L. Pick, and G. R. Fariello. 1974. Cognitive maps in children and men. *Child Development* 45 (3): 707–716.

Lynch, K. 1960. *The image of the city.* Cambridge, MA: MIT Press.

Mullet, K., and D. Sano. 1995. *Designing visual interfaces: Communication oriented techniques.* Englewood Cliffs, NJ: SunSoft Press.

Pardue, J. H., J. P. Landry, E. Kyper, and R. Lievano. 2009. Look-ahead and look-behind shortcuts in large item category hierarchies: The impact on search performance. *Interacting with Computers* 21:235–242.

Passini, R. 1980. Way-finding: A conceptual framework. *Man–Environment Systems* 10:22–30.

Pejtersen, A. M. 1980. Design of a classification scheme for fiction based on an analysis of actual user–librarian communication, and use of the scheme for control of librarians' search strategies. In *Theory and application of information research,* ed. O. Harbo and L. Kajberg. London: Mansell.

———. 1988. Search strategies and database design for information retrieval in libraries. In *Tasks, errors and mental models,* ed. L. P. Goodstein, H. B. Andersen and S. E. Olsen. Bristol, PA: Taylor & Francis.

———. 1992. The BookHouse: An icon based database system for fiction retrieval in public libraries. In *The marketing of library and information services 2,* ed. B. Cronin. London: ASLIB.

Rasmussen, J., A. M. Pejtersen, and L. P. Goodstein. 1994. *Cognitive systems engineering.* New York: John Wiley & Sons.

Siegel, A. W., and S. H. White. 1975. The development of spatial representations of large-scale environments. In *Advances in child development and behavior,* ed. H. W. Reese. New York: Academic Press.

Tolman, E. C. 1948. Cognitive maps in rats and men. *Psychological Review* 55 (4): 189–208.

Tufte, E. R. 1990. *Envisioning information.* Cheshire, CT: Graphics Press.

Watts-Perotti, J., and D. D. Woods. 1999. How experienced users avoid getting lost in large display networks. *International Journal of Human–Computer Interaction* 11 (4): 269–299.

Woods, D. D. 1984. Visual momentum: A concept to improve the cognitive coupling of person and computer. *International Journal of Man–Machine Studies* 21:229–244.

———. 1997. The theory and practice of representational design in the computer medium. Columbus: Ohio State University, Cognitive Systems Engineering Laboratory.

Woods, D. D., J. A. Wise, and L. F. Hanes. 1981. An evaluation of nuclear power plant safety parameter display systems. Paper read at the Human Factors Society 25th Annual Meeting, Santa Monica, CA.

Ziefle, M., and S. Bay. 2006. How to overcome disorientation in mobile phone menus: A comparison of two different types of navigation aids. *Human–Computer Interaction* 21:393–433.

# 16

# *Measurement*

## 16.1 Introduction

> Methodological issues are bread and butter to the working scientist but can be spinach to everyone else. (attributed to George Miller by Lachman, Lachman, and Butterfield 1979, p. 13)

The majority of this book has focused on the activities associated with the analysis of complex work domains and the ensuing processes of designing displays and interfaces for them. We have described concepts and analytical tools to assist in the former and principles of design that serve to streamline the latter. An equally important set of activities that have not been addressed directly are those associated with evaluation.

The issue is not simply the evaluation of a given display relative to conventional displays or to alternative designs in a "bakeoff" to identify the best option. Rather, the ultimate goal is the development of research programs that not only test displays but also evaluate the underlying theories and assumptions guiding the design. Thus, we are testing metaphysical assumptions (dyadic vs. triadic ontological positions), theories of human performance, theories of the domain, and general systems theories about the dynamics of the interactions between human, interfaces, and domains—as well as trying to answer the practical problem about at least incrementally improving the resilience of a specific system.

Thus, the evaluation process is jointly motivated by the desire to improve our basic understanding of cognitive systems and by the desire to design more effective interfaces that solve important practical problems (e.g., enhancing resiliency). We want to emphasize once again that theory is essential for generalizing from the laboratory to the field and from solutions in one domain to problems in another domain. Theory is essential to good design. Reciprocally, design challenges can provide acid tests for our theories. If our theories fail in meeting the practical challenges of improving resiliency in cognitive systems, then we have to wonder about the validity of our theories. Thus, we believe that the evaluation of interfaces is an important source of data for the science of cognitive systems.

## 16.2 Paradigmatic Commitments: Control versus Generalization

In Chapter 2 we introduced the concepts of a paradigm and the value and belief structures (i.e., the conventional rules of science) shared by researchers working within a paradigm. Throughout the book we have expressed our paradigmatic commitments with regard to design. The conventional rules are equally important during evaluation; they guide the choice of hypotheses, the choice of both independent and dependent variables, and the acceptable ways to parse the natural complexity. Ultimately, the question becomes how we can conduct meaningful scientific inquiry to develop effective decision support (including principles of design as well as specific displays and interfaces) while coping with the complexity of the real-world domains within which this decision support is to be used.

In short, how can we conduct research with the potential to generalize to real-world settings? Figure 16.1 provides a graphical representation to illustrate what we feel is a critical difference between the dyadic and triadic perspectives. Both graphs simultaneously relate critical dimensions: the complexity and fidelity of research settings, experimental control, and beliefs regarding the generalization of results.

### 16.2.1 Dyadic Commitments

From the conventional dyadic view (Figure 16.1a), generality is achieved by framing questions around fundamental information processing elements. Thus, the more we titrate our observations around fundamental particles, the more general will be the findings. Conversely, the more we incorporate the contextual peculiarities of particular domains, the less generalizable will be our conclusions. This is a very convenient belief, in that the more tightly we control our experiments (typically requiring low complexity and low fidelity), the more general will be the inferences that we can make (represented in Figure 16.1a by the arrows pointing from the upper right to the lower left).

### 16.2.2 Triadic Commitments

Contrast this conventional view with the triadic perspective (Figure 16.1b). This perspective—reflected perhaps most clearly in Brunswick's tradition of an ecological approach; see Brunswick (1956) and Kirlik (2006)—is that generality depends on the degree of match between the constraints in the research context (e.g., microworld experiment) and the constraints in the target context for the generalization (e.g., conclusions about ways to support command decision making).

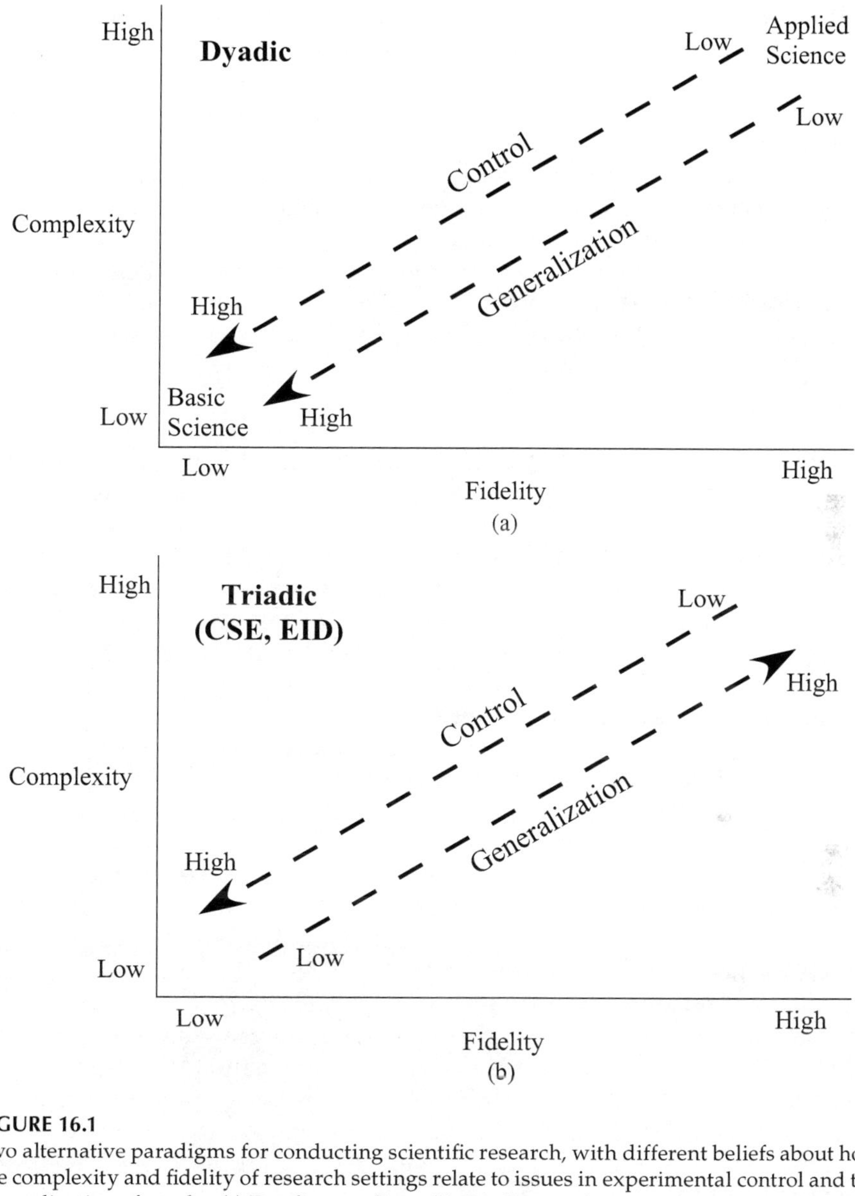


**FIGURE 16.1**
Two alternative paradigms for conducting scientific research, with different beliefs about how the complexity and fidelity of research settings relate to issues in experimental control and the generalization of results. (a) Dyadic paradigm. (b) Triadic paradigm.

From this tradition, it is important that the microworld be representative of the design context, in order to have any confidence that the results obtained can be used to guide design. With this perspective, the demands for generalization run against the demands for control. Experimental control is achieved through reducing complexity and abstracting away from contextual details. Generality is achieved through matching the complexity and incorporating the domain context into the experimental context.

Thus, for the triadic perspective, there is a tension between control and generalization (represented in Figure 16.1b by the arrows pointing in opposite directions) that is missing in the dyadic perspective. This *does not* mean that the value of carefully controlled experiments is not appreciated. However, this does mean that there is no free lunch! The results of carefully controlled experiments have to be considered as one component of a set of converging operations that explore the full space reflected in Figure 16.1. We cannot live exclusively in the lower left corner of this space if we hope to develop a comprehensive science of human experience.

### 16.2.3 Parsing Complexity

It is probably fair to say that, despite its early justification in general systems theory (e.g., Taylor 1911), human factors have classically taken a view consistent with the basic-applied distinction and that most scientific research has been conducted at the basic end of the scale (i.e., dyadic, reductionist) represented in Figure 16.1a. That is, there has been a strong tendency to break down complex phenomena into elementary information processing tasks and to assume that research with these elementary tasks allows both better control and higher generalizability.

These tasks are assumed to tap into "basic" information processing mechanisms. It is generally assumed that performance in complex tasks can be modeled as the sum of the appropriate basic elements (e.g., Card, Moran, and Newell 1983). As one researcher remarked, it is the same "head" in the laboratory and in the cockpit. This approach is generally paired with great skepticism about field studies and other naturalistic types of research due to concerns about control and about the narrow scope of generalization that might be possible from one natural context to another.

Recently, however, researchers such as Rasmussen, Hutchins, Woods, and Klein have been impressed by emergent properties that they have found when observing "cognition in the wild." There is a growing sense that performance in naturalistic settings has a situated dynamic that has not been well represented by the classical laboratory settings. There is a growing skepticism with models that describe complex phenomena as mere collections of elementary information processes. Thus, there is a move toward the conventions of the triadic paradigm.

This chapter will focus particularly on the challenge associated with parsing the complexity. In many respects, this is a pivotal issue separating the triadic and dyadic perspectives. The dyadic perspective parses the problem in a way that decouples mind from matter to build a general theory about how the various components in the head process information. The challenge for the triadic perspective is to parse the problem around functional relations over mind and matter that reflect particular functional dynamics. For example, the conventional dyadic approach treats memory as a process in the head. Wickens (1984) frames his discussion about memory in terms of the coding and storage of information in sensory, working, and long-term memory systems. On the other hand, Hutchins's (1995b) discussion of how a cockpit remembers presents memory as a process distributed over the work ecology.

The ultimate challenge is to parse the complexity in ways that are representative of the phenomenon of interest. If, for example, memory is viewed as a component in the head, then experiments organized around the information constraints of this component should be representative of the natural phenomenon. On the other hand, if memory depends in part on structure within the ecology (e.g., the layout of cockpit instruments), then preserving this structure may be critical in the design of experiments that will generalize to that domain. Thus, a foundational question with respect to measurement and evaluation is: what is the system?

## 16.3 What Is the System?

> I claim that many patterns of Nature are so irregular and fragmented, that, compared with Euclid—a term used in this work to denote all of standard geometry—Nature exhibits not simply a higher degree but an altogether different level of complexity ....
>
> The existence of these patterns challenges us to study those forms that Euclid leaves aside as being "formless," to investigate the morphology of the "amorphous." Mathematicians have disdained this challenge, however, and have increasingly chosen to flee from nature by devising theories unrelated to anything we can see or feel. (Mandelbrot 1983, p. 1)

Consistent with Mandelbrot's comments comparing the "cold" geometry of Euclid with the patterns of nature, there seems to be a growing dissatisfaction with the ability of classical experimental approaches to capture the complexity of activity "in the wild" (e.g., Hutchins 1995a). While many researchers have "disdained this challenge" and have fled from the apparently amorphous patterns of everyday work to study sterile laboratory tasks, a few have been plagued by a nagging fear that this research may not be representative.

The subject of this book—display and interface design—imposes a natural concern with regard to use in real work domains. In addressing this challenge, questions are raised about appropriate ways to measure the effectiveness of the support that they provide. There is a growing consensus that the cold geometry of context-free laboratory tasks and measurements (e.g., reaction time) is not capable of capturing the complexities of human performance in natural settings.

One of the first decisions researchers must make is to identify the phenomenon or system of interest. In Chapter 2 we described two alternative

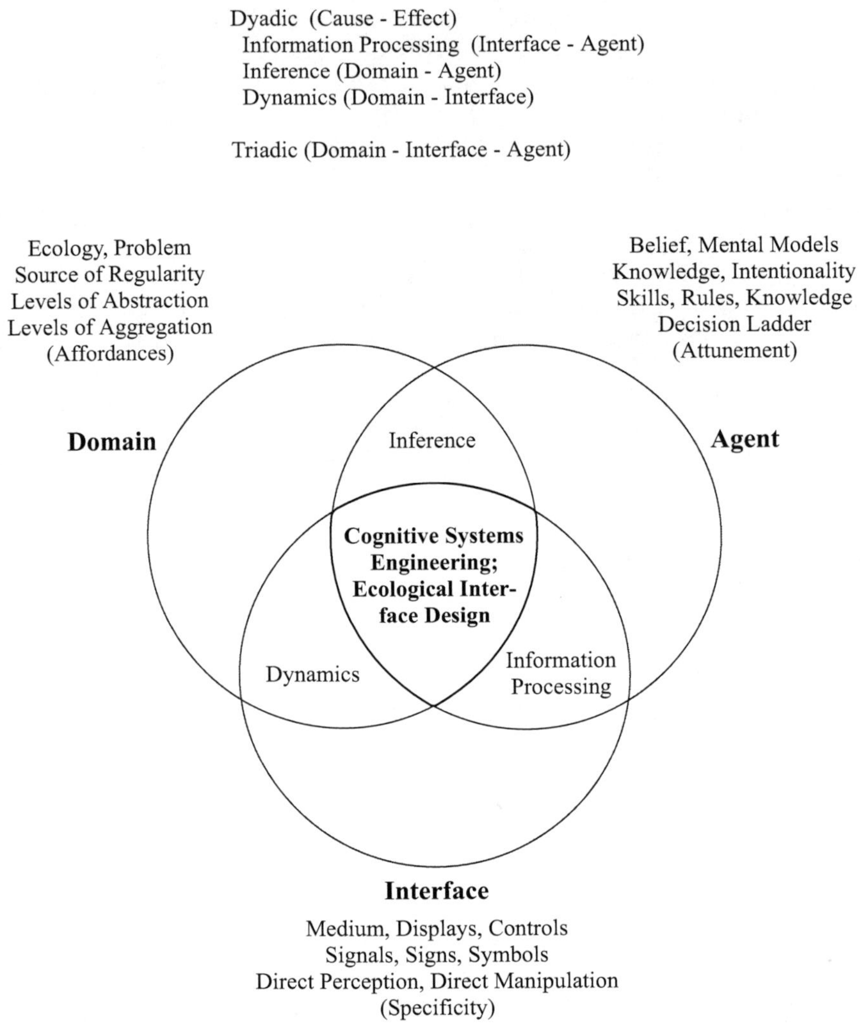


**FIGURE 16.2**
Cognitive systems engineering and ecological interface design is a triadic approach that lies at the intersection of mutually interacting constraints contributed by agents, domains, and interfaces.

paradigms that make different decisions in this regard: the dyadic (Saussure) and the triadic (Peirce) approaches. Figure 16.2 re-represents the three triadic system components first introduced in Chapter 2 (see Figures 2.3 and 6.2 [Chapter 6]) in Venn diagram form so that their mutual interactions are more explicit. As emphasized throughout the book, the triadic approach emphasizes the union of all three components (labeled cognitive systems engineering and ecological interface design in the figure). As a result, the triadic approach considers performance and action to be processes that are "situated" or "distributed." In terms of the present discussion, assessments of displays and interfaces will be meaningful only when constraints related to both the domain (i.e., situations) and the agent (awareness) are also explicitly incorporated into the evaluative setting.

In contrast to the triadic approach, with the dyadic paradigm the task or problem space is arbitrary. This literature creates a distinct impression that the system of interest is "in the human's head" and that the research goal is to characterize the internal limitations within isolated stages of information processing so that these limitations can be accommodated in the design of complex systems. The primary motivation for choosing one laboratory task or another is the ability to isolate specific stages of processing (e.g., encoding or decision making) or specific internal constraints (e.g., a bottleneck or resource limit). There is little discussion of how well such tasks represent the demands of natural situations. In terms of the present discussion, the evaluation of displays and interfaces focuses on the dyadic interaction between the agent and the interface.

The implication of a triadic perspective for the evaluation of display and interface designs is that we also need to consider how to measure both situations and awareness. Furthermore, we must measure them in a way in which we can index the fitness of one relative to the other. They are not two separate systems, but rather two facets of a single system. In the end, situation awareness is not a cognitive phenomenon that exists solely within the head of an agent; it is one that depends on the degree of fit between awareness and situations.

Thus, the situation is viewed as a critical component in the analysis, design, and evaluation of complex systems from the triadic paradigm. In fact, to a large extent, the evaluation of displays and interfaces should focus on the degree to which they support the agent in adapting to the problem or situation constraints. In the following sections we consider the measurement of both situations and awareness in greater detail. It is a complementary description of ideas presented in Chapters 3 and 4.

### 16.3.1 Measuring Situations

When considering measuring situations it is important to begin with fundamental lessons about the nature of information. The information value of an event (e.g., selecting a number from a jar) cannot be determined unless the possibilities (e.g., the other numbers in the jar) are specified. In simple choice decision tasks (e.g., Hick 1952; Hyman 1953), the other possibilities can be well defined using the number and probabilities of alternatives. However, how do you specify the possibilities when the task is controlling a nuclear power plant, flying a modern aircraft, or directing air operations during battle—much less when the question has to do with general purpose systems such as the Internet or the cell phone?

From the dyadic perspective, there is a natural tendency to extrapolate using the measures (i.e., probability-based information measures) that worked well for reaction time experiments where the possibilities were proscribed by the experimenter. This has stimulated initiatives to reduce events in a nuclear power plant and other complex situations to probabilities and durations that can be integrated using linear techniques like THERP (technique for human error rate prediction) or Monte Carlo style simulations such as MicroSAINT or ACT-R.

#### 16.3.1.1 The Abstraction (Measurement) Hierarchy

From the triadic perspective, there is great skepticism about whether the dynamics (or possibility space) of many natural situations can be captured using event probabilities or time-based measures alone. The alternative is to describe the constraints that shape the space of possibilities in terms of goals and values, general physical laws, organizational constraints, and specific physical properties. This perspective is well illustrated by Rasmussen's (1986) abstraction hierarchy, which is significant as one of the first clear specifications of the different classes of constraint that limit possibilities in natural work environments (see the detailed descriptions in Chapter 3). In relation to the focus of this chapter, it is important to recognize that measurement is a form of abstraction. Thus, the abstraction hierarchy is a statement about measurement. In fact, we suggest that it could easily have been termed a measurement hierarchy, where each level suggests different ways to index constraints within the work domain.

In the context of specifying or measuring situations, the abstraction hierarchy provides a useful guide for thinking about the various levels or types of constraints that shape the field of possibilities within a work domain. When considering how to measure situations, one must consider the need to describe the constraints in ways that reflect significant relations within and across the various levels. Vicente (1999) illustrates this very clearly with the DURESS example. One caution is that, although DURESS is a great illustration, it represents a task with fairly well defined goals and constraints. In many domains, the constraints will be far more amorphous, and discovering the right metrics to characterize the significant relations among the constraints is a significant challenge. But it is a challenge that must be engaged if there is to be any hope of understanding the computations involved in cognitive work.

Let us consider some of the levels associated with situation constraints and the issues associated with measuring them. First, consider goals and values. Even in simple laboratory tasks (e.g., those using reaction time or signal detection), it is evident that trade-offs between goals (e.g., speed vs. accuracy or hits vs. false alarms) have a significant role in shaping performance. Natural work domains are typically characterized by multiple goals and success often depends on balancing the demands associated with these goals (e.g., setting priorities or precedence).

An important question for measurement is how to index performance with respect to multiple goals so that the data can be integrated across the goal dimensions in a way that will reflect whether performance is satisfactory with regard to the aspirations for the system. Measures should allow some classification (satisfactory vs. unsatisfactory) or ordering (better or worse) of performance with respect to these aspirations. Note that unless weights or priorities are established for the various dimensions of performance, it is impossible to consider "optimality" or even whether performance is generally satisfactory.

Brungess's (1994) analysis of the SEAD (suppression of enemy air defenses) is an important example of someone who is explicitly wrestling with the problem of how to measure performance relative to goal constraints. For example, he writes:

> SEAD effectiveness in Vietnam was measured by counting destroyed SAM sites and radars. Applying that same criteria to SEAD technologies as used in Desert Storm yields a confused, possibly irrelevant picture. SEAD weapons and tactics evolution has outpaced the development of criteria to measure SEAD's total contribution to combat. (pp. 51–52)

The point Brungess is making is that overall measures of effectiveness that might have been valuable in one war may not be valuable in the next. He goes on to argue that, in an information age, the goal of SEAD is not to take out the missiles, but rather to "blind" the command and control systems that coordinate them. The lesson is that, as researchers, we cannot define ourselves around single general performance measures (e.g., "I study reaction time"). We need to choose measures that reflect the values of the domain of interest. If our quantitative evaluations are going to be useful, it will not be about numbers, but about values.

At another level, it should be quite obvious how general physical laws (e.g., thermodynamics or laws of motion) provide valuable insight into the possibilities of natural processes (e.g., feedwater regulation or vehicle control). These laws suggest what variables are important for specifying the state of the system (e.g., mass, energy, position, velocity). These variables are critical both to the researchers interested in describing the system and to the active control agents (whether human or automated) in terms of feedback (i.e., observability and controllability).

Note that for the variables to be useful in terms of feedback, they must be indexed in relation to both goals and control actions. That is, it must be possible to compare the information fed back about the current (and possibly future) states with information about the goals in a way that specifies the appropriate actions. Thus, questions about controllability and observability require indexes that relate goals, process states, and controls (e.g., Flach, Smith, et al. 2004). This is a clear indication of the need to choose measures that reflect relations within and across levels of the measurement hierarchy.

In addition to considering relations across levels in the situation measurement hierarchy, it is important not to lose sight of the fact that the measures should also help to reveal important relations to constraints on awareness. Remember that the system of interest includes both the problem and the problem solver. Understanding general physical laws can be very important in this respect because these laws suggest ways to organize information to allow humans with limited working memory to "chunk" multiple measures into a meaningful unit. This is a recurrent theme in the design of ecological displays: to use geometric relations to specify constraints (e.g., physical laws) that govern relations among state variables (see Chapters 8 through 11).

We have talked extensively about the abstraction hierarchy and its role in specifying the situation or problem to be solved (Chapter 3), so we will trust you to extrapolate to consider other aspects of measuring the situation. We hope that the general theme of measuring the situation using indexes that reveal relations across levels in the hierarchy and in relation to constraints on awareness is clear. For more discussion of other levels in the abstraction and measurement hierarchy, see Flach, Mulder, and van Paassan (2004).

### 16.3.2 Measuring Awareness

Dyadic approaches focus on identifying awareness dimensions that are independent of (or invariant across) situations. Thus, they tend to address issues such as perceptual thresholds, memory capacities, bandwidths or bottlenecks, and resource limits as attributes of an internal information processing mechanism. However, within this research literature, it is not difficult to find research that attests to the adaptive capacity of humans. For example, basic work on signal detection suggests that performance is relatively malleable as a function of the larger task context (e.g., expectancies and values). Even the sensitivity parameter (d′) is defined relative to signal and noise distributions. Thus, there is ample reason for skepticism about whether any attribute of human performance can be specified independently from the larger task context. Perhaps no single statement summarizes the dominant lessons from the last decade of research on cognition than this: *context matters!*

Whether or not it is possible to characterize constraints on human information processing that are independent of the task context, few can argue that humans are incredibly adaptive in their ability to meet the demands of natural situations. The triadic approach focuses on this adaptive capacity and this raises the question of measuring awareness relative to a domain (i.e., skill or expertise). There is little evidence to support the common belief that expert performance reflects innate talent. Rather, the evidence suggests that expert performance reflects skills acquired through extended, deliberate practice in a specific domain (Ericsson and Charness 1994). In fact, Ericsson and Charness conclude that "acquired skill can allow experts to circumvent basic capacity limits of short-term memory and of the speed of basic reactions, making potential limits irrelevant" (p. 731).

#### *16.3.2.1 Decision Ladder; SRK*

Again, Rasmussen (1986) was one of the first to recognize explicitly the flexibility of human information processing and to introduce a conceptual framework specifying important distinctions that must be addressed by any program to quantify human performance in natural contexts (i.e., the decision ladder and the SRK distinction between skill-, rule-, and knowledge-based processing). The decision ladder explicitly represents the shortcuts that might allow experts to circumvent basic capacity limitations (see Chapter 4). The SRK distinction provides a semiotic basis for relating the properties of the situation (e.g., consistent mapping) to the potential for utilizing the various shortcuts (see Chapter 5 and Flach and Rasmussen 2000).

Rasmussen (1986) illustrates how the decision ladder can be utilized to visualize qualitatively different strategies for fault diagnoses. This is clearly an important form of measurement that helps to index performance in relation to potential internal constraints on awareness and in relation to the demands of situations. Furthermore, it allows these strategies to be compared to normative models of diagnoses.

Consistent with the basic theory of information, it is important not simply to ask what experts know and what strategy they typically use; rather, we must explore the possibilities about what experts could know and about what strategies might be effective in a given situation. Note that the awareness of experts will be constrained by the types of representations to which they have been exposed. For example, pilots and aeronautical engineers utilize very different forms of representations for thinking about flying. Thus, there can be striking contrasts for how these different experts think about flight performance. Exploring the differing forms of awareness can be important for differentiating more and less productive ways for thinking about a problem such as landing safely (Flach et al. 2003). It is important to keep in mind that the best operators (e.g., pilots or athletes) often do not have the best explanations for how and why they do what they do.

Considering alternative representations across different experts can suggest possibilities for shaping awareness through the design of interfaces (as expressed throughout this book). As Hutchins's (1995a) work clearly illustrates, the choice of a specific technique for projecting the world onto the surface of a map has important implications for the cognitive processes involved in navigation. Again, this is a key theme behind the construct of ecological interface design: to shape the nature of awareness to facilitate information processing (issues that have been addressed in detail in many of the previous chapters). The point is not simply to match existing mental models, but rather to design representations that help shape the mental models to enhance awareness and resilience.

### 16.3.3 Measuring Performance

At the end of the day, one of the most important measurement challenges for display and interface evaluations is to be able to index the quality of performance (i.e., what constitutes "better" performance). Dyadic approaches prefer to focus on one dominant measure (e.g., time to completion or percent correct) to index quality. Even when there is clear evidence of the potential for trade-offs (e.g., speed vs. accuracy), dyadic-style research tends to frame the task to emphasize one dimension clearly (e.g., "go as fast as possible with zero errors").

These approaches typically assume that performance functions are monotonic (e.g., "faster is better"). This is usually generalized to research in human factors, where two designs might be evaluated in terms of which design produces a statistically significant advantage in response time. However, whether a statistically significant difference in response time leads to a practical gain in work performance is difficult to address with the dyadic approach. The general or CEO who asks whether the improved system will be worth the cost in terms of achieving the objectives important to him or her (e.g., greater safety or a competitive advantage) rarely gets a satisfactory answer.

#### 16.3.3.1 Multiple, Context-Dependent Indices

In the everyday world, there is rarely a single index of satisfaction. As discussed in the section on measuring situations, typically, multiple goals must be balanced; a good system should be efficient, accurate, safe, and not too expensive. This requires either multiple performance measures or at least an explicit integration of indexes associated with the various goals to yield a single score for ranking goodness or at least for distinguishing between satisfactory and unsatisfactory performance. Rarely are the quality indices monotonic. That is, success typically depends on responding at the right time (not too early or too late). At least for closed-loop systems, there will always be a stability boundary that limits the speed (i.e., gain) of response to stimuli. Thus, the quality function for response speed is rarely monotonic; a system that is too fast can become unstable (e.g., pilot-induced oscillations).

It is impossible to address questions about the right information, the right place, the right person, or the right time without considering the specific problem that is being solved (i.e., the work domain or task). "Right" is highly context dependent. It cannot be addressed by a dyadic-based research program that is designed to be context independent. This is an important motivation for a triadic approach to cognition and work: to specify the criteria for satisfying the demands of specific work domains.

In order to know whether a difference in response time is practically significant, it can be useful to compare this against landmarks that reflect the optimal or best-case situation. Here is where analytic control models (e.g., the optimal control model) or Monte Carlo simulations (e.g., Microsaint, ACT-R) can be very useful—not as models of human information processes, but rather as ways to explore the boundary conditions of performance. What is the best possible performance, assuming certain types of processes? How do changes at one step in a process (e.g., to speed or accuracy) or in properties of a sensor (e.g., signal-to-noise ratio) impact system performance? Where are the stability limits?

In this sense, the models are being used to explore boundaries (or limits) in the work space. These boundaries may provide important insights into what are realistic targets for improvement and into the practical value of specific improvements. In essence, these models can suggest normative landmarks against which to assess actual performance.

An example of a case where this type of insight might be useful is a recent initiative on the part of the air force to reduce the response time for executing dynamic targets (i.e., targets, typically threats, not specifically identified in the tactical mission plan) to single-digit minutes. This was motivated by the threat of mobile missile systems (e.g., SCUDs) that can fire a missile and then move to cover within about 10 minutes. Few have raised the question about whether this response time is realistic given the unavoidable lags associated with acquiring the necessary information and communicating with the weapons systems.

We fear that the blind pursuit of this single-digit-minute goal may lead to instabilities and unsatisfactory solutions to the overall goals of the air force. Rather than reacting faster, the solution to the SCUD missile problem may depend on improving the ability to predict or anticipate launches (e.g., Marzolf 2004). Thus, the solution to SCUD missiles may rest with the design of the air battle plan to include dedicated aircraft to patrol areas where launchers are likely to be hidden, with the authority to engage targets of opportunity when they arise, rather than to speed the dynamic targeting process for dealing with events not anticipated in the air battle plan.

#### *16.3.3.2 Process and Outcome*

Another important consideration for measuring performance is the distinction between process and outcome. In complex environments, an optimal process can still result in a negative outcome due to chance factors that may be completely beyond control. For example, a coach can call the perfect play that results in a touchdown and have it nullified by a penalty flag incorrectly thrown by a poor referee. Thus, it is important to include measures of process as well as measures of outcome. Also, it is important to have standards for measuring process as well as outcome. For example, most military organizations have doctrine that provides important standards for how processes should be conducted.

The astute reader should realize that as we talk about performance measurement, we are covering some of the same ground that was discussed in terms of measuring situations. We are talking about ends (goals and values) and means (processes). In classical dyadic approaches that define the system of interest in terms of what is inside the cognitive agent, the situation is typically treated as an independent variable and performance measures are treated as dependent variables. This creates the impression that these are different kinds of things. However, in natural work ecologies, understanding the situation requires consideration of both means and ends. Thus, using the abstraction hierarchy to think about situations will go a long way toward addressing questions about performance measures and should help to frame these questions in terms meaningful to the problem owners (those who have a stake in success).

#### *16.3.3.3 Hierarchically Nested*

Thus, consistent with our discussion about situations, we believe that, from a triadic framework, it is useful to think about a nested hierarchy of performance measures, where higher levels in the hierarchy reflect global criteria for success (e.g., how you know whether you are winning or losing), lower levels address subgoals and process measures that reflect the means to higher level goals (e.g., showing patterns of communication or organization), and still lower levels reflect the mechanics of action (e.g., decision and response latencies). It is important to keep in mind that the primary goal of measurement is to reveal the patterns of association between process and outcome. In other words, a key objective is to connect the microstructure associated with the design and organization of work activities to qualitative changes associated with global indexes of quality.

## 16.4 Synthetic Task Environments

In contrasting the dyadic and triadic approaches, our intent is not to eliminate basic research, but rather to make the case that this is only one element of a comprehensive research program. A research program that is exclusively framed in terms of basic, context-free tasks will not satisfy our goals to understand cognition in natural contexts or to inform the design of tools to support cognitive work. Thus, the goal of the triadic approach is to enrich the coupling between the laboratory and the natural world. It is in this context that we would like to suggest that research employing synthetic task environments or relatively high-fidelity microworlds can be an important means for bridging the gap between basic experimental research and natural cognition (e.g., Brehmer 1992; Brehmer, Leplat, and Rasmussen 1991; Dorner 1987; Rouse 1982–1983; Schiflett and Elliot 2000).

### 16.4.1 Not Just Simulation

We will use the term "synthetic task environment" to describe experimental situations where there is an explicit effort to represent the constraints of a natural work domain. This is in contrast to experimental settings designed around the parameters of a particular analytic model (e.g., choice reaction time or compensatory tracking) or designed to isolate a specific stage of an internal process (e.g., visual and memory search tasks). It is also in contrast to low-fidelity microworld research that attempts to represent the complexity of natural domains, without representing the constraints of specific actual domains (e.g., space fortress or other research using computer games). In a synthetic task, the work domain has to be more than a cover story. The task must be representative of some natural work—even though the implementation is synthetic (typically utilizing a simulation).

For example, research using a flight simulator may or may not satisfy our definition for synthetic task research. If the focus is on flight performance—perhaps in relation to a specific training protocol, to compare alternative interfaces, or to evaluate different procedures—then this is consistent with our definition of synthetic task research. However, if the focus is on cognitive workload and the flight task is essentially a manual control tracking task within a multiple task battery, then we would not consider this to be synthetic task research. Again, this does not mean that such research is not valuable; we simply want to emphasize that, for synthetic task research, the focus should be on the impact of specific natural constraints of the work on cognitive processes. The key to synthetic task research is *not* the use of a simulator, but rather the framing of research questions with respect to properties of the natural task or the natural work domain. The problem presented in the laboratory must be representative of the natural problem to which the results are intended to generalize.

### 16.4.2 Coupling between Measurement Levels

A second important facet of synthetic task environments is the ability to measure performance at multiple levels as discussed in previous sections of this chapter. The synthetic task environment should allow performance to be scored relative to global objectives (e.g., whether a landing was successful; whether the mission was completed successfully). It also should allow direct measures of the work processes at micro (e.g., time history of control and communication activities) and meso (e.g., system states, such as the actual flight path) levels. Finally, it should provide a basis for directly probing the underlying rational guiding action (e.g., through verbal protocols or after action reviews supported with playback of events).

One of the important goals for synthetic task research is to provide empirical data with respect to the coupling of global metrics (goals and values) and micrometrics (work activities and movement through the state space). The goal is to relate variations at one level of the measurement hierarchy empirically to variations at the other levels. For example, this may allow the earlier question about whether a significant difference in reaction time is practically significant with respect to the global intentions for the system to be addressed empirically: Do quantitative differences in response time to a particular class of events lead to increased probability of successfully completing the mission?

### 16.4.3 Fidelity

Another consideration with respect to synthetic task research is the question of fidelity. How much is enough? This is a bit tricky because this is one of the questions that we are asking when we frame questions around situations. What are the important constraints and how do they interact to shape performance? For this reason, the issue of fidelity can only be addressed iteratively. In general, it is best to start with as much fidelity as you can practically afford and to assume that it is not enough! The performance observed in synthetic tasks needs to be evaluated skeptically relative to generalizations to natural domains.

In our view, to be effective, a program of synthetic task research should be tightly coupled to naturalistic field studies. The patterns observed in the laboratory need to be compared to patterns observed in naturalistic settings. In this way, it may be possible to titrate down to identify critical constraints and interactions. The synthetic task observations allow more rigorous control and more precise measurement, but there is always the possibility that the patterns observed in the synthetic task are a result of the simulation and that they are not representative of the natural domain of interest. Ideally, however, synthetic task environments can improve our ability to see and quantify patterns obtained during more naturalistic observations.

It is also important to note that questions of fidelity should not be framed simply in terms of the simulation device. Consideration must be given to the participants of the research. Are they representative of the people who do this work in natural settings, in terms of knowledge, skill, motivation, etc.? Consideration also must be given to the task scenarios. Are the tasks representative of the work in the natural context in terms of probability of events, consequences, and organization? More specifically, are the experiences of the participants representative of experiences in the real work domain (e.g., in terms of motivation and stress)?

In order to bridge the gap between laboratory research and cognition in the world, synthetic task research will be most effective when the questions are driven by field observations of natural environments and when the multiple nested measures are motivated by (1) the values of the problem owners, (2) normative models of the work (e.g., information theory, control theory, queuing theory), and (3) basic theories of cognition. Currently, each of these three motivations has its champions and there seems to be a debate over which of these motivations is optimal. In our view, all three motivations are critical and none of them alone will meet our aspirations for a science of cognition. With respect to these three motivations, the synthetic task environment may provide a common ground to facilitate more productive coordination between the disparate constituencies across the basic and applied fields of cognitive science.

## 16.5 Conclusion

> Nature does exist apart from Man, and anyone who gives too much weight to any specific [ruler] ... lets the study of Nature be dominated by Man, either through his typical yardstick size or his highly variable technical reach. If coastlines are ever to become an object of scientific inquiry, the uncertainty concerning their lengths cannot be legislated away. In one manner or another, the concept of geographic length is not as inoffensive as it seems. It is not entirely "objective." The observer invariably intervenes in its definition. (Mandelbrot 1983, p. 27)

This quote from Mandelbrot reflects the difficulty in measuring a natural coastline: As the size of the ruler gets smaller, the length of the coastline can grow to infinity. If a simple attribute like "length of a coastline" creates this difficulty for measurement, how much more difficult is the problem when the nature that we are trying to measure involves human performance in complex work domains? In our view, perhaps this might be the biggest differentiator between dyadic and triadic approaches to cognition. The dyadic approach clings to the classical idea that it is possible to stand outside ourselves to measure cognition, work, or situation awareness objectively. The triadic approach believes that this is a myth (e.g., Flach, Dekker, and Stappers 2008).

Thus, it is important to recognize the inherent limits on any controlled scientific observation. The results will depend in part on properties of the phenomenon of interest and in part on the choices we make in designing the research environment. It is important to resist the temptation to become infatuated with a particular experimental methodology, setting, or task (whether microtask or specific synthetic task environment). No matter where we stand, no single perspective will be all encompassing with respect to nature's full complexity; no single perspective will be representative of the full space illustrated in Figure 16.1.

Whereas the dyadic approach sees the lower left corner of the space as a privileged perspective into the phenomenon, a triadic view cautions that *no single perspective is privileged.* There is no single critical experiment. Rather, we build understanding through a program of converging operations that takes the same care in sampling the fidelity/complexity space that the conventional dyadic approach takes in sampling subjects (e.g., see Kirlik 2006). A representative sample of situations is every bit as important as a representative sample of subjects.

Every measure, every level of description, and every perspective offers an opportunity to see some facet of nature, but hides other facets. Thus, understanding requires multiple measures, multiple levels of description, and/or multiple perspectives. In other words, the only way to eliminate or to "unconfound" the invariant of a specific measurement perspective from an invariant of nature is to measure from multiple perspectives. One is more confident in attributing an invariant to nature when that invariant is preserved over many changes of observation point. Note that this is not a special requirement for studying interface and display design. This will be true for any complex phenomenon in nature (e.g., weather systems or coastlines). By "complex," we simply mean a phenomenon that involves many interacting dimensions or degrees of freedom.

It is humbling to realize that nature/cognition cannot be reduced to reaction time and percent correct; to realize that the convenient measures (in terms of experimental control or analytic models) will not yield a complete picture; to realize that measures that work within the constraints of the ideals of Euclidean geometry do not do justice to the curves of natural coastlines. We get a distinct impression that the field of human–computer interaction is searching for a mythical holy grail—that is, a single framework (neuronets, neuroscience, chaos, etc.) and a specific measure (42, MRI, 1/f scaling, etc.) that will provide the key to the puzzle. We are skeptical.

Complex systems are difficult. They require multiple levels of measurement. An attractive feature of synthetic task environments is that they allow many measures (from micromeasures specifying the complete time histories of activity and state change to macromeasures specifying achievement relative to the intentions of operators and system designers). The problem is to make sense of all these data, weeding through the data to discover the patterns that allow insight, prediction, deeper understanding, and generalization.

Success in this search requires the intuitions available from normative systems theory (e.g., information, signal-detection, and control theory, computational and normative logic, nonlinear systems and complexity theory), from controlled laboratory research, and from naturalistic field observations. Again, none of these perspectives on research is privileged. We expect that if answers are to be discovered, they will be found at the intersection of these multiple perspectives. Thus, the value of synthetic task environments is to create common ground at the intersection of these various perspectives where we can constructively debate and test alternative hypotheses about the nature of cognitive systems.

Two alternative paradigms, or approaches, to conducting interface and display research in human factors were described in the introduction. The philosophical differences between these two paradigms fall into the realm of the conventional rules of science. This chapter attempts to reconcile the two alternative paradigms, much like what Neisser (1976) attempted to do for the field of experimental psychology years ago. Neisser encouraged cognitive psychologists to "make a greater effort to understand cognition as it occurs in the ordinary environment and in the context of natural purposeful activity. This would not mean an end to laboratory experiments, but a commitment to the study of variables that are ecologically important rather than those that are easily manageable" (p. 7).

In the end our message is very much the same. There appears to be no "free lunch" for human factors research; experimental control and complexity must be carefully considered and traded against each other and domain constraints must be analyzed and distilled into appropriate research settings. Research from the triadic paradigm is more difficult to conduct on a number of dimensions. However, doing so will improve the chances that human factors research will have an impact in actual applied settings and that it will feed back in ways that inform basic theories of human cognition.

## References

Brehmer, B. 1992. Human control of complex systems. *Acta Psychologica* 81:211–241.

Brehmer, B., J. Leplat, and J. Rasmussen. 1991. Use of simulation in the study of complex decision making. In *Distributed decision making: Cognitive models for cooperative work,* ed. J. Rasmussen, B. Brehmer and J. Leplat. London: John Wiley & Sons.

Brungess, J. R. 1994. *Setting the context. Suppression of enemy air defenses and joint war fighting in an uncertain world.* Maxwell AFB, AL: Air University Press.

Brunswick, E. 1956. *Perception and the representative design of psychological experiments,* 2nd ed. Berkeley: University of California Press.

Card, S. K., T. P. Moran, and A. Newell. 1983. *The psychology of human–computer interaction.* Hillsdale, NJ: Lawrence Erlbaum Associates.

Dorner, D. 1987. On the difficulties people have in dealing with complexity. In *New technology and human error,* ed. J. Rasmussen, K. Duncan and J. Leplat. New York: John Wiley & Sons.

Ericsson, K. A., and N. Charness, N. 1994. Expert performance: Its structure and acquisition. *American Psychologist* 48:725–747.

Flach, J. M., S. Dekker, and P. J. Stappers. 2008. Playing twenty questions with nature (The surprise version): Reflections on the dynamics of experience. *Theoretical Issues in Ergonomics Science* 9 (2): 125–145.

Flach, J. M., P. Jacques, D. Patrick, M. Amelink, M. M. van Paassen, and M. Mulder. 2003. A search for meaning: A case study of the approach-to-landing. In *Handbook of cognitive task design,* ed. E. Hollnagel, 171–191. Mahwah, NJ: Lawrence Erlbaum Associates.

Flach, J. M., M. Mulder, and M. M. van Paassen. 2004. The concept of the "situation" in psychology. In *A cognitive approach to situation awareness: Theory, measurement, and application,* ed. S. Banbury and S. Tremblay, 42–60. Aldershot, England: Ashgate.

Flach, J. M., and J. Rasmussen. 2000. Cognitive engineering: Designing for situation awareness. In *Cognitive engineering in the aviation domain,* ed. N. Sarter and R. Amalberti, 153–179. Mahwah, NJ: Lawrence Erlbaum Associates.

Flach, J. M., M. R. H. Smith, T. Stanard, and S. M. Dittman. 2004. Collision: Getting them under control. In *Theories of time to contact,* ed. H. Hecht and G. J. P. Savelsbergh, 67–91. Amsterdam: Elsevier, North-Holland.

Hick, W. E. 1952. On the rate of gain of information. *Quarterly Journal of Experimental Psychology* 4:11–26.

Hutchins, E. L. 1995a. *Cognition in the wild.* Cambridge, MA: MIT Press.

———. 1995b. How a cockpit remembers its speeds. *Cognitive Science* 19:265–288.

Hyman, R. 1953. Stimulus information as a determinant of reaction time. *Journal of Experimental Psychology* 45:188–196.

Kirlik, A. 2006. *Adaptive perspectives on human–technology interaction.* Oxford, England: Oxford University Press.

Lachman, R., J. L. Lachman, and E. C. Butterfield. 1979. *Cognitive psychology and information processing: An introduction.* Hillsdale, NJ: Lawrence Erlbaum Associates.

Mandelbrot, B. B. 1983. *The fractal geometry of nature.* New York: Freeman.

Marzolf, G. S. 2004. *Time-critical targeting: Predictive versus reactionary methods: An analysis for the future.* Maxwell AFB, AL: Air University Press.

Neisser, U. 1976. *Cognition and reality.* San Francisco, CA: W. H. Freeman and Company.

Rasmussen, J. 1986. *Information processing and human–machine interaction: An approach to cognitive engineering.* New York: Elsevier.

Rouse, W. B. 1982–1983. A mixed-fidelity approach to technical training. *Journal of Educational Technology Systems* 11:103–115.

Schiflett, S., and L. R. Elliot. 2000. Synthetic team training environments: Application to command and control aircrews. In *Aircrew training and assessment,* ed. J. H. F. O'Neill and D. Andrews. Mahwah, NJ: Lawrence Erlbaum Associates.

Taylor, F. W. 1911. *The principles of scientific management.* New York: Harper and Row.

Vicente, K. J. 1999. *Cognitive work analysis: Toward safe, productive, and healthy computer-based work.* Mahwah, NJ: Lawrence Erlbaum Associates.

Wickens, C. D. 1984. *Engineering psychology and human performance.* Columbus, OH: Merrill.

# 17

# *Interface Evaluation*

## 17.1 Introduction

> The crucial question is not "what" the actors are doing at the time of analysis, but "why," together with the alternatives for "how." (Rasmussen, Pejtersen, and Goodstein 1994, p. 31)

The complex issues involved in evaluation and generalization described in the previous chapter strongly suggest a need to categorize the various types of research that one might conduct in terms of complexity, fidelity, and control. Rasmussen et al. (1994) have devised a framework for evaluation from the cognitive systems engineering perspective that accomplishes this goal. This evaluative framework consists of five nested, hierarchical levels that are defined in terms of the "boundary conditions" or "constraint envelopes" that can be used to categorize alternative evaluative settings, as illustrated in Figure 17.1.

It might be useful to consider Figure 17.1 relative to Figure 16.1 in the previous chapter. In the reductionist strategy typically associated with the dyadic approach, control is greatest and generalizability is highest at the innermost boundary (level 1). However, the triadic approach maintains that, while control may be greatest at level 1, generalizability increases as one moves toward the outer boundaries (due to increasing representativeness of the measurement context). Thus, the triadic approach does not deny that basic research at level 1 can be an important component in a research program. However, evaluations must also be performed in contexts that incorporate more of the complexities that characterize real work domains. In contrast, the dyadic approach tends to dismiss the importance of evaluations at outer boundaries. We believe that only by looking across levels is it possible to relate the "how" of work with the "why" of work.

The purpose of this chapter is to provide some concrete examples that serve to make these concepts of evaluation, as well as the associated implications for system design, clearer. The chapter begins with a description of the cognitive systems engineering (CSE) framework for evaluation. This is followed by the discussion of a representative set of empirical evaluations that were conducted at different levels within this framework. This section also serves to provide concrete examples of a fundamental premise of the book: that the interactions

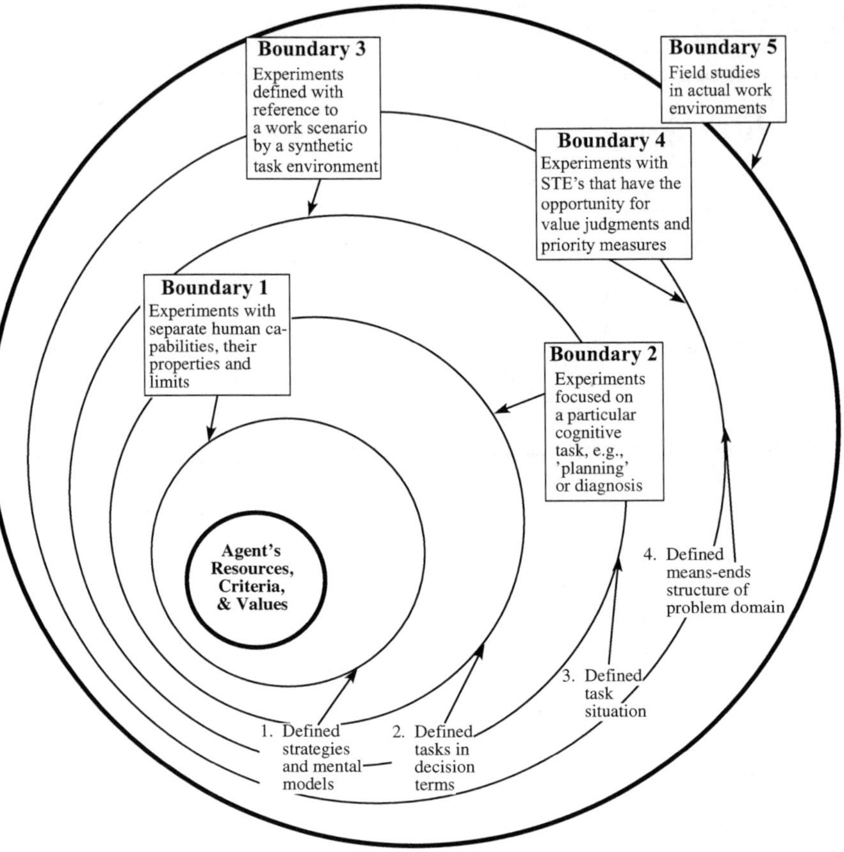


**FIGURE 17.1**
Rasmussen, Pejtersen, and Goodstein's cognitive systems engineering framework containing five levels of evaluation, each associated with different "constraint envelopes." (Adapted from Rasmussen, J., A. M. Pejtersen, and L. P. Goodstein. 1994. *Cognitive Systems Engineering.* New York: John Wiley & Sons. With permission.)

between the constraints contributed by domain, interface, and agent are fundamental considerations in display and interface design and evaluation. Finally, the results of these evaluations are examined for the degree of fit with regard to the belief structures of the two scientific paradigms.

## 17.2 Cognitive Systems Engineering Evaluative Framework

The CSE framework illustrated in Figure 17.1 provides a structured approach for devising complementary settings to evaluate the effectiveness of display and interface designs (Rasmussen et al. 1994). Each progressive step from inner to outer boundaries represents the incorporation of successively more of the constraints that characterize the underlying work domain. The defining characteristics of each boundary level will be considered in greater detail; the descriptions are tailored for display/interface evaluations and include representative examples.

### 17.2.1 Boundary 1: Controlled Mental Processes

Rasmussen et al. (1994) describe the evaluations performed at this level as controlled laboratory investigations. A defining characteristic of this level of evaluation is "the close relationship that often exists between the processes of the experimental equipment and the experimental task—this is not found in more complex simulation scenarios" (pp. 218–219). In terms of display and interface design, evaluations conducted at this level often examine basic perception–action capabilities of the agent with respect to specific design aspects of the interface—for example, the relationship between particular visual features that have been used to encode information into a display or graph and how well participants can extract or decode this information. Performance is usually measured in terms of the accuracy and the latency of responses.

High levels of experimental control can be maintained in boundary 1 evaluations. Because of the simplicity of these tasks, the strategies required for their completion are well-defined (and extremely limited). In Rasmussen and others' (1994) words:

> The formulation of the subject's instruction is at the procedural level and is very explicit. It serves to define the constraint boundary around the experimental situation and isolate it from (1) the general, personal knowledge background, and performance criteria of the subject, and (2) any eventual higher level considerations within the experimental domain itself. (p. 218)

An attractive aspect of boundary 1 evaluations is that performance at this level is most likely to conform to normative constraints. For example, performance tends to converge toward a normatively optimal level as skill increases in simple compensatory tracking tasks. Similarly, as mentioned in early chapters, performance in choice reaction time tasks to arbitrary stimuli tends to conform rather closely to simple models of communication channels. Thus, this research is important for validating general theories and principles (control theory or information theory) with respect to human performance. It is important to know that the predictions of information theory and of control theory apply when humans are the information channel or when they close the control loop.

Most of the research on display and interface design at this level has been performed from the dyadic perspective. The experimental tasks often have little "direct concern for the eventual contexts of the end users" (Rasmussen et al. 1994, p. 218). A prototypical example is the work of Cleveland and his colleagues (Cleveland 1985; Cleveland and McGill 1985), who systematically varied the visual features employed in alternative representations and ranked the participants ability to make discriminations using these visual features. Note that although it is less prevalent, research at this level can also be conducted from a triadic perspective (e.g., the research on emergent features described in Chapter 8). To qualify as a triadic evaluation, the experimental tasks need to be informed by aspects of the situation (see the discussions in Chapter 16, Section 16.3.1, and Chapter 17, Section 17.3).

### 17.2.2 Boundary 2: Controlled Cognitive Tasks

This level of evaluation assesses performance using experimental tasks designed to provide closer approximations of activities that characterize real work domains. The focus is on isolated "decision functions, such as diagnosis, goal evaluation, planning and/or the execution of planned acts" (Rasmussen et al. 1994, p. 219). These experimental tasks are typically more complex than those found at the previous level of evaluation. More fundamentally, to complete these tasks a participant must consider and act upon more than just the physical characteristics of the display alone. That is, the participant must obtain the information presented and then determine what this information means in the context of the task to be performed. Accuracy and latency may be used, but performance assessment may also involve additional types of measurements.

At this level of evaluation, a participant's general knowledge of the task and the particular strategies that they develop and employ will become more important (relative to boundary 1). Therefore, individual differences will have a more pronounced influence on the levels of performance that are obtained. As a result, some degree of experimental control will be lost relative to boundary level 1.

Prototypical examples of interface evaluations performed at this level can be found in MacGregor and Slovic (1986) and Goldsmith and Schvaneveldt (1984). MacGregor and Slovic employed a multicue probability judgment task, which required participants to consider multiple cues (age, total miles, fastest 10 km, time motivation) with varying degrees of diagnosticity to predict the amount of time runners would take to complete a marathon. The display formats used to present this information were varied and performance measures (e.g., achievement index) were calculated and analyzed using Brunswick's (1956) lens model. Research conducted at this level tends to tilt toward the triadic perspective since domain constraints are more likely to be incorporated into the experimental setting; however, this is by no means necessary (i.e., cover stories with relatively artificial domain constraints incorporated are often used).

### 17.2.3 Boundary 3: Controlled Task Situation

A synthetic task environment is required to conduct evaluations at boundary 3. These evaluations are "more complex experiments focused on actual task situations" that use "simulations … designed as replicas of actual work scenarios" (Rasmussen et al. 1994, pp. 222–223). These simulations may be causal or mathematical, but they must capture critical aspects of the work domain (i.e., complexity, fidelity). It follows that their development must be guided by work domain analyses. In contrast to the previous boundary levels, these evaluations are likely to incorporate a very direct link between the experimental tasks to be performed in the evaluation setting and the specific tasks that exist in a particular work domain.

Correspondingly, the experimental tasks to be completed at this level will be more complex than those encountered in the previous two boundary levels. The tasks will typically involve consideration of physical and functional characteristics of the domain, competing goals, limited resources to achieve these goals, and performance trade-offs. The measures that are used to assess performance will be defined by the domain itself and will therefore be relatively domain specific. At this level of evaluation, an individual's general knowledge, specific knowledge about the domain and task or tasks to be performed, and efficiency of the strategies that he or she employs will play a more important role in the findings that are obtained. All of these considerations point to the obvious: A larger degree of experimental control will be lost relative to boundary 2 and 1 evaluations.

Prototypical examples of display evaluations performed at this level include Moray, Lootsteen, and Pajak (1986), Mitchell and Miller (1986), and evaluations in this book (see Chapters 11, 14, and later in this chapter). For example, Mitchell and Miller (1986) simulated a flexible manufacturing plant; performance measurements included the number of units that were produced. Once again, the key to the use of synthetic task environments, from the triadic perspective, is that they are modeled on the constraints of real-world task domains (and not just games or microworld cover stories). Dyadic evaluations at this level are the exception rather than the rule.

### 17.2.4 Boundary 4: Complex Work Environments

Rasmussen et al. (1994) describe this evaluation boundary in the following fashion: "A more recent category of experiments has been focused on human problem solving behavior in complex simulated work environments in which the entire decision process is activated, including value formation, goal evaluation, and emotional factors" (p. 224). The synthetic task environments at this level could have a higher degree of fidelity (and complexity) than those at the previous level. However, this is not the primary distinguishing factor. Rather, it is critical that the evaluation is set up to determine the influence of the interface on the participant's goal formulation and performance criteria (i.e., value judgments and priorities).

Typically, the participant will be presented with a relatively open-ended task and will be free to formulate the task and goals on his or her own. For example, operators might be instructed to run a power plant in a full-scale simulator without explicit specification of values and priorities. In fact, this might involve a series of sessions with perhaps one or more of these sessions including faults that require the full range of cognitive processes, from recognizing that a problem is occurring to planning and carrying out an appropriate intervention that reflects value judgments (e.g., trade-offs between profit and safety).

### 17.2.5 Boundary 5: Experiments in Actual Work Environment

At this boundary, the system is evaluated using real practitioners in the actual domain (field studies). Published reports of this type of evaluation are relatively rare. Rasmussen and others' (1994) evaluation of an innovative library information retrieval system, the BookHouse, is an excellent example. This system assists librarians and patrons in the selection of books (fiction) from a library. This system was evaluated in a public library during a 6-month period. Numerous measures of performance were obtained, including "(a) a questionnaire, (b) online logging of all dialogue events (mouse clicks, etc.), (c) observation, and (d) interviewing by the librarians who (e) also kept a logbook with reports of user responses, system behavior, and so on" (p. 319). See also Hutchins's (1995) work on ship navigation.

### 17.2.6 Summary

The five boundaries in this framework span a range from controlled experiments to field evaluations. This has surface similarities to the basic and applied dimension described earlier, but it is conceptually distinct (i.e., consistent with the triadic paradigm). This is made clear in Rasmussen and others' (1994) description of the framework: "the innermost boundary [level 1] corresponds to the evaluation of actor-related issues in an environment that corresponds most closely to the traditions of experimental psychology. The remaining boundaries successively 'move' the context further from the actor to encompass more and more of the total work content" (p. 205). In essence, these boundary levels provide a structured means to parse the domain into meaningful subsets to evaluate the effectiveness of various aspects of display and interface design.

## 17.3 Representative Research Program; Multiboundary Results

In this section we will provide concrete examples of the framework and its use in evaluation. A research program that investigated issues in configural display design will be described (Bennett et al. 2000; Bennett, Payne, and Walters 2005; Bennett, Toms, and Woods 1993; Bennett and Walters 2001;

Bennett and Zimmerman 2001). A synthetic task environment (process control) was used in each of these empirical evaluations; this allowed a wide variety of performance measurements to be obtained simultaneously from multiple boundary levels of evaluation. The pattern of results obtained within and between levels of the framework will be described. The methodological "lessons learned" will be discussed, with an eye toward the larger issues in evaluation and generalization outlined in this and the previous chapter.

### 17.3.1 Synthetic Task Environment for Process Control

The development of the synthetic task environment was motivated by real-world concerns in a process control work domain. The description that follows will provide only a brief overview. See Woods and Roth (1988) and Roth and Woods (1988) for a detailed description of the rationale and development process, Haley and Woods (1988) for a description of the unique control theoretical aspects, and Bennett, Toms, and Woods (1993) for a concise description of the simulation.

The synthetic task environment simulates selected aspects of a nuclear power plant during the start-up phase. In the actual work domain, teams of operators are required to control energy (nuclear reactor), mass input to steam generators (feedwater), and mass output (from generators to turbine). The critical aspect is to maintain the indicated water level in the steam generators between high and low set-point boundaries (thereby avoiding automatic shutdown with high economic penalties).

Factors that contribute to the difficulty of the start-up task include poor information about system state, narrow performance windows, complex interactions (particularly time-delayed and counterintuitive thermodynamic effects), conflicting goals, and a need for intricate anticipation and coordination of the actions of the various operators. Successful completion of the start-up phase is sufficiently difficult that it could not be automated. The development of the synthetic task environment was motivated by concerns that were both economic (loss of substantial revenues) and practical (the full-scope simulations did not model the dynamics of the start-up phase faithfully; operators lacked practice and training on this infrequent event).

The synthetic task environment was not designed to replicate the full complexity of this difficult work domain. Rather, it was designed to capture the critical demands that are placed on human performance. The complex dynamics place a premium on a controller's ability to anticipate the effects of changes in plant state and the effects of various control actions on the indicated level of water within a steam generator. Compensatory responses must be made before the ultimate effect of these events on indicated level is seen. The controller needs to know what energy and mass factors have been introduced into the system in order to do this.

However, the dynamics and the poor state information make this assessment difficult. For example, determining whether a change in indicated level represents a longer term change in mass or a transitory energy effect (counterintuitive and time-delayed thermodynamic effects) can be quite difficult. Thus, performance at the feedwater/level task fundamentally revolves around separating the relative contributions of two independent functional processes on indicated level (separating long-term water mass effects from short-term energy effects—shrink and swell). These constitute the fundamental demands of the work domain that were faithfully replicated in the synthetic task environment.

### 17.3.2 Boundary Level 1 Evaluation

Bennett et al. (2000) performed an empirical study investigating issues in configural display design. Although simultaneous boundary 3 (i.e., system control, fault detection) and boundary 1 (accuracy and latency of information pickup) evaluations were conducted, the latter will be the focus of this discussion. The capability of a configural display to support an agent in obtaining domain information (both low-level data and high-level properties—domain constraints located at different levels of the abstraction hierarchy) was evaluated. One version of the configural display was augmented with several design techniques (the composite display, illustrated in Figure 9.12b in Chapter 9) with the potential to improve performance; the second configural display (the baseline display, illustrated in Figure 9.12a) did not have these techniques applied.

Two different experimental methodologies were used to collect data. With the visual methodology, the displays were visible during the completion of low-level data and high-level property probes, thus allowing an observer to use perceptual systems to extract information. With the memory methodology, the simulation was paused and the displays were removed from sight prior to the administration of a probe, thus requiring an observer to complete the probe from memory. In summary, the experimental design included two probe types (low-level data, high-level properties), two display types (composite, baseline), and two methodologies (visual, memory). The results of the evaluation, illustrated in Figure 17.2, provide diametrically opposed answers to the issues of display design that were posed, depending entirely upon the experimental methodology used.

There was a substantial and significant cost for completing low-level data probes when the memory methodology was used, as illustrated in the bottom of Figure 17.2. Low-level data probes took significantly longer to complete and were significantly less accurate than probes for high-level properties. (Note that pairs of data points in Figure 17.2 are labeled with the experimental manipulation that had the larger impact on performance). These findings suggest that the design techniques aimed to offset these potential costs had very little impact; they are very consistent with an interpretation based on principles of design that predict an inevitable cost for representing low-level

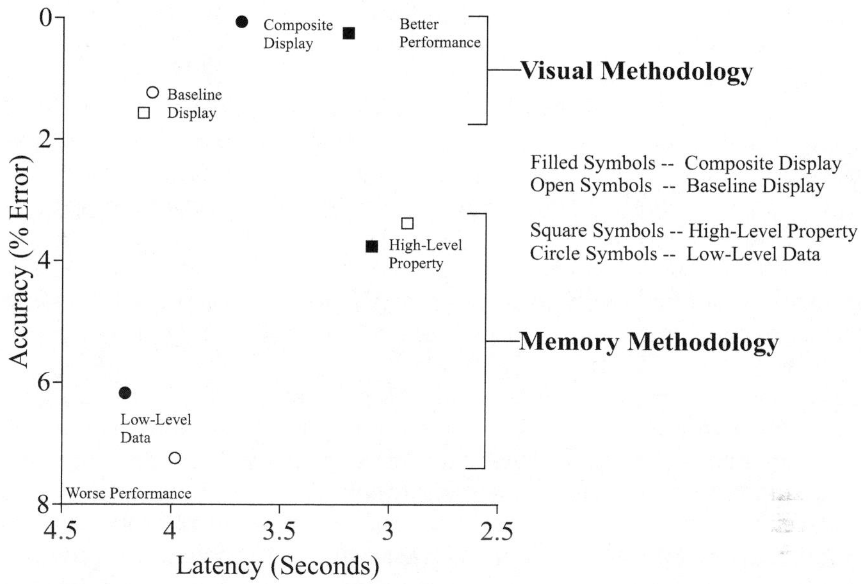


**FIGURE 17.2**
Diametrically opposed patterns of results obtained within a boundary level 1 evaluation. (Adapted with permission from Bennett, K. B. et al. 2000. *Human Factors* 42:287–298. Copyright 2000 by the Human Factors and Ergonomics Society. All rights reserved.)

data in a configural format (e.g., ungluing the parts of an object, integrality; see the associated discussion in Chapters 8 and 9).

The results for the visual methodology reveal a diametrically opposed pattern of results that requires a completely different interpretation. The display type, rather than the probe type, is the more important factor in determining performance. The design techniques aimed at offsetting potential costs were effective: The composite display produced significantly better performance than the baseline display for both types of probes. These results are consistent with a different theoretical perspective, which maintains that there is not an inherent cost for extracting low-level data from a configural display (e.g., principles based on configurality).

These results support Miller's observation that methodology is the "bread and butter of the working scientist." In this case, methodological choices alone determine the outcome. Either set of outcomes, conducted in isolation, would be interpreted as support for a particular theoretical orientation, divergent principles of design, and different design solutions. Both sets of outcomes have been replicated and there are very reasonable explanations when viewed from the triadic perspective. (Specifically, the methodological choices had an impact on the agent constraints that were relevant for performance in the setting.) See Chapter 9 for an interpretation of the results obtained with the visual methodology and Bennett et al. (2000) for an interpretation of the memory methodology.

The question is which set of findings should be used as the basis for principles of display design. The answer to this question lies in one's beliefs with regard to which set of results is most likely to generalize to real-world systems. From the triadic perspective, the whole point of ecological interface design is to leverage the powerful perception–action capabilities of the agent whenever possible and, conversely, to avoid situations where the agent is required to remember information and perform mental calculations. Thus, the results obtained with the visual methodology are preferable simply because they are more relevant for the kinds of activities that we are most interested in supporting.

Note that this experiment is a representative example of a boundary level 1 evaluation conducted within the triadic paradigm. Different display formats were evaluated with regard to the obtaining of meaningful information while interacting with a synthetic task environment. The different types of information came from different levels of the abstraction hierarchy (abstract function, physical function), and these variables and properties were previously identified in domain analyses as being critical to performance in the work domain. The results also validate the assumptions of the triadic perspective by illustrating the importance of considering the complex interactions that occur at the union of the three system components illustrated in Figure 16.2.

### 17.3.3 Boundary Levels 2 and 3

The second example draws from a line of research (Bennett and Zimmerman 2001; Bennett et al. 2005) that investigated the complementary benefits of configural and temporal displays (i.e., displays that present system state over time), as described in Chapter 10, Section 10.4.4. The potential benefits of a combined representational format are clear on logical, empirical (Pawlak and Vicente 1996), and practical (see the Bhopal example in Chapter 10) considerations. However, demonstrating this potential empirically proved to be a surprisingly difficult challenge. Bennett and Zimmerman (2001) summarized the results of the first study: "The results of the present experiment are not particularly encouraging for the time tunnels display design technique" (p. 198). Nor were the results of a number of subsequent evaluations that varied a number of critical parameters (e.g., the display was redesigned, alternative measures of control performance were used, experimental procedures were varied, etc.).

Bennett et al. (2005) eventually obtained empirical evidence supporting the time tunnel design technique. The determining factor in the success of this evaluation involved a change in the boundary level of evaluation. Experiment 1 used simultaneous evaluations at multiple levels of evaluation: boundary level 3 (system control, fault compensation), boundary level 2 (cognitive tasks, including fault detection and estimates of "true" system state), and boundary level 1 (obtaining and remembering the value of variables). In experiment 2 the boundary level 3 evaluation was dropped; participants did not control the system in real time. However, participants were still required to complete the same boundary level 2 cognitive tasks (see Figure 17.1) that were intimately tied to the work domain semantics; the stimuli were "snapshots" of actual system states that the same participants had generated in experiment 1.

Unlike previous evaluations, the boundary level 2 evaluation yielded significant results favoring a time tunnel display relative to two other display formats (a baseline configural display without temporal information and a configural display with temporal information presented separately in a traditional strip-chart or line-graph display—the trend display). The results favoring the time tunnel display are illustrated in Figure 17.3 (see Bennett et al., 2005, for the complete set of results). The display supported participants in determining the state of the system in the presence of counterintuitive thermodynamic effects and time delays. This is a capability that is critical to successful performance, as determined by both work domain analyses (Roth and Woods 1988; Woods and Roth 1988) and empirical evaluation (Bennett and Zimmerman 2001).

The question then becomes the following: "Why did the same participants performing the same tasks with the same displays portraying the same data produce significant performance differences in experiment 2 but not in

**Boundary 2 ($p <$)**

| | Non-Fault | | Reservoir Leak | | Stuck Valve | | |
|---|---|---|---|---|---|---|---|
| **(a) Time Tunnel Vs. Trend** | Signed Error | Absolute Error | Signed Error | Absolute Error | Signed Error | Absolute Error | Fault Certainty |
| Time After Onset (Sec) | | | | | | | |
| 1. 40 | | | | | | | 0.0004 |
| 2. 80 | 0.002 | 0.004 | | | 0.02 | 0.04 | |
| 3. 120 | 0.02 | 0.006 | 0.002 | 0.002 | | | |
| 4. 160 | | 0.002 | 0.006 | 0.009 | | 0.004 | |
| 5. 200 | | 0.05 | | | | | |
| **(b) Time Tunnel Vs. Baseline** | | | | | | | |
| Time After Onset (Sec) | | | | | | | |
| 1. 40 | | | | | | | |
| 2. 80 | | | | | | 0.02 | |
| 3. 120 | | 0.02 | | | | | |
| 4. 160 | 0.02 | 0.03 | | 0.02 | 0.007 | 0.007 | |
| 5. 200 | 0.002 | 0.002 | | 0.004 | 0.0008 | 0.0004 | |

**FIGURE 17.3**
Significant results indicating improved performance for the time tunnel display format in a boundary level 2 evaluation. (Adapted with permission from Bennett, K. B., M. Payne, and B. Walters. 2005. *Human Factors* 47:342–359. Copyright 2005 by the Human Factors and Ergonomics Society. All rights reserved.)

experiment 1?" Recall that the primary advantage of the time tunnel display technique relative to traditional configural displays lies in the presentation of temporal information (i.e., explicitly specifying the dynamic changes in domain variables and properties over time). Participants had full access to the unfolding temporal context in experiment 1 (i.e., their primary job was to monitor and control the process in real time). This allowed the participants to integrate internally the temporal information necessary for successful performance.

Thus, the external representation of temporal information in the time tunnel display was not effective because it was simply not needed. In contrast, access to the unfolding temporal context was eliminated in experiment 2. Under these circumstances, the external representation of temporal information provided by the time tunnel display became very important to successful performance.

At the outset of the project, we were very surprised at the lack of empirical research on temporal displays (Bennett et al. 2005): "Despite their intuitive appeal and widespread use, there has been surprisingly little research conducted on temporal displays in dynamic settings" (p. 342). Based on these findings, it appears that the reason why is related to the difficulties in decoupling the agent's capability to integrate temporal information from the potential benefits provided by a temporal representation.

There are several methodological messages from this example. First, it illustrates the advantages of conducting research with a synthetic task environment (STE). There is a very tight coupling between characteristics of the real work domain, characteristics of the STE, and behavioral measurements. Moreover, simultaneous measurements of performance at multiple boundary levels can be obtained. Much like the first example, it also validates the fundamental assumptions of the triadic approach with regard to the need to consider the complex interactions occurring between the three triadic components depicted in Figure 16.2. In this case, the adaptive nature of system constraints contributed by the agent played a major role in the experimental outcomes. Finally, the example demonstrates the increased complexity and loss of experimental control that occurs as one moves to higher boundary levels (see Figure 16.1 and associated discussion).

### 17.3.4 Boundary Levels 1 and 3

The third and final example discusses three studies (Bennett et al. 1993, 2000; Bennett and Walters 2001) in which simultaneous evaluations at two different boundary levels (1 and 3) were conducted. The potential for results to generalize between these two boundary levels was very high: The same basic configural display (see Figure 9.13 in Chapter 9), the same synthetic task environment, and the same participants (within experiments) were used.

The results of these experiments are illustrated in Figure 17.4. Each study and its display manipulations are listed in the leftmost column. The two boundary levels and the dependent measures contained within are listed in the center (boundary 1) and the rightmost (boundary 3) columns. Each cell formed by the intersection between display manipulations (rows) and dependent measures (columns) is a place-holder for a statistical test that was performed; only results that were significant ($p < 0.05$) are shown.

The boundary level 1 evaluations assessed the effect of various formats (e.g., propositional, alternative analogical displays) and design techniques (e.g., color coding, scales) on the ability of agents to obtain domain information. The issues, principles, and many of the specific manipulations have been described previously in Chapters 8, 9, and 10. It is fairly clear that these display manipulations had a significant impact on the capability of an agent to obtain domain information, given the numerous effects that appear in the boundary level 1 column in Figure 17.4.

The boundary level 3 evaluations assessed the effect of these same formats and techniques on the ability of agents to perform domain-related tasks (e.g., system control). It is clear that these evaluations produced substantially fewer significant results. However, each significant effect that occurred at this boundary level was also paired with at least one significant effect at boundary level 1. Therefore, each instance represents potential evidence to support the generalization of results between boundaries. A detailed analysis of these findings indicates that there is extremely limited evidence for generalization.

The significant results for the comparisons labeled 9 and 10 (see highlighted results in Figure 17.4) do not support generalization because the effect of the manipulation was in the opposite direction for the two boundary levels. As described in Chapter 9, the digital display (Figure 9.13f) improved performance at boundary 1 because the associated display constraints (a digital value) matched the task constraints exactly (provide a numerical estimate of the value of a variable). In contrast, the digital-only display imposed a truly severe set of constraints for performance at boundary 3, primarily because it did not contain the analog geometric format. Participants could not use powerful pattern recognition capabilities to complete domain tasks because the domain constraints (relationships, properties, goals, etc.) were not directly visible in the digital format. Instead, the participants were forced to derive the current system state mentally, using the digital values in conjunction with their conceptual knowledge about the system (see Bennett and Flach, 1992, and Bennett, Nagy, and Flach, 2006, for a more detailed discussion of similar considerations).

One might be tempted to conclude that the significant results for the comparisons labeled 11 (circled results with no fill) support generalization since the composite display (Figure 9.13e) was associated with performance advantages at both boundaries. However, the fundamental findings outlined in the previous paragraph suggest an alternative interpretation. The composite display contained both the analog configural display (associated with improved performance at boundary 3) and digital values (associated with improved performance at boundary 1).

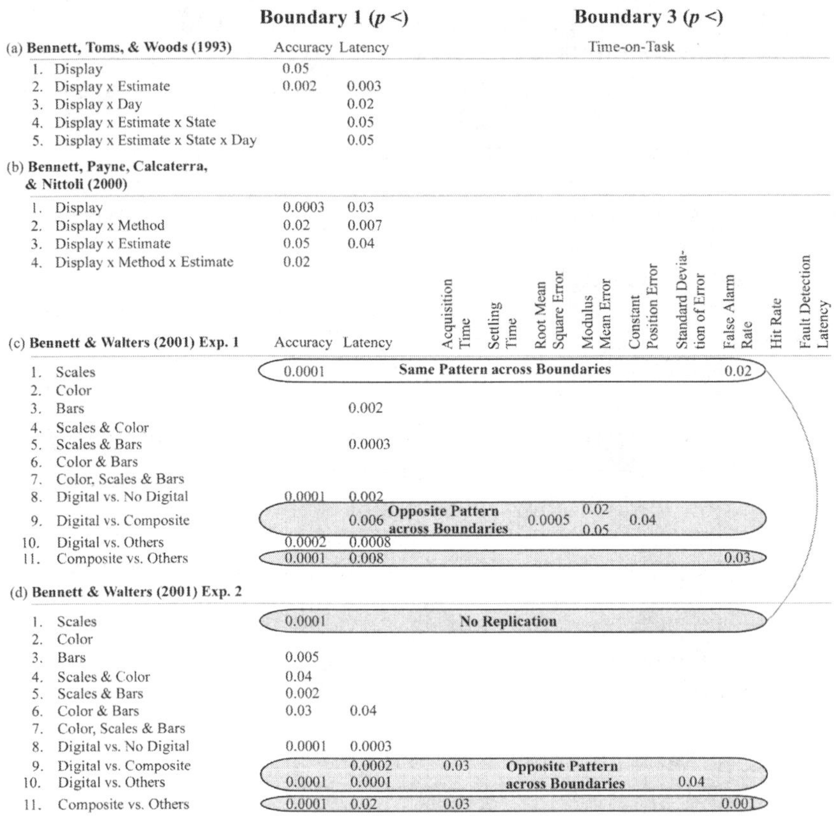


**FIGURE 17.4**
Patterns of performance between multiple boundary levels in multiple evaluations. (Adapted, in part, with permission from Bennett, K. B., and B. Walters. 2001. *Human Factors* 43 (3):415–434. Copyright 2001 by the Human Factors and Ergonomics Society. All rights reserved.)

Thus, the more likely interpretation is that participants were able to use each set of design features, independently to improve performance at the appropriate boundary. These are a very positive set of findings (see Chapter 9); they indicate that it is possible to combine propositional and analogical representations to improve performance at both boundaries (as opposed to separate displays for each type of task). But an interpretation of these results as support for the generalization of results between boundaries is clearly a stretch.

In the end, a single common finding indicated that a performance advantage obtained for a display manipulation in one boundary successfully generalized to another boundary. This is the effect labeled 1 in Figure 17.4c and it is highlighted with a white bounding oval. The presence of scales and gridlines in the display significantly increased accuracy while obtaining information (boundary 1) and significantly lowered the false alarm rate (boundary 3). However, the results were not replicated in experiment 2; the boundary level 3 finding was not significant.

The collective results of these three studies provide a fairly clear message: The evidence supporting the generalization of results between boundaries is quite meager. The overall lack of generalization between boundaries is particularly striking if one considers that the potential was quite high: five experiments, dozens of display manipulations, hundreds of similar statistical comparisons, and the same experimental subjects (within experiments) generating data.

## 17.4 Conventional Rules and Experimental Results

In this section we consider these results in light of the conventional rules (i.e., common beliefs shared by adherents of an approach) of the dyadic and triadic paradigms. As illustrated in Figures 16.1a and 16.1b, these two paradigms view the impact of complexity and fidelity on generalization in very different ways.

### 17.4.1 Match with Dyadic Conventional Rules

The results described in the previous section pose substantial difficulties for researchers who adhere to the conventional rules of a dyadic paradigm. To reiterate, dyadic approaches adopt a framework that is based on Saussure's model of semiotics (see Figure 2.1 in Chapter 2 and related discussion). Meaning is defined in terms of the relations between internal constructs (i.e., awareness, or agent constraints) and external representations (signifiers, or interface constraints). The third dimension represented in Figure 16.2—the ecology (i.e., situations, domain constraints)—is largely ignored, addressed in very general terms, or treated as a cover story.

A classic example of the dyadic approach applied to display and interface design is the proximity compatibility principle of Wickens and his colleagues (e.g., Wickens 1992; Wickens and Carswell 1995). As described in Chapter 9, the problem of display design is explicitly framed in terms of the relation between the sensory surfaces of an agent (i.e., perceptual proximity) and the internal concepts in the agent's mind (i.e., processing proximity). Domain constraints are considered only in general terms (e.g., computational or Boolean integration) and are conceptualized primarily in terms of their impact on cognitive processing (i.e., task proximity).

In contrast to the conventional beliefs of the dyadic paradigm, the pattern of results clearly indicates that domain constraints played a substantial role in the evaluation of display and interface designs. Very specific domain constraints (the complex system dynamics, changes over time, information associated with different levels in the abstraction hierarchy—low-level data, high-level properties, etc.) played a critical role in determining performance outcomes. These constraints interacted with both interface constraints (e.g., analog vs. digital, design techniques, temporal information) and agent constraints (visual attention, perception vs. memory, temporal integration, knowledge about the system). Meaningful interpretations of the results depended upon very specific and complex interactions between all three system components. This strongly suggests that a dyadic approach that does not explicitly consider the mutual interactions between all three components has fundamental and severe limitations as a scientific framework.

At a higher level, the lack of generalization between boundaries is particularly problematic for the dyadic paradigm. Recall that the dyadic approach is reductionistic in nature (see Figure 16.1a and the associated discussion). The higher order dynamics between the components illustrated in Figure 16.2 are considered to be a source of experimental confounds and uncontrolled variables. Simple experimental settings, devoid of the complexities of work domains, are devised. The search is for fundamental primitives of interaction between internal constructs (i.e., basic information processing capabilities and limitations of the agent) and external representations (i.e., variations in particular types of displays).

A fundamental belief of the dyadic paradigm is that the highest potential for generalization occurs when complexity and fidelity are low (i.e., at the lower left corner of Figure 16.1a). The belief is that these fundamental primitives can be extrapolated or combined to explain behavior in more complicated settings (i.e., they are basic truths that will generalize across all work domains). The lack of generalization between boundary levels 1 and 3 strongly suggests that these beliefs merely constitute false hope.

### 17.4.2 Match with Triadic Conventional Rules

In contrast, the results of these experiments are very consistent with the paradigmatic beliefs of the triadic approach. This paradigm adopts a framework that is based on Peirce's model of semiotics (see Figure 2.1). Thus, the core triadic components illustrated in Figure 16.2 (i.e., domain/ecology, interface, agent/awareness) are the critical determinants of performance both within a boundary and between boundaries; the degree of fit between the three sets of constraints will determine the levels of performance that are obtained.

At different points we have described the likelihood that generalization would occur in terms of complexity, fidelity, and control (Figure 16.1b) and the boundary levels of evaluation (Figure 17.1) as well. We can represent these related concerns in terms of a combined graph, as illustrated in Figure 17.5. Movement from inner to outer boundary levels involves the incorporation of more and more of the total constraint envelope of the system into an evaluative setting.

This forms the basis for interpretation of the results from the triadic perspective. An interface manipulation will produce superior performance *within* a particular boundary level only if the constraints that it contributes are well matched to those constraints contributed by domain and agent at that level (see previous discussions and Chapters 8 through 11). The generalization of results for an interface manipulation *between* two different boundaries depends upon the quality of the specific constraint mappings that occur *at each specific boundary level*. Thus, the farther apart the boundaries are, the less likely it is that a single display manipulation will simultaneously satisfy the combined domain and agent constraints that exist at two boundaries of evaluation.

The finding that display manipulations that produced significant results at the boundary level 1 evaluation did not generalize to the boundary level 3 setting is perfectly consistent with predictions derived from the triadic perspective. The display manipulations at boundary 1 (e.g., color coding, layering and separation, extenders) were aimed at increasing the capability of an agent to obtain domain-related information from an analog configural display. However, the capability to obtain this kind of information is necessary, but far from sufficient, for increased performance at

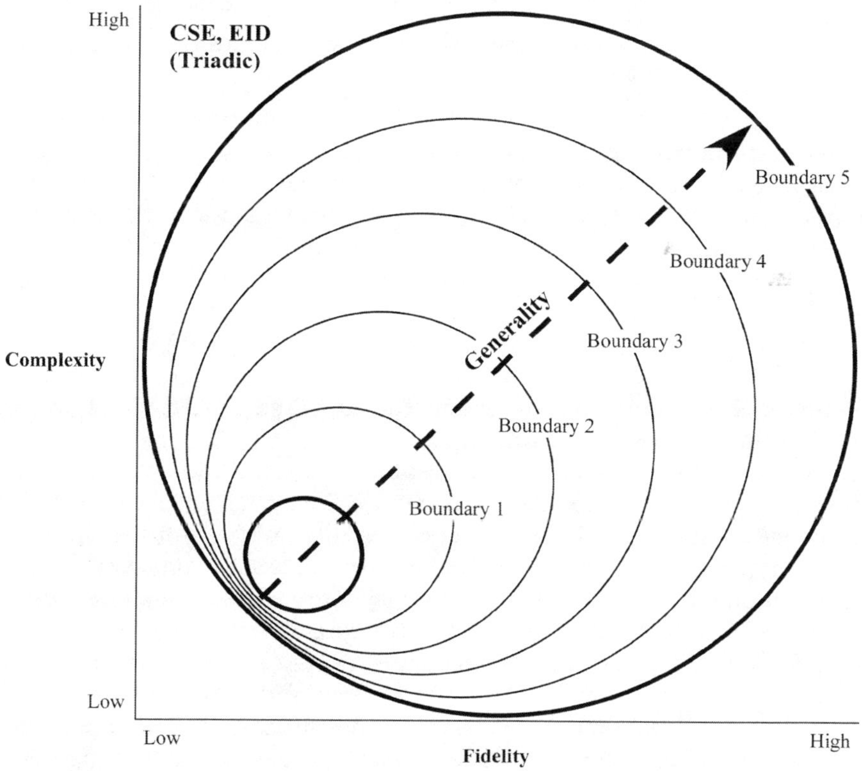


**FIGURE 17.5**
An illustration of the triadic view of generalization in the context of complexity, fidelity, control, and the boundary levels of the CSE evaluative framework (Figures 16.1 and 17.1 combined).

boundary level 3. Performance at this level depends upon a much wider set of domain (e.g., complex system dynamics) and agent (attunement, strategies) constraints.

The failure of these design interventions to generalize to boundary level 3, at least in terms of statistically significant results, simply reflects the fact that the interface constraints that they contributed were not particularly critical for performance at that boundary level. Compare this to the effect of removing the configural display altogether (i.e., the digital display, Figure 9.13f), which altered the display constraints in ways that were critical for performance at boundary level 3 (see detailed discussion in Section 17.3.4).

## 17.5 Broader Concerns Regarding Generalization

> For applied researchers it is absolutely essential that the results of studies generalize beyond the original research setting. If studies lack generalizability, then with each new operational setting one is forced to conduct research specific to that setting. Such an approach is expensive and inefficient. (Payne and Blackwell 1997, p. 330)

The ultimate goal of our analysis, design, and evaluation efforts is to produce display and interface interventions that prove useful in applied settings. In other words, it is imperative that our results generalize. The triadic approach maintains that the potential for generalization increases as the evaluative context becomes more specific relative to the underlying work domain; it emphasizes the role of very specific mappings between all three system components. All three evaluation examples in the previous section support this perspective, providing outcomes that were extremely context specific and very little evidence to support the generalization of results between evaluation boundaries. These considerations lead us to the concerns that are captured succinctly in the preceding quote.

Fortunately, it is not true that each interface design initiative must start from scratch; generalization is expected. The key factor that determines the potential for generalization is commonality in constraint boundaries: Successful design solutions (e.g., controls and displays) in one work domain will generalize to other work domains that share similar behavior-shaping constraints. For example, in Figure 10.3 (Chapter 10), the graphical arrangement of two bar graphs and a connecting line are used to represent the difference between commanded and actual flow and between mass in and mass out. This display will generalize to other process control applications, or to any domain where there is a constraint that requires comparisons between two values (e.g., account balances, supply/demand, etc.). Similarly, if a constraint requires precise values of variables or properties, the annotation of graphical representations with propositional representations will be effective. Many of the specific design solutions outlined in the book could be analyzed in a similar fashion.

If one moves beyond the simple definition of generalization as the production of statistically significant results, it becomes apparent that there are other ways in which design solutions can generalize. Consider the relationships between the evaluation boundaries depicted in Figures 17.1 and 17.5. These boundary levels are not isolated and independent. Rather, each inner boundary is nested within outer boundaries (i.e., circles representing inner boundary levels are placed entirely inside other circles representing outer boundary levels). These nested relationships suggest that each outer level builds upon inner levels—that successful solutions at inner boundary levels play a role in supporting performance at outer levels.

The implications for a revised definition of generalization can be more concrete by considering the previous examples. Obtaining domain-related information (boundary level 1, e.g., steam and feed flows, mass balance) is a fundamental activity that needs to be supported via effective design. The capability to obtain domain information is necessary to perform higher order domain tasks (e.g., boundary level 2, e.g., fault detection and diagnosis) even though it is not sufficient (i.e., successful performance depends upon a broader set of considerations than just obtaining information). Similarly, design features that support successful performance at the isolated tasks found at boundary level 2 are necessary, but not sufficient, to support performance at boundary level 3 (e.g., fault compensation requiring consideration of alternative resources, interconnections, goals, and functional trade-offs).

Thus, we would not necessarily expect the display manipulations that were effective at an inner boundary level to generalize to an outer boundary level in the sense that they produce statistically significant differences. However, getting the design right at inner boundary levels will generalize in the sense that it contributes to or supports performance at outer boundary levels. The framework suggests that evaluations that address the progressively larger sets of domain constraints will be required. It further suggests that improvements at the various boundaries can work together, producing display or interface designs with emergent properties that are greater than the sum of their parts.

Finally, from a broader perspective, the principles of analysis and design that have been described throughout the book will generalize: the CSE and ecological interface design (EID) frameworks are sufficiently general to be used to develop interfaces for any work domain. Modeling of work domain constraints in terms of the abstraction and aggregation hierarchy will serve to specify the affordances of the work domain. Modeling constraints in terms of modes of behavior (i.e., skill-based, rule-based, and knowledge-based behaviors) and information processing and states of knowledge (i.e., the decision ladder) will serve to specify the attunement of agents. Interfaces incorporating the principles of direct perception (i.e., analogical, metaphorical, and propositional representations) and direct manipulation (i.e., an intact perception–action loop) will provide decision-making and problem-solving support (i.e., interface constraints) with a high degree of specificity (i.e., matching domain and agent constraints).

## 17.6 Conclusion

> In short, we find that the framework suggested here can serve to resolve much of the standing controversy about the value of laboratory experiments for understanding "real life" work—a controversy caused by designers' frustrations concerning the lack of explicit descriptions of the boundary conditions of experiments in terms that could facilitate the transfer of results to their work contexts. (Rasmussen et al. 1994, p. 222)

We believe that the "standing controversy" and the "frustrations" referred to in this quote as well as the disconnect between research and application are a direct result of the fact that most researchers in human factors and related fields have historically adopted a dyadic approach to the study of display and interface design. It is somewhat ironic that the search for universally applicable principles of design has produced results that are, by and large, simply not applicable. Much like Rasmussen and his colleagues, we believe that the triadic approach "can resolve much of the standing controversy." However, adopting the paradigmatic commitments of the triadic approach entails facing a considerable set of challenges.

A fundamental paradigmatic commitment of the triadic approach is that each interface solution will require the coordination of diverse activities (e.g., analysis, design, and evaluation) guided by the characteristics of specific work domains. Detailed work domain analyses are required for the effective design of displays and interface solutions. Each new design effort presents a unique challenge for the development of decision-making and problem-solving support. Numerous examples throughout the book illustrate the extremely context-specific nature of display and interface design. These are challenges that can be met only through the completion of work domain analyses, the uncovering of context-specific constraints, and the development of specifically tailored design solutions. Although successful solutions from domains with similar constraints can be modified, overall there are no "cookie-cutter" solutions—no "cookbooks" or checklists that ensure effective design.

Work domain analyses are also critical in evaluation. The evaluation of displays and interfaces will require multiple evaluations at different boundary levels. A key concern lies in the parsing of the overall work domain into subsets that reside at each boundary level. Decisions with regard to how this parsing should proceed can only be informed by the process and products of a work domain analysis. The analytical tools that CSE provides (e.g., abstraction and aggregation hierarchies; decision ladder) can facilitate this process by serving as "lenses" to focus on the content that needs to be incorporated at each level.

The additional requirement to develop synthetic task environments is viewed as a key element of this approach. The need to evaluate analog configural displays in dynamic settings (see Section 8.4.4 in Chapter 8) is but one example. STEs facilitate the process of evaluation by bridging the gap between laboratory experiments and field studies. When developed properly, they incorporate key elements of the complexity encountered in real work domains and at the same time allow a degree of experimental control that is required for empirical evaluations. As described previously, they also allow performance measurements to be obtained at multiple levels and therefore alternative perspectives on the efficacy of design interventions to be obtained.

An inevitable consequence in adopting this multitiered approach to evaluation is illustrated quite clearly in Figure 16.1b and in the second and third examples in the previous section. There is a loss in experimental control as one incorporates more complexity and fidelity into the evaluation setting (i.e., moves toward outer boundaries). For example, in describing boundary level 3 evaluations, Vicente (1997) states: "This type of research sacrifices even more experimental control in order to determine whether results obtained under research of types 1 and 2 [boundary levels 1 and 2] prevail in the face of myriad additional factors that had not been addressed or were held constant" (p. 326).

In straightforward terms, the ability to detect significant differences between design interventions will be degraded as one moves to outer boundary levels. Thus, moving outward eventually requires a switch from empirical, outcome-based methodologies (focusing on averaged group performance) to more analytical, process-based evaluations (focusing on individual trajectories through a problem space).

In many ways it is unfortunate that the Saussurian, reductionistic, dyadic approach to display and interface design, where complexity and fidelity do not trade off against generalizability, has proven not to be viable. The Peircean, ecological, triadic approach obviously entails a substantially greater amount of work to develop and evaluate interfaces. Researchers and contract monitors who do not share these paradigmatic commitments may well view the latter approach as "expensive and inefficient." We view this simply as a fact of life. The difficulty of designing effective displays and interfaces is a far greater challenge than most people realize; the evidence supporting this claim is all around us in the form of interfaces that are difficult to learn and frustrating to use. We believe that CSE and EID make up a comprehensive and integrated framework that can be used to make this situation better. While there is more work involved, the end result will justify the additional time, energy, and resources required.

## References


Bennett, K. B., and J. M. Flach. 1992. Graphical displays: Implications for divided attention, focused attention, and problem solving. *Human Factors* 34:513–533.

Bennett, K. B., A. L. Nagy, and J. M. Flach. 2006. Visual displays. In *Handbook of human factors and ergonomics,* ed. G. Salvendy. New York: John Wiley & Sons.

Bennett, K. B., M. Payne, J. Calcaterra, and B. Nittoli. 2000. An empirical comparison of alternative methodologies for the evaluation of configural displays. *Human Factors* 42 (2): 287–298.

Bennett, K. B., M. Payne, and B. Walters. 2005. An evaluation of a "time tunnel" display format for the presentation of temporal information. *Human Factors* 47 (2): 342–359.

Bennett, K. B., M. L. Toms, and D. D. Woods. 1993. Emergent features and configural elements: Designing more effective configural displays. *Human Factors* 35:71–97.

Bennett, K. B., and B. Walters. 2001. Configural display design techniques considered at multiple levels of evaluation. *Human Factors* 43 (3): 415–434.

Bennett, K. B., and J. Zimmerman. 2001. A preliminary investigation of the time tunnels display design technique. *Displays* 22 (5): 183–199.

Brunswick, E. 1956. *Perception and the representative design of psychological experiments,* 2nd ed. Berkeley, CA: University of California Press.

Cleveland, W. S. 1985. *The elements of graphing data.* Belmont, CA: Wadsworth.

Cleveland, W. S., and R. McGill. 1985. Graphical perception and graphical methods for analyzing scientific data. *Science* 229:828–833.

Goldsmith, T. E., and R. W. Schvaneveldt. 1984. Facilitating multiple-cue judgments with integral information displays. In *Human factors in computer systems,* ed. J. C. Thomas and M. L. Schneider. Norwood, NJ: Lawrence Erlbaum Associates.

Haley, P. H., and D. D. Woods. 1988. Methods and apparatus for dynamic systems control. Patent number 4,770, 841, patent date Sept. 13.

Hutchins, E. L. 1995. *Cognition in the wild.* Cambridge, MA: MIT Press.

MacGregor, D., and P. Slovic. 1986. Graphic representation of judgmental information. *Human–Computer Interaction* 2:179–200.

Mitchell, C. M., and R. A. Miller. 1986. A discrete control model of operator function: A methodology for information display design. *IEEE Transactions on Systems, Man, and Cybernetics SMC* SMC-16:343–357.

Moray, N., P. Lootsteen, and J. Pajak. 1986. Acquisition of process control skills. *IEEE Transactions on Systems, Man, and Cybernetics* SMC-16:497–504.

Pawlak, W. S., and K. J. Vicente. 1996. Inducing effective operator control through ecological interface design. *International Journal of Human–Computer Studies* 44 (5): 653–688.

Payne, D. G., and J. M. Blackwell. 1997. Toward a valid view of human factors research: Response to Vicente (1997). *Human Factors* 39:329–331.

Rasmussen, J., A. M. Pejtersen, and L. P. Goodstein. 1994. *Cognitive systems engineering.* New York: John Wiley & Sons.

Roth, E. M., and D. D. Woods. 1988. Aiding human performance: I. Cognitive analysis. *Le Travail Humain* 51:39–64.

Vicente, K. J. 1997. Heeding the legacy of Meister, Brunswik, and Gibson: Toward a broader view of human factors research. *Human Factors* 39:323–328.

Wickens, C. D. 1992. *Engineering psychology and human performance,* 2nd ed. New York: Harper Collins.

Wickens, C. D., and C. M. Carswell. 1995. The proximity compatibility principle: Its psychological foundation and relevance to display design. *Human Factors* 37 (3): 473–494.

Woods, D. D., and E. M. Roth. 1988. Aiding human performance: II. From cognitive analysis to support systems. *Le Travail Humain* 51:139–171.

# 18

# *A New Way of Seeing?*

## 18.1 Introduction

> If you change the way you look at things, the things you look at change. (Dyer 2004)

While the database of a science grows incrementally, understanding is a far more nonlinear process that is prone to phase transitions (what Kuhn, 1962, called paradigm shifts). As the database grows, the working assumptions of the current paradigm begin to be questioned and new insights and organizations begin to emerge. Throughout this book, we have advocated for an alternative organizing framework or paradigm for thinking about human performance in general and about interface design in particular. It is not really a new perspective since many of the ideas can be traced back to early functionalist psychologists (e.g., James and Dewey) and philosophers (e.g., Peirce). More recently, the practical value of the insights of these early psychologists has been rediscovered by cognitive systems engineers concerned about the functionality of human–machine systems (e.g., Rasmussen and Vicente 1989).

Additionally, the basic significance of these ideas takes on a new power when coupled with the logic and analytic tools of dynamical systems theory (e.g., Juarrero 2002; Kelso 1995; Kugler and Turvey 1987). In this final chapter we first restate and summarize our case for a shift from a dyadic to a triadic stance on the meaning processing problem. Then we will review the specific implications for interface design. Finally, we will look down the road to describe some of the opportunities and challenges we see looming on the path ahead.

## 18.2 A Paradigm Shift and Its Implications

> [T]here is continuity in at least some of the facts, laws, and theories of a field. Some facts and laws are so well established that it would violate the rational rules of science to dispute them, and some theories so broadly supported that they survive paradigm shifts. However, the paradigm

> shift may result in re-interpretation of such facts, laws and theories. (Lachman, Lachman, and Butterfield 1979, pp. 37–38)

Figure 18.1 illustrates what we believe is a paradigm shift that is reflected in a triadic view of perception, action, cognition, and performance. The critical difference rests with the stance toward the ecology. The majority of researchers and developers would probably not deny the influence of ecological factors in ultimately shaping performance. However, the fundamental question is whether the ecological factors are integral or peripheral with regard to the science of human performance. The dyadic perspective, underlying the information processing approach to human performance, treats the ecological factors as extrinsic. In essence, the information processing approach has been formulated to focus almost exclusively on the dynamics within the head while the ecology is reduced to the role of a somewhat arbitrary stimulus that initiates the information process. From the triadic perspective, however, the ecology is viewed as an integral part of the system. A triadic perspective is concerned with the fit between what is in the head and (to paraphrase Mace 1977) what the head is inside of.

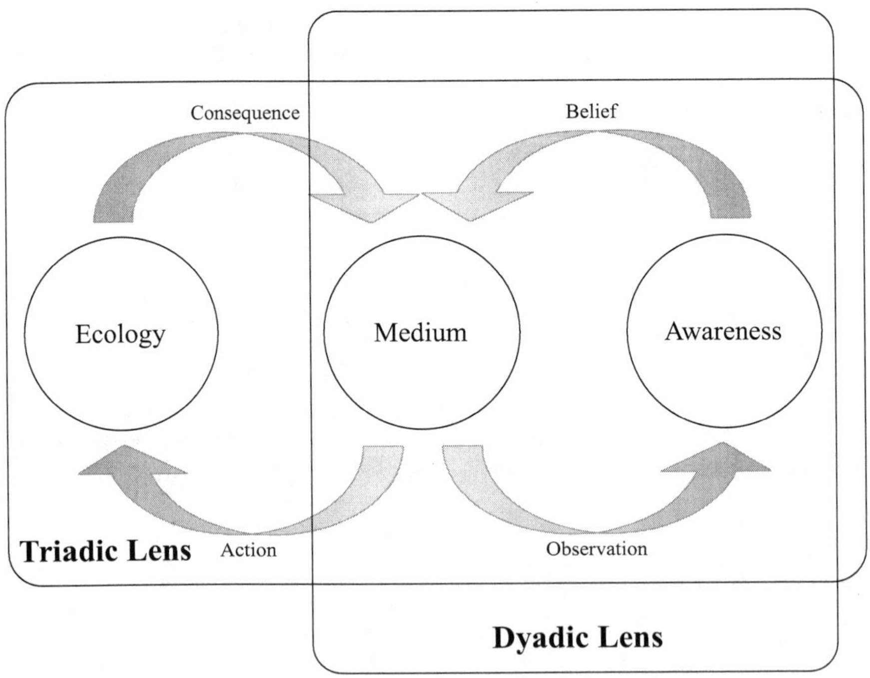


**FIGURE 18.1**
Two alternative ways of looking at the cognitive system. The dyadic lens treats the ecology as extrinsic to the system. The triadic lens includes the ecology as an intrinsic component of the cognitive system.

This essentially comes down to the ultimate ontological basis for meaning. In the dyadic approach, meaning is a product of information processing, an interpretation constructed in the head. In the triadic approach, meaning is an emergent property of the interaction between human and ecology. In the dyadic approach, the focus is on the logic underlying the internal computations and the standards of comparison are based in normative logic (induction and deduction) and mathematics (e.g., probability and economic models of choice). In the triadic approach, the focus is on the pragmatic consequences of decisions and actions and the standards of comparison are based on the normative logic of dynamical systems (information and control theory, abduction).

Note that while information and control theory played a significant role in the development of information processing theory (e.g., Miller, Galanter, and Pribram 1960), the significance has evolved to be a loose metaphor. The implications of closed-loop dynamics have been lost due to the seduction of simple causal explanations (e.g., Jagacinski and Flach 2003).

### 18.2.1 Chunking: A Representative Example of Reinterpretation

To appreciate how the same thing (or construct) looks different from the two distinct perspectives, consider the construct of *chunk*. In the dyadic approach, a chunk is created as a product of information processing. For example, experts in a domain are able to see and to remember more and to be more efficient decision makers due to the ability to chunk information. This ability of experts is typically attributed to superior knowledge (in the head) and the increased efficiency is attributed to the match with capacity limits (i.e., 7 ± 2 chunks, also in the head). Thus, for the dyadic approach, the concept of a chunk requires no consideration of factors residing outside the head. In this case, chunk is a purely *mental* phenomenon.

From a triadic perspective, however, another facet of the *chunk* construct becomes salient. The focus turns to how the organization in the head fits with structure in the ecology. In other words, the question now becomes how the associations in the head match with structure in the ecology. By "structure in the ecology," we mean that there are patterns or correlations among events that reduce uncertainty so that an observer who recognizes these associations can anticipate or predict events more efficiently than someone who is unaware of the associations. This provides an alternative explanation for why chunking leads to more efficient processing: It leverages natural constraints within the ecology to reduce uncertainty.

Thus, for the triadic approach, it is essential to understand both the structure in the ecology and the structure in the head in order to explain the efficiencies due to chunking. The efficiencies are a function of this interaction. It is the match between the associations in the head and the deep structure in the domain that allows the experts to think more productively than novices. This does not require that we deny the main effects of constraints in the head (limited capacity or limited resources), but it does mean that those constraints must be considered in relation to the interaction with the ecology in order to reliably predict performance. In other words, this means that the construct of a chunk is situated (e.g., Suchman 1987; Hutchins 1995) or embodied (Clark 1997).

### 18.2.2 Summary

In sum, the triadic approach is a more comprehensive approach to the problem of situation awareness that considers both the internal constraints on awareness and the external constraints on situations. Performance is considered always to be a joint function of the deep structure within the problem (e.g., work domain) and the perspicacity (e.g., knowledge, beliefs, and/or expectations) of the agent. The better the match between the deep structure and the beliefs of the agents is, the more stable the relation is expected to be (i.e., the agents will possess a higher degree of attunement).

In conclusion, it is important to understand that while we are advocating for a new way of looking at human performance, the adoption of this approach does not imply the dismissal of the wealth of empirical data that the previous paradigm has produced. The generation of psychologists who framed the information processing approach has produced valuable insights into the limits of human performance. In fact, it is the elegance of much of this work that attracted us to the field of experimental psychology and that eventually led us to explore the application of cognitive science to interface design. Additionally, there are important threads related to information theory and control theory that will continue to play a critical role in theories of human performance. One need only look at the content of Chapters 4, 5, 7, and 8 to see that we appreciate the value of these contributions.

What does change, however, is the interpretation of many of these findings. In essence, these data take on new significance when interpreted from a different set of paradigmatic beliefs. The previous discussion of chunking provides one example. A more extensive example is provided by the discussion of emergent features and display design described in Chapters 8 through 11. The findings related to internal constraints on awareness (i.e., form perception) are certainly important in their own right and no one would question the results of these elegant experiments. However, these findings take on new meaning when they are applied to display design. From the triadic perspective, the forms that are being perceived represent external constraints on situations, rather than just marks on a page or screen.

## 18.3 Interface Design

Thus, this sets up the context where the goal of interface design is to enhance the fit between situation and awareness. The implication of a triadic perspective for interface design is that the function of the interface is to enhance the correspondence between the deep structure of a particular problem or situation and the beliefs and expectations of the human agents. This requires that the designer consider the situation, to identify the deep structure, as well as the agents' awareness, to understand the conceptual organization.

One of the first questions that an interface designer should consider is the nature of the problem to be solved. We suggest that problems might be placed on a continuum from problems that are tightly constrained by invariant physical laws (e.g., aviation and nuclear power), on one end, to problems based on intentions, social conventions, or abstract logical relations (e.g., information databases such as libraries or the World Wide Web) at the other end of the continuum. Vicente (1999) referred to the former as *correspondence-driven domains* and to the later as *coherence-driven domains*.

### 18.3.1 Constraint Mappings for Correspondence-Driven Domains

In correspondence-driven domains, it is logical to focus first on the problem domain to identify the underlying physical, functional, and goal-related constraints and to consider representations that make those constraints salient to the human agents who must manage the process. In this case, the primary goal of the designer is to shape the human agents' awareness so that it effectively leverages the patterns in the domain. Thus, in correspondence-driven domains like process control (Chapter 10) and aviation (Chapter 11), work analysis focuses on the domain to attain a deep understanding of the physical process (e.g., the mass and energy balance relations). In such a domain, it is not only essential to consider the opinions of expert controllers (e.g., pilots), but also important to consider the perspectives of others who are knowledgeable about the underlying principles governing the process (e.g., aeronautical engineers).

Note that the different experts all have their own understandings and thus each may see things in a different way. Achieving direct perception/manipulation in these domains involves the development of analogical representations that directly reflect the associated situation constraints and are also compatible with awareness constraints. Rasmussen (1998) has recommended that exploring the graphical representations used by scientists and engineers can be a good source of inspiration for alternative designs for controlling physical processes. A good example of this is the Rankine cycle display proposed by Beltracchi (1987) as a component for controlling nuclear power processes.

Also, it is important to appreciate that the awareness of operators will be shaped by the interfaces with which they are most familiar. Thus, there will often be initial resistance and possibly lower performance as the result of a change in representation, even when a new interface is an objectively better representation of the deep structure of the domain. This is due to the mismatch between the representation and the operators' expectations. Any change in representation will often require some degree of training so that the operators can tune their awareness to leverage the benefits of the new representation fully. Again, stability will always depend on the interaction of situation and awareness.

A final point to consider with correspondence-driven domains is Ashby's (1956) law of requisite variety. The greatest danger of a dyadic approach is that in framing an interface around the user limitations, without any theory of the ecology, there is a significant risk that the problem will be trivialized. That is, there is a possibility that the representation will not reflect some aspects of the problem structure—that some aspect of the requisite variety will be hidden from view. Often the trivialization will not be apparent during routine operations when the operators are following predetermined procedures. But the trivialization may lead to ambiguity and confusion during faults where discovering solutions depends on a deep understanding of the process. It was the unanticipated variability associated with faults in nuclear power plants that originally inspired the construct of ecological interfaces (Rasmussen and Vicente 1989).

### 18.3.2 Constraint Mappings for Coherence-Driven Domains

At the other end of the continuum, in coherence-driven domains like libraries (Chapter 6) and interfaces to wireless data networks (Chapter 13), it is logical to begin with the awareness side of the equation—that is, to start with the experiences, desires, and expectations of the user population. In some respects, organization on the Internet or within a library is only limited by the imaginations of the designers who build the databases. Thus, books could be categorized by size, color, author, genre, etc. In this case, the challenge is to come up with a structure that matches the expectations and functional aspirations of the users. Often, this means creating common ground between the professionals who structure the database and the people who use it. In essence, the database is the ecology. However, it is not an ecology of immutable physical laws, but rather a categorical structure that can be changed to meet specific functional goals.

At the coherence-driven domain end of the continuum, the added value of a triadic perspective is not simply to consider the ecology as it exists (i.e., the conventional categorical structure), but also to consider the possible structures. Thus, with the BookHouse, the AMP classification system offered a structure for cataloguing fiction that better reflected the distinctions that made a difference to the reader experience. Thus, at the coherence end of the continuum, the design challenge is not simply to describe the ecological constraint, but in fact also may be to create or invent those constraints so that the constraints in the ecology (e.g., database) better reflect the intentions and expectations of the user population.

The fundamental danger at the coherence end of the continuum occurs when the interface is designed to reflect the narrow perspective of the technologists' experience. In these cases, the representation may be constrained in order to facilitate manufacturing or fabrication (e.g., Ford's famous dictum that the customer could have any color he wanted as long as it was black) or it is constrained to match specialized knowledge of the technologists or of narrowly trained experts (e.g., a library classification system designed for librarians).

Innovation in coherence-driven domains often seems to be associated with leveraging general skills and expectations of the user population. For example, the graphical user interface (GUI) allowed people to generalize expectations from broad experiences of interacting with space and physical objects to logical operations involving manipulation of data and information. One of the primary keys to the success of the iPhone (Chapter 13) was a shift from a narrow window utilizing a steep, hierarchically organized menu structure that placed heavy demands on memory for procedures to a wider window utilizing a flattened organization that better leveraged perceptual–motor skills to reduce memory load.

Also, a fundamental principle in the iPhone design philosophy is that common or important functions should be salient on the interface, whereas many other PDAs hide functionality in a hierarchical system of menus. In coherence-driven domains, the greatest returns on investment often come from shaping the databases or technologies to match preexisting skills and expectations of the user population.

Thus, as the nature of the problem domain shifts from coherence-driven domains (e.g., iPhone) toward more correspondence-driven domains (e.g., aviation), the emphasis of the work analysis will also shift from a *user-centered* focus in coherence-driven domains to a *use-centered* focus in the correspondence-driven domains. In either case, however, the goal for the interface design should be that the form of the representation should specify the functionally significant problem constraints.

In fact, both approaches are ultimately use centered in that the ultimate focus is on functionality. To reiterate, the key difference is that, in correspondence-driven domains, the functionally significant constraints will typically be immutable physical dynamics of the process being controlled. For coherence-driven domains, the functionally significant constraints will typically be the intentions of the users.

Most work domains fall in the middle regions of the continuum between the largely correspondence-driven problems and the largely coherence-driven problems. These include domains such as military command and control (Chapter 14), health care systems, and emergency management systems. In such domains, one must consider both the use-centered constraints and the user-centered constraints. A key challenge in this middle ground will be to distinguish aspects of the ecology that are immutable constraints and those that can be changed through more effective organization. We cannot rewrite the laws of mass and energy balance, but we can change the way books are catalogued or the way information is organized in a Web database. Such reorganization within the ecology may go hand in hand with the reorganization of how the work is represented in the interface—again as illustrated by the BookHouse example.

### 18.3.3 The Eco-Logic of Work Domains

Thus, creating and describing the ecology is an essential co-requisite of interface design if we are to build representations that lead to productive thinking. Rasmussen (1986; Rasmussen, Pejtersen, and Goodstein 1994) and Vicente (1999, 2003) have offered the abstraction hierarchy as a way of describing the use-centered side of the equation (i.e., the work domain constraints; see Chapter 3). This is a nested hierarchy with five levels linked through means–ends relations.

The labels of the different levels have shifted over the years. We think this might reflect changing characteristics of domains that have been explored. Early applications were at the correspondence end of the continuum, while more recent applications have explored domains closer to the coherence end of the continuum. We have also relied heavily on the abstraction hierarchy in our own work and find it to be a very useful framework for describing work domains (Amelink et al. 2005; Flach, Mulder, and van Paassen 2004). However, we feel that it is important not to get too hung up on the specific labels for the various levels but rather to choose labels that make sense for the domain of interest.

Our recommendation is not to begin work analysis with the goal of building an abstraction hierarchy. Rather, the goal should simply be to discover the deep structure of the work. An interesting comment from Rasmussen is that, typically, when he has entered a domain, many of the activities of experts seemed to be illogical or inefficient (relative to normative expectations). However, these initial impressions have often changed as he learned more about the specific domain. In fact, operations that initially seemed illogical or clumsy often appeared quite elegant once he understood the constraints of the domain better. The lesson here is to try to check your expectations at the door when you enter a new domain. Approach it with the humble attitude of a novice; watch and listen carefully in order to discover the *eco-logic* that shapes the work activities of domain experts.

As you learn more and more about the eco-logic of the work domain, you will need some way to organize and communicate your observations. At this point, the abstraction hierarchy might be considered as one option. But you should consider a wide range of options and pick the organization that makes the most sense to you and, perhaps, to others in the multidisciplinary team that might be involved in the design process.

### 18.3.4 Productive Thinking in Work Domains

We also recommend that the decision ladder be considered as a way to organize and represent both the awareness and situation sides of the equation

(Chapter 4). Traditionally, the decision ladder has been used to illustrate specific strategies of experts. However, we have found that this framework can be used to model not only the actual activities of experts, but also the possible strategies (see, for example, Naikar, Moylan, and Pearce 2006). Here, one might consider possible associations based on structure in the work domain that could *shortcut* the computational demands of higher cognitive processes in order to support more recognition-primed interactions. In other words, one might consider what patterns or consistencies in the work domain might be leveraged to help anticipate events, reducing the effective information processing load.

In sum, the bottom line for interface design is that the goal is to support what Wertheimer (1959) called *productive thinking*. It is clear from Wertheimer's work and a large body of work on the topic of problem solving that a change in representation can be the difference between a problem that is easy to solve (where the solution can be quickly recognized) and a problem that is difficult to solve (where the solution requires intensive thought or analysis). A good representation is one that reveals the *deep structure* of the problem. However, unless we can specify what we mean by deep structure, this prescription can be circular (i.e., the test of whether a representation reveals the deep structure is whether it leads to productive thinking). This is where formalisms like the abstraction hierarchy become important because they provide converging operations for defining what we mean by deep structure. In other words, the abstraction hierarchy allows a prescriptive formulation for identifying the deep structure of situations, as opposed to a post hoc explanation based on the ultimate impact on performance.

There are many advocates for the user-centered perspective, so for that reason we have put specific emphasis on the need to consider the use-centered constraints. In presenting this perspective, we may sometimes seem to be at odds with those advocating for a user-centered perspective, due to the dialectic "either–or" style of scientific discourse. However, we want to emphasize that it is not an issue of either–or; rather, we need to take a "both–and" approach. This is the essence of the triadic framework: that it is essential to consider *both* situation *and* awareness sides of the dynamical system.

## 18.4 Looking over the Horizon

Advances in information technologies have created enormous opportunities for alternative ways to organize and present information. Concepts such as Web 2.0 (O'Reilly 2005) and semantic computing (Berners Lee, Hendler, and Lassila 2001) are signs of increasing awareness of the need to move beyond the design of information processing systems to begin thinking about the design of meaning processing systems. This is an important challenge for cognitive scientists to test the values of their theories of human problem solving against the challenges to improve information technologies.

We believe the biggest obstacle that prevents the productive collaborations between cognitive scientists, engineers/technologists, and domain experts that are necessary to address these challenges is the dualist ontology that still dominates thinking, at least in Western-based cultures. There is still a dominant view that *mind* and *matter* reflect distinct realities: The field of mind is owned by the cognitive scientists and the field of matter is owned by the engineers/technologists. Thus, research tends to be organized around the main effects of mind and the main effects of matter, with the implicit assumption that these will eventually add up to a deeper understanding of human–machine systems or of *what matters.*

### 18.4.1 Mind, Matter, and What Matters

However, we believe that *what matters* is ultimately a function of the interactions between mind and matter. These interactions cannot be understood as the sum of the main effects. On the contrary, we believe that the main effects can only be fully appreciated in the context of the interactions that contain them. From the cognitive science side, many are coming to a similar conclusion. This is reflected in terms such as ecological psychology, situated cognition, and embodied cognition, as well as the growing interest in evolutionary psychology.

The more radical claim that we would like to make is that it is also important to consider the interactions with cognition on the matter side of the equation. This is particularly appreciated at the quantum level of observation, where the choices of scientists may have a role in creating the phenomena that result. As Wheeler (1980) has argued, the idea of an objective world "out there" apart from the human observers is a fiction. This was the ontological position of William James that he called radical empiricism.

Generally, the question of mind and matter is framed in a dialectic case of mind *or* matter and three ontological positions are generally considered:

Idealism: there is only mind.

Materialism: there is only matter.

Dualism: there are two distinct realities—one of mind and the other of matter.

With radical empiricism, William James was offering a fourth alternative. James argued that human experience is neither mind nor matter, but both. The distinction is that mind and matter are not two realities, but facets of a single reality. That human experience is a joint function of mind *and* matter. Thus, this ontological position demands a scientific framework that includes both the constraints typically associated with awareness (mind) and the constraints typically associated with the ecology (matter). This demands a single science of *what matters.*

### 18.4.2 Information Processing Systems

The information processing paradigm was framed in the context of the challenges of World War II, due to the advent of advanced technologies (e.g., radar, automatic weapons systems, and early computers) and the related sciences. This paradigm shift caused a renewed appreciation for the mental qualities of human experience, which had been neglected in the rise of behaviorism. In essence, there was a shift from a materialist stance to a dualist stance. Now, perhaps it is time to take one more step that involves more than just introducing a second science of mind. Perhaps, it is time to think about a single science integrating mind and matter—a single science of what matters.

Implicit in the information processing approach is a comparison between human and computer in which the human is seen as the weak link in the system (i.e., limited memory, inconsistent and unreliable with respect to following the rules of logic). However, the limitations of computers (e.g., Dreyfus 1992) have become quite evident (somewhat ironically) as a result of the incredible amount of time, effort, and funding that has been spent in attempting to make them act intelligently. Today, rather than lamenting the limitations of humans, the question is more often, "Why can't computers be more flexible and adaptive, like humans?"

In a world dominated by a computer image of mind, loss of the physical or action dimension seemed of little consequence. However, today, when there is an increased desire to put computers to work as robots and adaptive control systems, it becomes necessary to ground the information processing in the realities of a physical world so that a computer (i.e., robot) might walk through a cluttered environment or respond intelligently to unanticipated contingencies. Now the attention shifts from abstract logic to the pragmatic concerns of adapting to real ecologies. Now we see computer scientists shifting their attention from purely logical engines to abductive engines (Flach and Kakas 2000).

### 18.4.3 Meaning Processing Systems

The computer is not only moving into the physical world, but it is also increasingly integrated into other facets of human experience as the Internet and mobile computing reshape our social and economic experiences. This has huge implications for how people solve problems and for organizational and social sensemaking (e.g., Rochlin 1997; Weick 1995). Increasingly, organizations are becoming flatter and the sensemaking process is becoming increasingly distributed. The role of centralized planning in coordinating organizations and social systems is getting less important. Increasingly, social systems are achieving stable forms through self-organization. This, again, has huge implications for how we think about mind and matter and what we mean by a cognitive process.

This requires that we shift our image of humans from seeing them as *information processing systems* to seeing them as *meaning processing systems* or as components of *sensemaking organizations*. In the information processing view, the primary goal was to describe the processing limits and the sources of human error. In the meaning processing view, the primary goal will be to leverage fully the human capacity for creative problem solving. In the information processing view, the emphasis was on constraining humans to follow the plan or procedures. In the meaning processing view, the goal will be to support the human in adapting to complexity (i.e., unanticipated variability, contingencies not considered in the plans). In the old paradigm, the ideals were framed around accurately following the rules (e.g., the computer was the ideal). In the new paradigm, the challenge will be to rewrite the rules continually to keep pace with rapidly changing contingencies. In the new paradigm, the goal will be creative self-organization as reflected in dynamical systems.

We believe that those who determine the form of the representations that couple humans and information technologies will have a significant role in shaping the direction of the self-organizing processes. By changing the way people look at things, we will not only change the things at which they look, but also change the things they are!

## References

Amelink, M. H. J., M. Mulder, M. M. van Paassen, and J. M. Flach. 2005. Theoretical foundations for a total energy-based perspective flight-path display. *International Journal of Aviation Psychology* 15 (3): 205–231.

Ashby, R. 1956. *An introduction to cybernetics.* New York: Chapman and Hall.

Beltracchi, L. 1987. A direct manipulation interface for heat engines based upon the Rankine cycle. *IEEE Transactions on Systems, Man, and Cybernetics* SMC-17:478–487.

Berners Lee, T., J. Hendler, and O. Lassila. 2001. The semantic web. *Scientific American* 284 (5): 34–43.

Clark, A. 1997. *Being there: Putting brain, body, and world together again.* Cambridge, MA: MIT Press.

Dreyfus, H. L. 1992. *What computers still can't do*. Cambridge, MA: MIT Press.

Dyer, W. (2004). *The power of intention.* Carlsbad, CA: Hay House, Inc.

Flach, J. M., M. Mulder, and M. M. van Paassen. 2004. The concept of the "situation" in psychology. In *A cognitive approach to situation awareness: Theory, measurement, and application,* ed. S. Banbury and S. Tremblay. Aldershot, England: Ashgate.

Flach, P. A., and A. C. Kakas, eds. 2000. *Abduction and induction: Essays on their relation and integration.* Dordrecht, the Netherlands: Kluwer Academic Publishers.

Hutchins, E. L. 1995. *Cognition in the wild.* Cambridge, MA: MIT Press.
Jagacinski, R. J., and J. M. Flach. 2003. *Control theory for humans.* Mahwah, NJ: Lawrence Erlbaum Associates.
Juarrero, A. 2002. *Dynamics in action. Intentional behavior as a complex system.* Cambridge, MA: MIT Press.
Kelso, S. 1995. *Dynamic patterns: The self-organization of brain and behavior.* Cambridge, MA: MIT Press.
Kugler, P. N., and M. T. Turvey. 1987. *Information, natural law, and the self-assembly of rhythmic movement.* Hillsdale, NJ: Lawrence Erlbaum Associates.
Kuhn, T. 1962. *The structure of scientific revolutions.* Chicago, IL: University of Chicago Press.
Lachman, R., J. L. Lachman, and E. C. Butterfield. 1979. *Cognitive psychology and information processing: An introduction.* Hillsdale, NJ: Lawrence Erlbaum Associates.
Mace, W. M. 1977. James J. Gibson's strategy for perceiving: Ask not what's inside your head but what your head's inside of. In *Perceiving, acting and knowing,* ed. R. E. Shaw and J. Bransford. Hillsdale, NJ: Lawrence Erlbaum Associates.
Miller, G., E. Galanter, and K. Pribram. 1960. *Plans and the structure of behavior.* New York: Holt, Rinehart & Winston.
Naikar, N., A. Moylan, and B. Pearce. 2006. Analyzing activity in complex systems with cognitive work analysis: Concepts, guidelines, and case study for control task analysis. *Theoretical Issues in Ergonomics Science* 7:371–394.
O'Reilly, T. 2005. What is Web 2.0? Design patterns and business models for the next generation of software. http://oreilly.com/pub/a/web2/archive/what-is-web20.html
Rasmussen, J. 1986. *Information processing and human–machine interaction: An approach to cognitive engineering.* New York: Elsevier.
———. 1998. Ecological interface design for complex systems: An example: SEAD-UAV systems. Smorum, Denmark: HURECON.
Rasmussen, J., A. M. Pejtersen, and L. P. Goodstein. 1994. *Cognitive systems engineering.* New York: John Wiley & Sons.
Rasmussen, J., and K. J. Vicente. 1989. Coping with human errors through system design: Implications for ecological interface design. *International Journal of Man–Machine Studies* 31:517–534.
Rochlin, G. I. 1997. *Trapped in the net.* Princeton, NJ: Princeton University Press.
Suchman, L. A. 1987. *Plans and situated actions: The problem of human–machine communication.* New York: Cambridge University Press.
Vicente, K. J. 1999. *Cognitive work analysis: Toward safe, productive, and healthy computer-based work.* Mahwah, NJ: Lawrence Erlbaum Associates.
———. 2003. *The human factor.* New York: Routledge.
Weick, K. E. 1995. *Sensemaking in organizations.* Thousand Oaks, CA: Sage Publications.
Wertheimer, M. 1959. *Productive thinking.* New York: Harper and Row.
Wheeler, J. A. 1980. Frontiers of time. In *Problems in the foundations of physics,* ed. G. T. di Francia. Amsterdam: North-Holland.

# Index

## D

**I**

## J



## K

## Q



## R

## S